Molecular Systematics of Parasitic Helminths

Urusa Thaenkham • Kittipong Chaisiri •
Abigail Hui En Chan

Molecular Systematics of Parasitic Helminths

Springer

Urusa Thaenkham
Department of Helminthology,
Faculty of Tropical Medicine
Mahidol University
Bangkok, Thailand

Kittipong Chaisiri
Department of Helminthology,
Faculty of Tropical Medicine
Mahidol University
Bangkok, Thailand

Abigail Hui En Chan
Department of Helminthology,
Faculty of Tropical Medicine
Mahidol University
Bangkok, Thailand

ISBN 978-981-19-1785-1 ISBN 978-981-19-1786-8 (eBook)
https://doi.org/10.1007/978-981-19-1786-8

This Springer imprint is published by the registered company Springer Nature Singapore Pte Ltd.
The registered company address is: 152 Beach Road, #21-01/04 Gateway East, Singapore 189721, Singapore

Foreword

To date, parasitic helminth infection is still one of the important NTDs. Molecular and genomic technology should be applied to support the traditional method of using morphology as the main criteria for diagnosis and species identification, study species diversity, and enhance helminth taxonomy and molecular systematics. This book provides updated information on the fundamental concepts of molecular evolution of parasitic helminths, molecular systematics and relationship within nematodes, trematodes, and cestodes, and the best practice and recommendations for molecular studies on parasitic helminths.

Considering the need for continuous progression and the field of parasitic helminthology to remain relevant in this molecular era, this book provides useful and valuable information for students, researchers, and professionals alike. I believe that this book will help to contribute knowledge towards helminth molecular systematics and be an asset serving as building blocks for budding molecular parasitologists. It is, therefore, a great pleasure to present this book to the scientific community and interested readers.

25 January 2022 Jitra Waikagul

Preface

Systematics is the intersection of phylogeny and taxonomy, where phylogeny refers to an analysis of evolutionary or genealogical relationships, while taxonomy is the process of species delimitation, classification, and standard nomenclature. However, we sometimes face challenges in species identification based on morphological characters, particularly for cryptic lineages, homoplasy, morphological plasticity caused by environmental adaptation, etc. Recently, molecular systematics came into play due to molecular advances, where relationships among individuals and species based on the appropriate molecular genetic markers can be revealed. Progress in helminthic research is continually revealing species boundaries for the delimitation of many parasitic helminths. Although advances have been made in the molecular systematics and identification of parasitic helminths, their high diversity and variation are examples of obstacles hindering the analysis of robust phylogenetic relationships.

We aim to provide fundamental knowledge and information for molecular systematics research, focusing mainly on parasitic helminths (nematode, trematode, cestode). Firstly, we provide updated information on helminth diversity and classical taxonomy, along with molecular systematics involved in their diversity. The shreds of evidence of molecular systematics studies of parasitic nematodes, cestodes, and trematodes will be compiled and discussed regarding the utilities and pitfalls of the genetic marker used for various purposes. Secondly, evolutionary processes influencing parasite diversity and genetic adaptation, together with the fundamentals of molecular evolution, will be reviewed. These fundamentals of molecular evolution will be put into the context of parasitic helminths, which is helpful knowledge for understanding how species are evolutionarily related. Thirdly, pitfalls and challenges for helminth species identification will be elaborated on, including guidelines for parasitic helminth sample preparation and management for molecular identification and molecular systematics analysis. Lastly, further perspectives of utilizing genetic markers for molecular studies on parasitic helminths will be addressed in applications from the laboratory to fieldwork, such as DNA barcoding and the environmental DNA manipulation of parasitic helminths. We hope that undergraduate and postgraduate students will benefit from our book, especially those who require concepts in molecular systematics. We also hope that our book can be helpful for researchers desiring a guideline to select genetic markers and analyze DNA sequences for phylogenetic inferences of parasitic helminths.

Given the significance of the impacts of climate change on biodiversity and increasing anthropogenic activity in the coming years, the study on the molecular systematics of parasitic helminths is even more relevant. We face a race with time due to climate change to discover extant helminth species before they become extinct. Also, as the interface between humans and animals decreases, helminthic diseases with zoonotic potential can emerge or re-emerge, possibly leading to more severe disease outbreaks. With molecular tools being applied in this genomic era, the fundamentals, guidelines, and examples on the molecular systematics of parasitic helminths provided in our book will hopefully benefit a wide range of readers across various scientific disciplines.

Bangkok, Thailand

Urusa Thaenkham
Kittipong Chaisiri
Abigail Hui En Chan

Contents

Abbreviations

A	Adenine base
AFA	Alcohol-formalin-acetic acid
AFLP	Amplified fragment length polymorphism
AI	Artificial intelligence
ATP	Adenosine triphosphate
BI	Bayesian inference
BOLD	Barcode of Life Database
Bp	Base pairs
BSC	Biological species concept
C	Cytosine base
CBOL	Consortium for the Barcode of Life
CNS	Central nervous system
COI	Cytochrome *c* oxidase subunit 1 gene
COII	Cytochrome *c* oxidase subunit 2 gene
COIII	Cytochrome *c* oxidase subunit 3 gene
CRISPR	Clustered regularly interspaced short palindromic repeats
CSC	Cohesion species concept
cytB	Cytochrome b gene
DALYs	Disability-adjusted life years
ddNTP	Dideoxynucleotide
DNA	Deoxyribonucleic acid
dNTP	Deoxynucleotides
EcSC	Ecological species concept
eDNA	Environmental DNA
Ef-1 α	Elongation factor 1 alpha gene
EID	Emerging infectious diseases
ES	Excretory-secretory products
ESC	Evolutionary species concept
ESTs	Expressed sequence tags
EtBr	Ethidium bromide
ETS	External transcribed spacer region

F81 model	Felsenstein 1981 model
FAAL	Formaldehyde alcohol azocarmine lactophenol
FAO	Food and Agriculture Organization
G	Guanine base
GAA	Glacial acetic acid
GTR model	General time reversible model
HKY model	Hasegawa-Kishino-Yano model
HSPs	Heat shock proteins
IGS	Intergenic spacer region
IPCC	Intergovernmental Panel on Climate Change
ITS	Internal transcribed spacer region
ITS1	Internal transcribed spacer 1
ITS2	Internal transcribed spacer 2
JC model	Jukes and Cantor's one-parameter model
K2P model	Kimura's two-parameter model
Kb	Kilobases
LNR	Long non-coding region
MAFFT	Multiple alignments using fast Fourier transform
matK	Maturase K gene
Mb	Megabases
MCMC	Markov chain Monte Carlo
ME	Minimum evolution
ML	Maximum likelihood
MOTU	Molecular operational taxonomic unit
MP	Maximum parsimony
MRCA	Most recent common ancestor
mRNA	Messenger RNA
MSA	Multiple sequence alignment
MSC	Morphological species concept
mtDNA	Mitochondrial DNA
MUSCLE	Multiple sequence comparison by log-expectation
NCBI	National Center for Biotechnology Information
ND1 to *ND6*	Nicotinamide adenine dinucleotide 1 to 6
NGS	Next generation sequencing
NHM	Host-Parasite Database of Natural History Museum, London
NJ	Neighbor joining
NTDs	Neglected tropical diseases
NTS	Non-transcribed spacer
NUMTs	Nuclear insertions of mitochondrial sequences
OTUs	Operational taxonomic units
PAA	Population aggregation analysis
PCR	Polymerase chain reaction
PCR-RAPD	Polymerase chain reaction-random amplified polymorphic DNA

PCR-RFLP	Polymerase chain reaction-restriction fragment length polymorphism
PMC	PubMed Central
PSC	Phylogenetic species concept
qPCR	Quantitative real-time PCR
QTL	Quantitative trait loci
RADSeq	Restriction site-associated DNA sequencing
rbcL	Ribulose-bisphosphate carboxylase
rDNA	Ribosomal DNA
RNA	Ribonucleic acid
RNAi	RNA interference
rRNA	Ribosomal RNA gene
s.l.	Sensu lato
s.s.	Sensu stricto
SDS-PAGE	Sodium-dodecyl sulfate-polyacrylamide gel electrophoresis
SMRT	Single-molecule real-time sequencing
SNR	Short non-coding region
SSCP	Single-strand conformation polymorphism
SSR	Single sequence repeats
STH	Soil-transmitted helminth
T	Thymine base
TE	Transposable elements
tRNA	Transfer RNA
U	Uracil base
UPGMA	Unweighted pair group method with arithmetic means
USNPC	US National Parasite Collection
UV	Ultraviolet
WHO	World Health Organization

Chapter 1
Concept of Molecular Systematics

Abstract The biodiversity of metazoans, including parasitic helminths, has been explored and studied continuously. To characterize and interpret global diversity, systematics has been established as the branch of the foundation for biology. It provides basic evolutionary questions related to host-parasite association, life history, genetics, and disease transmission, which is the cornerstone of epidemiology. Systematics is the intersection of phylogeny, either evolutionary or genealogical relationships, and taxonomy, which is the process of species delimitation, classification, and standard nomenclature. As we know, parasitic helminths are a massive group of metazoans with high genetic diversity; this group of parasites lacks fossil records to support their classification strongly. Recently, molecular systematics, which uses genome information, particularly sequences and structures of DNA, RNA molecules, and amino acid chains, has been applied to taxonomy and demonstrates phylogenies. This chapter gives the definition and contribution of systematics to biology. The principle of molecular systematics and its application for conducting research, particularly in parasitic helminths, were also provided more in each chapter of this book.

1.1 What Is Molecular Systematics?

Generally, systematics contributes to major tasks such as (1) identifying organisms or taxonomy including a standard nomenclature, (2) discovering the ancestral relationships among organisms or phylogenetics, and (3) organizing information about the diversity of organisms into hierarchical systems or classification, including the process of species delimitation. However, species identification based on morphological character alone is problematic. Like other metazoans, adaptation and coevolution between host and parasites are constantly occurring, together with the complex life cycle of parasites to enhance their diversity. Moreover, the consequences of the environmental changes due to anthropogenic activity and ecological processes on host-parasite interaction influence adaptation to new environments, enhancing the parasites' phenotypic plasticity and adaptive radiation. Species delimitation is a challenging task, which requires data to verify its confidence by having

U. Thaenkham et al., *Molecular Systematics of Parasitic Helminths*,
https://doi.org/10.1007/978-981-19-1786-8_1

congruency across methods and tie to species concepts as the criteria for consideration.

Recently, we can easily obtain genome information. The cost of obtaining the target DNA sequences of molecular genetic markers and whole-genome sequence is dramatically decreased. There are several reasons that molecular data enhances the ability of systematics more than morphological information. Firstly, DNA sequences are strictly heritable entities. At the same time, morphological characters can be influenced by changing environmental factors throughout time. Secondly, the assessment of homological characters based on molecular data is more accessible than morphological characters, providing a more straightforward interpretation of the phylogenetic relationship. Thirdly, we have molecular characters obtained from various molecular genetic markers, typically partial or the entire length of a protein-coding gene or noncoding region from nuclear and mitochondrial genomes. The genetic information obtained represents differentiation in a relatively short period of time compared to morphological characters. Lastly, molecular genetic markers can be applied to deformed and partly degraded materials, including long-term preserved, formalin-fixed, and glycerine-embedded specimens by using the standard protocol or commercial kits for DNA extraction (Jeon & Eom, 2013; Jeon et al., 2011; Moradi et al., 2019; Schneider et al., 2008). Using molecular information for systematics and supporting morphological identification has become the standard work for indicting the phylogenetic relationships of parasitic helminths (Brooks, 1985; Brooks & Hoberg, 2001; Spakulova, 2002).

1.2 How Does Molecular Systematics Support Other Research Works?

Brooks and Hoberg (2001) mentioned the opportunities and obstacles of systematic parasitology in the twenty-first century. Interestingly, he addressed the significant point that while we have tried to eradicate parasitic diseases of humans and livestock and the effect of wild diversity is threatened with extinction in many regions, we often have incomplete information on managing what remains, particularly potential parasitic disease agents. They critically suggested that we face irreversible species extinction, which may impact the loss of socio-economic potential and evolutionary potential that has been the source of biotic recovery from ecological perturbations and environmental disasters on a global scale.

Until now, most phylogenetic relationship information has been emphasized at important parasites of humans and livestock (Brooks & Hoberg, 2001; Holterman, 2007; Morand, 2015). Many of them have been obtained from non-systematists who tried to construct the phylogenies to conduct their phylogenies using user-friendly programs based on gene or gene fragment sequences. Thus, the vast majority of those topologies presented in the literature may not depict robust phylogenetic relationships. A fully trained or know-how for molecular systematics is required, which will guide the research to assess sequence data for sampling bias and missing data, nucleotide substitution, and alignment problems to provide the most robust phylogeny. The principles and know-how of molecular systematics are necessary, particularly in tropical medicine and public health, to avoid producing poor phylogenetic trees with reliable data to be the consensus for other phylogenies.

Although morphology is the cornerstone of the systematics of parasitic helminths, genetic information can provide us with a capacity for identifying species, even for cryptic species, revealing phenotypic plasticity, linking life cycle stages, inferring phylogeny, and exploring geographic variation (Forsman, 2015; Nadler & De León, 2011; Nolan & Cribb, 2005). The framework as the best molecular practice for trematode systematics was provided by Blasco-Costa et al. (2016), which summarized the trends that applied molecular methods for systematics work from 2011 to 2015. They included research on species concept applied to trematodes in the molecular age, molecular genetic markers applied for systematics works, and also the best practice for molecular systematic analysis (Blasco-Costa et al., 2016). Molecular approaches and practices, especially using short fragments of DNA sequences, are advantageous in this -omic era for applying to DNA barcoding and metabarcoding. Next-generation sequencing technologies now offer more opportunities to explore many aspects of fundamental biology. Many complete mitochondrial genomes of parasitic helminths are now available, providing more robust phylogenetic relationships and better diagnostic markers, particularly closely related species in cryptic lineages.

1.3 Overview Topics

This book aims to provide fundamental knowledge, best practice, and guidance for molecular systematics research, mainly on parasitic helminths (nematodes, trematodes, cestodes). We also provided guidelines for sample preparation and management for molecular systematics and molecular identification. Additionally, we also delivered the procedure for research on molecular systematics and DNA taxonomy as the guideline to explore parasitic helminths. Finally, further perspectives on utilizing molecular genetic markers for molecular studies on parasitic helminths are addressed in applications from the laboratory to fieldwork, such as DNA barcoding and the environmental DNA manipulation of helminthic parasites. The chapters were arranged from the principle to the applications to support non-systematists who would like to conduct molecular systematics, particularly parasitic helminths, as the friendly handbook.

Chapter 2 provided general information on parasitic helminths, particularly the species reported as of medical and public health importance, including general information on other helminth-like parasites infecting humans. Essential information like biology, taxonomy, diversity, and classification is also reviewed in this chapter. A list of parasitic helminths and other parasites updated from the NHM Host-Parasite Database is provided in the Appendix. Nucleotide, protein sequences, and genome information from the public databases were congregated by extracting data from the NHM Host-Parasite Database and selected parameters from the National Center for Biotechnology Information (NCBI) database. The information obtained can serve as helpful information for helminth molecular systematics.

With the accelerating rate of disease emergence in wildlife, livestock, and humans, possibly attributed to global climate change, the topic of parasite diversity and its distribution is concerning.

Chapter 3 provided an overview of parasitic helminth diversity and how molecular systematics is involved. This chapter focused on helminth diversity by emphasizing parasitic helminths, which are reported to be distributed globally with complex relationships between hosts and parasites. This chapter also introduced how the evolutionary process allowed adaptation and coevolution between the parasite and their hosts and the factors that contributed to enhancing the genetic diversity of parasites. The genetic adaptation and variability of parasites during climate change (under unpredictable conditions) lead to species-specific adaptation of the parasites to environmental conditions through host-parasite interaction. Trends in molecular systematic studies in helminthic parasite diversity were also reviewed using the various molecular genetic markers from the past to present. Additionally, phylogenomics, which is the interface between molecular and evolutionary relationships involving genome data, is also mentioned. In defining the origins of parasitism in different lineages, understanding the assembly of hosts and parasites in various ecosystems, mapping the patterns of parasite diversification, and revealing the evolutionary patterns in developmental and other systems, phylogenomics is the trend for further work on molecular systematics.

To understand evolutionary relationships, fundamental knowledge of molecular evolution is essential. Evolutionary forces act upon the genome makeup of populations. An essential component of the evolutionary process is changing gene frequencies with time at the population level, which is critical to whether speciation occurs. Chapter 4 provided the principle of molecular evolution, emphasizing nucleotide sequences. Genomic diversity in Platyhelminthes and Nematoda is shown through parasitic helminth genomes. Principle information of genes and gene structures of eukaryotic organisms was mentioned, including the genetic codes of both nuclear and mitochondrial genes. Moreover, nucleotide and chromosomal mutation was elaborated on, as it is the engine of evolution to generate genetic variation. The dynamics of genes in populations under various evolutionary forces causes a change in allele frequencies. A new mutant variant can completely replace a previous one in the population at the molecular level by slightly affecting phenotypes and fitness. Under evolutionary forces, gene substitutions occur in succession, with one allele replacing another and replacing it in time with a newer allele, called the rate of gene substitution (the number of fixations of new alleles per unit time). Genetic polymorphisms of the population are thus seen under evolutionary forces.

As changes in nucleotide sequences are used in molecular evolutionary studies, the evolutionary rate used for reconstructing the evolutionary history of organisms has to be estimated. Two different sequences are needed to study the genetic relationship between the common ancestral and descendant sequences by using statistical methods as a model to detect evolutionary changes in DNA sequences. The series of nucleotide substitutions in new alleles arises and is fixed in the population as the nucleotide substitution model. The essential molecular information of parasitic helminths for molecular analysis like codon usage bias (the unequal usage of the synonymous codon) was also updated by the reporting of GC content, which is indicative of the observed variation in the effective number of codons, codon bias measure, and amino acid frequency differences. This chapter also mentioned the main factors that drive genetic change over time in the genome by observing different evolutionary time scales such as microevolution and macroevolution. Uncertainty of the continuity of micro- and macroevolutions was also indicated, and the paradox inferred by opponents of those two different evolutionary levels could be eliminated by a hierarchical application of the geometric-mean fitness principle (Simons, 2002).

As mentioned in Chap. 3, the genomes of parasites are affected by environmental changes, which in turn drive the evolutionary process. The evolution of parasite genomes is reflected in their widespread diversity, and the diversity can be transferred from one generation to another, making it possible to trace the lines of descent of a gene back through the generations. As the molecular genetic marker, mitochondrial DNA is mainly used to measure genetic variations and gene genealogies for parasitic helminths. Mitogenome data offers a source of markers for the systematics and population genetics of socioeconomically important parasitic helminths of humans and other animals. A list of updated complete mitochondrial genomes (mitogenomes) of parasitic helminths was compiled from the NCBI database in the Appendix, serving as essential information for diagnosis, epidemiology, and ecology of infections.

Through reproductive isolation, diverged lineages can arise, and genetic divergence is enhanced over time. However, determining species boundaries and discovering new species (species delimitation) remains a challenge. In Chap. 5, essential knowledge was provided to deal with the challenges of species delimitation. Species concepts, which are the criteria of species delimitation and the principle of speciation, particularly parasitic speciation, were mentioned. Additionally, the pitfalls and problematic issues for species identification based on morphological characteristics were discussed by focusing on cryptic species, species hybridization, homoplasy under convergent evolution, and phenotypic plasticity under adaptative evolution.

For species delimitation, utilizing DNA sequences as particular targets for PCR can represent proper resolution for molecular aspects, including molecular identification and molecular systematics. Chapter 6 provided general information on molecular genetic markers utilized in parasitic helminth research. General properties, utilities, and limitations of nuclear and mitochondrial genetic markers, including nuclear ribosomal RNA genes and spacer regions, nuclear protein-coding genes, mitochondrial protein-coding genes, mitochondrial ribosomal RNA genes, and whole mitochondrial genomes, were mentioned. This chapter also gives critical criteria for selecting proper molecular genetic markers for some parasitic helminths. To support beginners in molecular systematic research, practical steps for data mining from nucleotide databases, primer design, in silico PCR, general setup, condition, optimization, troubleshooting for PCR amplification, and post-PCR processes were detailed in Chap. 7. Guidelines for conducting polymerase chain reaction (PCR) and primer design were also provided.

Chapter 8 gave guidelines for analysis and selecting appropriate DNA sequence alignment methods and phylogenetic inferences. The basics of DNA sequence alignment and phylogenetic inferences were provided, including nucleotide substitution saturation, homogeneity between sequences, types of nucleotide substitution models, and phylogenetic trees. The measurement of phylogenetic tree confidence in traditional phylogenetics and Bayesian inference was explained. Additionally, other types of the phylogenetic tree were also mentioned, such as the secondary structure of the rRNA and the amino acid sequences of protein-coding genes. These can be the alternative characters for inferring evolutionary relationships. The practical steps for multiple sequence alignments and phylogenetic analysis, including selecting the proper outgroup, were demonstrated.

Before starting a molecular systematics study, morphological identification is essential for producing reliable data. Obtaining high-quality specimens through appropriate specimen preparation is crucial for downstream analysis. Well-prepared and high-quality specimens will also allow us to make comparisons between the morphological, functional, and ecological aspects of parasitic helminths. Chapter 9 provided standard protocols for high-quality sample preparation to ensure the specimens were maintained in the best condition for classical identification or molecular studies.

Chapters 10 to 12 focused on the molecular systematics of parasitic nematodes, trematodes, and cestodes. The best practice for molecular phylogenetic is the backbone of systematic works using the nuclear ribosomal gene, mitochondrial genomes, and whole genomes for medically important nematodes (in clades I, III, IV, and V), medically important trematodes (orders Diplostomida, Plagiorchiida), and medically important cestodes (orders Cyclophyllidea and Diphyllobothridea). The challenges and trends of future works on molecular systematics of those three groups of parasitic helminths were discussed.

The molecular approach has been implemented to support species delimitation for parasitic helminths by fulfilling the gaps in helminth species identification that morphology alone cannot do, particularly species in cryptic lineages. Chapter 13 demonstrated that molecular advances allowed DNA use for taxonomical studies, bringing about DNA taxonomy and barcoding. Examples of the molecular identification of the three groups of parasitic helminths and gaps in helminth DNA barcoding progress were detailed. Practical steps for DNA barcoding and molecular identification of helminths were given, followed by recommendations for application.

Chapter 14 demonstrated the implementation of molecular genetic makers from molecular systematics to DNA taxonomy and field applications. In recent years, a shift towards using next-generation sequencing (NGS) technologies has been prominent, and these molecular advances are playing a critical role in species identification, monitoring, and disease diagnosis for a massive variety of organisms. This chapter discussed the overview, advantages and applications, and examples of DNA metabarcoding for helminths. DNA metabarcoding has been applied to helminths in various contexts, with examples ranging from biodiversity and conservation, biomonitoring, and noninvasive detection to monitoring parasitic helminths. Further research and other field applications for helminth DNA metabarcoding were also suggested. Finally, concluding remarks and future perspectives on parasitic helminth research were proposed in Chap. 15. A glossary was provided at the end of this book to assist readers in easily understanding the contents discussed.

References

Blasco-Costa, I., Cutmore, S. C., Miller, T. L., & Nolan, M. J. (2016). Molecular approaches to trematode systematics: 'Best practice' and implications for future study. *Systematic Parasitology, 93*(3), 295–306. https://doi.org/10.1007/s11230-016-9631-2

Brooks, D. R. (1985). Phylogenetics and the future of helminth systematics. *The Journal of Parasitology, 71*, 719–727.

Brooks, D. R., & Hoberg, E. P. (2001). Parasite systematics in the 21st century: Opportunities and obstacles. *Trends in Parasitology, 17*(6), 273–275.

Forsman, A. (2015). Rethinking phenotypic plasticity and its consequences for individuals, populations and species. *Heredity, 115*(4), 276–284. https://doi.org/10.1038/hdy.2014.92

Holterman, M. (2007). *Phylogenetic relationships within the phylum Nematoda as revealed by ribosomal DNA, and their biological implications*. PhD thesis, Wageningen University.

Jeon, H. K., & Eom, K. S. (2013). Molecular approaches to *Taenia asiatica*. *The Korean Journal of Parasitology, 51*(1), 1–8. https://doi.org/10.3347/kjp.2013.51.1.1

Jeon, H. K., Kim, K. H., & Eom, K. S. (2011). Molecular identification of *Taenia* specimens after long-term preservation in formalin. *Parasitology International, 60*(2), 203–205. https://doi.org/10.1016/j.parint.2010.12.001

Moradi, M., Meamar, A. R., Akhlaghi, L., Roozbehani, M., & Razmjou, E. (2019). Detection and genetic characterization of *Echinococcus granulosus* mitochondrial DNA in serum and formalin-fixed paraffin embedded cyst tissue samples of cystic echinococcosis patients. *PLoS One, 14*(10), e0224501. https://doi.org/10.1371/journal.pone.0224501

Morand, S. (2015). (macro-) evolutionary ecology of parasite diversity: From determinants of parasite species richness to host diversification. *International Journal for Parasitology: Parasites and Wildlife, 4*(1), 80–87. https://doi.org/10.1016/j.ijppaw.2015.01.001

Nadler, S. A., & De León, G. P. P. (2011). Integrating molecular and morphological approaches for characterizing parasite cryptic species: Implications for parasitology. *Parasitology, 138*(13), 1688–1709.

Nolan, M. J., & Cribb, T. H. (2005). The use and implications of ribosomal DNA sequencing for the discrimination of digenean species. *Advances in Parasitology, 60*, 101–163.

Schneider, R., Gollackner, B., Edel, B., Schmid, K., Wrba, F., Tucek, G., Walochnik, J., & Auer, H. (2008). Development of a new PCR protocol for the detection of species and genotypes (strains) of *Echinococcus* in formalin-fixed, paraffin-embedded tissues. *International Journal for Parasitology, 38*(8–9), 1065–1071. https://doi.org/10.1016/j.ijpara.2007.11.008

Simons A.M. (2002) The continuity of microevolution and macroevolution. Journal of Evolutionary Biology. 15(5), 688-701. https://doi.org/10.1046/j.1420-9101.2002.00437.x

Spakulova, M. (2002). Current problems of helminth taxonomy: Species level reflections. *Helminthologia, 39*(2), 111–118.

Chapter 2
Parasitic Helminths of Medical and Public Health Importance

Abstract This chapter provided general information on parasitic helminths, including biology, taxonomic classification, a list of parasitic helminths of medical and public health importance globally, and their impacts on humans to address some fundamental questions on how many parasitic helminth species have ever been reported to infect humans and which parasite taxonomic groups are of scientific significance. Research interests in medical parasitology, molecular biology, health, and life science information were obtained mainly from the Host-Parasite Database of the Natural History Museum, London (NHM), and other published sources. This chapter also discussed the diversity of parasitic helminths and their relationships based on classical systematics and taxonomy.

2.1 General Information and Impacts of Parasitic Helminths

The word "helminth" is the common term (meaning "worm" in Greek) used to refer to wormlike eukaryotic, multicellular endoparasites of animals and humans. Helminths are known as one of the most common parasitic infections in animals and humans throughout the world. Helminths are of different sizes, ranging from very small, less than a millimeter, to several meters in length. They live as parasites within the host body's systems, e.g., the digestive, excretory, blood, respiratory, musculoskeletal, and nervous systems. Parasitic helminths represent important public health problems with significant economic impact worldwide, particularly in tropical and subtropical regions.

Parasitic helminths have coevolved with humans over thousands of years, with over one-quarter of the world's population estimated to be infected by helminths (Maizels et al., 2009; Ryan et al., 2020). There are a variety of impacts caused by helminths, from mild to potential disabilities or mortalities. The direct effects of helminths on their hosts include robbing effects, nutritional competition, and direct trauma (e.g., blood suckling, lumen obstruction, ulceration) (Crompton & Nesheim, 2002; Loukas et al., 2016; Ünal et al., 2020), whereas indirect injurious effects include the induction of secondary infection by pathogenic microbes and chronic

U. Thaenkham et al., *Molecular Systematics of Parasitic Helminths*,
https://doi.org/10.1007/978-981-19-1786-8_2

disorders such as cellular reactions, fibrosis, calcification, or malignant neoplasm (Brindley & Loukas, 2017; Prueksapanich et al., 2018; Reynolds et al., 2017). In most cases, helminths that parasitize the gastrointestinal tract of hosts tend not to result in severe symptoms or serious pathology, as compared to those parasitizing the extraintestinal system. For example, soil-transmitted helminths (STHs), minute intestinal flukes, or adult tapeworm infections in the gastrointestinal tract have a relatively high prevalence in endemic areas. However, infections are generally asymptomatic or present only mild symptoms such as weakness, nausea, abdominal pain, diarrhea, or constipation in the cases with mild or moderate infections (Chai & Jung, 2017; García et al., 2003; WHO, 2017). Unlike intestinal helminths, infection with tissue parasites can cause relatively severe pathological effects, usually with more prominent health impacts, mainly when parasites migrate to the vital organs such as the central nervous system (CNS) or eyes. For example, filarial nematodes causing lymphatic filariasis (or human elephantiasis) in Asia and ocular onchocerciasis (or river blindness) in sub-Saharan Africa can result in the patient's inability to live normally or be permanently disabled (Hotez et al., 2008; Zeldenryk et al., 2011). Some stages of parasites (usually larvae or eggs), ingested via contaminated water or intermediate hosts, can cause eosinophilic meningoencephalitis in humans through migration to the CNS, causing trauma and inflammation. These include helminths that cause cysticercosis, gnathostomiasis, angiostrongyliasis, toxocariasis, and schistosomiasis (Graeff-Teixeira et al., 2009; Ramirez-Avila et al., 2009). Long-term parasitic infection can produce irreversible functional changes in tissues, e.g., hyperplasia or cancer. The liver flukes *Opisthorchis viverrini* and *Clonorchis sinensis* are examples where chronic infections result in cholangiocarcinoma or bile duct cancer (Qian & Zhou, 2017; Sripa et al., 2011).

The negative impacts of parasitic helminths aside, there are also positive aspects of host-parasite interactions, where recent researches claim the beneficial impact of helminths on their hosts. Related to part of the "hygiene hypothesis," researchers note the positive role that parasitic infections play in the proper maturation of the host's immune system (Briggs et al., 2016; Caraballo, 2018). Helminth infections are known to modulate host immune responses in different ways, including allergic reactivity reduction and the suppression of inflammatory processes and immunopathology (La Flamme et al., 2003; McSorley et al., 2011; Summers et al., 2003). They have been recognized as potential modifiers of allergy, chronic inflammatory diseases, and auto-immune disorders, shedding light on an alternative medical treatment for these noncommunicable diseases (Sipahi & Baptista, 2017). Therefore, biological products derived from helminths such as antigenic proteins and excretory-secretory products (ES) can be used as potential new drugs.

Parasitic helminths of medical and public health importance are generally comprised of the two major phyla of flatworms and roundworms, i.e., Phylum Platyhelminthes (cestodes and digenean trematodes) and Phylum Nematoda (nematodes) (Brindley et al., 2009; Hotez et al., 2008). Parasitic helminths also include other organisms in the phylum Acanthocephala and the phylum Nematomorpha, as they are found occasionally parasitizing humans (Fig. 2.1). Here, we focus on Platyhelminthes and Nematoda as the main parasitic helminths of medical and public

Taxonomic rank	
Domain: Eukaryota (Chatton, 1925)	Helminths
Kingdom: Animalia Linnaeus, 1758	
Phylum: Platyhelminthes Claus, 1887	
Class: Cestoda Rudolphi, 1808	
Class: Trematoda Rudolphi, 1808	
Phylum: Nematoda Diesing, 1861	
Class: Chromadorea Inglis, 1983	
Class: Enoplea Inglis, 1983	
Phylum: Acanthocephala Koelreuter, 1771	Other parasites
Class: Archiacanthocephala Meyer, 1931	
Class: Eoacanthocephala Van Cleave, 1936	
Class: Palaeacanthocephala Meyer, 1931	
Phylum: Nematomorpha Vejdovsky, 1886	
Class: Gordioida Vejdovsky, 1886	

Fig. 2.1 Taxonomic rank of parasitic helminths and other helminth-like parasites known to infect humans

health importance. The morphological classification and identification of helminths depend on several features, including body shape, body cavity, integument, digestive system, reproductive system, attachment organs, eggs, host group, microhabitat (target organ inside the hosts), and life cycle. Common developmental stages of parasitic helminths can be roughly divided into the three main stages: egg, larva, and adult, with some further specialized stages in each group.

In general, helminths are unable to multiply in the hosts that other microorganisms like viruses, bacteria, and fungi can. Therefore, the level of parasitic burden (whether mild, moderate, or heavy) depends on the quantity of infective stages being uptaken by the host. Infective stages (larval forms) are either free-living or parasitize intermediate or paratenic hosts and are subsequently ingested by definitive hosts where helminths can develop further into adults to start sexual reproduction. The life cycles of parasitic helminths can be divided into two types: (1) direct life cycle (monoxenous)—intermediate host is not required—and (2) indirect life cycle (heteroxenous)—at least one or two intermediate hosts are required for biological development, usually asexual reproduction. Knowledge of helminth life cycles helps understand how parasites are transmitted to the host. The three modes of transmission include (1) oral route via consumption, (2) vector-borne transmission, and (3) skin penetration. Recognizing the developmental stages, life cycles, and transmission routes of parasitic helminths is an essential basis for understanding better the epidemiology, diagnosis, and pathology of helminthiasis and developing a better plan for disease prevention and control.

2.1.1 Cestode

Tapeworms, or cestodes, are internal parasites of vertebrate hosts, including humans. Making up a very diversified group, it is believed that almost 4000 species have been described, and nearly every vertebrate species on earth is parasitized by at least one or more species of cestodes (Schmidt, 1986). Adult worms are long and flat-segmented with bilateral symmetry, acoelomate, and no digestive tract. The body is divided into the head (scolex) and strobila, which comprise a short neck portion and a long part of a chain of segments or proglottids. The proglottids are divided into three types according to the segments' development and maturity of reproductive organs. The three types are the immature segment (reproductive organ absent), mature segment (both male and female reproductive organs fully developed), and gravid segment (containing fertilized eggs in the uterus). Proglottids develop progressively downward in the posterior direction, and the gravid segments are usually shed through fecal material into the environment. The scolex bears different forms of attachment organs for attaching to the host intestinal wall. The various scolex forms include four muscular suckers (sucking groove -bothria), specialized apical organs, and a rostellum unarmed or armed with rows of hooks. All cestodes are monoecious (hermaphrodites), having both male and female reproductive organs in the same individual.

The life cycle of cestodes is indirect, requiring one or two intermediate hosts for their development and reproduction. Species infecting humans utilize either the aquatic or terrestrial life cycles. The aquatic cycle is usually the strategy of the cestode species in the order Pseudophyllidea, and three hosts are involved in the cestode developmental process. Eggs are shed into the water with the coracidium (ciliated oncosphere embryo) developed and eventually hatched out. Either egg-containing coracidia or freely swimming coracidia are ingested by small aquatic arthropods (first intermediate host) and further develop into the next larval stage called procercoid. Once the infected arthropods are eaten by bigger aquatic predators (usually fishes), the procercoid develops further into a plerocercoid in the fish muscle (second intermediate host). The plerocercoid has an elongated body with a wormlike appearance living in the second intermediate host without forming a cyst ("solid-form metacestode"). This infective stage plerocercoid is then passed to a definitive host (i.e., appropriate vertebrate species and humans) through predation. In the definitive host, the ingested plerocercoid attaches to the host's gut and further develops into adult worms, completing the life cycle. For the terrestrial cycle, cestodes in the order Cyclophyllidea use this strategy to complete their developmental process. Embryonated eggs are passed around in the environment with a fully developed oncosphere (hexacanth embryo) inside. The infective eggs are ingested by arthropods and herbivores, which act as appropriate intermediate hosts. In the intermediate hosts, the oncosphere larvae hatch and penetrate through the intestinal mucosa to develop into either cysticercoid larvae in the coelom of arthropods or different types of cystic larvae in the tissues or internal organs of vertebrates. These cyst-form metacestodes consist of cysticercoid, a microscopic-sized metacestode in a small arthropod; cysticercus, a big fluid-filled bladder containing a protoscolex larva; and coenurus and hydatid cyst, a fluid-filled cyst with multiple protoscolices inside. Finally, when definitive hosts such as carnivores and humans ingest the metacestodes, the parasite excysts and attaches to the host's intestinal wall and develops further to reach the adult stage in the intestinal tract.

Examples of cestodes of medical and public health importance include the pig tapeworm *Taenia solium*, causing either taeniasis when adult cestodes occupy the human intestinal tract or cysticercosis when humans serve as intermediate hosts after ingesting *T. solium* eggs via contaminated food. Cysticercosis in humans results when the oncosphere larvae develop into metacestodes and encyst in muscles and internal organs. In cases where the metacestodes migrate to vital organs such as the CNS, more severe neurologic symptoms with a high impact on morbidity or even mortality result (Martins-Melo et al., 2016; Sorvillo et al., 2007).

Echinococcosis, or hydatid disease, is another cestode infection causing a serious public health problem in many areas of the world. Small cestode species of wild and domestic canids, e.g., *Echinococcus granulosus* and *E. multilocularis*, can be transmitted to humans by having close contact with dogs or exposure to dog feces. Like with *Taenia solium* cysticercosis, humans become infected through the accidental ingestion of food products contaminated with eggs. Oncosphere larvae develop into unilocular or multilocular hydatids encysting in tissues and several internal organs, including the liver, lung, spleen, kidney, heart, brain, and bone in humans (Martins-Melo et al., 2016).

As a member of the pseudophyllidean cestodes, the broad-fish tapeworm *Diphyllobothrium latum* is recognized as a fish-borne parasitic infection in humans. This cestode species is believed to be the longest parasite species infecting humans, growing to an estimated 30 ft in length (CDC, 2021). The tapeworm can infect any age group or gender but is frequently reported in middle-aged men with a habit of raw fish consumption. With the increasing popularity of raw fish menus such as sashimi and sushi, several cases of human diphyllobothriasis have emerged recently, particularly in urban and developed countries (Scholz et al., 2009). The clinical manifestation of the broad-fish tapeworm infection is relatively minor or asymptomatic.

2.1.2 Digenean Trematodes

Trematodes are classified in the phylum Platyhelminthes, class Trematoda, comprising parasitic flatworm species that usually occupy invertebrates and vertebrates as their intermediate and definitive hosts. Within the class Trematoda, two groups belong to the members, i.e., subclass Aspidogastrea and subclass Digenea. The former contains internal parasites of freshwater fishes, turtles, and mollusks, and they have a large attachment organ (holdfast) covering most of the ventral surface of their bodies. The latter is much larger, comprising over 2500 genera of internal parasitic flukes, which usually use mollusks as intermediate hosts and vertebrates as definitive hosts (Gibson et al., 2002).

The body of digenean flukes is dorsoventrally flattened, unsegmented, and covered by a tegument armed or unarmed with spines. They have one or two muscular suckers, i.e., the oral sucker is usually used for feeding, whereas the ventral or posterior sucker (acetabulum) is used mainly for attachment. The digestive system is incomplete, containing two blind bifurcated intestinal ceca along the lateral sides of the body. Almost all digenean flukes are hermaphrodite, except for the members of the order Strigeidida that are dioecious with prominent sexual dimorphism. The reproductive organs of digeneans usually comprise two or more testes but only a single ovary. The uterus is filled with eggs containing miracidium, which is a ciliated larva.

The life cycle of Digenea involves different larval stages that develop asexually (multiplication of germinal cells) in one or two mollusk intermediate hosts. These larval stages include miracidium, sporocyst, redia, cercaria, and finally metacercaria as the infective stage, before developing into sexually mature adult worms in definitive hosts. In brief, the development of the larval stages is as follows: after hatching from eggs, the swimming miracidium penetrates a suitable mollusk, the first intermediate host. The miracidium develops into sporocysts, rediae, and cercariae through a polyembryony mechanism in the first intermediate host, and cercariae are reproduced from either sporocyst or redia. The cercariae leave the first intermediate host to encyst as metacercariae in the second intermediate hosts, e.g., fishes, tadpoles, crustaceans, mollusks, aquatic vegetation, or solid objects in the water body. In some digenetic trematodes, particularly the order Strigeidida (e.g., blood flukes), the cercariae penetrate definitive hosts directly without needing a second intermediate host in the life cycle.

Here, some important human trematodiases are raised as examples. Schistosomiasis is a neglected tropical parasitic disease caused by blood fluke species in the genus *Schistosoma*, including some important species such as *Schistosoma haematobium, S. mansoni, S. japonicum, S. guineensis, S. intercalatum, and S. mekongi* (McManus et al., 2018). People in several areas of the tropics, i.e., sub-Saharan Africa, South America, Southeast Asia, and the Middle East, are affected. *Schistosoma* infection is considered the second most socioeconomically serious parasitic disease after malaria (Waknine-Grinberg et al., 2010). Transmission to humans occurs when the infective cercarial stage from snail intermediate hosts penetrates human skin and develops into mature adult worms in blood vessels. After fertilization, eggs from female worms are released and trapped in tissues and organs like the liver and urinary bladder, causing granulomatous and inflammatory responses in the organs where eggs are deposited. Chronic schistosomiasis can cause abdominal pain, enlarged liver, bloody stool, and increased liver fibrosis and cancer (Abdelghani et al., 2020; Calvisi, 2020). In addition, skin penetration of cercariae through the skin can result in a skin allergic reaction (rash) or cercarial dermatitis.

Food-borne trematode infections are also another important neglected parasitic disease and have recently been estimated to infect more than 85 million people worldwide (Sithithaworn et al., 2014). These parasites include the liver flukes, e.g., *Clonorchis sinensis, Opisthorchis viverrini,* and *Fasciola hepatica*; the intestinal flukes, e.g., *Fasciolopsis buski, Echinostoma* spp., *Haplorchis* spp.; and the lung flukes, e.g., *Paragonimus* spp. Consumption of traditional dishes and personal preference for raw or improperly cooked meat, mainly aquatic animals such as fishes, crustaceans, and snails, are the most important factors in determining trematode infection risk in humans.

2.1.3 Nematodes

Nematodes, or roundworms, are classified in the phylum Nematoda (or Nemathelminthes) and are believed to be highly diversified, falling behind only arthropods. They live either as free-living forms in various environments or as parasitic forms in plants, animals, and humans. In a survey of an animal biodiversity project, over 24,700 species of nematodes have been described (Hodda, 2011). These include 10,600 free-living species in freshwater, marine, and terrestrial habitats and over 15,000 parasitic nematode species of plants, vertebrates, and invertebrate animals (Hugot et al., 2001; Poulin & Morand, 2000). About 5% of living species of Nematoda are known to science, whereas the majority are waiting to be discovered and described (Morand et al., 2006). More information on parasite diversity will be more detailed in Chap. 3.

Nematodes are wormlike with an elongated, cylindrical, unsegmented body and bilateral symmetry. Adults vary in size, ranging from as tiny as about a millimeter (e.g., *Trichinella spiralis* and *Strongyloides stercoralis*), to 20–30 cm (e.g., *Ascaris lumbricoides*), or even up to a meter in length (e.g., *Dracunculus medinensis*). The body is covered with a thick collagenous cuticle layer varying in patterns and shapes, e.g., smooth, striation, ridge, spines, annulation, or punctuation. Nematodes undergo periodic molting through the shedding of their cuticle. The body cavity is a pseudocoelom (or false cavity), in which a fluid-filled cavity lies between the muscular layer and the digestive tract, in which all the visceral organs are suspended. The digestive system is complete, consisting of the mouth at the anterior part, esophagus, intestine, and anus located at the posterior end. There are different types of esophageal parts, which characteristically vary in shape and structure (e.g., rhabditiform, filariform, bulb, muscular-glandular, and stichosomal types), and are informative for nematode taxonomy and classification. The chemosensory organs, also known as amphids or phasmids, are present on either the anterior or posterior parts of the body.

Contrary to the digestive system, the respiratory and circulatory systems are not developed; sexual dimorphism is present in parasitic nematodes, where the male is usually smaller than the female, and the caudal end is coiled or curved. The male reproductive system comprises a single tubule of the testis, vas deferens, and seminal vesicle. Additionally, copulatory organs such as the spicule and bursa are present for members of the order Strongylida). On the other hand, the female reproductive system consists of the ovary, oviduct, seminal receptacle, uterus, vagina, and vulva. Females are able to lay eggs (oviparous) or have eggs with larvae that hatch in the reproductive tract before passing out (ovoviviparous), or the larvae can hatch directly (viviparous). Internal fertilization is used as the main strategy for sexual reproduction.

The life cycles of parasitic nematodes are either direct (not requiring an intermediate host) or indirect (complex), requiring one or more intermediate hosts. Six stages of nematode development are mainly applied to the majority, and the stages include eggs, four larval stages (L1, L2, L3, L4), and one adult stage. The L3 is usually the infective stage of several nematode groups. In some cases, the sexually immature adult worm is called L5 or subadult. Nematodes molt and shed their cuticles when developing from one stage to the next. In terms of transmission, there are several routes that the parasite can infect humans. The routes are skin penetration (e.g., hookworm and *S. stercoralis*); ingestion of infective egg (e.g., *Ascaris lumbricoides, Trichuris trichiura,* and *Enterobius vermicularis*), larvae in intermediate hosts (e.g., *Angiostrongylus cantonensis* and *Gnathostoma spinigerum*), and larvae encysted in muscle (e.g., *Trichinella spiralis*); arthropod vectors (e.g., *Brugia malayi, Wuchereria bancrofti* and *Onchocerca volvulus*); and inhalation of infective eggs contaminated in dust or airborne particles (e.g., *Enterobius vermicularis*) (Ghosh, 2018).

Examples of medically important nematodes include soil-transmitted helminths, a group of widespread intestinal nematode infections in humans. They include *Ascaris lumbricoides* (roundworm), *Ancylostoma duodenale* (hookworm), *Necator americanus* (hookworm), and *Trichuris trichiura* (whipworm). Although STH infection may not be fatal, it significantly impacts a child's nutritional impairment, growth, cognition, well-being, and welfare due to parasite-induced morbidity. Estimation revealed that the global burden of intestinal nematodes results in 39 million disability-adjusted life years (DALYs) lost, comparable with 35.7 million for malaria (Crompton & Nesheim, 2002).

Strongyloidiasis, caused by *S. stercoralis,* is endemic in tropical and subtropical regions (South America and Southeast Asia). *Strongyloides stercoralis* is known as the smallest nematode species (about 1–2.5 mm in length), causing disease in humans (Ghosh, 2018). *Strongyloides stercoralis* has a complex life cycle and can develop parasitic cycles in hosts and the free-living cycle in soil. Additionally, it can survive for long periods in the environment. Infection in humans occurs through contact with soil contaminated by the infective stage larva and usually causes mild symptoms or is asymptomatic. More importantly, immunocompromised patients are at high risk of developing *Strongyloides* hyperinfection syndrome with chronic infection. Hyperinfection results in a life-threatening situation when the larvae proliferate and disseminate to multiple organs, leading to organ failures and systemic sepsis (Greaves et al., 2013).

Trichinellosis is another infection (genus *Trichinella*, e.g., *T. spiralis, T. pseudospiralis, T. papuae,* and *T. britovi*) causing serious diseases in humans via the consumption of raw or improperly cooked meat from domestic pigs, wild boars, and other game or wildlife meat. After ingestion, adults reside and reproduce in the gastrointestinal tract, causing abdominal discomfort with mild to moderate symptoms such as nausea, vomiting, and diarrhea. Thereafter, the larvae migrate to encysted muscles, leading to severe muscle pain, inflammation, rash, fever, and eye swelling. The severity of infection largely depends on the number of worms ingested, in which case fatality can occur with heavy infection. This food-borne parasitic zoonosis occurs worldwide and is particularly endemic in East and Southeast Asia, the Pacific Islands, as well as sporadic incidences in Eastern Europe, the former USSR, and developed European countries such as France and Italy, where people consume raw horse meat (Bai et al., 2017; Kusolsuk et al., 2010; Pozio, 2001).

The various life cycle routes and transmission strategies, habitats, and ability to infect different host species are factors contributing to the species diversity of helminths and their cosmopolitan distribution, leading to the occurrence of helminthic diseases worldwide. Table 2.1 summarizes the key features of cestodes, trematodes, and nematodes in terms of their main morphological characteristics, life cycle and transmission routes, and important helminthic diseases. Although helminthic diseases are grouped under the neglected tropical disease (NTD) category, they still present a significant health and economic burden on many countries. Thus, the taxonomy of helminths is essential in their classification and identification to identify species that cause disease and detection and to understand disease transmission dynamics and distribution.

2.2 Diversity and Classification of Parasitic Helminths of Public Health Importance

Research and academic information on the diversity of parasitic helminths affecting human and public health concerns are available in classical and modern parasitology textbooks and research publications in scientific journals. However, these data are relatively sparse and not easily accessible. Here, a review of a list of parasitic helminths (Platyhelminthes and Nematoda) and other parasites (Acanthocephala and Nematomorpha) of medical and public health importance was obtained from the Host-Parasite Database of the Natural History Museum, London (Gibson et al., 2005). Data search was done via host species (i.e., *Homo sapiens* was specified in this case), and data was retrieved from all geographic locations. Taxonomic classification and synonyms of each parasite species were verified and updated via the NCBI Taxonomy Browser tool (https://www.ncbi.nlm.nih.gov/Taxonomy/Browser/wwwtax.cgi) and the Global Biodiversity Information Facility (GBIF, https://www.gbif.org/).

Search results revealed a high diversity of parasitic helminths and other wormlike parasites of humans, with 421 species reported in scientific publications since 1922 (Gibson et al., 2005). Platyhelminthes appeared as the majority with 228 species

Table 2.1 Summary of key features of cestodes, trematodes, and nematodes

Helminth	Cestodes	Trematodes	Nematodes
Key morphology	• Long, flat, segmented body • Bilateral symmetry • Acoelomate • No digestive system • Monoecious • Body divided into scolex and strobila	• Dorsoventrally flattened, unsegmented body • One or two muscular suckers • Incomplete digestive system • Monoecious (except *Schistosoma*)	• Elongated, cylindrical, unsegmented body • Body covered with cuticle • Pseudocoelom • Complete digestive system • Dioecious
Life cycle	• Indirect • Aquatic intermediate hosts (Pseudophyllidea) and terrestrial intermediate hosts (Cyclophyllidea)	• Indirect • Mollusk as first intermediate hosts and usually fishes or aquatic organisms as second intermediate hosts	• Indirect or direct • Various host types in terrestrial and aquatic environments
Development stages	*Pseudophyllidea*: Egg → coracidium → procercoid → pleurocercoid → adult *Cyclophyllidea*: Egg → cysticercoid → adult	Egg → miracidium → sporocyst → redia → cercaria → metacercaria → adult	Egg → L1 to L4 → adult
Examples of medically important species and transmission to humans	*Taenia solium*: Ingestion of egg or cysticercoid in undercooked pork *Echinococcus granulosus*: Ingestion of egg *Diphyllobothrium latum*: Ingestion of pleurocercoid in undercooked fish	*Schistosoma* sp.: Skin penetration of cercaria *Fish-borne* (*Opisthorchis viverrini*, *Clonorchis sinensis*, *Haplorchis* sp.): Ingestion of metacercaria in undercooked fish *Fasciola hepatica*: Ingestion of metacercaria on vegetables	*Soil-transmitted* (*Ascaris lumbricoides*, *Trichuris trichiura*, *Strongyloides stercoralis*, *Necator americanus*): Ingestion of egg or skin penetration of filariform larva *Vector-borne* (*Brugia malayi*, *Onchocerca volvulus*): Bite of arthropod with L3 *Food-borne* (*Angiostrongylus cantonensis*, *Gnathostoma spinigerum*, *Trichinella spiralis*): Ingestion of L3 in undercooked intermediate hosts

Table 2.2 Summarized taxon number and diversity of medically important helminths and other parasites (Data source: NHM Host-Parasite Database)

Phylum	Class	Order	Family	Genus	Species
Platyhelminthes	2	9	39	91	228
	Cestoda	4	10	27	71
	Trematoda	5	28	64	157
Nematoda	2	10	37	71	157
	Enoplea	4	7	11	21
	Chromadorea	6	30	60	136
Acanthocephala	3	5	6	8	10
Nematomorpha	1	2	3	9	26

documented, followed by Nematoda, Nematomorpha, and Acanthocephala with 157, 26, and 10 species reported in humans, respectively (Table 2.2).

2.2.1 Diversity and Classification of Parasitic Cestodes

Tapeworms or cestodes are classified in the class Cestoda alongside two other classes: Trematoda (digenean fluke) and Monogenea in the Neodermata clade (the group of obligate parasitic Platyhelminthes) of the phylum Platyhelminthes (Littlewood, 2008; Olson et al., 2001) (Fig. 2.2). All human parasitic tapeworms are classified into a single taxon: subclass Eucestoda. Eucestoda comprises several species belonging to the four orders, i.e., order Cyclophyllidea (with the six families: Anoplocephalidae, Davaineidae, Dipylidiidae, Hymenolepididae, Mesocestoididae, and Taeniidae), order Proteocephalidea (single family: Proteocephalidae), order Diphyllobothriidea or Pseudophyllidea (single family: Diphyllobothriidae) and Trypanorhyncha (two families: Hepatoxylidae and Tentaculariidae). Seventy-one species belonging to 27 genera have been documented in the databases.

2.2.2 Diversity and Classification of Parasitic Trematodes

Class Trematoda (Rudolphi, 1808) is divided into two subclasses, i.e., Aspidogastrea and Digenea (Fig. 2.2). All parasitic trematodes infecting humans are only from the latter subclass, whereas the former is a small taxon (around 80 valid species in 12 genera and 4 families), and they are known to parasitize mollusks, fishes, and turtles (Rohde, 2002). Subclass Digenea is obligate parasites in definitive vertebrate hosts and has a large diversity with over 18,000 species in about 2500 genera (Kostadinova & Pérez-del-Olmo, 2014). In the NHM Host-Parasite Database, 157 species in 64 genera of trematodes infecting humans have been documented. They belong to the five orders: Azygiida (one family: Isoparorchiidae),

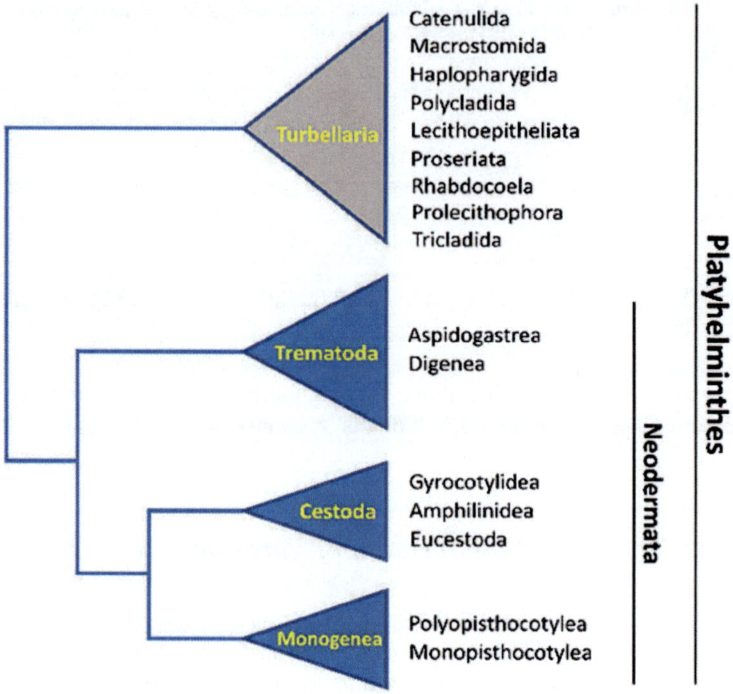

Fig. 2.2 Relationship of the phylum Platyhelminthes showing groups of obligate parasites in Neodermata (e.g., Cestoda and Trematoda) and free-living flatworms (Turbellaria) using morphological and molecular characters based on a review by Littlewood (2008)

Diplostomida (three families: Brachylaimidae, Cyathocotylidae, and Diplostomidae), Opisthorchiida (two families: Heterophyidae and Opisthorchiidae), Plagiorchiida (18 families: Achillurbainiidae, Cathaemasiidae, Dicrocoeliidae, Echinochasmidae, Echinostomatidae, Fasciolidae, Gastrothylacidae, Gymnophallidae, Himasthlidae, Lecithodendriidae, Microphallidae, Nanophyetidae, Notocotylidae, Paramphistomatidae, Phaneropsolidae, Philophthalmidae, Plagiorchiidae, and Troglotrematidae), Strigeidida (four families: Clinostomidae, Schistosomatidae, Spirorchiidae, and Strigeidae).

2.2.3 Diversity and Classification Nematodes

Nematodes have been classified based on recent genetic evidence to be closely related to arthropods (Phylum Arthropoda) in the Superphylum Ecdysozoa, a group of molting animals (Aguinaldo et al., 1997). Recent molecular phylogenies revealed that parasitic and free-living nematodes are comprised of the three monophyletic subclasses, i.e., Dorylaimia, Enoplia, and Chromadoria (Blaxter et al., 1998;

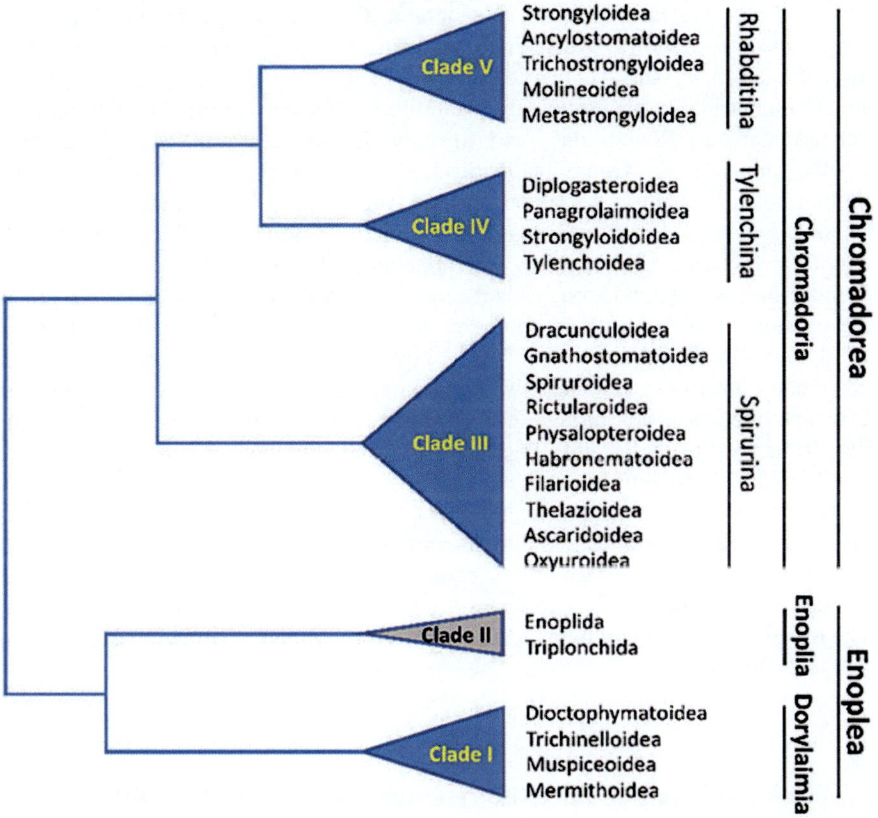

Fig. 2.3 Relationship of the phylum Nematoda using morphological and molecular characters from a compilation of studies. (Blaxter et al., 1998; De Ley & Blaxter, 2002; International Helminth Genomes Consortium, 2019; Morand et al., 2006) Medically important parasitic nematodes are in clades I, III, IV, and V, whereas nonhuman parasitic nematodes are in clade II

De Ley & Blaxter, 2002; International Helminth Genomes Consortium, 2019). The first two subclasses (clades I and II) are derived from the common ancestral class Enoplea, whereas subclass Chromadoria, comprising clades III, IV, and V, is separated. These five main clades, depicted in Fig. 2.3, are Dorylaimia (clade I with an example of order Trichinellida), Enoplia (clade II, where the majority of members are free-living in marine habitats), Spirurina (clade III with some examples in orders Ascaridida, Oxyurida, and Spirurida), Tylenchina (clade IV with various free-living, zooparasitic, and phytoparasitic nematodes), and Rhabditina (clade V with an example of order Strongylida) (Morand et al., 2006).

Nematode parasites of humans comprise members of four clades (i.e., clades I, III, IV, and V) (International Helminth Genomes Consortium, 2019). From the information obtained from the NHM Host-Parasite Database, 157 parasitic nematode species are known to infect humans, including species from the classes

Chromodorea and Enoplea. In Chromadorea, there are six orders involved, i.e., Ascaridida (six families: Anisakidae, Ascarididae, Oxyuridae, Pharyngodonidae, Raphidascarididae, and Toxocaridae), Camallanida (two families: Dracunculidae and Philometridae), Rhabditida (five families: Alloionematidae, Diplogasteridae, Panagrolaimidae, Rhabditidae, and Strongyloididae), Spirurida (eight families: Gnathostomatidae, Gongylonematidae, Onchocercidae, Physalopteridae, Rictulariidae, Setariidae, Tetrameridae, and Thelaziidae), Tylenchida (one family: Meloidogynidae), and Strongylida (eight families: Ancylostomatidae, Angiostrongylidae, Chabertiidae, Haemonchidae, Metastrongylidae, Molineidae, Syngamidae, and Trichostrongylidae). For class Enoplea, nematodes in this group are considered more primitive compared to those in Chromadorea (Blaxter et al., 1998; Schulze & Schierenberg, 2009). Four orders are recognized as parasitic nematodes infecting humans. These include Dioctophymatida (one family: Dioctophymatidae), Mermithida (one family: Mermithidae), Muspiceida (one family: Robertdollfusidae), and Trichinellida (four families: Anatrichosomatidae, Capillariidae, Trichinellidae, and Trichuridae).

As shown by the large species numbers, helminth diversity illustrates the complexities in the systematics and classification of helminths. In addition to classical taxonomy using morphological and ecological characters, molecular information through DNA sequences is becoming increasingly vital for helminth systematics. The following section will focus on the helminth molecular information available on public databases.

2.3 Major Taxa of Parasitic Helminths of Public Health Importance and Availability of Molecular Information on a Public Database

Focusing on the data extracted from the NHM Host-Parasite Database, we selected parameters from the National Center for Biotechnology Information (NCBI) database (https://www.ncbi.nlm.nih.gov) to accumulate information that is publicly available for parasitic helminths in humans. The data obtained represents the amount of available molecular information and scientific publications to consider the importance and tendency of parasitic helminth species in scientific, social, and economic interests. Here, numbers of Entrez records in the "Nucleotide" and "Protein" databases were derived from each human parasitic helminth (419 species) to record the entry number of available nucleotide sequence data and amino acid sequences of the helminth species (Fig. 2.4a, b). A number of complete and in-progress large-scale molecular projects studying the molecular biology of each helminth species (e.g., genome sequencing and assembly, transcriptome, metagenomic, annotation, expression, and mapping projects) were recorded through the "BioProject" database (Fig. 2.4c). In terms of available literature, many scientific publications related to

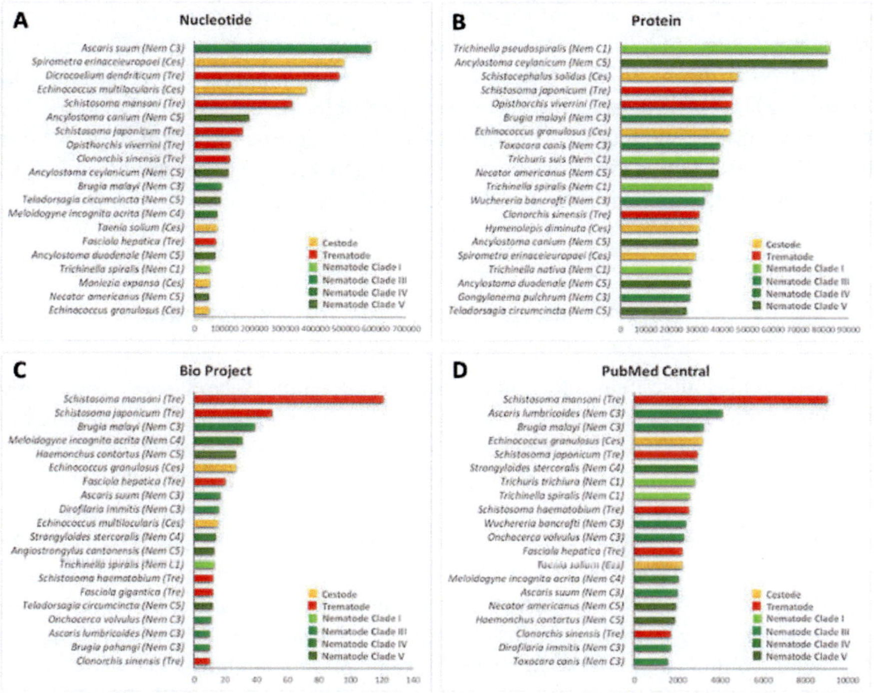

Fig. 2.4 The bar chart plots represent Entrez records of the Nucleotide (**a**), Protein (**b**), BioProject (**c**), and PubMed Central (**d**) databases of the top 20 parasitic helminth species infecting humans

each helminth species were collected from the "PubMed Central (PMC)" database, which is a digital archive of literature in life sciences journals (Fig. 2.4d).

As of October 2021, of the 419 species of helminths and other wormlike parasites (Acanthocephala and Nematomorpha), molecular data for 271 species (64.6% coverage—the proportion of species with molecular data available) was available in the NCBI database. Of these, there are 146 species of Platyhelminthes, i.e., 95 species of trematodes (60.5%) and 51 species of cestodes (71.8%); followed by 111 species of Nematoda (70.7%) and 7 species of Acanthocephala (87.5%), whereas the information is somewhat limited in Nematomorpha, where data is only available for 7 species (26.9%).

2.3.1 Molecular Information Available for Cestodes

Among cestodes, *Spirometra erinaceieuropaei* (causing sparganosis) have the most DNA sequences recorded in the nucleotide database (493,677 entries). Examples of the other major species include *E. multilocularis* (369,910 entries), *Taenia solium* (75,411 entries), *Moneizia expansa* (50,794 entries), and *E. granulosus* (49,676 entries). For the amino acid information, *Schistocephalus solidus* (46,057 entries), *E. granulosus* (42,701 entries), *Hymenolepis diminuta* (30,652 entries), *S. erinaceieuropaei* (29,255 entries), and *Dibothriocephalus* (or *Diphyllobothrium*) *latus* (20,071 entries) have the most data available in the protein database. For molecular project and journal published data, *E. granulosus* has the highest number of projects and publications in the BioProject and PMC databases (27 and 3158 entries, respectively) compared to other tapeworm species.

2.3.2 Molecular Information Available for Trematodes

For trematodes that can infect humans, *Dicrocoelium dendriticum*, commonly known as the lancet liver fluke, shows the highest number of DNA sequences recorded in the nucleotide database (479,012 entries). *Dicrocoelium dendriticum* are parasites of ruminants and can occasionally infect humans. The two blood flukes, i.e., *S. mansoni* (324,651 entries) and *S. japonicum* (160,073 entries), also have many DNA sequences available in the database. *Opisthorchis viverrini* (121,521 entries) and *Clonorchis sinensis* (117,834 entries), the two human liver flukes, also possess >100,000 DNA sequence entries, outcompeting the cattle liver fluke, *Fasciola hepatica*, which contains 72,022 entries. In terms of protein data, the following blood and tissue-infected flukes: *S. japonicum* (44,182 entries), *O. viverrini* (43,776 entries), *C. sinensis* (30,944 entries), and *Paragonimus westermani* (25,045 entries) are the major trematode species contributing to the protein database. The two blood flukes, *S. mansoni* and *S. japonicum*, also have the most significant number of molecular projects in the BioProject database (121 and 50 entries, respectively), surpassing the other parasitic helminth species in humans (Fig. 2.4c). Interestingly, *S. mansoni* is an outstanding species of scientific, public health, and socio-economic interests with incredible amounts of molecular information and associated projects, probably resulting in the most published species in scientific literature compared to other helminths of medical and public health importance (Fig. 2.4d).

2.3.3 Molecular Information Available for Nematodes

Among the human parasitic nematodes, the pig roundworm *Ascaris suum* that occasionally infects humans was the species with the most DNA sequences recorded in the nucleotide database (584,359 entries) (Fig. 2.4a). Compared to other nematodes, the sequence data available for *A. suum* is much greater, with 182,520 entries for the zoonotic hookworm species *Ancylostoma caninum*, 113,602 entries for *Ancylostoma ceylanicum*, and 90,323 entries for the lymphatic filarial nematode *Brugia malayi*. For the protein database, *T. pseudospiralis* (82,422 entries) and *A. ceylanicum* (81,728 entries) have the greatest amount of amino acid data available

(Fig. 2.4b). Other nematode species with a high amount of protein data include *B. malayi* (43,559 entries), *Toxocara canis* (39,104 entries), and *Trichuris suis* (38,673 entries). For BioProjects, *B. malayi* appeared as the main species among the other nematodes with 39 entries in the database, followed by the root-knot nematode *Meloidogyne incognita* (31 entries), which is an obligate plant-parasitic nematode rarely infecting humans (Bradbury & Speare, 2015); *Haemonchus contortus* (27 entries), the barbers pole worm of ruminants; the pig roundworm *A. suum* (17 entries); and *Dirofilaria immitis* (16 entries), the dog heartworm causing human pulmonary and ocular dirofilariasis (Dantas-Torres & Otranto, 2013; Pampiglione et al., 2009). Nematode species in the STH group, i.e., *A. lumbricoides* (4129 entries), *S. stercoralis* (2940 entries), and *T. trichiura* (2823 entries), are among the most published species in scientific journals, based on the information in the PMC database. In addition to the STH species, publications of some others such as *B. malayi* and *T. spiralis* reached >2500 entries in the database.

Based on the data accumulated in Fig. 2.4, there are differences in the amount of available data among the groups of parasitic helminths. For example, *S. mansoni* is the largest single contributor to the BioProject and PMC databases, and *T. pseudospiralis* and *A. ceylanicum* occupied most of the information in the protein database. These biases may reflect the actual situations of the helminthological field from the past to the present in terms of scientific, public health, and socio-economic priorities of helminthic infections in humans. The trend will undoubtedly be changed in the future as more attention is shifted to other understudied taxa (Johnston, 2006).

2.4 Other Helminth-Like Parasites Infecting Humans

In addition to parasitic helminths in the Platyhelminthes and Nematoda, other groups of wormlike parasites recorded infecting humans include Acanthocephala (thorny-headed worms) and Nematomorpha (horsehair worms).

2.4.1 Acanthocephala

The members of the phylum Acanthocephala are obligate parasites with complex life cycles, requiring at least two hosts (intermediate and definitive hosts), including invertebrates, fishes, reptiles, birds, mammals, and humans. Acanthocephala infection in humans is rare, but common foci have been documented in temperate zones such as European countries, China, and Japan (Lotfy, 2020; Schmidt, 1971; Tada et al., 1983). The worms are cylindrical, unsegmented, with the major characteristic of an invaginable proboscis armed with rows of spines. The parasites are dioecious, having the male and female reproductive organs in separate individuals. The entire phylum comprises four classes: Archiacanthocephala, Eoacanthocephala, Palaeacanthocephala, and Polyacanthocephala. At least ten species in the first three classes have been reported to infect humans, including *Macracanthorhynchus hirudinaceus*, *Macracanthorhynchus ingens*, and *Moniliformis moniliformis* from the class Archiacanthocephala; *Pallisentis celatus* from the class Eoacanthocephala; *Acanthocephalus rauschi*, *Pseudoacanthocephalus bufonis*, *Bolbosoma sp.*, *Plagiorhynchus sp.*, *Corynosoma validum*, and *Corynosoma strumosum* from the class Palaeacanthocephala (Anosike et al., 2000; Arizono et al., 2012; Counselman et al., 1989; Dingley & Beaver, 1985; Golvan, 1994; Haustein et al., 2010; Joe & Siang, 1959; Leng et al., 1983; Lotfy, 2020; Pradatsundarasar & Pechranond, 1965; Schmidt, 1971; Tada et al., 1983; Xu et al., 1998). Humans are infected by ingesting raw and undercooked arthropods, e.g., cockroaches, millipedes, and beetles for those acanthocephalans in the class Archiacanthocephala and raw fishes and crustaceans, mainly for worms in Palaeacanthocephala and Eoacanthocephala (Lotfy, 2020).

2.4.2 *Nematomorpha*

Nematomorpha, as obligate parasites, require arthropod intermediate hosts (e.g., crickets, grasshoppers, mantises, beetles, and other insects) during the juvenile stage, and live as free-living organisms in several aquatic habitats as adults. In addition to arthropods, Nematomorpha also parasitizes aquatic vertebrates (e.g., frogs and fishes), birds, mammals, and humans. However, it is uncommon for them to parasitize mammalian hosts. Parasitic worms in the phylum Nematomorpha are classified into two classes, Nectonematida and Gordioida, living in marine and freshwater habitats. Their threadlike elongated bodies are similar to nematodes and were initially grouped in the phylum Nematoda; they seem morphologically related to mermithid nematodes. However, the major difference between the adult stages of Nematomorpha and Nematoda is the degenerate gut (nonfunction) in Nematomorpha. Also, in contrast to nematodes, the head of the nematomorph larval stage bears rows of cuticular hooks and invaginable terminal stylets used to penetrate the host's tissue. These differences resulted in the separation of Nematomorpha from Nematoda (Veidovsky, 1886). In addition, phylogenetic analyses using nuclear small subunit sequences (SSU) confirmed the distinction between the two phyla, showing Nematomorpha and Nematoda as sister groups (Blaxter et al., 1998; Bleidorn et al., 2002; Giribet et al., 2007). Despite having a worldwide distribution, not more than 500 valid species are known to science (Poinar, 2010), indicating that this phylum still requires further investigation. In terms of public health importance, all horsehair worm species infecting humans are only from the class Gordioida. These include 26 worm species from Chordodidae, Parachordodidae, and Gordiidae (data from the NHM Host-Parasite Database). Evidence of nematomorph worms has been reported as accidental cases in several regions of the world (prominent foci in China, Japan, and Korea) from human excreta (i.e., vomit and feces) (Dexiang & Wenyuan, 1981; Hong et al., 2015; Lee et al., 2003; Yamada et al., 2012).

2.5 Conclusion

This chapter presents the diversity, classification, and relationships among the three major helminth groups. Major groups of helminths can be broadly categorized based on utilizing different life cycles, modes of transmission, reproductive strategies, morphological characters, and hosts. These characters have built the foundation for the helminth taxonomy and systematics. With more than 100,000 estimated species globally, the study of helminth systematics for understanding their diversity and distribution has important implications, especially for medically important parasitic helminths (Carlson et al., 2020). Helminth diversity encompassing both phenotypic and genetic variation can be attributed to evolutionary processes and interactions constantly occurring between environments, hosts, and helminths. The next chapter will introduce some of these evolutionary processes and adaptations behind the helminth diversity that we see today.

Appendix

Table 2.3 List of helminths and other parasites of medical and public health importance from the NHM Host-Parasite Database, London (Gibson et al., 2005)

Phylum	Class	Order	Family	Genus	Species
Acanthocephala	Palaeacanthocephala	Echinorhynchida	Echinorhynchidae	*Acanthocephalus*	*Acanthocephalus rauschi* (Schmidt, 1969)
Acanthocephala	Palaeacanthocephala	Echinorhynchida	Echinorhynchidae	*Pseudoacanthocephalus*	*Pseudoacanthocephalus bufonis* (Shipley, 1903)
Acanthocephala	Palaeacanthocephala	Polymorphida	Polymorphidae	*Bolbosoma*	*Bolbosoma* sp.
Acanthocephala	Palaeacanthocephala	Polymorphida	Polymorphidae	*Corynosoma*	*Corynosoma strumosum* (Rudolphi, 1802)
Acanthocephala	Palaeacanthocephala	Polymorphida	Polymorphidae	*Corynosoma*	*Corynosoma validum* (Van Cleave, 1953)
Acanthocephala	Palaeacanthocephala	Polymorphida	Plagiorhynchidae	*Plagiorhynchus*	*Plagiorhynchus* sp.
Acanthocephala	Archiacanthocephala	Moniliformida	Moniliformidae	*Moniliformis*	*Moniliformis moniliformis* (Bremser in Rudolphi, 1819)
Acanthocephala	Archiacanthocephala	Oligacanthorhynchida	Oligacanthorhynchidae	*Macracanthorhynchus*	*Macracanthorhynchus hirudinaceus* (Pallas, 1781)
Acanthocephala	Archiacanthocephala	Oligacanthorhynchida	Oligacanthorhynchidae	*Macracanthorhynchus*	*Macracanthorhynchus ingens* (Linstow, 1879)
Acanthocephala	Eoacanthocephala	Gyracanthocephala	Quadrigyridae	*Pallisentis*	*Pallisentis celatus* (Van Cleave, 1928)
Platyhelminthes	Cestoda	Cyclophyllidea	Anoplocephalidae	*Anoplocephala*	*Anoplocephala perfoliata* (Goeze, 1782)
Platyhelminthes	Cestoda	Cyclophyllidea	Anoplocephalidae	*Bertiella*	*Bertiella mucronata* (Meyner, 1895)
Platyhelminthes	Cestoda	Cyclophyllidea	Anoplocephalidae	*Bertiella*	*Bertiella satyri* (Blanchard, 1891)
Platyhelminthes	Cestoda	Cyclophyllidea	Anoplocephalidae	*Bertiella*	*Bertiella studeri* (Blanchard, 1891)
Platyhelminthes	Cestoda	Cyclophyllidea	Anoplocephalidae	*Moniezia*	*Moniezia expansa* (Rudolphi, 1805)
Platyhelminthes	Cestoda	Cyclophyllidea	Anoplocephalidae	*Inermicapsifer*	*Inermicapsifer cubensis* (Kouri, 1938)

(continued)

Table 2.3 (continued)

Phylum	Class	Order	Family	Genus	Species
Platyhelminthes	Cestoda	Cyclophyllidea	Anoplocephalidae	*Inermicapsifer*	*Inermicapsifer madagascariensis* (Davaine, 1869)
Platyhelminthes	Cestoda	Cyclophyllidea	Anoplocephalidae	*Mathevotaenia*	*Mathevotaenia symmetrica* (Baylis, 1927)
Platyhelminthes	Cestoda	Cyclophyllidea	Davaineidae	*Buginetta*	*Buginetta alouattae* (Baylis, 1947)
Platyhelminthes	Cestoda	Cyclophyllidea	Davaineidae	*Raillietina*	*Raillietina celebensis* (Janicki, 1902)
Platyhelminthes	Cestoda	Cyclophyllidea	Davaineidae	*Raillietina*	*Raillietina demerariensis* (Daniels, 1835)
Platyhelminthes	Cestoda	Cyclophyllidea	Davaineidae	*Raillietina*	*Raillietina asiatica* (von Linstow, 1901)
Platyhelminthes	Cestoda	Cyclophyllidea	Dipylidiidae	*Dipylidium*	*Dipylidium caninum* (Linnaeus, 1758)
Platyhelminthes	Cestoda	Cyclophyllidea	Hymenolepididae	*Drepanidotaenia*	*Drepanidotaenia lanceolata* (Bloch, 1782)
Platyhelminthes	Cestoda	Cyclophyllidea	Hymenolepididae	*Hymenolepis*	*Hymenolepis diminuta* (Rudolphi, 1819)
Platyhelminthes	Cestoda	Cyclophyllidea	Hymenolepididae	*Pseudanoplocephala*	*Pseudanoplocephala crawfordi* (Baylis, 1927)
Platyhelminthes	Cestoda	Cyclophyllidea	Hymenolepididae	*Rodentolepis*	*Rodentolepis fraterna* (Stiles, 1906)
Platyhelminthes	Cestoda	Cyclophyllidea	Hymenolepididae	*Rodentolepis*	*Rodentolepis nana* (Bilharz, 1851)
Platyhelminthes	Cestoda	Cyclophyllidea	Mesocestoididae	*Mesocestoides*	*Mesocestoides lineatus* (Goeze, 1782)
Platyhelminthes	Cestoda	Cyclophyllidea	Taeniidae	*Echinococcus*	*Echinococcus granulosus* (Batsch, 1786)
Platyhelminthes	Cestoda	Cyclophyllidea	Taeniidae	*Echinococcus*	*Echinococcus multilocularis* (Leuckart, 1863)

(continued)

Table 2.3 (continued)

Phylum	Class	Order	Family	Genus	Species
Platyhelminthes	Cestoda	Cyclophyllidea	Taeniidae	*Echinococcus*	*Echinococcus oligarthrus* (Diesing, 1863)
Platyhelminthes	Cestoda	Cyclophyllidea	Taeniidae	*Echinococcus*	*Echinococcus ortleppi* (Lopez-Neyra and Soler, 1943)
Platyhelminthes	Cestoda	Cyclophyllidea	Taeniidae	*Echinococcus*	*Echinococcus vogeli* (Rausch & Bernstein, 1972)
Platyhelminthes	Cestoda	Cyclophyllidea	Taeniidae	*Taenia*	*Taenia asiatica* (Eom & Rim, 1993)
Platyhelminthes	Cestoda	Cyclophyllidea	Taeniidae	*Taenia*	*Taenia multiceps* (Leske, 1780)
Platyhelminthes	Cestoda	Cyclophyllidea	Taeniidae	*Taenia*	*Taenia crassiceps* (Zeder, 1800)
Platyhelminthes	Cestoda	Cyclophyllidea	Taeniidae	*Taenia*	*Taenia hydatigena* (Pallas, 1766)
Platyhelminthes	Cestoda	Cyclophyllidea	Taeniidae	*Taenia*	*Taenia ovis* (Cobbold, 1869)
Platyhelminthes	Cestoda	Cyclophyllidea	Taeniidae	*Taenia*	*Taenia saginata* (Goeze, 1782)
Platyhelminthes	Cestoda	Cyclophyllidea	Taeniidae	*Taenia*	*Taenia serialis* (Gervais, 1847; Baillet, 1863)
Platyhelminthes	Cestoda	Cyclophyllidea	Taeniidae	*Taenia*	*Taenia solium* (Linnaeus, 1758)
Platyhelminthes	Cestoda	Cyclophyllidea	Taeniidae	*Taenia*	*Taenia taeniaeformis* (Batsch, 1786)
Platyhelminthes	Cestoda	Diphyllobothriidea	Diphyllobothriidae	*Adenocephalus*	*Adenocephalus pacificus* (Nybelin, 1931)
Platyhelminthes	Cestoda	Diphyllobothriidea	Diphyllobothriidae	*Dibothriocephalus*	*Dibothriocephalus dendriticus* (Nitzsch, 1824)
Platyhelminthes	Cestoda	Diphyllobothriidea	Diphyllobothriidae	*Dibothriocephalus*	*Dibothriocephalus latus* (Linnaeus, 1758)
Platyhelminthes	Cestoda	Diphyllobothriidea	Diphyllobothriidae	*Dibothriocephalus*	*Dibothriocephalus nihonkaiensis* (Yamane, Kamo, Bylund & Wikgren, 1986)
Platyhelminthes	Cestoda	Diphyllobothriidea	Diphyllobothriidae	*Diphyllobothrium*	*Diphyllobothrium cameroni* (Rausch, 1969)

(continued)

Table 2.3 (continued)

Phylum	Class	Order	Family	Genus	Species
Platyhelminthes	Cestoda	Diphyllobothriidea	Diphyllobothriidae	*Diphyllobothrium*	*Diphyllobothrium cordatum* (Leuckart, 1863) Gedoelst, 1911
Platyhelminthes	Cestoda	Diphyllobothriidea	Diphyllobothriidae	*Diphyllobothrium*	*Diphyllobothrium dalliae* (Rausch, 1956)
Platyhelminthes	Cestoda	Diphyllobothriidea	Diphyllobothriidae	*Diphyllobothrium*	*Diphyllobothrium elegans* (Krabbe, 1865) Meggitt, 1924
Platyhelminthes	Cestoda	Diphyllobothriidea	Diphyllobothriidae	*Diphyllobothrium*	*Diphyllobothrium giljacicum* (Rutkevich, 1937)
Platyhelminthes	Cestoda	Diphyllobothriidea	Diphyllobothriidae	*Diphyllobothrium*	*Diphyllobothrium hians* (Diesing, 1850)
Platyhelminthes	Cestoda	Diphyllobothriidea	Diphyllobothriidae	*Diphyllobothrium*	*Diphyllobothrium houghtoni*
Platyhelminthes	Cestoda	Diphyllobothriidea	Diphyllobothriidae	*Diphyllobothrium*	*Diphyllobothrium lanceolatum* (Krabbe, 1865) Cooper, 1921
Platyhelminthes	Cestoda	Diphyllobothriidea	Diphyllobothriidae	*Diphyllobothrium*	*Diphyllobothrium mansonoides* (Mueller, 1935)
Platyhelminthes	Cestoda	Diphyllobothriidea	Diphyllobothriidae	*Diphyllobothrium*	*Diphyllobothrium minus* (Cholodkovsky, 1916)
Platyhelminthes	Cestoda	Diphyllobothriidea	Diphyllobothriidae	*Diphyllobothrium*	*Diphyllobothrium nenzi* (Petrov, 1938)
Platyhelminthes	Cestoda	Diphyllobothriidea	Diphyllobothriidae	*Diphyllobothrium*	*Diphyllobothrium orcini* (Hatsushika & Shirouzu, 1990)
Platyhelminthes	Cestoda	Diphyllobothriidea	Diphyllobothriidae	*Diphyllobothrium*	*Diphyllobothrium scoticum* (Rennie & Reid, 1912)
Platyhelminthes	Cestoda	Diphyllobothriidea	Diphyllobothriidae	*Diphyllobothrium*	*Diphyllobothrium skrjabini* (Plotnikoff, 1933)
Platyhelminthes	Cestoda	Diphyllobothriidea	Diphyllobothriidae	*Diphyllobothrium*	*Diphyllobothrium stemmacephalum* (Cobbold, 1858)

(continued)

Table 2.3 (continued)

Phylum	Class	Order	Family	Genus	Species
Platyhelminthes	Cestoda	Diphyllobothriidea	Diphyllobothriidae	*Diphyllobothrium*	*Diphyllobothrium tungussicum* (Podyapolskaya and Gnedina, 1932)
Platyhelminthes	Cestoda	Diphyllobothriidea	Diphyllobothriidae	*Diphyllobothrium*	*Diphyllobothrium ursi* (Rausch, 1954)
Platyhelminthes	Cestoda	Diphyllobothriidea	Diphyllobothriidae	*Diplogonoporus*	*Diplogonoporus balaenopterae* (Lonnberg, 1892)
Platyhelminthes	Cestoda	Diphyllobothriidea	Diphyllobothriidae	*Diplogonoporus*	*Diplogonoporus brauni* (Leon, 1907)
Platyhelminthes	Cestoda	Diphyllobothriidea	Diphyllobothriidae	*Diplogonoporus*	*Diplogonoporus fukuokaensis* (Kamo & Miyazaki, 1970)
Platyhelminthes	Cestoda	Diphyllobothriidea	Diphyllobothriidae	*Diplogonoporus*	*Diplogonoporus grandis* (Blanchard, 1894)
Platyhelminthes	Cestoda	Diphyllobothriidea	Diphyllobothriidae	*Ligula*	*Ligula intestinalis* (Linnaeus, 1758; Gmelin, 1790)
Platyhelminthes	Cestoda	Diphyllobothriidea	Diphyllobothriidae	*Pyramicocephalus*	*Pyramicocephalus anthocephalus* (Monticelli, 1890)
Platyhelminthes	Cestoda	Diphyllobothriidea	Diphyllobothriidae	*Schistocephalus*	*Schistocephalus solidus* (Steenstrup, 1857)
Platyhelminthes	Cestoda	Diphyllobothriidea	Diphyllobothriidae	*Sparganum*	*Sparganum mansoni* (Cobbold, 1883)
Platyhelminthes	Cestoda	Diphyllobothriidea	Diphyllobothriidae	*Sparganum*	*Sparganum proliferum* (Ijima, 1905)
Platyhelminthes	Cestoda	Diphyllobothriidea	Diphyllobothriidae	*Spirometra*	*Spirometra erinaceieuropaei* (Rudolphi, 1819)
Platyhelminthes	Cestoda	Diphyllobothriidea	Diphyllobothriidae	*Spirometra*	*Spirometra mansoni* (Joyeux & Houdemer, 1927)
Platyhelminthes	Cestoda	Diphyllobothriidea	Diphyllobothriidae	*Spirometra*	*Spirometra mansonoides* (Mueller, 1935; Mueller, 1936)

(continued)

Table 2.3 (continued)

Phylum	Class	Order	Family	Genus	Species
Platyhelminthes	Cestoda	Diphyllobothriidea	Diphyllobothriidae	*Spirometra*	*Spirometra theileri* (Baer, 1925)
Platyhelminthes	Cestoda	Diphyllobothriidea	Diphyllobothriidae	*Spirometra*	*Spirometra ranarum* (Meggitt, 1925)
Platyhelminthes	Cestoda	Proteocephalidea	Proteocephalidae	*Ritacestus*	*Ritacestus ritaii* (Verma, 1926)
Platyhelminthes	Cestoda	Trypanorhyncha	Hepatoxylidae	*Hepatoxylon*	*Hepatoxylon trichiuri* (Holten, 1802)
Platyhelminthes	Cestoda	Trypanorhyncha	Tentaculariidae	*Nybelinia*	*Nybelinia surmenicola* (Okada in Dollfus, 1929)
Platyhelminthes	Trematoda	Azygiida	Isoparorchiidae	*Isoparorchis*	*Isoparorchis hypselobagri* (Billet, 1898)
Platyhelminthes	Trematoda	Diplostomida	Brachylaimidae	*Brachylaima*	*Brachylaima cribbi* (Butcher, 2003)
Platyhelminthes	Trematoda	Diplostomida	Cyathocotylidae	*Paracoenogonimus*	*Paracoenogonimus ovatus* (Katsurada, 1914)
Platyhelminthes	Trematoda	Diplostomida	Cyathocotylidae	*Prohemistomum*	*Prohemistomum vivax* (Sonsino, 1893) Azim, 1933
Platyhelminthes	Trematoda	Diplostomida	Diplostomidae	*Alaria*	*Alaria americana* (Hall and Wigdor, 1918)
Platyhelminthes	Trematoda	Diplostomida	Diplostomidae	*Alaria*	*Alaria marcianae* (La Rue, 1917)
Platyhelminthes	Trematoda	Diplostomida	Diplostomidae	*Diplostomum*	*Diplostomum spathaceum* (Rudolphi, 1819) Olsson, 1876
Platyhelminthes	Trematoda	Diplostomida	Diplostomidae	*Fibricola*	*Fibricola cratera* (Barker and Noll, 1915)
Platyhelminthes	Trematoda	Diplostomida	Diplostomidae	*Neodiplostomum*	*Neodiplostomum seoulensis* (Seo, Rim, and Lee, 1964)
Platyhelminthes	Trematoda	Opisthorchiida	Heterophyidae	*Apophallus*	*Apophallus donicus* (Skrjabin & Lindtrop, 1919)
Platyhelminthes	Trematoda	Opisthorchiida	Heterophyidae	*Ascocotyle*	*Ascocotyle* (*Phagicola*) *longa* (Ransom, 1920)

(continued)

Table 2.3 (continued)

Phylum	Class	Order	Family	Genus	Species
Platyhelminthes	Trematoda	Opisthorchiida	Heterophyidae	*Centrocestus*	*Centrocestus armatus* (Tanabe, 1922; Price, 1932)
Platyhelminthes	Trematoda	Opisthorchiida	Heterophyidae	*Centrocestus*	*Centrocestus caninus* (Leiper, 1913)
Platyhelminthes	Trematoda	Opisthorchiida	Heterophyidae	*Centrocestus*	*Centrocestus formosanus* (Nishigori, 1924)
Platyhelminthes	Trematoda	Opisthorchiida	Heterophyidae	*Centrocestus*	*Centrocestus kurokawai* (Kurokawa, 1935)
Platyhelminthes	Trematoda	Opisthorchiida	Heterophyidae	*Cryptocotyle*	*Cryptocotyle lingua* (Creplin, 1825; Fischoeder, 1903)
Platyhelminthes	Trematoda	Opisthorchiida	Heterophyidae	*Diorchitrema*	*Diorchitrema amplicaecale* (Katsuta, 1932)
Platyhelminthes	Trematoda	Opisthorchiida	Heterophyidae	*Haplorchis*	*Haplorchis microrchis* (Matsuda, 1932)
Platyhelminthes	Trematoda	Opisthorchiida	Heterophyidae	*Haplorchis*	*Haplorchis pleurolophocerca* (Sonsino, 1896)
Platyhelminthes	Trematoda	Opisthorchiida	Heterophyidae	*Haplorchis*	*Haplorchis pumilio* (Looss, 1896)
Platyhelminthes	Trematoda	Opisthorchiida	Heterophyidae	*Haplorchis*	*Haplorchis taichui* (Nishigori, 1924)
Platyhelminthes	Trematoda	Opisthorchiida	Heterophyidae	*Haplorchis*	*Haplorchis vanissimus* (Africa, 1938)
Platyhelminthes	Trematoda	Opisthorchiida	Heterophyidae	*Haplorchis*	*Haplorchis yokogawai* (Katsuta, 1932)
Platyhelminthes	Trematoda	Opisthorchiida	Heterophyidae	*Heterophyes*	*Heterophyes brevicaeca* (Africa and Garcia 1935)
Platyhelminthes	Trematoda	Opisthorchiida	Heterophyidae	*Heterophyes*	*Heterophyes dispar* (Looss, 1902)
Platyhelminthes	Trematoda	Opisthorchiida	Heterophyidae	*Heterophyes*	*Heterophyes aequalis* (Looss, 1902)
Platyhelminthes	Trematoda	Opisthorchiida	Heterophyidae	*Heterophyes*	*Heterophyes heterophyes* (Siebold, 1853)

(continued)

Table 2.3 (continued)

Phylum	Class	Order	Family	Genus	Species
Platyhelminthes	Trematoda	Opisthorchiida	Heterophyidae	*Heterophyes*	*Heterophyes nocens* (Onji & Nishio, 1916)
Platyhelminthes	Trematoda	Opisthorchiida	Heterophyidae	*Heterophyopsis*	*Heterophyopsis continua* (Onji & Nishio, 1916)
Platyhelminthes	Trematoda	Opisthorchiida	Heterophyidae	*Metagonimus*	*Metagonimus katsuradai* (Izumi, 1935)
Platyhelminthes	Trematoda	Opisthorchiida	Heterophyidae	*Metagonimus*	*Metagonimus miyatai* (Saito, Chai, Kim, Lee & Rim, 1997)
Platyhelminthes	Trematoda	Opisthorchiida	Heterophyidae	*Metagonimus*	*Metagonimus takahashii* (Suzuki in Takahashi, 1929)
Platyhelminthes	Trematoda	Opisthorchiida	Heterophyidae	*Metagonimus*	*Metagonimus yokogawai* (Katsurada, 1912)
Platyhelminthes	Trematoda	Opisthorchiida	Heterophyidae	*Procerovum*	*Procerovum calderoni* (Africa & Garcia, 1935)
Platyhelminthes	Trematoda	Opisthorchiida	Heterophyidae	*Procerovum*	*Procerovum varium* (Onji & Nishio, 1916)
Platyhelminthes	Trematoda	Opisthorchiida	Heterophyidae	*Pygidiopsis*	*Pygidiopsis summa* (Onji & Nishio, 1916)
Platyhelminthes	Trematoda	Opisthorchiida	Heterophyidae	*Stellantchasmus*	*Stellantchasmus falcatus* (Onji & Nishio, 1916)
Platyhelminthes	Trematoda	Opisthorchiida	Heterophyidae	*Stictodora*	*Stictodora fuscatum* (Onji & Nishio, 1916; Yamaguti, 1958)
Platyhelminthes	Trematoda	Opisthorchiida	Heterophyidae	*Stictodora*	*Stictodora lari* (Yamaguti, 1939)
Platyhelminthes	Trematoda	Opisthorchiida	Opisthorchiidae	*Clonorchis*	*Clonorchis sinensis* (Cobbold, 1875)
Platyhelminthes	Trematoda	Opisthorchiida	Opisthorchiidae	*Metorchis*	*Metorchis albidus* (Braun, 1893)
Platyhelminthes	Trematoda	Opisthorchiida	Opisthorchiidae	*Metorchis*	*Metorchis conjunctus* (Cobbold, 1860) Looss, 1899

(continued)

Table 2.3 (continued)

Phylum	Class	Order	Family	Genus	Species
Platyhelminthes	Trematoda	Opisthorchiida	Opisthorchiidae	*Opisthorchis*	*Opisthorchis felineus* (Rivolta, 1884)
Platyhelminthes	Trematoda	Opisthorchiida	Opisthorchiidae	*Opisthorchis*	*Opisthorchis guayaquilensis* (Rodríguez M., Gomez Lince & Montalvan C., 1949)
Platyhelminthes	Trematoda	Opisthorchiida	Opisthorchiidae	*Opisthorchis*	*Opisthorchis noverca* (Braun, 1902)
Platyhelminthes	Trematoda	Opisthorchiida	Opisthorchiidae	*Opisthorchis*	*Opisthorchis tenuicollis* (Rudolphi, 1819; Stiles & Hassall, 1896)
Platyhelminthes	Trematoda	Opisthorchiida	Opisthorchiidae	*Opisthorchis*	*Opisthorchis viverrini* (Poirier, 1886)
Platyhelminthes	Trematoda	Opisthorchiida	Opisthorchiidae	*Pseudamphistomum*	*Pseudamphistomum aethiopicum* (Pierantoni, 1942)
Platyhelminthes	Trematoda	Opisthorchiida	Opisthorchiidae	*Pseudamphistomum*	*Pseudamphistomum truncatum* (Rudolphi, 1819)
Platyhelminthes	Trematoda	Plagiorchiida	Cathaemasiidae	*Cathaemasia*	*Cathaemasia cabrerai* (Jueco and Monzon, 1984)
Platyhelminthes	Trematoda	Plagiorchiida	Echinostomatidae	*Artyfechinostomum*	*Artyfechinostomum mehrai* (Jain, 1957; Jain, 1960
Platyhelminthes	Trematoda	Plagiorchiida	Echinostomatidae	*Artyfechinostomum*	*Artyfechinostomum oraoni*
Platyhelminthes	Trematoda	Plagiorchiida	Echinostomatidae	*Artyfechinostomum*	*Artyfechinostomum sufrartyfex* (Lane, 1915)
Platyhelminthes	Trematoda	Plagiorchiida	Echinochasmidae	*Echinochasmus*	*Echinochasmus caninus* (Verma, 1935; Baschkirova, 1947)
Platyhelminthes	Trematoda	Plagiorchiida	Echinochasmidae	*Echinochasmus*	*Echinochasmus fujianensis* (Chen, 1992)
Platyhelminthes	Trematoda	Plagiorchiida	Echinochasmidae	*Echinochasmus*	*Echinochasmus japonicus* (Tanabe, 1926)

(continued)

Table 2.3 (continued)

Phylum	Class	Order	Family	Genus	Species
Platyhelminthes	Trematoda	Plagiorchiida	Echinochasmidae	*Echinochasmus*	*Echinochasmus jiufoensis* (Liang and Ke, 1988)
Platyhelminthes	Trematoda	Plagiorchiida	Echinochasmidae	*Echinochasmus*	*Echinochasmus liliputanus* (Looss, 1896) Odhner, 1910
Platyhelminthes	Trematoda	Plagiorchiida	Echinochasmidae	*Echinochasmus*	*Echinochasmus perfoliatus* (Ratz, 1908)
Platyhelminthes	Trematoda	Plagiorchiida	Echinostomatidae	*Echinoparyphium*	*Echinoparyphium recurvatum* (Linstow, 1873)
Platyhelminthes	Trematoda	Plagiorchiida	Echinochasmidae	*Echinostoma*	*Echinostoma angustitestis* (Wang, 1977)
Platyhelminthes	Trematoda	Plagiorchiida	Echinostomatidae	*Echinostoma*	*Echinostoma cinetorchis* (Ando & Ozaki, 1923)
Platyhelminthes	Trematoda	Plagiorchiida	Echinostomatidae	*Echinostoma*	*Echinostoma echinatum* (Zeder 1803)
Platyhelminthes	Trematoda	Plagiorchiida	Echinostomatidae	*Echinostoma*	*Echinostoma hortense* (Asada, 1926)
Platyhelminthes	Trematoda	Plagiorchiida	Echinostomatidae	*Echinostoma*	*Echinostoma ilocanum* (Garrison, 1908)
Platyhelminthes	Trematoda	Plagiorchiida	Echinostomatidae	*Echinostoma*	*Echinostoma macrorchis* (Ando & Ozaki, 1923)
Platyhelminthes	Trematoda	Plagiorchiida	Echinostomatidae	*Echinostoma*	*Echinostoma malayanum* (Leiper, 1911)
Platyhelminthes	Trematoda	Plagiorchiida	Echinostomatidae	*Echinostoma*	*Echinostoma miyagawai* (Ishii, 1932)
Platyhelminthes	Trematoda	Plagiorchiida	Echinostomatidae	*Echinostoma*	*Echinostoma paraulum* (Dietz, 1909)
Platyhelminthes	Trematoda	Plagiorchiida	Echinostomatidae	*Echinostoma*	*Echinostoma revolutum* (Froelich, 1802)

(continued)

Table 2.3 (continued)

Phylum	Class	Order	Family	Genus	Species
Platyhelminthes	Trematoda	Plagiorchiida	Echinostomatidae	*Euparyphium*	*Euparyphium jassyense* (Leon and Ciurea, 1922)
Platyhelminthes	Trematoda	Plagiorchiida	Echinostomatidae	*Hypoderaeum*	*Hypoderaeum conoideum* (Bloch, 1782)
Platyhelminthes	Trematoda	Plagiorchiida	Echinostomatidae	*Isthmiophora*	*Isthmiophora melis* (Schrank, 1788; Luhe, 1909)
Platyhelminthes	Trematoda	Plagiorchiida	Himasthlidae	*Acanthoparyphium*	*Acanthoparyphium kurogamo* (Yamaguti, 1939)
Platyhelminthes	Trematoda	Plagiorchiida	Himasthlidae	*Acanthoparyphium*	*Acanthoparyphium tyosenense* (Yamaguti, 1939)
Platyhelminthes	Trematoda	Plagiorchiida	Himasthlidae	*Himasthla*	*Himasthla muehlensi* (Vogel, 1933)
Platyhelminthes	Trematoda	Plagiorchiida	Fasciolidae	*Fasciola*	*Fasciola gigantica* (Cobbold, 1855)
Platyhelminthes	Trematoda	Plagiorchiida	Fasciolidae	*Fasciola*	*Fasciola hepatica* (Linnaeus, 1758)
Platyhelminthes	Trematoda	Plagiorchiida	Fasciolidae	*Fasciolopsis*	*Fasciolopsis buski* (Lankester, 1857; Looss, 1899)
Platyhelminthes	Trematoda	Plagiorchiida	Philophthalmidae	*Philophthalmus*	*Philophthalmus gralli* (Mathis & Leger, 1910)
Platyhelminthes	Trematoda	Plagiorchiida	Philophthalmidae	*Philophthalmus*	*Philophthalmus lacrymosus* (Braun, 1902)
Platyhelminthes	Trematoda	Plagiorchiida	Philophthalmidae	*Philophthalmus*	*Philophthalmus lucipetus* (Rudolphi, 1819)
Platyhelminthes	Trematoda	Plagiorchiida	Philophthalmidae	*Philophthalmus*	*Philophthalmus palpebrarum* (Looss, 1899)
Platyhelminthes	Trematoda	Plagiorchiida	Achillurbainiidae	*Poikilorchis*	*Achillurbainia congolensis* (Fain and Vandepitte, 1957)
Platyhelminthes	Trematoda	Plagiorchiida	Achillurbainiidae	*Achillurbainia*	*Achillurbainia nouveli* (Dollfus, 1939)

(continued)

Table 2.3 (continued)

Phylum	Class	Order	Family	Genus	Species
Platyhelminthes	Trematoda	Plagiorchiida	Achillurbainiidae	*Achillurbainia*	*Achillurbainia recondita* (Travassos, 1942)
Platyhelminthes	Trematoda	Plagiorchiida	Dicrocoeliidae	*Dicrocoelium*	*Dicrocoelium dendriticum* (Rudolphi, 1819)
Platyhelminthes	Trematoda	Plagiorchiida	Dicrocoeliidae	*Dicrocoelium*	*Dicrocoelium hospes* (Looss, 1907)
Platyhelminthes	Trematoda	Plagiorchiida	Dicrocoeliidae	*Eurytrema*	*Eurytrema pancreaticum* (Janson, 1889)
Platyhelminthes	Trematoda	Plagiorchiida	Nanophyetidae	*Nanophyetus*	*Nanophyetus salmincola* (Chapin, 1926)
Platyhelminthes	Trematoda	Plagiorchiida	Troglotrematidae	*Paragonimus*	*Paragonimus africanus* (Voelker and Vogel, 1965)
Platyhelminthes	Trematoda	Plagiorchiida	Troglotrematidae	*Paragonimus*	*Paragonimus bangkokensis* (Miyazaki and Vajrasthira, 1967)
Platyhelminthes	Trematoda	Plagiorchiida	Troglotrematidae	*Paragonimus*	*Paragonimus caliensis* (Little, 1968)
Platyhelminthes	Trematoda	Plagiorchiida	Troglotrematidae	*Paragonimus*	*Paragonimus heterotremus* (Chen and Hsia, 1964)
Platyhelminthes	Trematoda	Plagiorchiida	Troglotrematidae	*Paragonimus*	*Paragonimus kellicotti* (Ward, 1908)
Platyhelminthes	Trematoda	Plagiorchiida	Troglotrematidae	*Paragonimus*	*Paragonimus macrorchis* (Chen, 1962)
Platyhelminthes	Trematoda	Plagiorchiida	Troglotrematidae	*Paragonimus*	*Paragonimus mexicanus* (Miyazaki and Ishii, 1968)
Platyhelminthes	Trematoda	Plagiorchiida	Troglotrematidae	*Paragonimus*	*Paragonimus miyazakii* (Kamo et al., 1961)
Platyhelminthes	Trematoda	Plagiorchiida	Troglotrematidae	*Paragonimus*	*Paragonimus ohirai* (Miyazaki, 1939)
Platyhelminthes	Trematoda	Plagiorchiida	Troglotrematidae	*Paragonimus*	*Paragonimus rudis* (Diesing, 1850)

(continued)

Table 2.3 (continued)

Phylum	Class	Order	Family	Genus	Species
Platyhelminthes	Trematoda	Plagiorchiida	Troglotrematidae	*Paragonimus*	*Paragonimus skrjabini* (Chen, 1959)
Platyhelminthes	Trematoda	Plagiorchiida	Troglotrematidae	*Paragonimus*	*Paragonimus uterobilateralis* (Voelker and Vogel, 1965)
Platyhelminthes	Trematoda	Plagiorchiida	Troglotrematidae	*Paragonimus*	*Paragonimus westermani* (Kerbert, 1878)
Platyhelminthes	Trematoda	Plagiorchiida	Troglotrematidae	*Paragonimus*	*Paragonimus xiangshanensis* (He et al., 1995)
Platyhelminthes	Trematoda	Plagiorchiida	Gymnophallidae	*Gymnophalloides*	*Gymnophalloides seoi* (Lee, Chai & Hong, 1993)
Platyhelminthes	Trematoda	Plagiorchiida	Lecithodendriidae	*Prosthodendrium*	*Prosthodendrium glandulosum* (Looss, 1896)
Platyhelminthes	Trematoda	Plagiorchiida	Lecithodendriidae	*Prosthodendrium*	*Prosthodendrium molenkampi* (Lie Kian Joe, 1951)
Platyhelminthes	Trematoda	Plagiorchiida	Lecithodendriidae	*Prosthodendrium*	*Prosthodendrium obtusum* (Loos, 1896)
Platyhelminthes	Trematoda	Plagiorchiida	Phaneropsolidae	*Phaneropsolus*	*Phaneropsolus spinicirrus* (Kaewkes, 1991)
Platyhelminthes	Trematoda	Plagiorchiida	Microphallidae	*Microphallus*	*Microphallus brevicaeca* (Africa and Garcia, 1935)
Platyhelminthes	Trematoda	Plagiorchiida	Microphallidae	*Microphallus*	*Microphallus minus* (Ouchi, 1928)
Platyhelminthes	Trematoda	Plagiorchiida	Gastrothylacidae	*Fischoederius*	*Fischoederius elongatus* (Poirier, 1883)
Platyhelminthes	Trematoda	Plagiorchiida	Notocotylidae	*Notocotylus*	*Notocotylus stagnicolae* (Herber, 1942)
Platyhelminthes	Trematoda	Plagiorchiida	Paramphistomatidae	*Gastrodiscoides*	*Gastrodiscoides hominis* (Lewis & McConnall, 1876)

(continued)

Table 2.3 (continued)

Phylum	Class	Order	Family	Genus	Species
Platyhelminthes	Trematoda	Plagiorchiida	Paramphistomatidae	*Watsonius*	*Watsonius watsoni* (Conyngham, 1904)
Platyhelminthes	Trematoda	Plagiorchiida	Plagiorchiidae	*Plagiorchis*	*Plagiorchis harinasutai* (Radomyos et al., 1989)
Platyhelminthes	Trematoda	Plagiorchiida	Plagiorchiidae	*Plagiorchis*	*Plagiorchis javensis* (Sandgrolmn, 1940)
Platyhelminthes	Trematoda	Plagiorchiida	Plagiorchiidae	*Plagiorchis*	*Plagiorchis muris* (Tanabe, 1922)
Platyhelminthes	Trematoda	Plagiorchiida	Plagiorchiidae	*Plagiorchis*	*Plagiorchis philippinensis* (Sandground, 1940)
Platyhelminthes	Trematoda	Strigeidida	Clinostomidae	*Clinostomum*	*Clinostomum complanatum* (Rudolphi, 1814; Braun, 1899)
Platyhelminthes	Trematoda	Strigeidida	Schistosomatidae	*Austrobilharzia*	*Austrobilharzia terrigalensis* (Johnston, 1917)
Platyhelminthes	Trematoda	Strigeidida	Schistosomatidae	*Austrobilharzia*	*Austrobilharzia variglandis* (Miller & Northup, 1926)
Platyhelminthes	Trematoda	Strigeidida	Schistosomatidae	*Bilharziella*	*Bilharziella polonica* (Kowalewski, 1895)
Platyhelminthes	Trematoda	Strigeidida	Schistosomatidae	*Gigantobilharzia*	*Gigantobilharzia huttoni* (Leigh, 1953; Leigh, 1955)
Platyhelminthes	Trematoda	Strigeidida	Schistosomatidae	*Gigantobilharzia*	*Gigantobilharzia sturniae* (Tanabe, 1948)
Platyhelminthes	Trematoda	Strigeidida	Schistosomatidae	*Heterobilharzia*	*Heterobilharzia americana* (Prince 1929)
Platyhelminthes	Trematoda	Strigeidida	Schistosomatidae	*Orientobilharzia*	*Orientobilharzia cheni* (Hsu, 1951)
Platyhelminthes	Trematoda	Strigeidida	Schistosomatidae	*Orientobilharzia*	*Orientobilharzia turkestanicum* (Skrjabin, 1913)
Platyhelminthes	Trematoda	Strigeidida	Schistosomatidae	*Schistosoma*	*Schistosoma bovis* (Bilharz, 1852)

(continued)

Table 2.3 (continued)

Phylum	Class	Order	Family	Genus	Species
Platyhelminthes	Trematoda	Strigeidida	Schistosomatidae	*Schistosoma*	*Schistosoma guineensis* (Pagès, 2003)
Platyhelminthes	Trematoda	Strigeidida	Schistosomatidae	*Schistosoma*	*Schistosoma haematobium* (Bilharz, 1852)
Platyhelminthes	Trematoda	Strigeidida	Schistosomatidae	*Schistosoma*	*Schistosoma intercalatum* (Fisher, 1934)
Platyhelminthes	Trematoda	Strigeidida	Schistosomatidae	*Schistosoma*	*Schistosoma japonicum* (Katsurada, 1904)
Platyhelminthes	Trematoda	Strigeidida	Schistosomatidae	*Schistosoma*	*Schistosoma malayensis* (Greer, Ow-Yang & Yong, 1988)
Platyhelminthes	Trematoda	Strigeidida	Schistosomatidae	*Schistosoma*	*Schistosoma mansoni* (Sambon, 1907)
Platyhelminthes	Trematoda	Strigeidida	Schistosomatidae	*Schistosoma*	*Schistosoma mattheei*
Platyhelminthes	Trematoda	Strigeidida	Schistosomatidae	*Schistosoma*	*Schistosoma mekongi* (Voge, Bruckner & Bruce, 1978)
Platyhelminthes	Trematoda	Strigeidida	Schistosomatidae	*Schistosoma*	*Schistosoma rodhaini* (Brumpt, 1931)
Platyhelminthes	Trematoda	Strigeidida	Schistosomatidae	*Schistosoma*	*Schistosoma sinensium* (Bao 1958)
Platyhelminthes	Trematoda	Strigeidida	Schistosomatidae	*Schistosoma*	*Schistosoma spindale* (Montgomery, 1906)
Platyhelminthes	Trematoda	Strigeidida	Schistosomatidae	*Schistosomatium*	*Schistosomatium douthitti* (Cort, 1928)
Platyhelminthes	Trematoda	Strigeidida	Schistosomatidae	*Trichobilharzia*	*Trichobilharzia brevis* (Basch, 1966)
Platyhelminthes	Trematoda	Strigeidida	Schistosomatidae	*Trichobilharzia*	*Trichobilharzia dayushuensis* (Xu, Jiang, and Yao, 1997)
Platyhelminthes	Trematoda	Strigeidida	Schistosomatidae	*Trichobilharzia*	*Trichobilharzia franki* (Muller and Kimmig, 1994)

(continued)

Table 2.3 (continued)

Phylum	Class	Order	Family	Genus	Species
Platyhelminthes	Trematoda	Strigeidida	Schistosomatidae	*Trichobilharzia*	*Trichobilharzia gigantica* (Xu, Jiang, and Yao, 1997)
Platyhelminthes	Trematoda	Strigeidida	Schistosomatidae	*Trichobilharzia*	*Trichobilharzia guandongensis* (Tsai et al., 1979)
Platyhelminthes	Trematoda	Strigeidida	Schistosomatidae	*Trichobilharzia*	*Trichobilharzia jianensis* (Lui et al., 1977)
Platyhelminthes	Trematoda	Strigeidida	Schistosomatidae	*Trichobilharzia*	*Trichobilharzia ocellata* (La Valette, 1855)
Platyhelminthes	Trematoda	Strigeidida	Schistosomatidae	*Trichobilharzia*	*Trichobilharzia paoi* (Pao and Yung 1957)
Platyhelminthes	Trematoda	Strigeidida	Schistosomatidae	*Trichobilharzia*	*Trichobilharzia parocellata* (Islam & Copeman 1986)
Platyhelminthes	Trematoda	Strigeidida	Schistosomatidae	*Trichobilharzia*	*Trichobilharzia physellae* (Talbot, 1936)
Platyhelminthes	Trematoda	Strigeidida	Schistosomatidae	*Trichobilharzia*	*Trichobilharzia pingnana* (Xu, Jiang, and Yao, 1997)
Platyhelminthes	Trematoda	Strigeidida	Schistosomatidae	*Trichobilharzia*	*Trichobilharzia stagnicolae* (Talbot, 1936)
Platyhelminthes	Trematoda	Strigeidida	Schistosomatidae	*Trichobilharzia*	*Trichobilharzia szidati* (Neuhaus, 1952)
Platyhelminthes	Trematoda	Strigeidida	Schistosomatidae	*Trichobilharzia*	*Trichobilharzia zhongshani* (Tsai et al., 1979)
Platyhelminthes	Trematoda	Strigeidida	Spirorchiidae	*Spirorchis*	*Spirorchis haematobius* (Stunkard, 1922)
Platyhelminthes	Trematoda	Strigeidida	Strigeidae	*Cotylurus*	*Cotylurus japonicus* (Ishii, 1932)
Nematomorpha	Gordioida	Chordodea	Chordodidae	*Beatogordius*	*Beatogordius inesae* (Cavalieri, 1961)

(continued)

Table 2.3 (continued)

Phylum	Class	Order	Family	Genus	Species
Nematomorpha	Gordioida	Chordodea	Chordodidae	*Beatogordius*	*Beatogordius raphaelis* (Camerano, 1893)
Nematomorpha	Gordioida	Chordodea	Chordodidae	*Chordodes*	*Chordodes capensis* (Camerano, 1895)
Nematomorpha	Gordioida	Chordodea	Chordodidae	*Chordodes*	*Chordodes skorikowi* (Camerano, 1903)
Nematomorpha	Gordioida	Chordodea	Chordodidae	*Dacochordodes*	*Dacochordodes bacescui* (Capuse, 1966)
Nematomorpha	Gordioida	Chordodea	Chordodidae	*Neochordodes*	*Neochordodes colombianus* (Faust & Ramos, 1960)
Nematomorpha	Gordioida	Chordodea	Chordodidae	*Pseudogordius*	*Pseudogordius tanganyikae* (Yeh & Jordan, 1957)
Nematomorpha	Gordioida	Chordodea	Parachordodidae	*Gordionus*	*Gordionus alpestris* (Villot, 1884)
Nematomorpha	Gordioida	Chordodea	Parachordodidae	*Gordionus*	*Gordionus chinensis* (Villot, 1874)
Nematomorpha	Gordioida	Chordodea	Parachordodidae	*Gordionus*	*Gordionus violaceus* (Baird, 1853)
Nematomorpha	Gordioida	Chordodea	Parachordodidae	*Gordionus*	*Gordionus wolterstorffii* (Camerano 1888)
Nematomorpha	Gordioida	Chordodea	Parachordodidae	*Parachordodes*	*Parachordodes pustulosus* (Baird, 1853)
Nematomorpha	Gordioida	Chordodea	Parachordodidae	*Parachordodes*	*Parachordodes tolosanus* (Dujardin, 1842)
Nematomorpha	Gordioida	Gordea	Gordiidae	*Gordius*	*Gordius aquaticus* (Linnaeus 1758)
Nematomorpha	Gordioida	Gordea	Gordiidae	*Gordius*	*Gordius gesneri* (Heinze, 1937)
Nematomorpha	Gordioida	Gordea	Gordiidae	*Gordius*	*Gordius ogatai* (Inoue, 1979)
Nematomorpha	Gordioida	Gordea	Gordiidae	*Gordius*	*Gordius perronciti* (Camerano, 1887)
Nematomorpha	Gordioida	Gordea	Gordiidae	*Gordius*	*Gordius reddyi* (Singh and Rao, 1967)

(continued)

Table 2.3 (continued)

Phylum	Class	Order	Family	Genus	Species
Nematomorpha	Gordioida	Gordea	Gordiidae	*Gordius*	*Gordius robustus* (Leidy, 1851)
Nematomorpha	Gordioida	Gordea	Gordiidae	*Gordius*	*Gordius setiger* (Schneider, 1866)
Nematomorpha	Gordioida	Gordea	Gordiidae	*Gordius*	*Gordius villoti* (Rosa, 1882)
Nematomorpha	Gordioida	Chordodea	Gordiidae	*Paragordius*	*Paragordius areolatus* (Linstow, 1906)
Nematomorpha	Gordioida	Chordodea	Gordiidae	*Paragordius*	*Paragordius cinctus* (Linstow, 1906)
Nematomorpha	Gordioida	Chordodea	Gordiidae	*Paragordius*	*Paragordius esavianus* (Carvalho, 1942)
Nematomorpha	Gordioida	Chordodea	Gordiidae	*Paragordius*	*Paragordius tricuspidatus* (Dufour 1828)
Nematomorpha	Gordioida	Chordodea	Gordiidae	*Paragordius*	*Paragordius varius* (Leidy, 1851)
Nematoda	Chromadorea	Rhabditida	Diplogasteridae	*Diplogaster*	*Diplogaster coronata* (Cobb, 1893; Cobb, 1913)
Nematoda	Chromadorea	Rhabditida	Panagrolaimidae	*Halicephalobus*	*Halicephalobus deletrix* (Anderson & Bemrick, 1965)
Nematoda	Chromadorea	Rhabditida	Panagrolaimidae	*Turbatrix*	*Turbatrix aceti* (Muller, 1783)
Nematoda	Chromadorea	Rhabditida	Rhabditidae	*Mesorhabditis*	*Mesorhabditis graciliformis* (Goffart, 1935)
Nematoda	Chromadorea	Rhabditida	Rhabditidae	*Pelodera*	*Pelodera strongyloides* (Schneider, 1860)
Nematoda	Chromadorea	Rhabditida	Rhabditidae	*Pelodera*	*Pelodera teres* (Schneider, 1866)
Nematoda	Chromadorea	Rhabditida	Rhabditidae	*Rhabditella*	*Rhabditella axei* (Cobbold, 1884)
Nematoda	Chromadorea	Rhabditida	Rhabditidae	*Rhabditella*	*Rhabditella pseudoelongata* (Micoletzky, 1913; Andrássy, 1983)
Nematoda	Chromadorea	Rhabditida	Rhabditidae	*Rhabditis*	*Rhabditis niellyi* (Blanchard, 1855)

(continued)

Table 2.3 (continued)

Phylum	Class	Order	Family	Genus	Species
Nematoda	Chromadorea	Rhabditida	Rhabditidae	*Rhabditis*	*Rhabditis pellio* (Schneider, 1866; Bütschli, 1873)
Nematoda	Chromadorea	Rhabditida	Rhabditidae	*Rhabditis*	*Rhabditis terricola* (Dujardin, 1845)
Nematoda	Chromadorea	Rhabditida	Rhabditidae	*Rhabditis*	*Rhabditis usui* (Watanabe, 1927)
Nematoda	Chromadorea	Rhabditida	Rhabditidae	*Rhabditoides*	*Rhabditoides inermis* (Schneider, 1866)
Nematoda	Chromadorea	Rhabditida	Alloionematidae	*Cheilobus*	*Cheilobus quadrilabiatus* (Cobb, 1924)
Nematoda	Chromadorea	Rhabditida	Strongyloididae	*Strongyloides*	*Strongyloides cebus* (Darling, 1911)
Nematoda	Chromadorea	Rhabditida	Strongyloididae	*Strongyloides*	*Strongyloides fuelleborni* (Von Linstow, 1905)
Nematoda	Chromadorea	Rhabditida	Strongyloididae	*Strongyloides*	*Strongyloides myopotami* (Artigas & Pacheco, 1933)
Nematoda	Chromadorea	Rhabditida	Strongyloididae	*Strongyloides*	*Strongyloides papillosus* (Wedl, 1856)
Nematoda	Chromadorea	Rhabditida	Strongyloididae	*Strongyloides*	*Strongyloides planiceps* (Rogers, 1943)
Nematoda	Chromadorea	Rhabditida	Strongyloididae	*Strongyloides*	*Strongyloides procyonis* (Little, 1966)
Nematoda	Chromadorea	Rhabditida	Strongyloididae	*Strongyloides*	*Strongyloides ransomi* (Schwartz & Alicata, 1930)
Nematoda	Chromadorea	Rhabditida	Strongyloididae	*Strongyloides*	*Strongyloides stercoralis* (Bavay, 1876)
Nematoda	Chromadorea	Rhabditida	Strongyloididae	*Strongyloides*	*Strongyloides westeri* (Ihle, 1917)
Nematoda	Chromadorea	Ascaridida	Anisakidae	*Anisakis*	*Anisakis pegreffii* (Campana-Rouget and Biocca, 1955)
Nematoda	Chromadorea	Ascaridida	Anisakidae	*Anisakis*	*Anisakis physeteris* (Baylis, 1923)
Nematoda	Chromadorea	Ascaridida	Anisakidae	*Anisakis*	*Anisakis simplex* (Rudolphi, 1809)

(continued)

Table 2.3 (continued)

Phylum	Class	Order	Family	Genus	Species
Nematoda	Chromadorea	Ascaridida	Anisakidae	*Contracaecum*	*Contracaecum osculatum* (Rudolphi, 1802)
Nematoda	Chromadorea	Ascaridida	Anisakidae	*Pseudoterranova*	*Pseudoterranova decipiens* (Krabbe, 1878)
Nematoda	Chromadorea	Ascaridida	Ascarididae	*Ascaris*	*Ascaris lumbricoides* (Linnaeus, 1758)
Nematoda	Chromadorea	Ascaridida	Ascarididae	*Ascaris*	*Ascaris suum* (Goeze, 1782)
Nematoda	Chromadorea	Ascaridida	Ascarididae	*Baylisascaris*	*Baylisascaris procyonis* (Stefanski & Zarnowski, 1951)
Nematoda	Chromadorea	Ascaridida	Ascarididae	*Lagochilascaris*	*Lagochilascaris minor* (Leiper, 1909)
Nematoda	Chromadorea	Ascaridida	Ascarididae	*Lagochilascaris*	*Lagochilascaris sprenti* (Bowman, 1983)
Nematoda	Chromadorea	Ascaridida	Ascarididae	*Parascaris*	*Parascaris equorum* (Goeze, 1782)
Nematoda	Chromadorea	Ascaridida	Ascarididae	*Toxascaris*	*Toxascaris leonina* (Linstow, 1902)
Nematoda	Chromadorea	Ascaridida	Raphidascarididae	*Hysterothylacium*	*Hysterothylacium aduncum* (Rudolphi, 1802)
Nematoda	Chromadorea	Ascaridida	Toxocaridae	*Toxocara*	*Toxocara canis* (Werner, 1782)
Nematoda	Chromadorea	Ascaridida	Toxocaridae	*Toxocara*	*Toxocara cati* (Schrank, 1788)
Nematoda	Chromadorea	Ascaridida	Toxocaridae	*Toxocara*	*Toxocara pteropodis* (Baylis, 1936)
Nematoda	Chromadorea	Ascaridida	Toxocaridae	*Toxocara*	*Toxocara vitulorum* (Goeze, 1782)
Nematoda	Chromadorea	Ascaridida	Oxyuridae	*Enterobius*	*Enterobius gregorii* (Hugot, 1983)
Nematoda	Chromadorea	Ascaridida	Oxyuridae	*Enterobius*	*Enterobius vermicularis* (Linnaeus, 1758)
Nematoda	Chromadorea	Ascaridida	Oxyuridae	*Oxyuris*	*Oxyuris equi* (Schrank, 1788)
Nematoda	Chromadorea	Ascaridida	Oxyuridae	*Passalurus*	*Passalurus ambiguus* (Rudolphi, 1819)

(continued)

Table 2.3 (continued)

Phylum	Class	Order	Family	Genus	Species
Nematoda	Chromadorea	Ascaridida	Oxyuridae	*Syphacia*	*Syphacia obvelata* (Rudolphi, 1802)
Nematoda	Chromadorea	Ascaridida	Pharyngodonidae	*Spauligodon*	*Spauligodon blydeensis* (Hering-Hagenbeck, 2001)
Nematoda	Chromadorea	Camallanida	Dracunculidae	*Dracunculus*	*Dracunculus medinensis* (Linnaeus, 1758)
Nematoda	Chromadorea	Camallanida	Philometridae	*Philometra*	*Philometra sanguinea* (Rudolphi, 1819)
Nematoda	Chromadorea	Spirurida	Onchocercidae	*Brugia*	*Brugia beaveri* (Ash and Little, 1964)
Nematoda	Chromadorea	Spirurida	Onchocercidae	*Brugia*	*Brugia ceylonensis* (Jayewardene, 1962)
Nematoda	Chromadorea	Spirurida	Onchocercidae	*Brugia*	*Brugia guyanensis* (Orihel, 1964)
Nematoda	Chromadorea	Spirurida	Onchocercidae	*Brugia*	*Brugia malayi* (Brug, 1927)
Nematoda	Chromadorea	Spirurida	Onchocercidae	*Brugia*	*Brugia pahangi* (Buckley & Edeson, 1956)
Nematoda	Chromadorea	Spirurida	Onchocercidae	*Brugia*	*Brugia timori* (Partono et al., 1977)
Nematoda	Chromadorea	Spirurida	Onchocercidae	*Mansonella*	*Mansonella perstans*
Nematoda	Chromadorea	Spirurida	Onchocercidae	*Dirofilaria*	*Dirofilaria conjunctivae* (Addario, 1885)
Nematoda	Chromadorea	Spirurida	Onchocercidae	*Dirofilaria*	*Dirofilaria immitis* (Leidy, 1856)
Nematoda	Chromadorea	Spirurida	Onchocercidae	*Dirofilaria*	*Dirofilaria louisianensis* (Faust, Thomas, and Jones, 1941)
Nematoda	Chromadorea	Spirurida	Onchocercidae	*Dirofilaria*	*Dirofilaria magalhaesi* (Blanchard 1896)
Nematoda	Chromadorea	Spirurida	Onchocercidae	*Dirofilaria*	*Dirofilaria magnilarvatum* (Price, 1959)
Nematoda	Chromadorea	Spirurida	Onchocercidae	*Dirofilaria*	*Dirofilaria ochmanni* (Fuelleborn 1908)

(continued)

Table 2.3 (continued)

Phylum	Class	Order	Family	Genus	Species
Nematoda	Chromadorea	Spirurida	Onchocercidae	*Dirofilaria*	*Dirofilaria repens* (Railliet & Henry, 1911)
Nematoda	Chromadorea	Spirurida	Onchocercidae	*Dirofilaria*	*Dirofilaria spectans* (Freitas & Lent, 1949)
Nematoda	Chromadorea	Spirurida	Onchocercidae	*Dirofilaria*	*Dirofilaria striata* (Molin, 1858)
Nematoda	Chromadorea	Spirurida	Onchocercidae	*Dirofilaria*	*Dirofilaria tenuis* (Chandler, 1942)
Nematoda	Chromadorea	Spirurida	Onchocercidae	*Dirofilaria*	*Dirofilaria ursi* (Yamaguti, 1941)
Nematoda	Chromadorea	Spirurida	Onchocercidae	*Mansonella*	*Mansonella ozzardi* (Manson, 1897)
Nematoda	Chromadorea	Spirurida	Onchocercidae	*Mansonella*	*Mansonella rodhaini* (Peel and Chardome, 1946)
Nematoda	Chromadorea	Spirurida	Onchocercidae	*Mansonella*	*Mansonella streptocerca* (Macfie & Corson, 1922)
Nematoda	Chromadorea	Spirurida	Onchocercidae	*Meningonema*	*Meningonema peruzzii* (Orihel and Esslinger, 1973)
Nematoda	Chromadorea	Spirurida	Onchocercidae	*Onchocerca*	*Onchocerca cervicalis* (Railliet & Henry, 1910)
Nematoda	Chromadorea	Spirurida	Onchocercidae	*Onchocerca*	*Onchocerca dewittei japonica* (Uni et al., 2001)
Nematoda	Chromadorea	Spirurida	Onchocercidae	*Onchocerca*	*Onchocerca gutturosa* (Neumann, 1910)
Nematoda	Chromadorea	Spirurida	Onchocercidae	*Onchocerca*	*Onchocerca volvulus* (Leuckart, 1894)
Nematoda	Chromadorea	Spirurida	Onchocercidae	*Wuchereria*	*Wuchereria bancrofti* (Cobbold, 1877)
Nematoda	Chromadorea	Spirurida	Onchocercidae	*Wuchereria*	*Wuchereria lewisi* (Schacher, 1969)
Nematoda	Chromadorea	Spirurida	Setariidae	*Dipetalonema*	*Dipetalonema arbuta* (Highby, 1943)
Nematoda	Chromadorea	Spirurida	Setariidae	*Dipetalonema*	

(continued)

Table 2.3 (continued)

Phylum	Class	Order	Family	Genus	Species
					Dipetalonema sprenti (Anderson 1953)
Nematoda	Chromadorea	Spirurida	Setariidae	*Setaria*	*Setaria equina* (Abildgaard, 1789)
Nematoda	Chromadorea	Spirurida	Setariidae	*Setaria*	*Setaria labiatopapillosa* (Alessandrini, 1838)
Nematoda	Chromadorea	Spirurida	Gnathostomatidae	*Gnathostoma*	*Gnathostoma binucleatum* (Almeyda–Artigas, 1991)
Nematoda	Chromadorea	Spirurida	Gnathostomatidae	*Gnathostoma*	*Gnathostoma doloresi* (Tubangui, 1925)
Nematoda	Chromadorea	Spirurida	Gnathostomatidae	*Gnathostoma*	*Gnathostoma hispidum* (Fedstchenko, 1872)
Nematoda	Chromadorea	Spirurida	Gnathostomatidae	*Gnathostoma*	*Gnathostoma malaysiae* (Miyazaki and Dunn, 1965)
Nematoda	Chromadorea	Spirurida	Gnathostomatidae	*Gnathostoma*	*Gnathostoma nipponicum* (Yamaguti, 1941)
Nematoda	Chromadorea	Spirurida	Gnathostomatidae	*Gnathostoma*	*Gnathostoma procyonis* (Chandler, 1942)
Nematoda	Chromadorea	Spirurida	Gnathostomatidae	*Gnathostoma*	*Gnathostoma spinigerum* (Owen, 1836)
Nematoda	Chromadorea	Spirurida	Gnathostomatidae	*Gnathostoma*	*Gnathostoma turgidum* (Stossich, 1902)
Nematoda	Chromadorea	Spirurida	Tetrameridae	*Tetrameres*	*Tetrameres fissipina* (Diesing, 1861)
Nematoda	Chromadorea	Spirurida	Physalopteridae	*Abbreviata*	*Abbreviata caucasica* (Linstow, 1902)
Nematoda	Chromadorea	Spirurida	Rictulariidae	*Rictularia*	

(continued)

Table 2.3 (continued)

Phylum	Class	Order	Family	Genus	Species
					Rictularia lucifugus (Douvres, 1956)
Nematoda	Chromadorea	Spirurida	Gongylonematidae	*Gongylonema*	*Gongylonema pulchrum* (Molin, 1857)
Nematoda	Chromadorea	Spirurida	Thelaziidae	*Spirocerca*	*Spirocerca lupi* (Rudolphi, 1809)
Nematoda	Chromadorea	Spirurida	Thelaziidae	*Thelazia*	*Thelazia californiensis* (Price, 1930)
Nematoda	Chromadorea	Spirurida	Thelaziidae	*Thelazia*	*Thelazia callipaeda* (Railliet & Henry, 1910)
Nematoda	Chromadorea	Tylenchida	Meloidogynidae	*Meloidogyne*	*Meloidogyne incognita acrita* (Chitwood, 1949)
Nematoda	Chromadorea	Strongylida	Ancylostomatidae	*Ancylostoma*	*Ancylostoma brasiliense* (De Faria, 1910)
Nematoda	Chromadorea	Strongylida	Ancylostomatidae	*Ancylostoma*	*Ancylostoma canium* (Ercolani, 1859; Hall, 1913)
Nematoda	Chromadorea	Strongylida	Ancylostomatidae	*Ancylostoma*	*Ancylostoma ceylanicum* (Looss 1911)
Nematoda	Chromadorea	Strongylida	Ancylostomatidae	*Ancylostoma*	*Ancylostoma duodenale* (Dubini, 1843)
Nematoda	Chromadorea	Strongylida	Ancylostomatidae	*Ancylostoma*	*Ancylostoma japonica* (Fukuda & Katsurada, 1925)
Nematoda	Chromadorea	Strongylida	Ancylostomatidae	*Ancylostoma*	*Ancylostoma malayanaum* (Alessandrini, 1905)
Nematoda	Chromadorea	Strongylida	Ancylostomatidae	*Ancylostoma*	*Ancylostoma tubaeforme* (Zeder, 1800)
Nematoda	Chromadorea	Strongylida	Ancylostomatidae	*Bunostomum*	*Bunostomum phlebotomum* (Railliet, 1900)
Nematoda	Chromadorea	Strongylida	Ancylostomatidae	*Bunostomum*	*Bunostomum trigonocephalum* (Rudolphi, 1808)

(continued)

Table 2.3 (continued)

Phylum	Class	Order	Family	Genus	Species
Nematoda	Chromadorea	Strongylida	Ancylostomatidae	*Necator*	*Necator americanus* Stiles, 1902
Nematoda	Chromadorea	Strongylida	Ancylostomatidae	*Uncinaria*	*Uncinaria stenocephala* (Railliet, 1884)
Nematoda	Chromadorea	Strongylida	Angiostrongylidae	*Angiostrongylus*	*Angiostrongylus cantonensis* (Chen, 1935)
Nematoda	Chromadorea	Strongylida	Angiostrongylidae	*Angiostrongylus*	*Angiostrongylus costaricensis* (Morera & Cespedes, 1971)
Nematoda	Chromadorea	Strongylida	Angiostrongylidae	*Angiostrongylus*	*Angiostrongylus malaysiensis* (Bhaibulaya and Cross, 1971)
Nematoda	Chromadorea	Strongylida	Metastrongylidae	*Metastrongylus*	*Metastrongylus apri* (Gmelin, 1790)
Nematoda	Chromadorea	Strongylida	Molineidae	*Nematodirus*	*Nematodirus abnormalis* (May, 1920)
Nematoda	Chromadorea	Strongylida	Chabertiidae	*Cyclodontostomum*	*Cyclodontostomum purvisi* (Adams, 1933)
Nematoda	Chromadorea	Strongylida	Syngamidae	*Mammomonogamus*	*Mammomonogamus laryngeus* (Railliet, 1899)
Nematoda	Chromadorea	Strongylida	Syngamidae	*Mammomonogamus*	*Mammomonogamus nasicola* (Linstow, 1899)
Nematoda	Chromadorea	Strongylida	Chabertiidae	*Oesophagostomum*	*Oesophagostomum aculeatum* (Linstow, 1879)
Nematoda	Chromadorea	Strongylida	Chabertiidae	*Oesophagostomum*	*Oesophagostomum apiostomum* (Willach 1891)
Nematoda	Chromadorea	Strongylida	Chabertiidae	*Oesophagostomum*	*Oesophagostomum bifurcum* (Creplin, 1849)
Nematoda	Chromadorea	Strongylida	Chabertiidae	*Oesophagostomum*	*Oesophagostomum stephanostomum*
Nematoda	Chromadorea	Strongylida	Chabertiidae	*Ternidens*	*Ternidens deminutus* (Railliet & Henry, 1905)

(continued)

Table 2.3 (continued)

Phylum	Class	Order	Family	Genus	Species
Nematoda	Chromadorea	Strongylida	Haemonchidae	*Haemonchus*	*Haemonchus contortus* (Rudolphi, 1803)
Nematoda	Chromadorea	Strongylida	Haemonchidae	*Marshallagia*	*Marshallagia marshalli* (Ransom, 1907)
Nematoda	Chromadorea	Strongylida	Haemonchidae	*Mecistocirrus*	*Mecistocirrus digitatus* (Linstow, 1906)
Nematoda	Chromadorea	Strongylida	Haemonchidae	*Teladorsagia*	*Teladorsagia circumcincta* (Stadelman, 1894)
Nematoda	Chromadorea	Strongylida	Haemonchidae	*Ostertagia*	*Ostertagia ostertagi* (Stiles, 1892)
Nematoda	Chromadorea	Strongylida	Trichostrongylidae	*Trichostrongylus*	*Trichostrongylus affinis* (Graybill, 1924)
Nematoda	Chromadorea	Strongylida	Trichostrongylidae	*Trichostrongylus*	*Trichostrongylus axei* (Cobbold, 1879)
Nematoda	Chromadorea	Strongylida	Trichostrongylidae	*Trichostrongylus*	*Trichostrongylus brevis* (Otsuru, 1962)
Nematoda	Chromadorea	Strongylida	Trichostrongylidae	*Trichostrongylus*	*Trichostrongylus calcaratus* (Ransom, 1911)
Nematoda	Chromadorea	Strongylida	Trichostrongylidae	*Trichostrongylus*	*Trichostrongylus capricola* (Ransom, 1907)
Nematoda	Chromadorea	Strongylida	Trichostrongylidae	*Trichostrongylus*	*Trichostrongylus colubriformis* (Giles, 1892)
Nematoda	Chromadorea	Strongylida	Trichostrongylidae	*Trichostrongylus*	*Trichostrongylus lerouxi* (Biocca, Chabaud & Ghadirian, 1974)
Nematoda	Chromadorea	Strongylida	Trichostrongylidae	*Trichostrongylus*	*Trichostrongylus orientalis* (Jimbo, 1914)
Nematoda	Chromadorea	Strongylida	Trichostrongylidae	*Trichostrongylus*	*Trichostrongylus probolurus* (Railliet, 1896)

(continued)

Table 2.3 (continued)

Phylum	Class	Order	Family	Genus	Species
Nematoda	Chromadorea	Strongylida	Trichostrongylidae	*Trichostrongylus*	*Trichostrongylus skrjabini* (Kalantarian, 1928)
Nematoda	Chromadorea	Strongylida	Trichostrongylidae	*Trichostrongylus*	*Trichostrongylus vitrinus* (Looss, 1905)
Nematoda	Enoplea	Dioctophymatida	Dioctophymatidae	*Dioctophyme*	*Dioctophyme renale* (Goeze, 1782)
Nematoda	Enoplea	Mermithida	Mermithidae	*Agamomermis*	*Agamomermis hominis* (Leidy 1850)
Nematoda	Enoplea	Mermithida	Mermithidae	*Agamomermis*	*Agamomermis restiformis* (Leidy, 1880)
Nematoda	Enoplea	Mermithida	Mermithidae	*Mermis*	*Mermis nigrescens* (Dujardin, 1842)
Nematoda	Enoplea	Muspiceida	Robertdollfusidae	*Haycocknema*	*Haycocknema perplexum* (Spratt, Beveridge, Andrews & Dennett, 1999)
Nematoda	Enoplea	Trichinellida	Anatrichosomatidae	*Anatrichosoma*	*Anatrichosoma cutaneum* (Allen, 1960)
Nematoda	Enoplea	Trichinellida	Anatrichosomatidae	*Anatrichosoma*	*Anatrichosoma cynomolgi* (Smith and Chitwood, 1954)
Nematoda	Enoplea	Trichinellida	Capillariidae	*Paracapillaria*	*Paracapillaria philippinensis* (Chitwood, Velasqueza & Salazar, 1968)
Nematoda	Enoplea	Trichinellida	Capillariidae	*Calodium*	*Calodium hepaticum* (Bancroft, 1893)
Nematoda	Enoplea	Trichinellida	Capillariidae	*Capillaria*	*Capillaria aerophila* (Creplin, 1839)
Nematoda	Enoplea	Trichinellida	Capillariidae	*Pearsonema*	*Pearsonema plica* (Rudolphi, 1819)
Nematoda	Enoplea	Trichinellida	Trichinellidae	*Trichinella*	*Trichinella britovi* (Pozio, la-Rosa, Murrell & Lichtenfels, 1992)

(continued)

Table 2.3 (continued)

Phylum	Class	Order	Family	Genus	Species
Nematoda	Enoplea	Trichinellida	Trichinellidae	*Trichinella*	*Trichinella murrelli* (Pozio & La Rosa 2000)
Nematoda	Enoplea	Trichinellida	Trichinellidae	*Trichinella*	*Trichinella nativa* (Britov & Boev 1972)
Nematoda	Enoplea	Trichinellida	Trichinellidae	*Trichinella*	*Trichinella nelsoni* (Britov & Boev 1972)
Nematoda	Enoplea	Trichinellida	Trichinellidae	*Trichinella*	*Trichinella papuae* (Pozio, Owen, La-Rosa, Sacchi, Rossi & Corona, 1999)
Nematoda	Enoplea	Trichinellida	Trichinellidae	*Trichinella*	*Trichinella pseudospiralis* (Garkavi, 1972)
Nematoda	Enoplea	Trichinellida	Trichinellidae	*Trichinella*	*Trichinella spiralis* (Owen 1835)
Nematoda	Enoplea	Trichinellida	Trichuridae	*Trichuris*	*Trichuris suis* (Schrank, 1788)
Nematoda	Enoplea	Trichinellida	Trichuridae	*Trichuris*	*Trichuris trichiura* (Linnaeus, 1771)
Nematoda	Enoplea	Trichinellida	Trichuridae	*Trichuris*	*Trichuris vulpis* (Frölich, 1789)

Gibson, D. I., Bray, R. A., & Harris, E. A. (2005). Host-Parasite Database of the Natural History Museum, London. Natural History Museum. Retrieved 21 October

References

Abdelghani, E., Zerpa, R., Iliescu, G., & Escalante, C. P. (2020). Schistosomiasis and liver disease: Learning from the past to understand the present. *Clinical Case Reports, 8*(8), 1522–1526. https://doi.org/10.1002/ccr3.2922

Aguinaldo, A. M., Turbeville, J. M., Linford, L. S., Rivera, M. C., Garey, J. R., Raff, R. A., & Lake, J. A. (1997). Evidence for a clade of nematodes, arthropods and other moulting animals. *Nature, 387*(6632), 489–493. https://doi.org/10.1038/387489a0

Anosike, J. C., Njoku, A. J., Nwoke, B. E., Okoro, O. U., Okere, A. N., Ukaga, C. N., & Adimonye, R. N. (2000). Human infections with *Moniliformis moniliformis* (Bremser 1811) Travassos 1915 in South-Eastern Nigeria. *Annals of Tropical Medicine and Parasitology, 94*(8), 837–838. https://doi.org/10.1080/00034980020020192

Arizono, N., Kuramochi, T., & Kagei, N. (2012). Molecular and histological identification of the acanthocephalan *Bolbosoma* cf. *capitatum* from the human small intestine. *Parasitology International, 61*(4), 715–718. https://doi.org/10.1016/j.parint.2012.05.011

Bai, X., Hu, X., Liu, X., Tang, B., & Liu, M. (2017). Current research of trichinellosis in China. *Frontiers in Microbiology, 8*, 1472. https://doi.org/10.3389/fmicb.2017.01472

Blaxter, M. L., De Ley, P., Garey, J. R., Liu, L. X., Scheldeman, P., Vierstraete, A., Vanfleteren, J. R., Mackey, L. Y., Dorris, M., Frisse, L. M., Vida, J. T., & Thomas, W. K. (1998). A molecular evolutionary framework for the phylum Nematoda. *Nature, 392*(6671), 71–75. https://doi.org/10.1038/32160

Bleidorn, C., Schmidt-Rhaesa, A., & Garey, J. R. (2002). Systematic relationships of Nematomorpha based on molecular and morphological data. *Invertebrate Biology, 121*(4), 357–364. Retrieved from http://www.jstor.org/stable/3227168

Bradbury, R. S., & Speare, R. (2015). Passage of *Meloidogyne* eggs in human stool: Forgotten, but not gone. *Journal of Clinical Microbiology, 53*(4), 1458–1459. https://doi.org/10.1128/JCM.03384-14

Briggs, N., Weatherhead, J., Sastry, K. J., & Hotez, P. J. (2016). The hygiene hypothesis and its inconvenient truths about helminth infections. *PLoS Neglected Tropical Diseases, 10*(9), e0004944. https://doi.org/10.1371/journal.pntd.0004944

Brindley, P. J., & Loukas, A. (2017). Helminth infection-induced malignancy. *PLoS Pathogens, 13*(7), e1006393. https://doi.org/10.1371/journal.ppat.1006393

Brindley, P. J., Mitreva, M., Ghedin, E., & Lustigman, S. (2009). Helminth genomics: The implications for human health. *PLoS Neglected Tropical Diseases, 3*(10), e538. https://doi.org/10.1371/journal.pntd.0000538

Calvisi, D. F. (2020). *Schistosoma mansoni* and hepatocellular carcinoma: Is it all about c-Jun and signal transducer and activator of transcription 3? *Hepatology, 72*(2), 375–378. https://doi.org/10.1002/hep.31392

Caraballo, L. (2018). The tropics, helminth infections and hygiene hypotheses. *Expert Review of Clinical Immunology, 14*(2), 99–102. https://doi.org/10.1080/1744666x.2018.1424543

Carlson, C. J., Dallas, T. A., Alexander, L. W., Phelan, A. L., & Phillips, A. J. (2020). What would it take to describe the global diversity of parasites? *Proceedings of the Biological Sciences, 287*(1939), 20201841. https://doi.org/10.1098/rspb.2020.1841

CDC. (2021). *Parasites-Diphyllobothrium infection*. Retrieved October 10, 2021.

Chai, J. Y., & Jung, B. K. (2017). Fishborne zoonotic heterophyid infections: An update. *Food and Waterborne Parasitology, 8-9*, 33–63. https://doi.org/10.1016/j.fawpar.2017.09.001

Counselman, K., Field, C., Lea, G., Nickol, B., & Neafie, R. (1989). *Moniliformis moniliformis* from a child in Florida. *The American Journal of Tropical Medicine and Hygiene, 41*(1), 88–90.

Crompton, D. W., & Nesheim, M. C. (2002). Nutritional impact of intestinal helminthiasis during the human life cycle. *Annual Review of Nutrition, 22*, 35–59. https://doi.org/10.1146/annurev.nutr.22.120501.134539

Dantas-Torres, F., & Otranto, D. (2013). Dirofilariosis in the Americas: A more virulent *Dirofilaria immitis*? *Parasites & Vectors, 6*(1), 288. https://doi.org/10.1186/1756-3305-6-288

De Ley, P., & Blaxter, M. (2002). Systematic position and phylogeny. In *The biology of nematodes* (pp. 1–30). Taylor & Francis.

Dexiang, W., & Wenyuan, Y. (1981). *Parachordodes* sp. (Nematomorpha) human infestation of the lower urinary tract: The first case report in China. *Acta Academiae Medicinae Wuhan, 1*(2), 40–45. https://doi.org/10.1007/BF02857074

Dingley, D., & Beaver, P. C. (1985). *Macracanthorhynchus ingens* from a child in Texas. *The American Journal of Tropical Medicine and Hygiene, 34*(5), 918–920. https://doi.org/10.4269/ajtmh.1985.34.918

García, H. H., Gonzalez, A. E., Evans, C. A. W., Gilman, R. H., & Cysticercosis Working Group in P. (2003). *Taenia solium* cysticercosis. *Lancet (London, England), 362*(9383), 547–556. https://doi.org/10.1016/S0140-6736(03)14117-7

Ghosh, S. K. (2018). In S. K. Ghosh (Ed.), *Paniker's textbook of medical parasitology* (8th ed.). Jaypee Brothers Medical Publishers.

Gibson, D. I., Jones, A., & Bray, R. A. (2002). *Keys to the Trematoda* (Vol. 1). CABI.

Gibson, D. I., Bray, R. A., & Harris, E. A. (2005). Host-parasite database of the Natural History Museum, London. Natural History Museum. Retrieved October 21.

Giribet, G., Dunn, C. W., Edgecombe, G. D., & Rouse, G. W. (2007). A modern look at the animal tree of life. *Zootaxa, 1668*(1), 61–79.

Golvan, Y. J. (1994). Nomenclature of the Acanthocephala. *Research and Reviews in Parasitology, 54*(3), 135–205.

Graeff-Teixeira, C., da Silva, A. C. A., & Yoshimura, K. (2009). Update on eosinophilic meningoencephalitis and its clinical relevance. *Clinical Microbiology Reviews, 22*(2), 322–348. https://doi.org/10.1128/CMR.00044-08

Greaves, D., Coggle, S., Pollard, C., Aliyu, S. H., & Moore, E. M. (2013). *Strongyloides stercoralis* infection. *British Medical Journal, 347*, f4610. https://doi.org/10.1136/bmj.f4610

Haustein, T., Lawes, M., Harris, E., & Chiodini, P. L. (2010). An eye-catching acanthocephalan. *Clinical Microbiology and Infection, 16*(6), 787–788. https://doi.org/10.1111/j.1469-0691.2009.02896.x

Hodda, M. (2011). Phylum Nematoda Cobb 1932. In Z. Q. Zhang (Ed.), *Animal biodiversity: An outline of higher-level classification and survey of taxonomic richness* (p. 3148). Zootaxa.

Hong, E.-J., Sim, C., Chae, J.-S., Kim, H.-C., Park, J., Choi, K.-S., Yu, D.-H., Yoo, J.-G., & Park, B.-K. (2015). A horsehair worm, *Gordius* sp. (Nematomorpha: Gordiida), passed in a canine feces. *The Korean Journal of Parasitology, 53*(6), 719–724. https://doi.org/10.3347/kjp.2015.53.6.719

Hotez, P. J., Brindley, P. J., Bethony, J. M., King, C. H., Pearce, E. J., & Jacobson, J. (2008). Helminth infections: The great neglected tropical diseases. *The Journal of Clinical Investigation, 118*(4), 1311–1321. https://doi.org/10.1172/JCI34261

Hugot, J.-P., Baujard, P., & Morand, S. (2001). Biodiversity in helminths and nematodes as a field of study: An overview. *Nematology, 3*(3), 199–208. https://doi.org/10.1163/156854101750413270

International Helminth Genomes Consortium. (2019). Comparative genomics of the major parasitic worms. *Nature Genetics, 51*(1), 163–174. https://doi.org/10.1038/s41588-018-0262-1

Joe, L. K., & Siang, T. K. (1959). Human intestinal helminths obtained from autopsies in Djakarta, Indonesia. *The American Journal of Tropical Medicine and Hygiene, 8*(4), 518–523. https://doi.org/10.4269/ajtmh.1959.8.518

Johnston, D. A. (2006). Genomes and genomics of parasitic flatworms. In A. G. Maule & N. J. Marks (Eds.), *Parasitic flatworms molecular biology, biochemistry, immunology and physiology* (pp. 37–80). CAB International.

Kostadinova, A., & Pérez-del-Olmo, A. (2014). The systematics of the Trematoda. In R. Toledo & B. Fried (Eds.), *Digenetic trematodes* (pp. 21–44). Springer. https://doi.org/10.1007/978-1-4939-0915-5_2

Kusolsuk, T., Kamonrattanakun, S., Wesanonthawech, A., Dekumyoy, P., Thaenkham, U., Yoonuan, T., Nuamtanong, S., Sa-Nguankiat, S., Pubampen, S., Maipanich, W., Panitchakit,

J., Marucci, G., Pozio, E., & Waikagul, J. (2010). The second outbreak of trichinellosis caused by *Trichinella papuae* in Thailand. *Transactions of the Royal Society of Tropical Medicine and Hygiene, 104*(6), 433–437. https://doi.org/10.1016/j.trstmh.2009.12.005

La Flamme, A. C., Ruddenklau, K., & Bäckström, B. T. (2003). Schistosomiasis decreases central nervous system inflammation and alters the progression of experimental autoimmune encephalomyelitis. *Infection and Immunity, 71*(9), 4996–5004. https://doi.org/10.1128/IAI.71.9.4996-5004.2003

Lee, K. J., Bae, Y. T., Kim, D. H., Deung, Y. K., Ryang, Y. S., Im, K. I., & Yong, T. S. (2003). *Gordius* worm found in a three year old girl's vomitus. *Yonsei Medical Journal, 44*(3), 557–560. https://doi.org/10.3349/ymj.2003.44.3.557

Leng, Y. J., Huang, W. D., & Liang, P. N. (1983). Human infection with *Macracanthorhynchus hirudinaceus* Travassos, 1916 in Guangdong Province, with notes on its prevalence in China. *Annals of Tropical Medicine and Parasitology, 77*(1), 107–109. https://doi.org/10.1080/00034983.1983.11811681

Littlewood, D. T. (2008). Platyhelminth systematics and the emergence of new characters. *Parasite, 15*(3), 333–341. https://doi.org/10.1051/parasite/2008153333

Lotfy, W. (2020). Neglected rare human parasitic infections: Part III: Acanthocephaliasis. *Parasitologists United Journal, 13*(3), 145–150. https://doi.org/10.21608/puj.2020.47335.1090

Loukas, A., Hotez, P. J., Diemert, D., Yazdanbakhsh, M., McCarthy, J. S., Correa-Oliveira, R., Croese, J., & Bethony, J. M. (2016). Hookworm infection. *Nature Reviews. Disease Primers, 2*(1), 16088. https://doi.org/10.1038/nrdp.2016.88

Maizels, R. M., Pearce, E. J., Artis, D., Yazdanbakhsh, M., & Wynn, T. A. (2009). Regulation of pathogenesis and immunity in helminth infections. *The Journal of Experimental Medicine, 206*(10), 2059–2066. https://doi.org/10.1084/jem.20091903

Martins-Melo, F. R., Ramos, A. N., Jr., Cavalcanti, M. G., Alencar, C. H., & Heukelbach, J. (2016). Neurocysticercosis-related mortality in Brazil, 2000–2011: Epidemiology of a neglected neurologic cause of death. *Acta Tropica, 153*, 128–136. https://doi.org/10.1016/j.actatropica.2015.10.011

McManus, D. P., Dunne, D. W., Sacko, M., Utzinger, J., Vennervald, B. J., & Zhou, X.-N. (2018). Schistosomiasis. *Nature Reviews. Disease Primers, 4*(1), 13. https://doi.org/10.1038/s41572-018-0013-8

McSorley, H. J., Gaze, S., Daveson, J., Jones, D., Anderson, R. P., Clouston, A., Ruyssers, N. E., Speare, R., McCarthy, J. S., Engwerda, C. R., Croese, J., & Loukas, A. (2011). Suppression of inflammatory immune responses in celiac disease by experimental hookworm infection. *PLoS One, 6*(9), e24092. https://doi.org/10.1371/journal.pone.0024092

Morand, S., Bouamer, S., & Hugot, J. P. (2006). Nematodes. In S. Morand, B. R. Krasnov, & R. Poulin (Eds.), *Micromammals and macroparasites: From evolutionary ecology to management* (pp. 63–79). Springer-Verlag.

Olson, P. D., Littlewood, D. T., Bray, R. A., & Mariaux, J. (2001). Interrelationships and evolution of the tapeworms (Platyhelminthes: Cestoda). *Molecular Phylogenetics and Evolution, 19*(3), 443–467. https://doi.org/10.1006/mpev.2001.0930

Pampiglione, S., Rivasi, F., & Gustinelli, A. (2009). Dirofilarial human cases in the Old World, attributed to *Dirofilaria immitis*: A critical analysis. *Histopathology, 54*(2), 192–204. https://doi.org/10.1111/j.1365-2559.2008.03197.x

Poinar, G. O. (2010). Chapter 9—Nematoda and nematomorpha. In J. H. Thorp & A. P. Covich (Eds.), *Ecology and classification of north American freshwater invertebrates* (3rd ed., pp. 237–276). Academic Press. https://doi.org/10.1016/B978-0-12-374855-3.00009-1

Poulin, R., & Morand, S. (2000). The diversity of parasites. *The Quarterly Review of Biology, 75*(3), 277–293. https://doi.org/10.1086/393500

Pozio, E. (2001). New patterns of *Trichinella* infection. *Veterinary Parasitology, 98*(1–3), 133–148. https://doi.org/10.1016/s0304-4017(01)00427-7

Pradatsundarasar, A., & Pechranond, K. (1965). Human infection with the acanthocephalan *Macracanthorhynchus hirudinaceus* in Bangkok: Report of a case. *The American Journal of Tropical Medicine and Hygiene, 14*(5), 774–776. https://doi.org/10.4269/ajtmh.1965.14.774

Prueksapanich, P., Piyachaturawat, P., Aumpansub, P., Ridtitid, W., Chaiteerakij, R., & Rerknimitr, R. (2018). Liver fluke-associated biliary tract cancer. *Gut Liver, 12*(3), 236–245. https://doi.org/10.5009/gnl17102

Qian, M. B., & Zhou, X. N. (2017). Global burden of cancers attributable to liver flukes. *The Lancet Global Health, 5*(2), e139. https://doi.org/10.1016/s2214-109x(16)30301-1

Ramirez-Avila, L., Slome, S., Schuster, F. L., Gavali, S., Schantz, P. M., Sejvar, J., & Glaser, C. A. (2009). Eosinophilic meningitis due to *Angiostrongylus* and *Gnathostoma* species. *Clinical Infectious Diseases, 48*(3), 322–327. https://doi.org/10.1086/595852

Reynolds, L. A., Redpath, S. A., Yurist-Doutsch, S., Gill, N., Brown, E. M., van der Heijden, J., Brosschot, T. P., Han, J., Marshall, N. C., Woodward, S. E., Valdez, Y., Borchers, C. H., Perona-Wright, G., & Finlay, B. B. (2017). Enteric helminths promote *Salmonella* coinfection by altering the intestinal metabolome. *The Journal of Infectious Diseases, 215*(8), 1245–1254. https://doi.org/10.1093/infdis/jix141

Rohde, K. (2002). Subclass Aspidogastrea Faust & Tang, 1936. In *Keys to the Trematoda* (Vol. 1, pp. 5–14). CAB International.

Ryan, S. M., Eichenberger, R. M., Ruscher, R., Giacomin, P. R., & Loukas, A. (2020). Harnessing helminth-driven immunoregulation in the search for novel therapeutic modalities. *PLoS Pathogens, 16*(5), e1008508. https://doi.org/10.1371/journal.ppat.1008508

Schmidt, G. D. (1971). Acanthocephalan infections of man, with two new records. *The Journal of Parasitology, 57*(3), 582–584.

Schmidt, G. D. (1986). *CRC handbook of tapeworm identification*. CRC.

Scholz, T., Garcia, H. H., Kuchta, R., & Wicht, B. (2009). Update on the human broad tapeworm (genus *Diphyllobothrium*), including clinical relevance. *Clinical Microbiology Reviews, 22*(1), 146–160. https://doi.org/10.1128/CMR.00033-08

Schulze, J., & Schierenberg, E. (2009). Embryogenesis of *Romanomermis culicivorax*: An alternative way to construct a nematode. *Developmental Biology, 334*(1), 10–21. https://doi.org/10.1016/j.ydbio.2009.06.009

Sipahi, A. M., & Baptista, D. M. (2017). Helminths as an alternative therapy for intestinal diseases. *World Journal of Gastroenterology, 23*(33), 6009–6015. https://doi.org/10.3748/wjg.v23.i33.6009

Sithithaworn, P., Sripa, B., Kaewkes, S., Nawa, Y., & Haswell, M. R. (2014). 53 - food-borne trematodes. In J. Farrar, P. J. Hotez, T. Junghanss, G. Kang, D. Lalloo, & N. J. White (Eds.), *Manson's tropical infectious diseases* (23rd ed., pp. 726–736.e722). W.B. Saunders. https://doi.org/10.1016/B978-0-7020-5101-2.00054-6

Sorvillo, F. J., DeGiorgio, C., & Waterman, S. H. (2007). Deaths from cysticercosis, United States. *Emerging Infectious Diseases, 13*(2), 230–235. https://doi.org/10.3201/eid1302.060527

Sripa, B., Bethony, J. M., Sithithaworn, P., Kaewkes, S., Mairiang, E., Loukas, A., Mulvenna, J., Laha, T., Hotez, P. J., & Brindley, P. J. (2011). Opisthorchiasis and *Opisthorchis*-associated cholangiocarcinoma in Thailand and Laos. *Acta Tropica, 120*(Suppl 1), S158–S168. https://doi.org/10.1016/j.actatropica.2010.07.006

Summers, R. W., Elliott, D. E., Qadir, K., Urban, J. F., Jr., Thompson, R., & Weinstock, J. V. (2003). *Trichuris suis* seems to be safe and possibly effective in the treatment of inflammatory bowel disease. *The American Journal of Gastroenterology, 98*(9), 2034–2041. https://doi.org/10.1111/j.1572-0241.2003.07660.x

Tada, I., Otsuji, Y., Kamiya, H., Mimori, T., Sakaguchi, Y., & Makizumi, S. (1983). The first case of a human infected with an acanthocephalan parasite, *Bolbosoma* sp. *The Journal of Parasitology, 69*(1), 205–208.

Ünal, E., Arslan, S., Onur, M. R., & Akpinar, E. (2020). Parasitic diseases as a cause of acute abdominal pain: Imaging findings. *Insights Into Imaging, 11*(1), 86. https://doi.org/10.1186/s13244-020-00892-5

Veidovsky, F. (1886). Zur Morphologie der Gordiiden. *Wissenschaftliche Zoologie, 43*, 369–433.

Waknine-Grinberg, J. H., Gold, D., Ohayon, A., Flescher, E., Heyfets, A., Doenhoff, M. J., Schramm, G., Haas, H., & Golenser, J. (2010). *Schistosoma mansoni* infection reduces the incidence of murine cerebral malaria. *Malaria Journal, 9*(1), 5. https://doi.org/10.1186/1475-2875-9-5

WHO. (2017). *Guideline: preventive chemotherapy to control soil-transmitted helminth infections in at-risk population groups*. World Health Organization.

Xu, L., Jiang, Z., & Yao, M. (1998). Survey of human parasite species in China (continued). *Zhongguo Ji Sheng Chong Xue Yu Ji Sheng Chong Bing Za Zhi, 16*(5), 388–393.

Yamada, M., Tegoshi, T., Abe, N., & Urabe, M. (2012). Two human cases infected by the horsehair worm, *Parachordodes* sp (Nematomorpha: Chordodidae), in Japan. *The Korean Journal of Parasitology, 50*(3), 263–267. https://doi.org/10.3347/kjp.2012.50.3.263

Zeldenryk, L. M., Gray, M., Speare, R., Gordon, S., & Melrose, W. (2011). The emerging story of disability associated with lymphatic filariasis: A critical review. *PLoS Neglected Tropical Diseases, 5*(12), e1366. https://doi.org/10.1371/journal.pntd.0001366

Chapter 3
Overview of Parasitic Helminth Diversity: How Molecular Systematics Is Involved

Abstract Chapter 2 focused on helminth diversity by emphasizing parasitic helminths, which are reported to be distributed globally with complex relationships between hosts and parasites. This chapter presented how the evolutionary process allowed adaptation and coevolution between the parasite and their hosts. The life-cycle complexity of helminths enhances their genetic diversity and host–parasite assemblage, allowing for the coevolution of the parasites with their host. With the accelerating rate of disease emergence in wildlife, livestock, and humans possibly attributed to global climate change, the topic of parasite diversity and its distribution is concerning. Moreover, the number of parasite taxa has dramatically decreased due to extinction. Around ten thousand helminthic parasites in vertebrates were recently re-estimated using the largest database of host–parasite association of the Natural History Museum in London (NHM) and the US National Parasite Collection (USNPC). Only 15% of those estimated species were already described, mainly by morphology. This chapter also mentioned evolutionary factors contributing to parasitic helminth diversity and their genetic adaptation. These factors were exemplified, mainly based on molecular systematics studies, to demonstrate how molecular systematics is involved in parasitic helminth diversity.

3.1 Parasite Diversity in the Global Environmental Change

The consequences of environmental changes, mostly due to human anthropogenic activity and natural ecological processes on host–parasite interaction, have directly influenced the capacity of parasites to adapt to the new environment (Aleuy & Kutz, 2020). In a highly changing environment, metazoans, including helminths, can adapt themselves to produce various phenotypes across various environments while still expressing under a single genotype, so-called phenotypic plasticity. Parasitic helminth populations exposed to wildly different environments, such as different host species and host immune systems, may contribute to phenotypic variation (Hildebrand et al., 2015; Schneider & Meyer, 2017). Moreover, rapid environmental changes may drive the rapid evolution of new species through phenotypic segregation into different ecological niches (Choi et al., 2020; Schneider & Meyer, 2017).

U. Thaenkham et al., *Molecular Systematics of Parasitic Helminths*,
https://doi.org/10.1007/978-981-19-1786-8_3

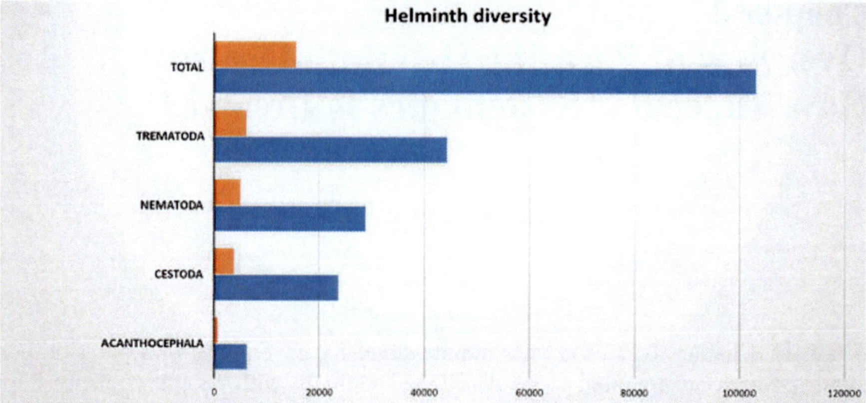

	Acanthocephala	Cestoda	Nematoda	Trematoda	Total
■ Number of species described	684.53	3799.84	4903.48	6196.68	15461.7
■ Number of species re-estimated	6223	23,749	28844	44262	103078

Fig. 3.1 Helminth diversity based on the number of species re-estimated and described. The number of species was estimated using "big data" sources for biodiversity research from the National History Museum in London (NHM) and the US National Parasite Collection (USNPC) (Carlson et al., 2020)

Contrarily, helminthic parasites that have adapted to similar environments or ecological niches are affected by convergent evolution, in which different species independently evolve to possess similar traits (Brooks, 1985; Spakulova, 2002).

The species diversity of parasitic helminths has been discovered, characterized, and classified like other organisms. The systematics of parasites also allows the exploration of host–parasite association, life history, and patterns of infection and transmission (Zarlenga et al., 2014b). Although without fossil records to support the evolution of extant parasite species, the systematics of helminths can predict species diversity and evolutionary history from extant organisms (Hoberg & Brooks, 2013; Zarlenga et al., 2014b). Before genetic information came into the forefront, helminthic parasites were classified only based on morphological characters, leading to confusion due to morphological ambiguity (Poulin & Presswell, 2016). After 2000, a paradigm shift toward utilizing morphological and genetic markers for helminth taxonomy was widely applied through phylogenetic analyses (Scholz, 2001).

Estimates of helminth species diversity have been difficult due to the lack of taxonomists and systematists after the 1960s (Brooks & Hoberg, 2001; Carlson et al., 2020; Poulin, 2014). Carlson et al. (2020) mentioned that the total global helminth is re-estimated to be 100,000 to 300,000 species of endoparasites, including possible cryptic species. Figure 3.1 compares the number of species that have been re-estimated for helminths (Cestoda, Nematoda, Trematoda, and Acanthocephala) parasitizing various vertebrate classes with the number of species described. They also deliberated significant problems of cryptic species, which consist of a fraction of undescribed species because they are morphologically indistinguishable. Carlson et al. (2020) suggested the overall richness of cryptic species was on average 2.6 cryptic species per species of acanthocephalan, 2.4 per species of cestode, 1.2 per species of nematode, and 3.1 per species of trematode. Also, they advised that many publications may be biased toward cryptic species estimation, resulting in underestimation (Perez-Ponce de Leon & Poulin, 2018).

Perez-Ponce de Leon and Poulin (2018) proposed that the more extraordinary occurrence of cryptic diversity among trematodes may be due to either the traditional taxonomic approach for species identification or biological features such as their mode of reproduction, the complexity of their life cycle requiring intermediate hosts, and their frequently lacking fast-evolving hard structure to serve as reliable and taxonomically informative morphological traits. Misidentifications based on morphologically indistinguishable characters may also overlook cryptic species of trematodes due to the lack of proper molecular identification using a molecular genetic marker with high species-level resolution (e.g., mitochondrial protein-coding genes) (Blouin, 2002; de Leon & Nadler, 2010; Perez-Ponce de Leon & Poulin, 2018; Vilas et al., 2005).

3.2 Evolution of Parasitic Helminths

Parasitic helminths are invertebrates and are large macroparasites characterized by their general external shape and host organ infection. As mentioned in Chap. 2, parasitic helminths are divided into flatworms (flukes and tapeworms in the phylum Platyhelminthes) and roundworms (in the phylum Nematoda) (Castro, 1996). Flatworms are principally endoparasites that coevolved with human and animal hosts (Littlewood et al., 1999; Zarowiecki & Berriman, 2015). Nematodes parasitize a large spectrum of hosts, including animals and plants, and have evolved into several lineages (Blaxter & Koutsovoulos, 2015; Blaxter et al., 1998). Parasitic helminths were hypothesized to have evolved from free-living worms and subsequently adapted to parasitism according to the diversity of hosts and their niches. Parasitic helminths simplified their external morphology by reducing pigmentation and sensory organs but evolved strategies to increase their reproductive output through host invasion and evasion of the host immune system (Zarowiecki & Berriman, 2015).

As Poulin and Morand (2000) mentioned, there are two main reasons we should understand parasite diversity. Firstly, parasitism allowed us to efficiently study the process of evolution as the ideal biological model for studying ecological specialization, speciation mechanisms, and diversification as compared to free-living helminths. Secondly, many helminthic parasites are important to medical and veterinary science to understand biological phenomena such as host transfers and subsequent speciation (Poulin & Morand, 2000). Studies have discussed the diversity of helminths, particularly the global diversity and distribution of parasite richness in various hosts that are under the impact of climatological forces of anthropogenic origin and environmental agitation (Brooks & Hoberg, 2013; Brooks & Hoberg, 2007; Dallas et al., 2018; Dobson et al., 2008b).

Recent sudden and extreme environmental changes can have various consequences for host–parasite dynamics and parasite communities (Deutsch et al., 2008). Some genotypes may survive and distribute in the new environment, while others may not be able to adjust themselves to survive the changing environments, leading to the disappearance of the population. Allele frequencies in the population can be changed from generation to generation under natural selection (Aleuy & Kutz, 2020; Helmuth et al., 2005). Colles et al. (2009) discussed that specialist species might be at risk and are unable to survive in varying environmental conditions as compared to generalists because of the small initial size of local populations, restricted geographic range, and the limited utilization of resources and habitats (Colles et al., 2009).

The environment of organisms, including parasites, is usually variable in space and time. The environment of parasites is principally biotic, in which parasites constantly interact with their host. Parasites have evolved strategies to possess advantages for survival. Constantly interacting between parasites and hosts suggests that local adaptation is the principal process for coevolution (Kaltz & Shykoff, 1998). Local adaptation of parasites refers to the ability of parasites to perform better in or from hosts from the same environment/niches (Gandon & Michalakis, 2002). The transmission of parasites to their hosts is related to the parasite environment, which can be either homogeneous or heterogeneous (McCoy, 2003). In a homogenous environment, parasites have similar probabilities of transmitting to each host type existing in the same geographical area. Genetic differentiation/ isolation of parasites infecting different host types can occur under strong enough selection pressure from the host immune response and finally produce homogamy.

Conversely, a heterogeneous environment is when a host has different resources/ niches within a local environment (McCoy, 2003). It can also arise in host genotypes, host races, host species, and the abundance of host individuals (Gandon & Van Zandt, 1998). Transmission of parasites in the heterogeneous environment can result in partial or complete isolation of the parasite population and a limited probability of transmission back to the parental population, supporting the genetic divergence between populations (Fig. 3.2). Kaltz and Shykoff (1998) suggested that the parasite population should be locally adapted for increased fitness on sympatric rather than allopatric hosts. Compared to their hosts, local adaptation of parasites is favored by relatively short generation time, high fecundity, and a high migration rate (gene flow) (Kaltz & Shykoff, 1998). These evolutionary forces enable novel alleles to be introduced into the local parasite population at a high rate, enhancing the coevolution race and quickly overcoming new host resistance. The introduction of novel alleles can allow parasites to counteract the host defense response (Kaltz & Shykoff, 1998; Loker & Hofkin, 2015).

Although parasites may frequently be theorized to be locally adapted to their host (Kaltz & Shykoff, 1998; McCoy, 2003), many empirical studies demonstrated local maladaptation (Kaltz et al., 1999; Oppliger et al., 1999). Maladaptation occurs when parasites migrate less than hosts, and the parasites are less able to infect sympatric than allopatric hosts (Gandon et al., 1996). One example is provided by Oppliger et al. (1999), who studied the susceptibility of lizard *Gallotia galloti* to

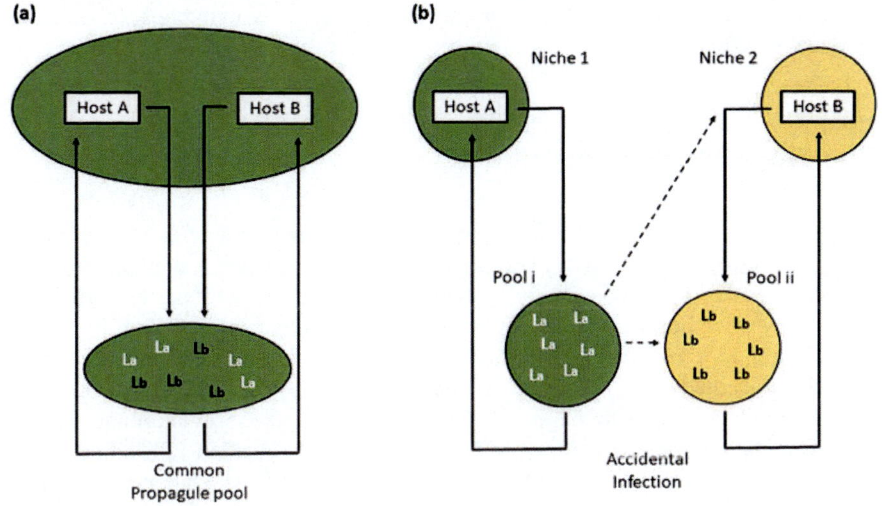

Fig. 3.2 Transmission of parasites between sympatric hosts in (**a**) homogeneous and (**b**) heterogeneous environments. In (**a**), propagules (larvae (L) from host A and host B) released into the environment are mixed in a common propagule pool. In contrast, in (**b**), propagules released from host types use different niches within the local environment, are isolated into propagule pools (pool i and pool ii), and have a limited transmission probability. Isolation can cause genetic divergence between populations by the physical barrier due to the distance between hosts (considered allopatric speciation). The dashed arrows indicate limited transmission probability between hosts A and B or their larvae (McCoy, 2003)

hemogregarine blood parasites transmitted by mites (Oppliger et al., 1999). The migration of the lizard is faster than the parasites because its offspring typically disperse from their natal site soon after hatching and without contact with their parents, which is the potential carrier of blood parasites. This may cause a slow down in the rate of hemogregarine migration between lizard populations. However, local adaptation may not constantly occur in the host-parasite system because the migration rate between host and parasite is sometimes comparable. Prugnolle et al. (2006) discussed environmental and genetic factors to explain the variations in susceptibility/infectivity between *Biomphalaria glabrata* and *Schistosoma mansoni* populations (Prugnolle et al., 2006). There was no evidence of local adaptation, for *S. mansoni* and *B. glabrata*, because of a similar migration rate between host and parasite.

Gandon and Michalakis (2002) analyzed host-parasite coevolution using the metapopulation model and studied the effects of population size, mutation rates, migration rates for both host and parasites, and parasite generation time. The results showed that all evolutionary factors, particularly mutation and migration, promote local adaptation. Additionally, they revealed that a shorter parasite generation time may not always favor parasite local adaptation, particularly in parasite populations with limited genetic variability. A short generation time may increase the strength of selection and decrease the parasite's ability to adapt to an evolving host (Gandon & Michalakis, 2002).

3.3 Host–Parasite Coevolution

Hosts and parasites can interact for many generations under evolutionary processes, leading to various host-parasite evolutionary histories. However, extant patterns and characteristics of the host-parasite system might not be evolutionarily correlated, particularly when hosts and parasites evolve independently (Loker & Hofkin, 2015). Coevolution is a reciprocal evolutionary change, a dynamic microevolutionary process concerned with changes in allele frequencies within a population or among populations due to major evolutionary forces. These evolutionary forces include mutation, genetic drift, gene flow, and natural selection (Criscione et al., 2005; Gandon & Michalakis, 2002; MacIntosh & Frias, 2017). Under coevolution, parasite-induced reduction in host fitness enhances the selection for host resistance mechanisms. Conversely, novel host defenses raise selection on the parasite. Changes in gene frequencies in host and parasite populations can occur during coevolution, thus maintaining polymorphism in crucial genes, affecting parasite infectivity, host resistance, host searching, and avoidance behavior by the host (Woolhouse et al., 2002). These changes in gene frequencies can influence the parasite's virulence and transmissibility, potentially impacting medical and veterinary perspectives.

Under environmental change, hosts are under selective pressure to resist parasites, while parasites are selected to overcome host defenses. Mostowy and Engelstadter (2011) constructed a mathematical model to indicate the effect of environmental change on host-parasite coevolutionary outcomes by assuming that host–parasite interaction is mediated by the genetic basis of host-parasite interaction with matching-allele and gene-for-gene models. The matching-allele model represents interactions between the host immune system and antigenic parasites, explicitly matching the host to infect it (Mostowy & Engelstadter, 2011). The gene-for-gene model is the type of interaction that represents the situation where the host needs to recognize specific effectors of the parasites to launch its defense; hence, matching is equivalent to resistance. This study revealed potential consequences of environmental variation in coevolution and the importance of characterizing genotype-by-genotype-by-environment interactions in natural host-parasite systems. The environment changes the direction of selection between the two species, which may effectively alter the selective pressure and level of specialism in the population by selecting antagonistic interaction specificity.

Direct coevolutionary evidence for the operation of antagonistic interactions between host and parasite is rare, particularly for long-term temporal dynamics, wherein the interaction gives rise to a continual natural selection for adaptation and counter-adaptation. The Red Queen hypothesis proposes antagonistic coevolution between interacting species selected to maintain outcrossing, which allows

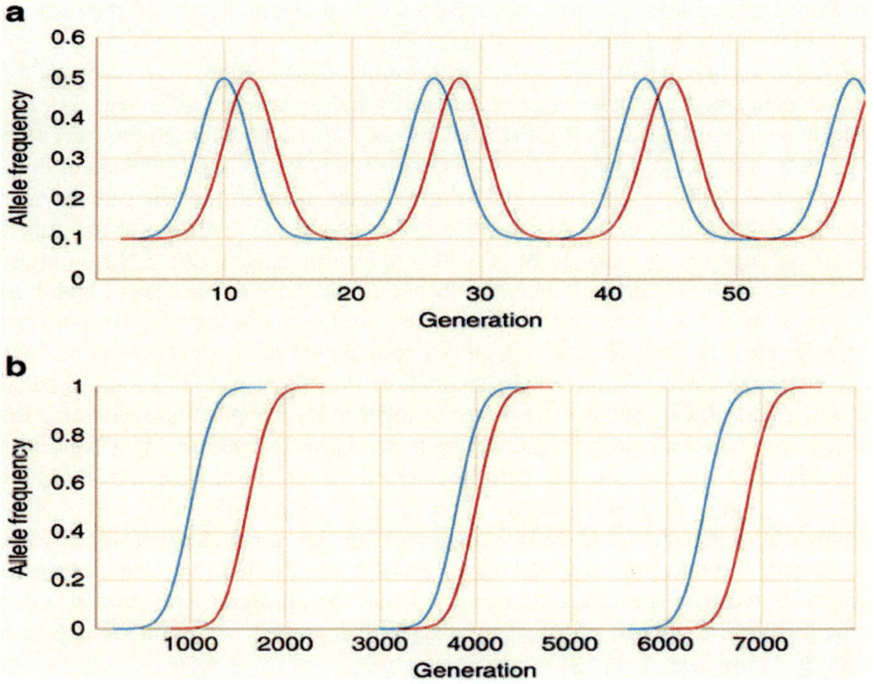

Fig. 3.3 Allele frequency changes driven by Red Queen coevolution. In (**a**), Red Queen dynamics is a dynamic equilibrium of coevolution acting on host immune alleles (blue) and pathogen virulence alleles (red). The equilibrium is maintained by balancing selection acting on existing polymorphism over relatively short evolutionary time scales. In (**b**), the Red Queen arms race is a recurrent bout of positive selection in host and pathogen operating on new polymorphisms arising through mutation (Lighten et al., 2017)

producing offspring with rare genotypes, whereas prolonged periods of uniparental reproduction propagate genetically uniform lineages. The Red Queen dynamics balance different variants at loci related to host defense and parasite offense, operating to fluctuate allele frequencies under negative frequency-dependent selection (balancing selection) (Fig. 3.3a) (Bernatchez & Landry, 2003; Lighten et al., 2017; Tellier et al., 2014; Thrall et al., 2016).

As the strength of selection is almost impartial, the Red Queen dynamics is a general model of host–parasite coevolution. The Red Queen dynamic contrasts with the evolutionary arms race, where the arms race dynamic is biased with interactions between predator and prey (Loker & Hofkin, 2015). The arms race dynamic is a series of selective sweeps when coevolving species develop a positive selection for antagonized traits. The possible outcome of coevolution between host-parasite interaction occurs when a selective sweep replaces a new trait over an old one and it becomes fixed in the parasite population. The host population also progresses by fixing a new host trait in order to counter the parasite, replacing the old trait with an antagonistic response. At any time, there may be polymorphisms for the relevant traits in both host and parasite populations (Fig. 3.3b).

Variations in population sizes shape coevolution dynamics because it is one of the critical parameters determining genetic diversity and the intensity of genetic drift and selection (Papkou et al., 2016). Parasites reduce host fitness through their negative effects on host survival and reproduction, particularly when large numbers of hosts are infected. Reducing host abundance decreases the niche size for the parasite. One example is the effect on the reproductive cycles of a wild population of red grouse (*Lagopus lagopus scoticus*) infected with *Trichostrongylus tenuis* (Hudson et al., 1998), leading to reciprocally induced changes in the population sizes of both host and parasite. On the other hand, parasites usually have superior adaptive potential compared to their hosts because of their large population sizes, short generation time, and high frequencies of horizontal gene transfer (Woolhouse et al., 2002). Simultaneously, parasites are subject to extreme changes in their population size during the interaction with a single host and during transmission between hosts. Changes in population size induced by host–parasite interactions can substantially affect genetic diversity, genetic drift, and selection, potentially making nonidentical coevolutionary dynamics. Gokhale et al. (2013) suggested that the combination of stochasticity and changing population size can enhance fast allele fixation (consistent recurrent selective sweep or arms race dynamics), while the constant population size can produce patterns typical of negative frequency-dependent selection (Red Queen dynamic) (Gokhale et al., 2013).

3.4 The Evolution of Virulence and Transmission

Factors affecting the parasite's population size are important for understanding how virulence evolves in parasites and the likelihood of the host suffering from a heavy infection, reducing host fitness. Loker and Hofkin (2015) suggested that understanding virulence requires (1) a general theoretical framework to speculate on the evolution of virulence and (2) peculiarities of any particular host–parasite system being considered which may resist generalization and require more specific consideration (Loker & Hofkin, 2015).

Generally, we understand the virulence of a parasite when first associated with a host. Through adaptation, parasites accommodate hosts over time and develop a state of avirulence to prolong their coexistence with their host. This was evidenced by the discovery of *Paragonimus westermani* and *Capillaria philippinensis* found infecting mummies in Korea (Shin et al., 2009). The evolution of virulence is embraced mainly by the trade-off hypothesis, which recognizes virulence and transmission to a new host as linked traits (Alizon et al., 2009). Local adaptation of parasites is one of the potential outcomes of coevolutionary host-parasite interaction, where parasites and hosts are sympatric (Gandon & Michalakis, 2002; Kaltz & Shykoff, 1998), and can potentially lead to divergence in parasite populations and subsequently speciation. These two research areas are used to study how the environment alters virulence-transmission relationships. Phylodynamics, a new approach that emphasizes how interactions between parasite epidemiological and evolutionary processes shape parasite phylogenies, can be applied to reconstructing population history based on genetic sequence information from a single gene sequence through deep sequencing methods (Ho & Shapiro, 2011; Kao et al., 2014; Pybus & Rambaut, 2009). This new approach has been applied as a framework for examining transmission dynamics between viruses and hosts (Gupta et al., 2020; Turner et al., 2021; Webster et al., 2017).

Turner et al. (2021) reviewed the consensus on how virulence and transmission are related by three interrelated approaches for host-parasite systems. These three approaches are the evolutionary trade-off theory, local adaptation, and parasite phylodynamics. They highlighted questions for future research in virulence-transmission relationships and addressed how the environment contributes to virulence-transmission relationships between parasites and their hosts. The outstanding question is, for example, the contribution between environmental conditions and the evolution of parasite traits or life-history strategies. The question also addresses how various selection pressures and the environment gradient affect the parasite's survival. We should also be concerned about the degree of linking genetic traits and phenotypic plasticity of the parasite, which is relevant to their life-history strategy. Finally, the variation in environmental conditions across a parasitic range can be used to comprehend how environmental conditions alter virulence and transmission traits and host–parasite dynamics (Turner et al., 2021). Questions remain as to how to solve and better arrest the mechanisms and forces leading to the evolution of virulence and transmission under changing environmental conditions, especially with global climate change and land use transformation occurring rapidly. Understanding the relationship between the parasite's virulence and transmission will enhance our ability to predict disease outbreaks in a changing environment.

3.5 Parasite Diversity Enhanced by the Parasitic Life Cycles Along the Evolutionary Process

Basically, host-parasite interactions have been concerned based on two different views; the parasite-centered view rooted in Darwinian evolution and the host-centered view rooted in orthogenesis (Brooks & Hoberg, 2007; Brooks & McLennan, 1993; Fenton & Pedersen, 2005). Brooks and Hoberg (2007) concluded from those different views that parasites are ecological specialists with specific microhabitat requirements (preferred infection site) and modes of transmission from their host through the environment (life cycle pattern). The host-centered tradition is considered a rare event because the coevolution of hosts provides a safeguard against emerging infectious diseases (EID), and it is still questionable whether host species would want to capture more parasites. The parasite-centered tradition is considered different, as it assumes that the parasite infection is not related to the host evolution as the parasite prefers to inhibit all susceptible hosts with their contact at any place and time.

Changes in geographical distribution or environmental conditions associated with parasite transmission should produce immediate changes in the host range of parasites. Parasites colonize and persist in the new environment by using novel resources or forming a new association with other species, resulting in accompanying suites of traits when they encounter another new condition; this is so-called ecological fitting (Agosta & Klemens, 2008; Brooks & Ferrao, 2005; Brooks et al., 2006a, b). Ecological fitting relates to the concept of exaptation (shift in the function of a trait during evolution) and the positive role of phenotypic plasticity in evolution. Agosta and Klemens (2008) discussed that ecological fitting could be either a precursor to evolutionary diversity or a maintainer of the evolutionary stasis, serving as an important concept to understand species association and evolution of biodiversity during climate change episodes.

In the context of climate change, the impact of extreme and sudden environmental changes on a parasite's performance, such as structural and behavioral adaptations for survival strategies, life-history traits, and adaptive ecological mechanisms of direct and indirect life cycles, has been discussed. Parasitic nematodes are highly diverse (+23,000 species infecting vertebrates) and present in terrestrial, freshwater, and marine ecosystems (Aleuy & Kutz, 2020; Blaxter & Koutsovoulos, 2015; Dobson et al., 2008a). Environmental conditions affecting the development and survival of parasitic nematodes vary, depending on stage-specific adaptations of eggs and/or larval stages interacting with their intermediate hosts and the environment, which are different between the direct and indirect life cycles (Aleuy & Kutz, 2020; Molnar et al., 2013). In the indirect life cycle, the egg/larval stages allow the nematode to avoid temperature and humidity extremes, protecting them due to the thermoregulatory behavior of their intermediate host, which can increase parasite fitness under certain conditions. However, they may be at high risk of extinction under climate change compared to nematodes utilizing a direct life cycle because multiple, precise, and obligate interactions with their intermediate host are required (Dobson et al., 2008b; Harvell et al., 2002).

Aleuy and Kutz (2020) discussed the role of complex life history and ecological traits in parasites, such as the presence of intermediate hosts, phenotypic plasticity in development traits, local adaptation to key environmental factors, and parasite fecundity and longevity. They are considered in host-parasite interactions, which are generated by adapting to environmental changes from the past to the present (Fig. 3.4). They also confirmed the effects of environmental change by phenotypic plasticity and local adaptation on the freezing temperature of *Marshallagia marshalli* (gastrointestinal nematode infecting various artiodactyl species). The nematodes survived by adapting their life-history traits by developing the third larval stage in an egg to increase their fitness advantage, protecting the first stage larvae from freezing conditions. The study revealed that plasticity in essential traits could allow short-term modification in the developmental pathway, thus extending the potential window for the selection of evolutionary adaptation. Natural selection can promote such plastic traits to be expressed in many individuals within just a few

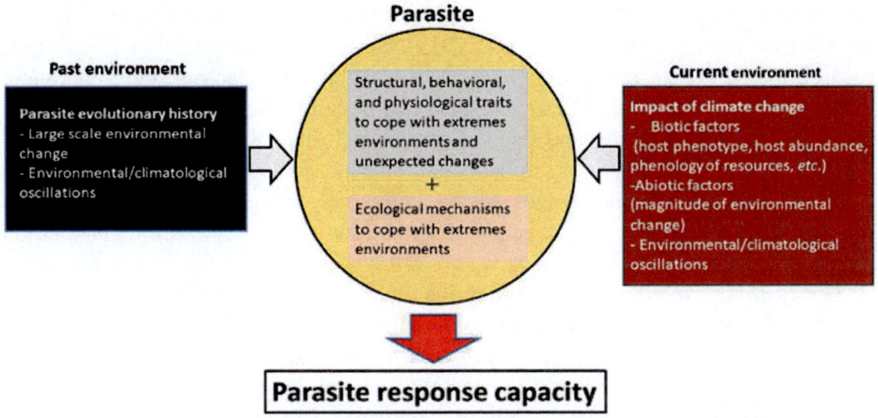

Fig. 3.4 Influential factors of parasites from past to current environments. Parasites had been affected by environmental changes and climatological oscillations in their evolutionary history. Impacts of climate change like host phenology, host abundance, phenology resources, the magnitude of environmental change, and environmental/climatological oscillations also affect the parasites' current environment. These changes influence the parasites' capacity to respond to environmental changes by changing structural, behavioral, and physiological traits (for example, resistant structures in eggs and larvae) and ecological mechanisms to cope with extreme environments (for example, properties of the parasite life cycle, phenotypic plasticity in specific adaptation to extremes, and local adaptation) (Aleuy & Kutz, 2020)

generations. The developmental plasticity manifested by parasitic nematodes might facilitate both short- and long-term patterns such as phenotypic and evolutionary responses under the current regime of accelerating climate and environmental perturbation (Aleuy & Kutz, 2020).

3.6 Genetic Adaptation and Variability of the Parasites

As we know, environmental unpredictability during climate change leads to species-specific adaptation of parasites to environmental conditions through host-parasite interaction. Under unpredictable conditions, parasites can survive by producing different phenotypes in response to environmental variation, although they are of a single genotype (phenotypic plasticity). Phenotypic plasticity occurs along with adaptive evolution and is expected to play an important role in species responses to sudden environmental change by the various phenotypes existing in the population (Hoffmann & Sgro, 2011; Parmesan, 2006). Additionally, Kelly (2019) suggested that genetic variation in plastic responses to the environment (phenotypes determined by genetics and influenced by the environment; G $\times$ E) might be an important predictor of species vulnerabilities to climate-driven decline or extinction. Plasticity already plays an essential role in allowing organisms to maintain fitness across heterogeneous environments. A key question is what factors maintain genetic variation for plasticity within and between natural populations and why these factors vary across traits and species (Kelly, 2019). Several hypotheses have been made regarding factors that maintain high levels of plasticity. The factors tend to maintain genetic variation, including large effective population size, high mutation rate, spatially varying selection, and relaxed selection (Kelly, 2019). The genetic variation is driven by spatially varying selection and local adaptation to environmental gradients. Environmental changes act on individual organisms over evolutionary time by linking current phenotypic plasticity and long-term evolutionary adaptation to climate change.

Under environmental changes, adaptive evolution, which is the consequence of natural selection, can be rapid and potentially helpful for species under stressful conditions (Hoffmann & Sgro, 2011). Adaptive evolution enables species to fit their traits to the environment. Schmalhausen (1949) recognized that mutations are likely to be expressed differently in various environments (Schmalhausen, 1949). Genetic variation in phenotypic plasticity can result, as these mutations disrupt the most advantageous norm (genotype-environment interaction) (Via & Lande, 1985). Through investigating the effects of extreme environmental change on cryptic genetic variation and phenotypic plasticity, Salinas et al. (2019) showed that cryptic genetic variation has slight or no effect on phenotypic variation, so-called hidden variation (Salinas et al., 2019). However, cryptic genetic variation is a critical component of the adaptive capacity of populations under new environmental conditions. The genetic variation, in turn, generates heritable phenotypic variation (Paaby & Rockman, 2014). Therefore, cryptic genetic variation aids in our understanding of how species rapidly adapted to novel environments in the past. Kelly (2019) suggested that if adaptation occurs through the modification of plasticity responses, environmentally responsive target genes are expected to be diverse (Kelly, 2019). The expression levels of these target genes might be altered, increasing the divergence between populations (Levis et al., 2017).

Host-parasite interactions are the main part of ecosystems that vary with the ecological settings and changes that occur within them (Brunner & Eizaguirre, 2016; Wolinska & King, 2009). The consequences of the parasite transmission pathway, the interaction between hosts and parasites, and the ecological setting when hosts' population structure changes depend on the specific characteristics of the disease triangle (Brunner & Eizaguirre, 2016; Budria & Candolin, 2014). The disease triangle comprises the host community, parasite community, and reciprocal interactions with the environment. Host factors contain the variation of life-history traits, immunity, and abundance, while parasite factors comprise the parasite transmission rate, ability to complete its life cycle, and parasite virulence. Factors from the host and parasite are affected by environmental interactions, where environmental interactions are the initial impacts affecting hosts and parasites (Brunner & Eizaguirre, 2016).

The process of parasite speciation depends on reducing the gene flow of different hosts (reciprocally of different parasite populations). In the case of low gene flow, the populations in each environmental condition have to adapt themselves to occupy the specific niche of an adaptive landscape. Divergent selection can occur within the population, as the population has experienced different selective pressures to promote genetic differentiation (Brunner & Eizaguirre, 2016; Crispo, 2008; Futuyma, 2010; Turelli et al., 2001). The promotion of certain traits can occur through local adaptation and increased fitness. Speciation can result if reproductive isolation develops via behavioral barriers and/or genetic incompatibilities (Coyne & Orr, 2004).

Brunner and Eizaguirre (2016) discussed that host-parasite interactions can alter speciation mechanisms if they affect Darwinian fitness by ensuring a gene pool for the next generations to survive and reproduce in their environment. A recurrent selective sweep of alleles in the population through the evolutionary arms race mechanism may lead to faster local adaptation and population divergence (Brunner & Eizaguirre, 2016). Parasite speciation should be expected when narrow adaptive peaks for each parasite population promote local adaptation and decrease gene flow, which favors the emergence of co-locally adapted gene complexes and species (Greischar & Koskella, 2007; Summers et al., 2003). Therefore, specialist parasites should be more likely to speciate than generalist parasites (Johnson et al., 2003). Although parasite speciation is a major evolutionary force contributing to host evolution, this factor has only recently been conceptualized (Brunner & Eizaguirre, 2016).

3.7 Trends of Studies in Parasite Diversity Involved by Molecular Systematics: Past to the Present

According to evolutionary processes, coevolved parasites and their hosts result in the buildup of parasite diversity. Diversity can be either morphological or genetic. Defining the species diversity of parasites, including parasitic helminths, is essential for studying epidemiology, developing species detection and discrimination methods, developing management practices to limit transmission, and designing treatment regimes to diminish or eliminate infections. Studies on species diversity have mainly been conducted on morphological characters for over 200 years. Species-level identification has relied on comparative morphology and often depends on examining fully developed adult worms. Structural morphology and the physiological and behavioral characteristics of the parasite population are used. The characteristics of the parasite may influence host relationships, and information on environmental differences among different host species helps study the systematics of parasitic helminths. Haley (1962) suggested that host-dependent variations in parasitic helminths should be performed to assess better the taxonomic stability of characters used in classification (Haley, 1962). For helminths, the analysis of adult structure is the principal basis for classification. Information using larval structure, life history, physiology, and host relationships is helpful when available. However, the use of morphology is limited, especially when parasites are under evolutionary processes from climate change episodes.

Since 2000, the molecular-based polymerase chain reaction (PCR) method has enabled the building up of the DNA sequence database of helminths from a particular DNA target, which is known as a molecular genetic marker. DNA sequences of neutral genetic markers can provide a high-resolution phylogenetic relationship among helminths and confirm confounding status within the cryptic lineage. DNA sequencing platforms are continually developing, producing higher data quantities with better quality. These developments can ultimately improve our ability to accurately and precisely measure genetic variation (Pareek et al., 2011). The DNA sequences of parasitic helminths, focusing on the groups of human and veterinary parasites, can allow for molecular taxonomy, phylogenetic systematics, and population genetics, which can then aid in solving epidemiological problems (molecular epidemiology) (Foxman & Riley, 2001).

Unique DNA sequences or molecular genetic markers are used to characterize either neutral variation or adaptive genetic variations. Neutral genetic markers have been applied for research on population structure (Luikart et al., 2003; Soto-Cerda & Cloutier, 2013), and numerous methods have been developed to identify species. Studies using these genetic markers enable monitoring of the geographical range of the taxon, the presence of hybrids, and the emergence of disease or invasive species (Schwartz et al., 2007). Adaptive (non-neutral) makers, on the other hand, are adaptive genes that directly influence the fitness or the regions of the genome responding to selective forces stemming from interactions with an environment, hosts, or other parasites (Kirk & Freeland, 2011). Both neutral and adaptive molecular genetic markers have been used to study mass drug administration programs and the evolution of drug resistance in helminths (Lustigman et al., 2012).

Early studies on helminth genetic variations employed isoenzyme markers of parasites existing among geographic isolates of single species or between species in parasitic helminths—for example, the study with *Schistosoma mansoni* (Coles, 1970), *Trichinella* spp. (Flockhart et al., 1982), and *Haemonchus contortus* (Bentounsi & Cabaret, 1999). Since Mullis developed the PCR technique in 1983, genetic variations of parasitic helminths could be revealed by molecular genetic markers by fingerprinting approaches. The genetic diversity of organisms can be investigated without requiring prior sequence information. Examples include the PCR-random amplified polymorphic DNA (PCR-RAPD), amplified fragment length polymorphism (AFLP), and the single-strand conformation polymorphism (SSCP) approach (Gasser & Chilton, 2001; Gasser et al., 2006).

PCR-RAPD, which is the random amplification of DNA segments using short primers composed of an arbitrary nucleotide sequence, is useful for species identification and population genetics (Rishi & McManus, 1987). Using PCR-RAPD, Eom et al. (2002) demonstrated that *Taenia asiatica* is sympatrically distributed with *T. saginata* and *T. solium* in humans in mainland China (Eom et al., 2002). Thaenkham et al. (2012) also applied the PCR-RAPD to reveal the two main genetic clusters of *Angiostrongylus cantonensis* populations across Thailand (Thaenkham et al., 2012), serving as a platform for further studies using mitochondrial protein-coding genes to confirm the presence of *A. cantonensis* and its closely related species *A. malaysiensis* (Dusitsittipon et al., 2017, 2018; Rodpai et al., 2016). *Fasciola hepatica* and *F. gigantica* were also discriminated against based on the genetic polymorphisms of the RAPD profiles and are helpful for species diagnosis and epidemiological surveys in *Fasciola* endemic areas (Ramadan et al., 2010; Ramadan & Saber, 2004).

AFLP is a genetic marker for genotyping based on amplifying specific restriction fragments from the genome. Although the AFLP method has not been extensively used, it can solve many important issues as it allows the polymorphic sites in the eukaryotic genomes, including parasitic helminths (Restrepo et al., 2018). The population structure and variation of *Oesophagostomum bifurcum* in Ghana were studied using AFLP, allowing for the differentiation of isolates according to host origins (human and non-human primates) (De Gruijter et al., 2005). AFLP was also applied for *Necator americanus*, revealing at least two genetically distinct groups, suggesting that these genetic variations may represent a species complex (de Gruijter et al., 2006). Nejsum et al. (2005) used AFLP to analyze the population structure of *Ascaris suum*, where there was low geographic variation among isolates, suggesting that *A. suum* is an active zoonosis in Denmark (Nejsum et al., 2005). Genetic variations of *Trichinella nativa* (Mikkonen et al., 2005) and *Wuchereria bancrofti* (Patra et al., 2007) were also detected using AFLP. Although AFLP is useful for studying genetic variations, it is currently not widely applied because of the rapidly growing number of sequence-derived markers and a rising number of commercial next-generation sequencing (NGS) service providers, enhancing the accumulation of sequences in databases. However, AFLP is still advantageous in the era of NGS as it requires a lower cost for studying genetic variations (Restrepo et al., 2018). Small-sized laboratories can implement AFLP without requiring complex bioinformatic data processing. Nowadays, a technique known as restriction site-associated DNA sequencing (RADSeq) can be applied for access to a particular subset of the genome. RADseq combines restriction analysis, PCR amplification, and massively parallel sequencing (Andrews et al., 2016; Baird et al., 2008; Davey et al., 2011) and has been implemented to investigate anthelmintic resistant mutations in nematode genomes such as *H. contortus* and *Teladorsagia circumcincta*, which are important nematodes infecting the abomasum of small ruminants (Choi et al., 2017; Luo et al., 2017).

The SSCP is one of the analytical applications of the mutation scanning technique and is a powerful and cost-effective electrophoretic tool for the direct analysis of genetic variation, particularly when large numbers of samples are required (Gasser, 1997, 2006; Gasser et al., 2006). This technique can be used to display point mutations for small (100–300) amplicons and is powerful in detecting mutation/variation in the population. SSCP is helpful as a sensitive analytical and diagnostic tool for molecular investigations of unicellular and multicellular parasites, including helminths (Gasser, 2006). This method has been implemented among helminths to study their systematics, ecology, epidemiology, population genetics, and control. PCR-based SSCP analysis of the internal transcribed spacer 1 (ITS1) and internal transcribed spacer 2 (ITS2) rDNA regions has been employed effectively to identify parasites to species or strains where morphological characters are unreliable (Gasser & Chilton, 2001; Gasser & Monti, 1997). The third-stage larvae (L3) of the two species of porcine nodule worm, *Oesophagostomum dentatum,* and *O. quadrangulatum*, were distinguished, and the results showed the ability of the SSCP method to detect either the mixture between the two species of *Oesophagostomum* or detect pure *O. dentatum* in the laboratory-maintained cultures of L3 (Gasser et al., 1998b). The SSCP method was also applied to identify and distinguish between seven different hookworm species (*Ancylostoma caninum, A. tubaeforme, A. ceylanicum, A. duodenale, Uncinaria stenocephala, Bunostomum trigonocephalum*, and *Necator americanus*) (Gasser et al., 1998a). For molecular identification of cryptic species of helminthic parasites (Anderson et al., 1998; Gasser, 1999), SSCP-based identification of *Pseudoterranova decipiens* sensu lato (s.l) and its five sibling species in the *P. decipiens* complex (i.e., *P. krabbei, P. decipiens* sensu stricto (s.s), *P. bulbosa, P. azarasi*, and *P. cattani*) from different seal and sea lion hosts and geographic origins was performed. The SSCP-based analysis of ITS2 amplicons showed an unequivocal differentiation among the species in the *P. decipiens* complex. Additionally, the robustness of the SSCP method was supported through confirmation using DNA sequence data (Zhu et al., 2002). However, DNA sequence-based techniques through PCR have gradually replaced the SSCP method as enzymatic amplification of DNA can now be achieved in vitro from minute amounts of materials.

With the accumulation of DNA sequences, molecular phylogenetics and related fields have had an immediate and far-reaching impact on biology. Recent advances in molecular biology provide novel tools to address evolutionary, ecological, and taxonomic questions. Variations in DNA sequences can now be observed with a level of accuracy that was previously impossible, as the nucleotide level is often not visible at the phenotypic level. DNA-based markers have many advantages over phenotypic markers in that they are highly heritable, are relatively easy to assay, and are not affected by the environment. Generally, the target DNA using the molecular genetic marker should possess enough genetic variation to allow differentiation at least to the species level. If possible, the genetic variation of the target DNA region should be able to simultaneously discriminate at both the interspecific (species) and intraspecific (strain/ population) levels (Chan et al., 2021). Useful molecular genetic markers are present in various target regions in both nuclear and mitochondrial genomes, including nuclear ribosomal DNA (rDNA), repetitive DNA elements (microsatellite loci), mitochondrial protein-coding genes, and mitochondrial ribosomal RNA genes (Chan et al., 2021; Hu et al., 2004; López-Flores & Garrido-Ramos, 2012; Patwardhan et al., 2014). Zarlenga et al. (2014b) suggested no consensus as to what constitutes appropriate discovery methods and analytical approaches for defining species, particularly in the context of cryptic diversity (Zarlenga et al., 2014a).

Chan et al. (2021) assessed the suitability of mitochondrial and nuclear DNA genetic markers for molecular identification and systematics based on GenBank's DNA sequences by estimating genetic distances within and among species for nematodes, trematodes, and cestodes. They concluded that the nuclear and mitochondrial rRNA genes were the best possible molecular genetic markers for helminth molecular systematics, whereas the mitochondrial protein-coding genes were suitable for molecular identification for those three groups of parasitic helminths. Cutoff genetic distance values per taxonomic level on mitochondrial rDNA genetic markers based on the *K-mean* algorithm were generated, and they revealed that a general gauge of genetic distances might not be adequate for all helminths. Therefore, genetic distances should be considered separately for each helminth taxa (Chan et al., 2021).

The utility of suitable genetic markers can be evidenced by the robustness of phylogenetic relationships based on DNA sequence variation. The evolutionary divergence of taxonomic taxa can be visualized by constructing phylogenetic trees based on the order in which these evolutionary events transpired, which diverged from common ancestry. As mentioned in Chap. 1, systematics considers a historical pattern of species diversification arising from common ancestry. The inclusion of genomic and molecular data in assessments of relationships within and among parasitic groups has dramatically expanded across all helminth taxa, particularly parasitic helminths. Examples include molecular-based phylogenies for the phylum Nematoda using a single gene (small nuclear subunit ribosomal RNA gene-18S rRNA gene) by dividing the phylum into five major clades based on 53 species (Blaxter et al., 1998), and the molecular systematics of trematodes using the complete small and large subunit ribosomal DNA sequences for 163 Digenea taxa (Olson et al., 2003).

Phylogenetic applications allow the development of hypotheses for relationships of lineages, species, and higher taxa. Single-gene or multigene studies have been utilized extensively to develop phylogenetic inference of various parasitic helminth species, for example, *Trichinella* (Korhonen et al., 2016; Pozio et al., 2009), *Schistosoma* (Bowles et al., 1995; Webster et al., 2006), *Taenia* (Hoberg, 2006; Hoberg et al., 2000; Nakao et al., 2002), *Anisakis* (Mattiucci & Nascetti, 2006; Valentini et al., 2006), *Opisthorchis* (Park, 2007; Pomaznoy et al., 2013), *Metagonimus* ((Pornruseetairatn et al., 2016; Shumenko et al., 2017), and *Angiostrongylus* (Chan et al., 2020; Eamsobhana et al., 2010; Tokiwa et al., 2012), etc. However, the problem of varying evolutionary rates among DNA target regions can occur using a specific genetic marker. This issue can be resolved through phylogenomics by expanding the comparison to whole genomes or genomic features, as phylogenomics can often overcome variances at the microscale, reducing the effect of misalignments and horizontal gene transfer and missing data.

3.8 Phylogenomics: Trend to Study Deep Evolutionary Relationship

Phylogenomics is the interface between molecular and evolutionary relationships, which has been used to refer to the analysis involving genome data and evolutionary reconstructions. This approach becomes theoretically practicable to elucidate ancient evolutionary relationships or deep evolutionary relationships (Young & Gillung,

2019). Phylogenomics draws information by comparing entire genomes or large portions of genomes and has two major goals, including (1) to infer the phylogenetic relationships between taxa and understanding molecular evolution mechanisms and (2) to use phylogenomic comparisons of multispecies to infer putative functions for DNA or protein sequences (Young & Gillung, 2019). Modern phylogenomic analyses take advantage of hundreds to thousands of loci across the genome, more enormous than traditional Sanger sequencing datasets. The size of the genome datasets significantly reduces the impact of stochastic error and data availability, offering great promise for resolving historically inconsistent nodes in the tree of life. Phylogenomics was applied to mitochondrial genomes where recently, Kern et al. (2020) elucidated the mitogenome phylogenetics of nematodes and suggested that mitochondrial genomes are a promising molecular genetic marker for nematode systematics (Kern et al., 2020). The International Helminth Genomes Consortium (2019) investigated the genome diversity of 56 nematodes and 25 platyhelminth species using genome-wide analysis. The phylogenomic relationships were constructed based on a partitioned analysis of a concatenated data matrix of 21,649 amino acid sites from 202 single-copy orthologous proteins in at least 23 species (International Helminth Genomes, 2019). The phylogenomic relationships obtained revealed similar relationships among the nematode clades from the nuclear 18S rRNA gene (Blaxter et al., 1998). Blaxter and Koutsovoulos (2015) recommended that phylogenomic can resolve nematode systematics by defining the origins of parasitism in different lineages, understanding the assembly of hosts and parasites in various ecosystems, mapping the patterns of parasite diversification, and revealing the evolutionary patterns in developmental and other systems (Blaxter & Koutsovoulos, 2015). More details on the molecular evolution of parasitic helminths will be presented in Chap. 4.

3.9 Conclusion

As mentioned in the chapter, coevolution between host and parasites allowed helminths to adapt themselves in the evolutionary process. Changes in the environment have influenced phenotypic plasticity, which is a problematic issue for morphological identification and systematics. The genetic background of parasites has been changed due to evolutionary forces which can alter virulence-transmission relationships between hosts and parasites. Through evolution, coevolved parasites and their hosts build up parasite diversity in morphological and genetic characters. Molecular advances enabled the use of various molecular genetic markers for a particular target to improve our ability to accurately and precisely measure genetic variation. Genetic markers allow the visualization of phylogenetic relationships and historical diversification patterns originating from the common ancestor, which supports systematics. Recently, modern phylogenomics enabled us to study deep evolutionary relationships by taking advantage of the hundreds to thousands of loci across genomes. However, before we thoroughly move to the genome era for parasitic helminths, molecular genetic markers from the target regions are helpful for molecular systematics and molecular identification.

References

Agosta, S. J., & Klemens, J. A. (2008). Ecological fitting by phenotypically flexible genotypes: Implications for species associations, community assembly and evolution. *Ecology Letters, 11*(11), 1123–1134. https://doi.org/10.1111/j.1461-0248.2008.01237.x

Aleuy, O. A., & Kutz, S. (2020). Adaptations, life-history traits and ecological mechanisms of parasites to survive extremes and environmental unpredictability in the face of climate change. *International Journal for Parasitology: Parasites and Wildlife, 12*, 308–317. https://doi.org/10.1016/j.ijppaw.2020.07.006

Alizon, S., Hurford, A., Mideo, N., & Van Baalen, M. (2009). Virulence evolution and the trade-off hypothesis: History, current state of affairs and the future. *Journal of Evolutionary Biology, 22*(2), 245–259.

Anderson, T. J., Blouin, M. S., & Beech, R. N. (1998). Population biology of parasitic nematodes: Applications of genetic markers. *Advances in Parasitology, 41*, 219–283.

Andrews, K. R., Good, J. M., Miller, M. R., Luikart, G., & Hohenlohe, P. A. (2016). Harnessing the power of RADseq for ecological and evolutionary genomics. *Nature Reviews Genetics, 17*(2), 81–92.

Baird, N. A., Etter, P. D., Atwood, T. S., Currey, M. C., Shiver, A. L., Lewis, Z. A., Selker, E. U., Cresko, W. A., & Johnson, E. A. (2008). Rapid SNP discovery and genetic mapping using sequenced RAD markers. *PLoS One, 3*(10), e3376.

Bentounsi, B., & Cabaret, J. (1999). Analysis of helminth genetic data: Comparative examples with *Haemonchus contortus* isozymes using exact tests or resampling procedures. *Parasitology Research, 85*(10), 855–857.

Bernatchez, L., & Landry, C. (2003). MHC studies in nonmodel vertebrates: What have we learned about natural selection in 15 years? *Journal of Evolutionary Biology, 16*(3), 363–377.

Blaxter, M., & Koutsovoulos, G. (2015). The evolution of parasitism in Nematoda. *Parasitology, 142*(Suppl 1), S26–S39. https://doi.org/10.1017/S0031182014000791

Blaxter, M. L., De Ley, P., Garey, J. R., Liu, L. X., Scheldeman, P., Vierstraete, A., Vanfleteren, J. R., Mackey, L. Y., Dorris, M., & Frisse, L. M. (1998). A molecular evolutionary framework for the phylum Nematoda. *Nature, 392*(6671), 71–75.

Blouin, M. S. (2002). Molecular prospecting for cryptic species of nematodes: Mitochondrial DNA versus internal transcribed spacer. *International Journal for Parasitology, 32*(5), 527–531.

Bowles, J., Blair, D., & McManus, D. (1995). A molecular phylogeny of the human schistosomes. *Molecular Phylogenetics and Evolution, 4*(2), 103–109.

Brooks, D. R. (1985). Historical ecology: A new approach to studying the evolution of ecological associations. *Annals of the Missouri Botanical Garden, 72*, 660–680.

Brooks, D. R., & Ferrao, A. L. (2005). The historical biogeography of coevolution: Emerging infectious diseases are evolutionary accidents waiting to happen. *Journal of Biogeography, 32*(8), 1291–1299. https://doi.org/10.1111/j.1365-2699.2005.01315.x

Brooks, D. R., & Hoberg, E. P. (2001). Parasite systematics in the 21st century: Opportunities and obstacles. *Trends in Parasitology, 17*(6), 273–275.

Brooks, D. R., & Hoberg, E. P. (2007). How will global climate change affect parasite-host assemblages? *Trends in Parasitology, 23*(12), 571–574. https://doi.org/10.1016/j.pt.2007.08.016

Brooks, D., & Hoberg, E. (2013). The emerging infectious disease crisis and pathogen pollution: A question of ecology and evolution. In *The balance of nature and human impact* (pp. 215–229). Cambridge University Press.

Brooks, D. R., & McLennan, D. A. (1993). Comparative study of adaptive radiations with an example using parasitic flatworms (Platyhelminthes: Cercomeria). *The American Naturalist, 142*(5), 755–778.

Brooks, D. R., León-Règagnon, V., McLennan, D. A., & Zelmer, D. (2006a). Ecological fitting as a determinant of the community structure of platyhelminth parasites of anurans. *Ecology, 87*(sp7), S76–S85.

Brooks, D. R., McLennan, D. A., León-Règagnon, V., & Hoberg, E. (2006b). Phylogeny, ecological fitting and lung flukes: Helping solve the problem of emerging infectious diseases. *Revista mexicana de biodiversidad, 77*(2), 225–233.

Brunner, F. S., & Eizaguirre, C. (2016). Can environmental change affect host/parasite-mediated speciation? *Zoology, 119*(4), 384–394.

Budria, A., & Candolin, U. (2014). How does human-induced environmental change influence host-parasite interactions? *Parasitology, 141*(4), 462–474.

Carlson, C. J., Dallas, T., Alexander, L., Phelan, A., & Phillips, A. (2020). What would it take to describe the global diversity of parasites? *Proceedings of the Royal Society B, 287*(1939), 1841.

Castro, G. (1996). Helminths: Structure, classification, growth, and development. In S. Baron (Ed.), *Medical microbiology* (4th ed.). University of Texas Medical Branch at Galveston.

Chan, A. H. E., Chaisiri, K., Dusitsittipon, S., Jakkul, W., Charoennitiwat, V., Komalamisra, C., & Thaenkham, U. (2020). Mitochondrial ribosomal genes as novel genetic markers for discrimination of closely related species in the *Angiostrongylus cantonensis* lineage. *Acta Tropica, 211*, 105645.

Chan, A. H. E., Chaisiri, K., Saralamba, S., Morand, S., & Thaenkham, U. (2021). Assessing the suitability of mitochondrial and nuclear DNA genetic markers for molecular systematics and species identification of helminths. *Parasites & Vectors, 14*(1), 233. https://doi.org/10.1186/s13071-021-04737-y

Choi, Y.-J., Bisset, S. A., Doyle, S. R., Hallsworth-Pepin, K., Martin, J., Grant, W. N., & Mitreva, M. (2017). Genomic introgression mapping of field-derived multiple-anthelmintic resistance in *Teladorsagia circumcincta*. *PLoS Genetics, 13*(6), e1006857.

Choi, Y. J., Fontenla, S., Fischer, P. U., Le, T. H., Costabile, A., Blair, D., Brindley, P. J., Tort, J. F., Cabada, M. M., & Mitreva, M. (2020). Adaptive radiation of the flukes of the family Fasciolidae inferred from genome-wide comparisons of key species. *Molecular Biology and Evolution, 37*(1), 84–99. https://doi.org/10.1093/molbev/msz204

Coles, G. (1970). A comparison of some isoenzymes of *Schistosoma mansoni* and *Schistosoma haematobium*. *Comparative Biochemistry and Physiology, 33*(3), 549–558.

Colles, A., Liow, L. H., & Prinzing, A. (2009). Are specialists at risk under environmental change? Neoecological, paleoecological and phylogenetic approaches. *Ecology Letters, 12*(8), 849–863. https://doi.org/10.1111/j.1461-0248.2009.01336.x

Coyne, J. A., & Orr, H. A. (2004). *Speciation* (Vol. 37). Sinauer Associates.

Criscione, C. D., Poulin, R., & Blouin, M. S. (2005). Molecular ecology of parasites: Elucidating ecological and microevolutionary processes. *Molecular Ecology, 14*(8), 2247–2257. https://doi.org/10.1111/j.1365-294X.2005.02587.x

Crispo, E. (2008). Modifying effects of phenotypic plasticity on interactions among natural selection, adaptation and gene flow. *Journal of Evolutionary Biology, 21*(6), 1460–1469.

Dallas, T. A., Aguirre, A. A., Budischak, S., Carlson, C., Ezenwa, V., Han, B., Huang, S., & Stephens, P. R. (2018). Gauging support for macroecological patterns in helminth parasites. *Global Ecology and Biogeography, 27*(12), 1437–1447. https://doi.org/10.1111/geb.12819

Davey, J. W., Hohenlohe, P. A., Etter, P. D., Boone, J. Q., Catchen, J. M., & Blaxter, M. L. (2011). Genome-wide genetic marker discovery and genotyping using next-generation sequencing. *Nature Reviews Genetics, 12*(7), 499–510.

De Gruijter, J., Gasser, R., Polderman, A., Asigri, V., & Dijkshoorn, L. (2005). High resolution DNA fingerprinting by AFLP to study the genetic variation among *Oesophagostomum bifurcum* (Nematoda) from human and non-human primates from Ghana. *Parasitology, 130*(2), 229–237.

de Gruijter, J. M., Polderman, A. M., Dijkshoorn, L., Roberts, H., Ziem, J., Kunwar, C. B., & Gasser, R. B. (2006). AFLP fingerprinting for the analysis of genetic diversity within *Necator americanus*. *Molecular and Cellular Probes, 20*(5), 317–321.

de Leon, G. P., & Nadler, S. A. (2010). What we don't recognize can hurt us: A plea for awareness about cryptic species. *The Journal of Parasitology, 96*(2), 453–464. https://doi.org/10.1645/GE-2260.1

Deutsch, C. A., Tewksbury, J. J., Huey, R. B., Sheldon, K. S., Ghalambor, C. K., Haak, D. C., & Martin, P. R. (2008). Impacts of climate warming on terrestrial ectotherms across latitude. *Proceedings of the National Academy of Sciences, 105*(18), 6668–6672.

Dobson, A., Lafferty, K. D., Kuris, A. M., Hechinger, R. F., & Jetz, W. (2008a). Colloquium paper: Homage to Linnaeus: How many parasites? How many hosts? *Proceedings of the National Academy of Sciences of the United States of America, 105*(Suppl 1), 11482–11489. https://doi.org/10.1073/pnas.0803232105

Dobson, A., Lafferty, K. D., Kuris, A. M., Hechinger, R. F., & Jetz, W. (2008b). Homage to Linnaeus: How many parasites? How many hosts? *Proceedings of the National Academy of Sciences, 105*(Supplement 1), 11482–11489.

Dusitsittipon, S., Criscione, C. D., Morand, S., Komalamisra, C., & Thaenkham, U. (2017). Cryptic lineage diversity in the zoonotic pathogen *Angiostrongylus cantonensis*. *Molecular Phylogenetics and Evolution, 107*, 404–414.

Dusitsittipon, S., Criscione, C. D., Morand, S., Komalamisra, C., & Thaenkham, U. (2018). Hurdles in the evolutionary epidemiology of *Angiostrongylus cantonensis*: Pseudogenes, incongruence between taxonomy and DNA sequence variants, and cryptic lineages. *Evolutionary Applications, 11*(8), 1257–1269.

Eamsobhana, P., Lim, P. E., Solano, G., Zhang, H., Gan, X., & Yong, H. S. (2010). Molecular differentiation of *Angiostrongylus* taxa (Nematoda: Angiostrongylidae) by cytochrome c oxidase subunit I (*COI*) gene sequences. *Acta Tropica, 116*(2), 152–156.

Eom, K. S., Jeon, H.-K., Kong, Y., Hwang, U. W., Yang, Y., Li, X., Xu, L., Feng, Z., Pawlowski, Z. S., & Rim, H.-J. (2002). Identification of *Taenia asiatica* in China: Molecular, morphological, and epidemiological analysis of a Luzhai isolate. *Journal of Parasitology, 88*(4), 758–764.

Fenton, A., & Pedersen, A. B. (2005). Community epidemiology framework for classifying disease threats. *Emerging Infectious Diseases, 11*(12), 1815.

Flockhart, H. A., Harrison, S. E., Dobinson, A., & James, E. (1982). Enzyme polymorphism in *Trichinella*. *Transactions of the Royal Society of Tropical Medicine and Hygiene, 76*(4), 541–545.

Foxman, B., & Riley, L. (2001). Molecular epidemiology: Focus on infection. *American Journal of Epidemiology, 153*(12), 1135–1141.

Futuyma, D. J. (2010). Evolutionary constraint and ecological consequences. *Evolution: International Journal of Organic Evolution, 64*(7), 1865–1884.

Gandon, S., & Michalakis, Y. (2002). Local adaptation, evolutionary potential and host–parasite coevolution: Interactions between migration, mutation, population size and generation time. *Journal of Evolutionary Biology, 15*(3), 451–462.

Gandon, S., & Van Zandt, P. A. (1998). Local adaptation and host–parasite interactions. *Trends in Ecology & Evolution, 13*(6), 214–216. https://doi.org/10.1016/s0169-5347(98)01358-5

Gandon, S., Capowiez, Y., Dubois, Y., Michalakis, Y., & Olivieri, I. (1996). Local adaptation and gene-for-gene coevolution in a metapopulation model. *Proceedings of the Royal Society of London. Series B: Biological Sciences, 263*(1373), 1003–1009.

Gasser, R. B. (1997). Mutation scanning methods for the analysis of parasite genes. *International Journal for Parasitology, 27*(12), 1449–1463.

Gasser, R. B. (1999). PCR-based technology in veterinary parasitology. *Veterinary Parasitology, 84*(3–4), 229–258.

Gasser, R. B. (2006). Molecular tools—Advances, opportunities and prospects. *Veterinary Parasitology, 136*(2), 69–89. https://doi.org/10.1016/j.vetpar.2005.12.002

Gasser, R. B., & Chilton, N. B. (2001). Applications of single-strand conformation polymorphism (SSCP) to taxonomy, diagnosis, population genetics and molecular evolution of parasitic nematodes. *Veterinary Parasitology, 101*(3–4), 201–213.

Gasser, R. B., & Monti, J. R. (1997). Identification of parasitic nematodes by PCR-SSCP of ITS-2 rDNA. *Molecular and Cellular Probes, 11*(3), 201–209.

Gasser, R. B., Monti, J. R., Bao-Zhen, Q., Polderman, A. M., Nansen, P., & Chilton, N. B. (1998a). A mutation scanning approach for the identification of hookworm species and analysis of population variation. *Molecular and Biochemical Parasitology, 92*(2), 303–312.

Gasser, R. B., Woods, W. G., & Bjørn, H. (1998b). Distinguishing *Oesophagostomum dentatum* from *Oesophagostomum quadrispinulatum* developmental stages by a single-strand conformation polymorphism method. *International Journal for Parasitology, 28*(12), 1903–1909.

Gasser, R. B., Hu, M., Chilton, N. B., Campbell, B. E., Jex, A. J., Otranto, D., Cafarchia, C., Beveridge, I., & Zhu, X. (2006). Single-strand conformation polymorphism (SSCP) for the analysis of genetic variation. *Nature Protocols, 1*(6), 3121–3128.

Gokhale, C. S., Papkou, A., Traulsen, A., & Schulenburg, H. (2013). Lotka–Volterra dynamics kills the Red Queen: Population size fluctuations and associated stochasticity dramatically change host-parasite coevolution. *BMC Evolutionary Biology, 13*(1), 1–10.

Greischar, M. A., & Koskella, B. (2007). A synthesis of experimental work on parasite local adaptation. *Ecology Letters, 10*(5), 418–434.

Gupta, P., Robin, V., & Dharmarajan, G. (2020). Towards a more healthy conservation paradigm: Integrating disease and molecular ecology to aid biological conservation. *Journal of Genetics, 99*(1), 1–26.

Haley, A. J. (1962). Role of host relationships in the systematics of helminth parasites. *The Journal of Parasitology, 48*, 671–678.

Harvell, C. D., Mitchell, C. E., Ward, J. R., Altizer, S., Dobson, A. P., Ostfeld, R. S., & Samuel, M. D. (2002). Climate warming and disease risks for terrestrial and marine biota. *Science, 296*(5576), 2158–2162.

Helmuth, B., Kingsolver, J. G., & Carrington, E. (2005). Biophysics, physiological ecology, and climate change: Does mechanism matter? *Annual Review of Physiology, 67*, 177–201.

Hildebrand, J., Adamczyk, M., Laskowski, Z., & Zalesny, G. (2015). Host-dependent morphology of *Isthmiophora melis* (Schrank, 1788) Luhe, 1909 (Digenea, Echinostomatinae)—Morphological variation vs. molecular stability. *Parasites & Vectors, 8*, 481. https://doi.org/10.1186/s13071-015-1095-8

Ho, S. Y., & Shapiro, B. (2011). Skyline-plot methods for estimating demographic history from nucleotide sequences. *Molecular Ecology Resources, 11*(3), 423–434.

Hoberg, E. P. (2006). Phylogeny of *Taenia*: Species definitions and origins of human parasites. *Parasitology International, 55*, S23–S30.

Hoberg, E. P., & Brooks, D. R. (2013). Episodic processes, invasion and faunal mosaics in evolutionary and ecological time. In K. Rohde (Ed.), *The balance of nature and human impact* (pp. 199–213). Cambridge University Press.

Hoberg, E. P., Jones, A., Rausch, R. L., Eom, K. S., & Gardner, S. L. (2000). A phylogenetic hypothesis for species of the genus *Taenia* (Eucestoda: Taeniidae). *Journal of Parasitology, 86*(1), 89–98.

Hoffmann, A. A., & Sgro, C. M. (2011). Climate change and evolutionary adaptation. *Nature, 470*(7335), 479–485. https://doi.org/10.1038/nature09670

Hu, M., Chilton, N. B., & Gasser, R. B. (2004). The mitochondrial genomics of parasitic nematodes of socio-economic importance: Recent progress, and implications for population genetics and systematics. *Advances in Parasitology, 56*(03), 133–212.

Hudson, P. J., Dobson, A. P., & Newborn, D. (1998). Prevention of population cycles by parasite removal. *Science, 282*(5397), 2256–2258. https://doi.org/10.1126/science.282.5397.2256

International Helminth Genomes, C. (2019). Comparative genomics of the major parasitic worms. *Nature Genetics, 51*(1), 163–174. https://doi.org/10.1038/s41588-018-0262-1

Johnson, K. P., Adams, R. J., Page, R. D., & Clayton, D. H. (2003). When do parasites fail to speciate in response to host speciation? *Systematic Biology, 52*(1), 37–47. https://doi.org/10.1080/10635150390132704

Kaltz, O., & Shykoff, J. A. (1998). Local adaptation in host–parasite systems. *Heredity, 81*(4), 361–370.

Kaltz, O., Gandon, S., Michalakis, Y., & Shykoff, J. A. (1999). Local maladaptation in the anther-smut fungus *Microbotryum violaceum* to its host plant Silene latifolia: Evidence from a cross-inoculation experiment. *Evolution, 53*(2), 395–407.

Kao, R. R., Haydon, D. T., Lycett, S. J., & Murcia, P. R. (2014). Supersize me: How whole-genome sequencing and big data are transforming epidemiology. *Trends in Microbiology, 22*(5), 282–291.

Kelly, M. (2019). Adaptation to climate change through genetic accommodation and assimilation of plastic phenotypes. *Philosophical Transactions of the Royal Society of London. Series B, Biological Sciences, 374*(1768), 20180176. https://doi.org/10.1098/rstb.2018.0176

Kern, E. M. A., Kim, T., & Park, J.-K. (2020). The mitochondrial genome in nematode Phylogenetics. *Frontiers in Ecology and Evolution, 8*, 250. https://doi.org/10.3389/fevo.2020.00250

Kirk, H., & Freeland, J. R. (2011). Applications and implications of neutral versus non-neutral markers in molecular ecology. *International Journal of Molecular Sciences, 12*(6), 3966–3988. https://doi.org/10.3390/ijms12063966

Korhonen, P. K., Pozio, E., La Rosa, G., Chang, B. C., Koehler, A. V., Hoberg, E. P., Boag, P. R., Tan, P., Jex, A. R., & Hofmann, A. (2016). Phylogenomic and biogeographic reconstruction of the *Trichinella* complex. *Nature Communications, 7*(1), 1–8.

Levis, N. A., Serrato-Capuchina, A., & Pfennig, D. W. (2017). Genetic accommodation in the wild: Evolution of gene expression plasticity during character displacement. *Journal of Evolutionary Biology, 30*(9), 1712–1723.

Lighten, J., Papadopulos, A. S. T., Mohammed, R. S., Ward, B. J., Baillie, L., Bradbury, I. R., Hendry, A. P., Bentzen, P., & van Oosterhout, C. (2017). Evolutionary genetics of immunological supertypes reveals two faces of the red queen. *Nature Communications, 8*(1), 1294. https://doi.org/10.1038/s41467-017-01183-2

Littlewood, D., Rohde, K., Bray, R., & Herniou, E. (1999). Phylogeny of the Platyhelminthes and the evolution of parasitism. *Biological Journal of the Linnean Society, 68*(1–2), 257–287.

Loker, E., & Hofkin, B. (2015). *Parasitology: A conceptual approach*. Garland Science.

López-Flores, I., & Garrido-Ramos, M. (2012). The repetitive DNA content of eukaryotic genomes. *Repetitive DNA, 7*, 1–28.

Luikart, G., England, P. R., Tallmon, D., Jordan, S., & Taberlet, P. (2003). The power and promise of population genomics: From genotyping to genome typing. *Nature Reviews Genetics, 4*(12), 981–994.

Luo, X., Shi, X., Yuan, C., Ai, M., Ge, C., Hu, M., Feng, X., & Yang, X. (2017). Genome-wide SNP analysis using 2b-RAD sequencing identifies the candidate genes putatively associated with resistance to ivermectin in *Haemonchus contortus*. *Parasites & Vectors, 10*(1), 1–10.

Lustigman, S., Geldhof, P., Grant, W. N., Osei-Atweneboana, M. Y., Sripa, B., & Basanez, M. G. (2012). A research agenda for helminth diseases of humans: Basic research and enabling technologies to support control and elimination of helminthiases. *PLoS Neglected Tropical Diseases, 6*(4), e1445. https://doi.org/10.1371/journal.pntd.0001445

MacIntosh, A. J. J., & Frias, L. (2017). Coevolution of hosts and parasites. In *The international encyclopedia of primatology* (pp. 1–8). https://doi.org/10.1002/9781119179313.wbprim0395

Mattiucci, S., & Nascetti, G. (2006). Molecular systematics, phylogeny and ecology of anisakid nematodes of the genus *Anisakis* Dujardin, 1845: An update. *Parasite, 13*(2), 99–113.

McCoy, K. D. (2003). Sympatric speciation in parasites—What is sympatry? *Trends in Parasitology, 19*(9), 400–404. https://doi.org/10.1016/s1471-4922(03)00194-6

Mikkonen, T., Koort, J., Björkroth, K., & Sukura, A. (2005). Testing of amplified fragment length polymorphism (AFLP) technique as a tool for molecular epidemiology of *Trichinella nativa*. *Veterinary Parasitology, 132*(1–2), 19–22.

Molnar, P. K., Dobson, A. P., & Kutz, S. J. (2013). Gimme shelter--the relative sensitivity of parasitic nematodes with direct and indirect life cycles to climate change. *Global Change Biology, 19*(11), 3291–3305. https://doi.org/10.1111/gcb.12303

Mostowy, R., & Engelstadter, J. (2011). The impact of environmental change on host-parasite coevolutionary dynamics. *Proceedings of the Biological Sciences, 278*(1716), 2283–2292. https://doi.org/10.1098/rspb.2010.2359

Nakao, M., Okamoto, M., Sako, Y., Yamasaki, H., Nakaya, K., & Ito, A. (2002). A phylogenetic hypothesis for the distribution of two genotypes of the pig tapeworm *Taenia solium* worldwide. *Parasitology, 124*(6), 657–662.

Nejsum, P., Frydenberg, J., Roepstorff, A., & Davis Parker, J. E. (2005). Population structure in *Ascaris suum* (Nematoda) among domestic swine in Denmark as measured by whole genome DNA fingerprinting. *Hereditas, 142*(2005), 7–14.

Olson, P. D., Cribb, T. H., Tkach, V. V., Bray, R. A., & Littlewood, D. T. J. (2003). Phylogeny and classification of the Digenea (Platyhelminthes: Trematoda). *International Journal for Parasitology, 33*(7), 733–755. https://doi.org/10.1016/s0020-7519(03)00049-3

Oppliger, A., Vernet, R., & Baez, M. (1999). Parasite local maladaptation in the Canarian lizard *Gallotia galloti* (Reptilia: Lacertidae) parasitized by haemogregarian blood parasite. *Journal of Evolutionary Biology, 12*(5), 951–955.

Paaby, A. B., & Rockman, M. V. (2014). Cryptic genetic variation: Evolution's hidden substrate. *Nature Reviews. Genetics, 15*(4), 247–258. https://doi.org/10.1038/nrg3688

Papkou, A., Gokhale, C. S., Traulsen, A., & Schulenburg, H. (2016). Host-parasite coevolution: Why changing population size matters. *Zoology (Jena, Germany), 119*(4), 330–338. https://doi.org/10.1016/j.zool.2016.02.001

Pareek, C. S., Smoczynski, R., & Tretyn, A. (2011). Sequencing technologies and genome sequencing. *Journal of Applied Genetics, 52*(4), 413–435. https://doi.org/10.1007/s13353-011-0057-x

Park, G.-M. (2007). Genetic comparison of liver flukes, *Clonorchis sinensis* and *Opisthorchis viverrin*i, based on rDNA and mtDNA gene sequences. *Parasitology Research, 100*(2), 351–357.

Parmesan, C. (2006). Ecological and evolutionary responses to recent climate change. *Annual Review of Ecology, Evolution, and Systematics, 37*(1), 637–669. https://doi.org/10.1146/annurev.ecolsys.37.091305.110100

Patra, K., Ramu, T., Hoti, S., Pragasam, G. S., & Das, P. (2007). Identification of a molecular marker for genotyping human lymphatic filarial nematode parasite *Wuchereria bancrofti. Experimental Parasitology, 116*(1), 59–65.

Patwardhan, A., Ray, S., & Roy, A. (2014). Molecular markers in phylogenetic studies—a review. *Journal of Phylogenetics & Evolutionary Biology, 2014.*

Perez-Ponce de Leon, G., & Poulin, R. (2018). An updated look at the uneven distribution of cryptic diversity among parasitic helminths. *Journal of Helminthology, 92*(2), 197–202. https://doi.org/10.1017/S0022149X17000189

Pomaznoy, M., Tatkov, S., Katokhin, A., Afonnikov, D., Babenko, V., Furman, D., Brusentsov, I., Belavin, P., Najakshin, A., & Guselnikov, S. (2013). Adult *Opisthorchis felineus* major protein fractions deduced from transcripts: Comparison with liver flukes *Opisthorchis viverrini* and *Clonorchis sinensis. Experimental Parasitology, 135*(2), 297–306.

Pornruseetairatn, S., Kino, H., Shimazu, T., Nawa, Y., Scholz, T., Ruangsittichai, J., Saralamba, N. T., & Thaenkham, U. (2016). A molecular phylogeny of Asian species of the genus *Metagonimus* (Digenea)—Small intestinal flukes—Based on representative Japanese populations. *Parasitology Research, 115*(3), 1123–1130.

Poulin, R. (2014). Parasite biodiversity revisited: Frontiers and constraints. *International Journal for Parasitology, 44*(9), 581–589.

Poulin, R., & Morand, S. (2000). The diversity of parasites. *The Quarterly Review of Biology, 75*(3), 277–293.

Poulin, R., & Presswell, B. (2016). Taxonomic quality of species descriptions varies over time and with the number of authors, but unevenly among parasitic taxa. *Systematic Biology, 65*(6), 1107–1116. https://doi.org/10.1093/sysbio/syw053

Pozio, E., Hoberg, E., La Rosa, G., & Zarlenga, D. S. (2009). Molecular taxonomy, phylogeny and biogeography of nematodes belonging to the *Trichinella* genus. *Infection, Genetics and Evolution, 9*(4), 606–616.

Prugnolle, F., de Meeus, T., Pointier, J. P., Durand, P., Rognon, A., & Theron, A. (2006). Geographical variations in infectivity and susceptibility in the host-parasite system *Schistosoma mansoni/Biomphalaria glabrata*: No evidence for local adaptation. *Parasitology, 133*(Pt 3), 313–319. https://doi.org/10.1017/S0031182006000412

Pybus, O. G., & Rambaut, A. (2009). Evolutionary analysis of the dynamics of viral infectious disease. *Nature Reviews Genetics, 10*(8), 540–550.

Ramadan, N. I., & Saber, L. M. (2004). Detection of genetic variabiltiy in nonhuman isolates of *Fasciola hepatica* and *Fasciola gigantica* by the RAPD-PCR technique. *Journal of the Egyptian Society of Parasitology, 34*(2), 679–689.

Ramadan, I., Saber, L. M., Abd El Latif, M. M., Abdalla, N. A., & Ragab, H. M. (2010). Molecular genetic approach by using the RAPD-PCR technique for detection of genetic variability in non-human isolates of *Fasciola*. *Journal of American Science, 3*, 88–96.

Restrepo, C. M., Llanes, A., & Lleonart, R. (2018). Use of AFLP for the study of eukaryotic pathogens affecting humans. *Infection, Genetics and Evolution, 63*, 360–369.

Rishi, A., & McManus, D. (1987). Genomic cloning of human *Echinococcus granulosus* DNA: Isolation of recombinant plasmids and their use as genetic markers in strain characterization. *Parasitology, 94*(2), 369–383.

Rodpai, R., Intapan, P. M., Thanchomnang, T., Sanpool, O., Sadaow, L., Laymanivong, S., Aung, W. P., Phosuk, I., Laummaunwai, P., & Maleewong, W. (2016). *Angiostrongylus cantonensis* and *A. malaysiensis* broadly overlap in Thailand, Lao PDR, Cambodia and Myanmar: A molecular survey of larvae in land snails. *PLoS One, 11*(8), e0161128.

Salinas, S., Irvine, S. E., Schertzing, C. L., Golden, S. Q., & Munch, S. B. (2019). Trait variation in extreme thermal environments under constant and fluctuating temperatures. *Philosophical Transactions of the Royal Society B, 374*(1768), 20180177.

Schmalhausen, I. (1949). *Factors of evolution*. Blakiston.

Schneider, R. F., & Meyer, A. (2017). How plasticity, genetic assimilation and cryptic genetic variation may contribute to adaptive radiations. *Molecular Ecology, 26*(1), 330–350. https://doi.org/10.1111/mec.13880

Scholz, T. (2001). Taxonomy of helminths after the year 2000. *Helminthologia, 38*(3), 155–164.

Schwartz, M. K., Luikart, G., & Waples, R. S. (2007). Genetic monitoring as a promising tool for conservation and management. *Trends in Ecology & Evolution, 22*(1), 25–33.

Shin, D. H., Chai, J., Park, E., Lee, W., Lee, H., Lee, J., Choi, Y., Koh, B., Park, J., & Oh, C. (2009). Finding ancient parasite larvae in a sample from a male living in late 17th century Korea. *Journal of Parasitology, 95*(3), 768–771.

Shumenko, P., Tatonova, Y., & Besprozvannykh, V. (2017). *Metagonimus suifunensis* sp. n. (Trematoda: Heterophyidae) from the Russian southern Far East: Morphology, life cycle, and molecular data. *Parasitology International, 66*(1), 982–991.

Soto-Cerda, B. J., & Cloutier, S. (2013). Outlier loci and selection signatures of simple sequence repeats (SSRs) in flax (*Linum usitatissimum* L.). *Plant Molecular Biology Reporter, 31*(4), 978–990.

Spakulova, M. (2002). Current problems of helminth taxonomy: Species level reflections. *Helminthologia, 39*(2), 111–118.

Summers, K., McKeon, S., Sellars, J., Keusenkothen, M., Morris, J., Gloeckner, D., Pressley, C., Price, B., & Snow, H. (2003). Parasitic exploitation as an engine of diversity. *Biological Reviews, 78*(4), 639–675.

Tellier, A., Moreno-Gamez, S., & Stephan, W. (2014). Speed of adaptation and genomic footprints of host-parasite coevolution under arms race and trench warfare dynamics. *Evolution, 68*(8), 2211–2224. https://doi.org/10.1111/evo.12427

Thaenkham, U., Pakdee, W., Nuamtanong, S., Maipanich, W., Pubampen, S., Sa-Nguankiat, S., & Komalamisra, C. (2012). Population structure of *Angiostrongylus cantonensis* (Nematoda: Metastrongylidae) in Thaialnd based on PCR-RAPD markers. *The Southeast Asian Journal of Tropical Medicine and Public Health, 43*(3), 567.

Thrall, P. H., Barrett, L. G., Dodds, P. N., & Burdon, J. J. (2016). Epidemiological and evolutionary outcomes in gene-for-gene and matching allele models. *Frontiers in Plant Science, 6*, 1084.

Tokiwa, T., Harunari, T., Tanikawa, T., Komatsu, N., Koizumi, N., Tung, K.-C., Suzuki, J., Kadosaka, T., Takada, N., & Kumagai, T. (2012). Phylogenetic relationships of rat lungworm, *Angiostrongylus cantonensis*, isolated from different geographical regions revealed widespread multiple lineages. *Parasitology International, 61*(3), 431–436.

Turelli, M., Barton, N. H., & Coyne, J. A. (2001). Theory and speciation. *Trends in Ecology & Evolution, 16*(7), 330–343.

Turner, W. C., Kamath, P. L., van Heerden, H., Huang, Y.-H., Barandongo, Z. R., Bruce, S. A., & Kausrud, K. (2021). The roles of environmental variation and parasite survival in virulence–transmission relationships. *Royal Society Open Science, 8*(6), 210088.

Valentini, A., Mattiucci, S., Bondanelli, P., Webb, S. C., Mignucci-Giannone, A. A., Colom-Llavina, M. M., & Nascetti, G. (2006). Genetic relationships among *Anisakis* species (Nematoda: Anisakidae) inferred from mitochondrial cox2 sequences, and comparison with allozyme data. *Journal of Parasitology, 92*(1), 156–166.

Via, S., & Lande, R. (1985). Genotype-environment interaction and the evolution of phenotypic plasticity. *Evolution, 39*(3), 505–522.

Vilas, R., Criscione, C. D., & Blouin, M. S. (2005). A comparison between mitochondrial DNA and the ribosomal internal transcribed regions in prospecting for cryptic species of platyhelminth parasites. *Parasitology, 131*(Pt 6), 839–846. https://doi.org/10.1017/S0031182005008437

Webster, B. L., Southgate, V. R., & Littlewood, D. T. J. (2006). A revision of the interrelationships of *Schistosoma* including the recently described *Schistosoma guineensis*. *International Journal for Parasitology, 36*(8), 947–955.

Webster, J. P., Borlase, A., & Rudge, J. W. (2017). Who acquires infection from whom and how? Disentangling multi-host and multi-mode transmission dynamics in the 'elimination'era. *Philosophical Transactions of the Royal Society, B: Biological Sciences, 372*(1719), 20160091.

Wolinska, J., & King, K. C. (2009). Environment can alter selection in host–parasite interactions. *Trends in Parasitology, 25*(5), 236–244.

Woolhouse, M. E., Webster, J. P., Domingo, E., Charlesworth, B., & Levin, B. R. (2002). Biological and biomedical implications of the coevolution of pathogens and their hosts. *Nature Genetics, 32*(4), 569–577.

Young, A. D., & Gillung, J. P. (2019). Phylogenomics — Principles, opportunities and pitfalls of big-data phylogenetics. *Systematic Entomology, 45*(2), 225–247. https://doi.org/10.1111/syen.12406

Zarlenga, D., Hoberg, E., & Detwiler, J. (2014a). Diversity and history as drivers of helminth systematics and biology. In *Helminth infections and their impact on global public health* (pp. 1–28). Springer.

Zarlenga, D., Hoberg, E., Rosenthal, B., Mattiucci, S., & Nascetti, G. (2014b). Anthropogenics: Human influence on global and genetic homogenization of parasite populations. *The Journal of Parasitology, 100*(6), 756–772. https://doi.org/10.1645/14-622.1

Zarowiecki, M., & Berriman, M. (2015). What helminth genomes have taught us about parasite evolution. *Parasitology, 142*(Suppl 1), S85–S97. https://doi.org/10.1017/S0031182014001449

Zhu, X., D'amelio, S., Palm, H., Paggi, L., George-Nascimento, M., & Gasser, R. (2002). SSCP-based identification of members within the *Pseudoterranova decipiens* complex (Nematoda: Ascaridoidea: Anisakidae) using genetic markers in the internal transcribed spacers of ribosomal DNA. *Parasitology, 124*(6), 615–623.

Chapter 4
Molecular Evolution of Parasitic Helminths

Abstract Molecular evolution is a scientific discipline that encompasses two areas of study, including macromolecule evolution and understanding the evolutionary history of organisms through reconstruction using genes through phylogenetics. The first area, the evolution of macromolecules, refers to the characterization of changes in genetic materials (DNA and RNA sequences) and proteins over time through the rates and patterns of such changes. The second area, molecular phylogenetics, deals with the evolutionary history of organisms inferred from molecular data and the methodology of phylogenetic tree reconstruction. While these two areas of molecular evolution are independent, they are intimately related. Phylogenies are essential for determining the order of changes in molecular characters, whereas the pattern and rate of evolutionary changes of molecules are crucial for reconstructing the evolutionary history of organisms. This chapter provided essential knowledge in molecular evolution, emphasizing on the parasitic helminth genome. Topics included are gene and gene structure, genetic codes, and mutations generating genetic variations through micro- and macroevolution. Additionally, this chapter also mentioned genome evolution under parasitic adaptation, which is affected by evolutionary forces and mitochondrial genes used for population genetics of parasitic helminths.

4.1 Principle of Molecular Evolution

Knowledge of molecular evolution is powerful for understanding evolution, as it concerns changes in the frequency of genes in a population. Some characteristics may give the individual an advantage over other individuals, allowing these advantageous genes or characteristics to be passed on to their offspring. Evolutionary mechanisms drive a new variant of a gene to spread through a population. Higgs and Attwood (2005) suggested two essential factors defining evolution, and they are (1) error-prone self-replication and (2) variation in success at self-replication. They described that evolutionary mechanisms must have the ability to replicate them-

U. Thaenkham et al., *Molecular Systematics of Parasitic Helminths*,
https://doi.org/10.1007/978-981-19-1786-8_4

selves to increase the number of copies, which can occur during the lifetime of organisms. The replication process can either occur through self-replication of genes or by undergoing asexual reproduction (e.g., in bacteria). If self-replication is error-prone, it means that the copies are not always identical to the originals. Errors are essential for evolution during gene replication, as there would be no further evolution (no variation) if DNA replication occurred perfectly. However, errors in DNA replication do not occur too frequently as too many errors limit the traits that will be passed on to the next generation. Variations in self-replication successes can exponentially increase organisms if they are not limited by food or predation in their ecological niches.

Generally, if all organisms can survive and multiply at the same rate, there will not be any change in the frequency of variants, and no evolution will occur. In evolutionary biology, the selection of some variants will occur, and natural selection amplifies the variants with high fitness rather than those with lower fitness. However, two different variants of an equal population size might not remain exactly equal, even though they have equal fitness. Gene variant numbers generally fluctuate randomly in a population through random genetic drift, through the occurrence of alleles (variant forms of a gene) that will increase and decrease by chance. These variations due to chance mean that evolution can occur even without natural selection and is referred to as the neutral theory of molecular evolution (Higgs & Attwood, 2005).

In molecular evolution, evolutionary forces act upon the genome makeup of populations. An essential component of the evolutionary process is the change in gene frequencies with time at the population level. In evolutionary population genetics, studies are interested in determining how the frequency of mutant alleles will change over time under various evolutionary forces. Also, evolutionary population genetic studies are concerned with estimating how fast and how frequent the replacement process is. Whenever changes occur at the nucleotide level, it might slightly affect the organism's phenotype and fitness. Genetic polymorphism within populations and the distribution pattern of polymorphism can be characterized to investigate driving forces in molecular evolution in the population.

4.2 Nucleotide Sequences

Nucleotide sequences are the hereditary information of all living organisms, and two types of nucleic acids are of key importance in cells. The two types of nucleic acids are deoxyribonucleic acid (DNA) and ribonucleic acid (RNA). Hydrogen bonds between pairs of nucleotides join two DNA chains throughout their length to form a double-stranded helix. The DNA sequence contains a deoxyribose sugar (pentose sugar), a phosphate group, and a purine or pyrimidine base. The sugar and phosphate moieties, which are covalently linked in tandem by asymmetrical 5′–3′ phosphodiester bonds, form the DNA backbone. Carbons in deoxyribose sugar are conventionally numbered from 1 to 5, and the phosphate groups are linked to carbons 3 and 5. The DNA molecule is polarized by having a phosphoryl radical (–P) on the 5′ carbon of the terminal nucleotide called the 5′ (five prime) end and a free hydroxyl radical (–OH) on the 3′ carbon of the terminal nucleotide called the 3′ (three prime) end. The direction of the phosphodiester bond is conventionally written from the 5′end (upstream) to the 3′end (downstream) to locate a particular nucleotide in DNA sequences easily. A purine base always pairs with a pyrimidine base through the formation of hydrogen bonds. Adenine pairs with thymine with two hydrogen bonds (weak bond), while guanine pairs with cytosine with three hydrogen bonds (strong bond). The complementary pairing of two nucleotides in a double helix is indicated by the colon between bases (G:C and A:T), so-called canonical base pairs. RNA is another type of nucleic acid and differs from DNA by having a ribose instead of deoxyribose as its backbone sugar moiety. Also, the uracil (U) base is used in place of thymine. The length of a single-stranded nucleic sequence is measured in base pairs (bp) based on the number of nucleotides. When the single-stranded nucleic sequence contains thousands and millions of base pairs, they are indicated as kilobases (Kb) and megabases (Mb).

4.3 Genomes

In recent years, evolutionary studies at the whole genome level have become possible, such as comparing the genomes of different species to investigate which homologous genes are shared and separated by a speciation event. Most genomes contain many open reading frames that are supposed to be genes. Detecting similarities between sequences from different species may be challenging because of the large amount of evolutionary change. Sequences that are from different species can be referred to as orthologous sequences. Gene families, which are duplicated genes, are contained in the genome. These duplicated genes often have slightly different functions or expression sites within the organisms. Gene duplication events result in paralogous sequences, causing sequence divergences. Also, gene duplications can occur independently in different lineages, resulting in a single gene (in one species) homologous to all genes in the gene family (in the other species).

In 1950, Barbara McClintock discovered and described transposable elements (TEs) in genomic DNA and suggested that TEs can influence genome evolution (Piskurek & Jackson, 2012). TEs are repetitive genomic sequences that can move from one chromosomal location to another and can be subdivided into two major groups according to their mobilization mechanism. In the first group, TEs move by reverse transcription of an RNA intermediate termed retrotransposons or class I elements. In the second group, transposition occurs via a DNA intermediate, termed *DNA transposons* or class II elements (Bowen & Jordan, 2002). TEs can also replicate themselves to enhance the dynamics between elements and their host for evolutionary success. This replication strategy to enhance their success has given TEs the term “selfish sequences” or “genomic parasites.” To date, the number of parasite genomes is continually increasing, along with the number of identified TEs

(Thomas et al., 2010). Along with molecular advances, bioinformatic analyses can establish sets of homologous genes between species groups, and we can understand how those genes are allocated in their positions (Higgs & Attwood, 2005).

4.3.1 Parasitic Helminth Genome

In the past 6 years, the number of published genomes of helminthic parasites has increased dramatically, and research interest in functional helminth genomics is quickly gaining speed. For example, studies investigate how gene families modulate host immune responses, how parasites successfully invade the host, specific lineages in core metabolism, and protein families historically targeted for drug development (Consortium, 2019). Obtaining genome sequences from parasitic nematodes (roundworms) and platyhelminths (flatworms) is important, as the parasites impede economic development through human disability and economic loss from the livestock and crop industries (Charlier et al., 2014; Jones et al., 2013). Recently, the International Helminth Genomes Consortium (2019) conducted whole-genome analysis by combining 36 published genomes with new assemblies for 31 nematodes and 14 platyhelminth species into a large genome comparison of parasitic and nonparasitic worms. The results showed genomic diversity in both phyla and a relatively slight variation in gene set completeness (coefficient of variation, c. v. = 0.15) among the nematodes and Platyhelminthes, despite variation in assembly contiguity (c.v. = 8.5). Genome sizes within each phylum vary greatly, ranging from 42 to 700 Mb in nematodes and 104 to 1259 Mb in Platyhelminthes. The genome size variation is attributed mainly to noncoding elements, including repetitive sequences and non-repetitive DNA. Major factors driving genome size variation in increasing order are (1) long terminal repeat transposons, (2) simple repeats, (3) assembly quality, (4) DNA transposable elements, (5) total length of introns, and (6) low complexity sequences (Consortium, 2019).

4.4 Genes and Gene Structure

Traditionally, a gene is a region of DNA that contains the necessary information to make RNAs and protein molecules. Recent molecular findings, however, have altered our perception of genes. A gene is a sequence of genomic DNA or RNA essential for a specific function. According to its functions, the nucleotide sequences may not be required to undergo DNA translation or transcription (Li & Graur, 1999). Three types of genes are recognized: protein-coding, RNA-specifying, and untranscribed genes. The protein-coding genes, essential for a specific function, are transcribed from RNA. The RNA-specifying genes are also essential for a specific function, but they are transcribed and do not undergo translation. For example, many nuclear transfer RNA (tRNA)-specifying genes possess very short introns. All eukaryotic tRNA-specific genes contain an internal transcriptional start signal recognized by RNA polymerase III that transcribes essential noncoding RNAs, including the entire pool of tRNAs, and the 5S ribosomal RNA other small RNAs (Abascal-Palacios et al., 2018). Untranscribed genes, which do not undergo transcription, are likely independent of structural genes. Although several types have been tentatively identified, what we know of untranscribed genes is less than the other types of genes.

For protein-coding genes, protein synthesis processes include transcription, RNA processing, translation, and protein synthesis. The synthesis of proteins forms the foundation for the phenotypic characters of organisms. Firstly, transcription is RNA synthesis using DNA as a template and is carried out by the enzyme RNA polymerase. Transcription of single gene sequences is typically a few thousand base pairs in length and occurs in the nucleus. The double-stranded DNA helix is separated by a helicase, allowing RNA polymerase to temporarily bind to the promoter region of one of the two strands of DNA. The strand that is transcribed is the template strand. Transcription then proceeds by RNA synthesis from the 5′end to 3′ end. Transcription ends when RNA polymerase reaches the terminator region, where the promoter and terminator regions serve as signals to start and end the transcription process. The promoter (5′ flanking) and terminator (3′ flanking) regions are untranscribed regions of the gene, aiding gene regulation through determining the initiation, tempo, timing, and tissue specificity of the transcription process. In eukaryotes, the promoter region consists of the following signals: (1) TATA box located 19–27 bp upstream of the transcription start point, (2) CAAT box further upstream, and (3) one or more copies of GC box surrounding the CAAT box. Generally, the CAAT and GC boxes control the initial binding of the RNA polymerase, while the TATA box controls the start point of transcription for reading and decoding. The 3′ flanking region contains signals for the termination of the transcription process.

RNA processing occurs after the precursor messenger RNA (pre-mRNA) is obtained from the transcription process. Different pre-mRNAs are produced at any time in the cell, which differs significantly from each other in length and molecular weight. The pre-mRNA must be processed in several ways to produce a functional mRNA by making modifications at the 5′ and 3′ ends. The 5′ end is modified by adding guanine (G) and a protein complex (cap-binding complex), while the 3′ end is modified by adding 60–200 “adenine” (A) as the poly-A tail using *poly-A binding proteins*. The modified ends of the mRNA protect the mRNA from being quickly destroyed into individual nucleotides by nucleases. After modification of the 3′ and 5′ ends of the mRNA, the pre-mRNA sequences undergo splicing by the spliceosome enzyme. Introns, which are not needed sequences, are spliced out, leaving only pieces of exons containing protein-coding information for translation downstream. The fully processed mRNA is then transported from the nucleus through nucleopores in the nuclear membrane to the cytoplasm.

4.4.1 Genetic Codes

Protein synthesis occurs after transcription, using mRNA as a template. The mRNA is translated into primary amino acids using tRNA mediators. The tRNA is a small molecule, usually 70–90 nucleotides in length. Each amino acid has at least one type of tRNA assigned to it. Twenty specific enzymes called *aminoacyl-tRNA synthetases* couple each amino acid to the 3′ end of specific tRNA. Translation involves the sequential recognition of adjacent nonoverlapping triplets of nucleotides, called a codon, and a complementary sequence of those three nucleotides called an anticodon in the tRNA. Each tRNA has different anticodons and is different from one another in many other parts of the sequence. However, they all have the same cloverleaf secondary structure.

The genetic code is a form of sequences obtained from the four nucleotide bases (adenine, guanine, thymine, cytosine) that becomes information in sequences for 20 types of amino acids. Each amino acid is represented by groups of three bases, called codons, in the mRNA sequence. A total of 64 (4^3) codons can be made with the four nucleotide bases. Most amino acids have more than one codon that codes for them as the number of codons (64 codons) is greater than the number of amino acids (20) required. In the universal genetic code, 61 codons code for specific amino acids (sense codons), while the remaining three are stop codons (UAA, UAG, and UGA). Stop codons serve as signals to terminate the translation process. Standard genetic codes are used in most prokaryotic genomes and the nuclear genomes of most eukaryotes. However, several different genetic codes are reassigned in mitochondrial genomes and differ from standard ones. These include AUA as methionine, UGA as tryptophan, AGA, and AGG as a stop codon.

4.5 Mutation

DNA sequences are typically copied precisely during DNA replication, which is a semiconservative process. The semiconservative process indicates that the original double strand is replicated to give two double strands, where each strand contains one of the original strands and a newly synthesized complementary strand (by DNA polymerase III). Mutations can occur from errors in either DNA replication or repair, serving as a source for genetic variation and novel sequences for evolution. Moreover, mutagens, radiation, and chemicals can damage DNA, causing mutations.

Mutations occur in either somatic or germline cells and can be passed on to offspring. In eukaryotic organisms, somatic mutations are typically disregarded in an evolutionary context and are denoted by this term in germ cells. Typically, mutation refers to a permanently heritable change in the nucleotide sequence of a gene or a chromosome. The simplest mutation type is a point mutation, where another nucleotide replaces one nucleotide base. Several adjacent nucleotides can also be affected by insertions or deletions to one or a few nucleotides, resulting in a frameshift mutation. The chromosome structure can also be altered due to chromosomal mutation. Mutations in genes can also occur, but their occurrence is comparatively lower than chromosomal mutations. However, chromosomal mutations can sometimes be lethal.

4.5.1 Point Mutation (Substitution)

Point mutations occur by the mispairing of bases during DNA replication, resulting in nucleotide substitution (one nucleotide is replaced for another). Each DNA chain is a linear polynucleotide consisting of four kinds of nucleotides. The nucleotides are adenine (A) and guanine (G)—purines—and thymine (T) and cytosine (C)—pyrimidines. A transition occurs when a purine base is substituted for another purine base (or pyrimidine to another pyrimidine base), while a transversion occurs when a purine is substituted for a pyrimidine base (or vice versa).

When point mutations occur in protein-coding regions, they are classified according to the translation product. A synonymous or silent mutation results if the mutation causes no change in the amino acid being coded. On the other hand, if the mutation causes an amino acid change, it is a non-synonymous or missense mutation. Amino acid change due to non-synonymous nucleotides is called *a* replacement. This is called a nonsense mutation when the codon is changed to a stop codon due to a mutation causing termination of DNA translation results. The premature ending of the translation process leads to a truncated protein being produced. Synonymous mutation occurs mainly at the third nucleotide position of the codon (70% of all possible nucleotide changes). In comparison, all nucleotide changes and 96% of changes occur at the second and first positions of the codon, respectively, resulting in a non-synonymous mutation.

For animal mitochondrial genomes, the transition to transversion ratio is about 15 to 20. Also, some nucleotides are more mutable than others; for example, G and C tend to mutate more frequently than A and T in the nuclear DNA of mammals. Mutations are argued to occur either randomly or nonrandomly regarding their effect on the fitness of organisms (Baer et al., 2007; Cheverud, 1984; Hall, 1990). The relative frequencies of different types of nucleotide substitutions causing mutations in a random protein-coding sequence are shown in Fig. 4.1.

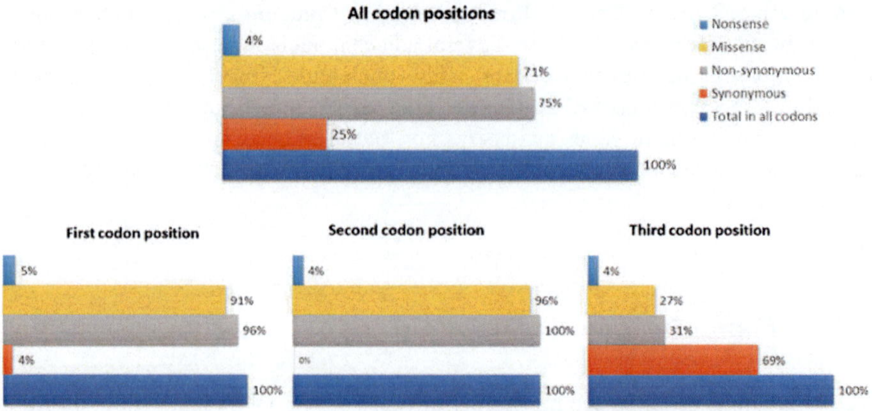

Fig. 4.1 Relative frequencies of different mutational substitutions in the random nuclear protein-coding sequences. The universal genetic code has 61 sense codons, and each of the sense codons can mutate into nine other codons to be a single nucleotide change. Therefore, there are 549 ($61 \times 9 = 549$) total possible substitution mutations of all codon positions, 183, 183, and 183 in first, second, and third codon positions, respectively (Li & Graur, 1999)

4.5.2 *Frameshift Mutations*

There are three types of frameshift mutations: deletion, insertion, and duplication. Deletion is the removal of nucleotide(s), while insertion is the addition of nucleotide (s), and both processes change the number of base pairs of the gene. Insertion and deletion are collectively called indel. For duplication, nucleotides are copied once or several times. Frameshift mutations alter the gene's open reading frame, which changes the initial amino acid sequence.

4.5.3 *Chromosomal Mutation*

Chromosomal mutations are alterations in the chromosome structure or chromosome number. They are caused by errors in crossing over during meiosis, altering chromosomal segments. There are four types of chromosomal structure alteration, including intrachromosomal deletion, gene duplication, inversion, and translocation. Intrachromosomal deletions are removals of chromosome segments, while gene duplication occurs when there are increasing copies of a particular allele due to duplication. For inversion, the orientation of the chromosome segment is changed. Inversion involves changes to very long stretches of DNA that are around hundreds or thousands of nucleotides in length. Translocation is the interchange of chromosomes between nonhomologous chromosomes by breaking one chromosome segment and fusing it to another chromosome.

4.6 Recombination

Recombination refers to the exchange of DNA strands, producing a new nucleotide rearrangement. Recombination differs from a mutation, where a permanent and heritable change in the nucleotide sequence of a gene or chromosome results. Recombination creates new sequence variants that are not seen when naturally preparing gametes. Recombination brings large-scale rearrangements to the genomes of organisms, leading to evolution, whereas mutation produces new alleles, introducing genetic variability to a particular population. Two types of homologous recombination can occur: crossing over (reciprocal recombination) and gene conversion (nonreciprocal recombination). Reciprocal recombination involves the exchange of homologous sequences between homologous chromosomes, producing new combinations of adjacent sequences and retaining both variants simultaneously.

On the other hand, nonreciprocal recombination involves the uneven replacement of one sequence by another. Nonreciprocal recombination results in the loss of one of the variant sequences involved in the recombination event. Both homologous recombination types involve a molecular intermediate called the Holliday structure, which forms stretches of mismatched double-stranded DNA called heteroduplex. Mismatches in the heteroduplex are recognized and removed by cellular enzymes. Two double-stranded DNA molecules then become separated into four strands to exchange segments of genetic information. The complementary strand is used as a template to repair the heteroduplex to fill up the gap made by DNA polymerase.

4.7 Dynamics of Genes in Populations

The evolutionary process causes a change in gene frequencies with time in populations. Population genetics deals with genetic changes within populations to understand molecular evolution and determine how a mutant allele‘s frequency changes with time under various evolutionary forces. However, in terms of the organism's phenotypes and fitness, molecular changes have only a tiny effect on them. A new mutant variant can completely replace a previous one in the population at the molecular level. The sequence variants at the same locus (allele) can be measured as allele frequencies. If sequence variations are available in many different individuals in the population, more than one allele is present in a polymorphic locus or polymorphism. A single copy of each locus in the genome is called a haploid, for example, in the most prokaryotic and mitochondrial genomes, whereas two copies of each locus from paired chromosomes (diploid) are generally found in many eukaryotic organisms. The two sequences at a given locus in an individual are copies of the same homozygous allele, while different alleles are heterozygous.

4.7.1 Changing Allele Frequencies

More than one allele may be present in a population at a locus. For example, allele frequencies can be measured, assuming the haploid population size is N individuals and alleles A_1 and A_2 are present at a particular locus. The number of allele copies for allele A_1 is n_1, while allele A_2 in n_2. Then, the allele frequencies are n_1/N and n_2/N for A_1 and A_2, respectively. $n_1 + n_2$ is equal to be N, and $n_1/N + n_2/N$ is equal to be 1. The set of all alleles at all loci existing in a population is called the gene pool.

A new mutation becomes significant in a population by increasing the allele frequencies and fixing them. If new mutant alleles are increasing at a low frequency, such mutations will have a negligible effect on evolutionary history. Other significant factors affecting population allele frequencies are natural selection and random genetic drift. Natural selection has been considered the driving force in evolution, while random genetic drift plays an essential role in evolution at the molecular level.

4.7.2 Natural Selection

The principle of natural selection states that mutant alleles with higher fitness will increase in number relative to those with lower fitness in a population. This mechanism is defined as the differential survival and reproduction of individuals due to differences in phenotype, which are genetically distinct within a population. Differential reproduction is caused by differences among individuals, such as mortality, fertility, fecundity, mating success, and the viability of the offspring. Thus, genetic variations among individuals are related to reproductive success. These factors for selection may lead to changes in allele frequencies over time. Most mutations are deleterious *mutations,* where fitness is reduced compared to the original when mutations are introduced into a functioning gene sequence. Deleterious mutations are usually rare in a large population because of natural selection acting against them. Also, other mutations may only have a negligible effect on the function of a gene that has no apparent consequences for the individual. However, minor selective effects are significant on the evolutionary time scale. Natural selection tends to keep the frequencies of slightly deleterious mutations at low levels over many generations, and this is also known as a negative or *purifying selection*. In some instances, a new mutation arises that increases some individuals' fitness, and the mutation may be as fit as the best allele in a population. Such a mutation is called positive or advantageous selection.

4.7.3 Random Genetic Drift

As mentioned above, natural selection is not the only factor that can cause changes in allele frequencies. Changes can also occur by chance without any direction (positive or negative). Random genetic drift is another factor that brings about random changes in allele frequencies affected by population sizes (Gillespie, 1991). The allele frequencies change from generation to generation by random direction at any point in time. The critical factor in producing random fluctuations in allele frequencies is the random sampling of gametes in the process of reproduction. However, only a minute fraction of gametes succeed in developing into adults. The alleles (existing genetic variants) in the offspring are a sample of those in the parents with a chance for survival or reproduction. Genetic variation is thus reduced and causes an initially increasing number of rare alleles to be fixed in the population. Allele frequency in the random genetic drift can reach either 0 or 1, from either loss or fixation of the allele in a population. Random genetic drift will occur unless there is a constant input of alleles into the population by mutation, migration, or polymorphism actively maintained by a balancing selection.

4.7.4 Effective Population Size

Population size is an essential parameter in population biology, and the census population size, *N*, is defined as the total number of individuals in a population. However, not all individuals take part in reproduction. The population size in the evolutionary process is different from the census size. In population genetics, we are concerned with the effective population size, denoted by N_e, as the number of individuals in a population contributing offspring to the next generation. Population size can be measured by simply counting the number of adults in a locality. In a population of size *N*, there will be 2 *N* genes. The interpretation of *N* for the theoretical equation is that *N* has been correctly measured when the chance of drawing two copies of the same gene is $(1/2\ N)^2$.

Effective population size is susceptible to differences in the number of reproductively active females versus males. When there are different numbers of each sex in a population, N_e's effective population size can be estimated as N_e *is* $4N_mN_f/(N_m + N_f)$, where N_m is the number of males and N_f is the number of females. For example, if there is 1 male and 9 females, $N_e = 3.6$, if there is 1 male and 1000 females, $N_e = 4$. These examples show an imbalanced sex ratio, reducing the effective population size.

4.7.5 Gene Substitution

Gene substitution is when a mutant allele completely replaces the predominant or wild-type allele in a population. In this process, a mutant allele arises in a single copy in a population and becomes fixed after a certain number of generations. However, not all mutants reach fixation because most of them are lost after a few generations. The probability (fixation probability) affects the chance that all the alleles will eventually be the ancestor within a population at that locus. The time it takes for a new allele to become fixed is called the fixation time. In the next paragraph, we shall identify the factors that affect the speed with which a new mutant allele replaces an old one in a population.

Gene substitutions occur in succession, with one allele replacing another and replacing it in time with a newer allele called the fixation time of gene substitution (the number of fixations of new alleles per unit time). If neutral mutations occur at a rate of u per gene per generation, then the mutant numbers arising at a locus in a diploid population of size N is $2Nu$ per generation. The fixation probability for each of these mutations is $1/(2\ N)$. The substitution rate of neutral alleles is obtained by multiplying the total number of mutations by the probability of their fixation, estimated by $K = 2Nu(1/2\ N) = u$. Thus, the rate of substitution is equal to the mutation rate. This result can be intuitively understood by observing a large population. The number of mutations that arise every generation is high, but each mutation's fixation probability is low. In contrast, the number of mutations that arise in every generation is low in a small population. For neutral mutations, the substitution rate is independent of the population size.

4.7.6 Genetic Polymorphism

A population is monomorphic at a locus only if one allele is at the locus. When two or more alleles coexist in the population, they are polymorphic. However, a locus is commonly defined as polymorphic only if the frequency of the most common allele is less than 99%. If one of the alleles has a very high frequency (more than 99%), it is impossible to observe the other alleles in a sample unless the sample size is enormous.

The simplest way to measure the extent of polymorphism in a population is to compute the average proportion of polymorphic loci (P) by dividing the number of polymorphic loci by the total number of sampled loci. For example, if 5 of the 20 loci are polymorphic, then $P = 5/20 = 0.25$. However, this measurement is dependent on the number of individuals studied. The smaller the sample size, the more difficult it is to identify polymorphic loci.

A more appropriate measure of polymorphism in a population is the average number of nucleotide differences per site between any two randomly chosen DNA sequences. This measure is called nucleotide diversity and is denoted by Π (pi). It

	Number of sequences	1	2	3	4	5	6	7	8	9	10	11	12	13	14	15	16	17	18	19	20	21	22	23	24	25	26	27	28	29	30	Frquency (x)
Sequence 1	2	A	G	C	A	G	C	T	A	A	A	G	C	A	G	C	A	G	G	T	G	C	T	G	C	A	G	C	A	G	T	2/10 =0.2
Sequence 2	5	A	G	C	C	G	C	T	A	A	A	G	C	A	C	C	A	G	G	T	G	C	T	G	C	A	G	C	A	G	T	5/10 =0.5
Sequence 3	2	A	G	C	C	G	C	T	A	A	T	G	C	A	G	C	A	G	G	T	G	C	G	G	C	A	G	A	A	G	T	2/10 =0.2
Sequence 4	1	A	G	C	T	G	C	T	A	A	A	G	C	A	T	C	A	G	G	T	G	C	A	G	C	A	G	G	A	G	T	1/10 =0.1

Fig. 4.2 Nucleotide sequences of population. The 30-nucleotide position of 10 sequence numbers shows five different nucleotide positions in shaded boxes. The nucleotide frequency of each sequence is calculated in the last column. Herein we have: $\pi_{12} = 2/30$, $\pi_{13} = 4/30$, $\pi_{14} = 4/30$, $\pi_{23} = 4/30$, $\pi_{24} = 4/30$, and $\pi_{34} = 5/30$

can be calculated by $n/(n - 1)\Sigma\, x_i x_j \pi_{ij}$ where x_i and x_j are the frequencies of the *i*th and *j*th types of DNA sequences, respectively. π_{ij} is the proportion of different nucleotides between the *i*th and *j*th types. For example, considering the sequences in Fig. 4.2, nuclcotide diversity is 0.042.

4.8 Nucleotide Substitution Models

In the DNA sequence, one nucleotide is substituted for another during a time period in the evolutionary process. Changes in nucleotide sequences are used in molecular evolutionary studies to estimate the evolutionary rate and reconstruct the evolutionary history of organisms. To detect evolutionary changes in DNA sequences, two different sequences are needed for comparison to study the genetic relationship between the common ancestral and descendent sequences by using statistical methods. The nucleotide substitution process is usually prolonged, as a series of nucleotide substitutions in new alleles arise as a single individual mutation increases their allele frequencies and ultimately becomes fixed in the population. The nucleotide substitution model of evolution describes the rates of change of fixed mutations among sequences and constitutes the basis of evolutionary analysis of genetic data at the molecular level.

The first substitution model corrected the effects of multiple nucleotide substitutions, such as Jukes and Cantor's one-parameter (JC) model and Kimura's two-parameter (K2P) model (Jukes & Cantor, 1969; Kimura, 1980). However, they are not applied in evolutionary analysis until other substitution models are generated to reduce bias in phylogenetic inferences (Lemmon & Moriarty, 2004; Minin et al., 2003; Posada & Crandall, 2001). The selection of the best-fit substitution model is a crucial stage in the pipeline of phylogenetic inferences (Posada & Crandall, 2001). Recently, substitution models have been routinely used in diverse areas of evolutionary biology. Many substitution models exist, and new models emerge to mimic the evolution of complex real datasets (Keane et al., 2006). Trends of substitution models are divided into DNA substitution, codon substitution, and amino acid substitution models.

Firstly, the simplest model to mimic the DNA substitution process is described by Jukes and Cantor (JC, 1969). This model considers one rate of change between all nucleotides and equal nucleotide frequencies. However, changes between bases with an equal chemical nature (transition) are more common than those with different chemical structures (transversion). In 1980, Kimura motivated the substitution model as a two-parameter model (K80), where the rate of change differs between transitions and transversions (Kimura, 1980). In 1981, Felsenstein (F81) extended the JC model to include different nucleotide frequencies, which can also appear due to the physicochemical properties of nucleotides and the operation of natural selection. Many models were later developed by incorporating extensions to those original models, for example, the Hasegawa-Kishino-Yano (HKY) model (Hasegawa et al., 1985). The most complex model is the general time-reversible model (GTR) (Tavaré, 1986), which incorporates different rates for every change and different nucleotide frequencies. Additionally, a proportion of invariable sites (+I) (Shoemaker & Fitch, 1989) and the rate of variation across sites (+G) (Yang, 1994) can be incorporated into any model. The GTR+G + I model usually fits real data better than other alternative models (Sumner et al., 2012), suggesting the complexity of evolutionary processes. The GTR model presents undesirable mathematical properties for its application as a Markov model in phylogenetics (Gatto et al., 2006; Sumner et al., 2012). Selecting the most complex substitution model is essential to improve the fit to mimic the actual nucleotide substitutions occurring (Arenas, 2015; Jayaswal et al., 2011). Methods to select the best-fit substitution model for reconstructing the phylogenetic relationship will be mentioned in Chap. 8.

Secondly, the codon substitution models consider that sites within codons evolve at different rates and should not be equally treated (Shapiro et al., 2006). Evolutionary inferences based on the modeling of codon evolution are more robust than those derived from empirical amino acid models (Arenas, 2015; Seo & Kishino, 2008). Mutations at the codon level can be classified as synonymous (silent mutation) and non-synonymous (amino-acid replaced), which provides a measure of selective pressure at the molecular level (molecular evolution). The first codon substitution models included the non-synonymous and synonymous substitution rates, *dN* and *dS*, respectively (Muse & Gaut, 1994), or *dN/dS* ratio (Goldman & Yang, 1994). *dN/dS* ratio should be estimated from samples of different populations (Kryazhimskiy & Plotkin, 2008; Pellissier, 2015) and is commonly estimated in evolutionary biology for testing hypotheses related to selective pressure (Arenas, 2015; Yang & Bielawski, 2000). A *dN/dS* larger than 1 indicates that substitutions in the protein-coding gene were enriched for altered amino acid states, suggesting diversifying (positive) selection. On the other hand, a *dN/dS* less than or equal to 1 can be interpreted as purifying (negative) selection and neutral evolution, respectively.

The emergence of mutation-selection models that attempt to integrate complex selection patterns into the mutation process is promising and outperforms simple neutral models (Lawrie et al., 2011). The interplay of mutation biases and weak selection can be complex, where constrained sites can evolve faster than neutral sites (Lawrie et al., 2011; McVean & Charlesworth, 1999). Considering these effects in substitution models, identifying functional regions and fitting them to the data can be improved through mutation-selection models (Lawrie et al., 2011).

Lastly, amino acid substitution models intend to mimic the evolution of protein data, which is crucial for testing various hypotheses such as selection toward novel proteins, the rate of protein evolution, the role of protein function on protein evolution, and evolutionary aspects of protein-protein interaction networks (Arenas, 2015; Liberles et al., 2012; Qian et al., 2011; Shapiro et al., 2006). Substitution models of amino acid evolution are classified into two major groups: (1) empirical models based on large protein databases for their implementation in phylogenetic software and (2) parametric models based on parameters described protein evolution (Anisimova, 2012; Halabi et al., 2021; Kosiol et al., 2007; Liò & Goldman, 1998; Shapiro et al., 2006). An empirical amino acid substitution model consists of a 20×20 matrix of exchangeability rates and 20 amino acids, while parametric substitution models are based on protein folding constraints. However, these models are not well-established in the evolutionary analysis of protein data because the commonly used evolutionary frameworks implement likelihood functions that are not site-dependent (Arenas, 2015; Buckley et al., 2001; Cankara, 2020).

4.9 Codon Usage Bias across Species in Parasitic Helminths

Codon usage bias is the unequal usage of the synonymous codon, which varies widely between species and between different regions of a genome within a single species (Bernardi, 1993; Vicario et al., 2007). Factors influencing codon usage include selection for translational accuracy and efficiency (Salim & Cavalcanti, 2008) and GC bias at a genomic level in prokaryotes and at an isochoric or regional level in vertebrates (Ermolaev, 2001; Palidwor et al., 2010; Xia et al., 2010). The GC bias is a major influence on codon bias in both eukaryotic (Nabiyouni et al., 2013; Sharp et al., 2010) and prokaryotic genomes (Lopez & Philippe, 2001; Qin et al., 2004), resulting in a close association between GC% at the third codon position called GC3 (Bernardi et al., 1997) and GC bias (genomic GC% in prokaryotes or isochoric GC% in mammals). All amino acids allow GC-changing synonymous substitutions in the third position. The substitutions have led to the use of synonymous G/C-ending codons, which should increase in frequency with increasing GC bias, while usage of A/T-ending codons should decrease (Palidwor et al., 2010; Xia et al., 2010).

Lamolle et al. (2019) performed a comparative genome-wide analysis of 22 species representative of the main clades and lifestyles of the phylum Platyhelminthes. Their study selected a set of 700 orthologous genes conserved in all species for measuring changes in GC content, codon, and amino acid usage in orthologous positions. They found that the GC values of the third codon position spanned over a wide range, and the codon usage bias values can discriminate between two distinctive clusters within freshwater turbellarians, cestodes, and trematodes. Figure 4.3 shows that the hierarchical clustering of codon usage data differs from the

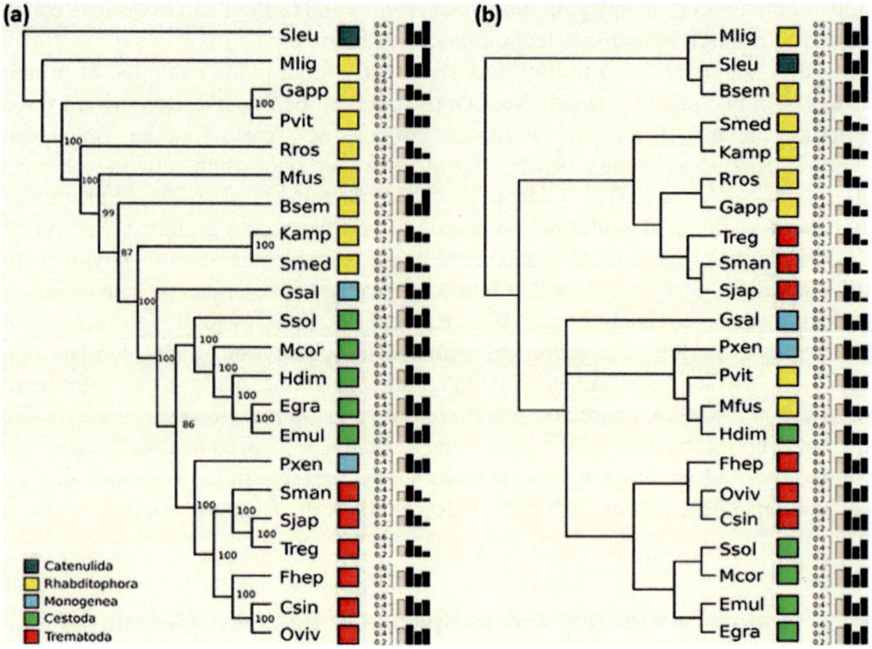

Fig. 4.3 GC composition variation across Platyhelminthes. (**a**) Maximum likelihood phylogenetic tree constructed from a set of 700 orthologous genes with SH-like branch support values for each node. (**b**) Hierarchical clustering based on the GC contents of relatively synonymous codons (RSCU). Histograms reflect overall GC content in all orthologous (gray bar) and codon positions (black bars). The colored boxes represent the classes of parasites: Sleu (*Stenostomum leucops*), Mlig (*Macrostomum lignano*), Gapp (*Geocentrophora applanata*), Pvit (*Prostheceraeus vittatus*), Rros (*Rhynchomesostoma rostratum*), Kamp (*Kronborgia amphipodicola*), Mfus (*Monocelis fusca*), Bsem (*Botrhioplana semperi*), Smed (*Schmidtea mediterranea*), Gsal (*Gyrodactilus salaris*), Pxen (*Protopolystoma xenopodis*), Sman (*Schistosoma mansoni*), Sjap (*Schistosoma japonicum*), Treg (*Trichobilharzia regenti*), Fhep (*Fasciola hepatica*), Csin (*Clonorchis sinensis*), Oviv (*Opistorchis viverrini*), Ssol (*Schistocephalus solidus*), Mcor (*Mesocestoides corti*), Hdim (*Hymenolepis diminuta*), Egra (*Echinococcus granulosus*), Emul (*Echinococcus multilocularis*) (Lamolle et al., 2019)

phylogenetic relationships constructed from a set of 700 orthologous genes. They detected a synonymous codon usage bias that was more extreme, containing GC-poor or GC-rich genomes. These biases impact the preferred amino acids encoded by codons biased for GC content and are associated with non-synonymous substitutions at orthologous positions.

Codon changes between *Schistosoma mansoni* and *Fasciola hepatica* as the representative species were analyzed for comparison, and the results indicated different codon frequencies at the conserved codons, with both synonymous and non-synonymous substitutions (Lamolle et al., 2019) (Fig. 4.4). No clear correlation of GC bias between parasitic lifestyles or evolutionary closeness was observed, and the results suggested that synonymous and non-synonymous changes provide

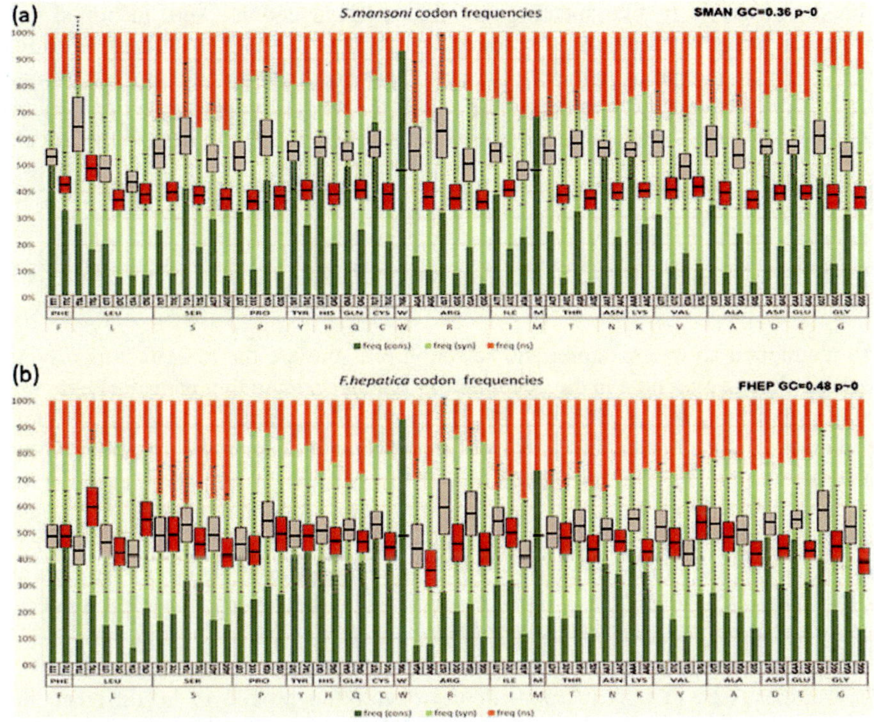

Fig. 4.4 Codon changes between *S. mansoni* (**a**) and *F. hepatica* (**b**). The substitution state for each codon in pairwise alignment is represented in bars: the dark green bar (conserved codons), the pale green bar (synonymous substitutions), and the orange bar (non-synonymous changes) (Lamolle et al., 2019)

evidence that mutation and selection forces drive the diverse coding strategies for flatworms.

Mitreva et al. (2006) studied codon usage within the diverse phylum Nematoda by analyzing 265,494 expressed sequence tags (ESTs) from 30 nematode species, the complete genomes of *Caenorhabditis elegans* and *C. briggsae*. A total of 25, 871,325 codons were analyzed, generating a comprehensive codon usage table. The codon usage similarity in Nematoda persists over a wide range of genera but is rapidly reduced within each clade. Nematodes in clade IV, comprising *Globodera*, *Meloidogyne*, *Pristionchus*, and *Strongyloides*, have the most highly derived codon usage patterns. The major factor affecting differences in codon usage between species is the coding sequence GC content, which varies from 32% to 51% in nematodes. The coding GC content, particularly at the third position in the codon, can explain much of the observed variation in the effective number of codons ($R = 0.70$) and measure codon bias. Codon usage is also affected by neighboring nucleotides. The coding GC content at the third position (GC3) correlates strongly with the estimated noncoding genomic GC content ($R = 0.92$). On examining abundant clusters in five species, candidate optimal codons were identified that might be performed in highly expressed transcripts. The GC content is indicative of the observed variation in the effective number of codons, codon bias measure, and amino acid frequency differences. The results are suggestive that directional mutation pressure drives codon usage in the examined nematode genomes (Mitreva et al., 2006).

4.10 Continuity of Micro- and Macroevolution

As mentioned above, mutation and selection pressure are the main factors driving genetic change over time in the genomes of organisms, including parasitic helminths (Gilleard, 2006; Wit & Gilleard, 2017). Two evolutionary processes in different evolutionary time scales are used to observe genetic change: microevolution and macroevolution. Firstly, microevolution is the process of evolution as observed through direct observation and experiment in only a few generations. In contrast, macroevolution is the process of evolution as observed through paleontology, archaeology, and history and involves tens or many more generations (Erwin, 2000; Simons, 2002). These more dramatic evolution events (speciation) occur rarely and slowly and are not directly observable in the present.

There have been debates over the process of micro- and macroevolution. Simpson (1944) discussed that if micro- and macroevolution proved to be different, the numerous studies of microevolution would become relatively unimportant and would have little value in the study of whole evolution (Simpson, 1944). Later, Folguera and Lombardi (2012) suggested new perspectives, showing the dependence of macroevolution and microevolution through different strategies. Another persistent debate in evolutionary biology is over the continuity of microevolution and macroevolution. Opponents of these processes argue that the time scales for micro- and macroevolution are disconnected, and thus the two processes should not be inferred together (Folguera & Lombardi, 2012).

Simons (2002) argued that the paradox inferred by opponents could be eliminated by a hierarchical application of the geometric-mean fitness principle, which is invoked only within the limited context of microevolution in response to environmental variance. The geometric-mean fitness principle implies that well-adapted genotypes are eliminated. On the premise of the temporal structure of environmental variance, selective extinction, and clade-level heritability, the evolutionary outcome of major environmental change may be viewed as identical in principle to the outcome of minor environmental fluctuations over the short term. Trend reversals are thus recognized as a fundamental property of selection operating at any phylogenetic level that occurs in response to event severities of any magnitude over all time scales.

A long-term genotype-level strategy maximizing lineage geometric-mean fitness in stochastic environments is called bet-hedging. This strategy decreases individual fitness variance, despite also lowering arithmetic mean fitness, finally leading to low variation in fitness (Haaland et al., 2019). The bet-hedging perspective occurs when organisms suffer decreased fitness in their typical conditions in exchange for increased fitness in stressful conditions (Simons, 2002). Selection for the bet-hedging strategy is not considered the norm of how natural selection operates. The principle of bet-hedging should be considered relevant to a broad range of micro-evolutionary studies and may also be applied hierarchically to the macroevolutionary time scales and all phylogenetic levels. In this perspective, environmental variance is treated as a potentially continuous variable, spanning from minor fluctuations to disastrous events, where event frequency and severity are inversely related (Heininger, 2015; Simons, 2002).

4.11 Genome Evolution Under Parasitic Adaptation

As mentioned in Chap. 3, the genomes of parasites are affected by environmental changes, which in turn drive the evolutionary process. The enigma of genotypic-phenotypic diversity and the biodiversity evolution of genes and genomes has been discussed since 1975 by the Institute of Evolution, University of Haifa. They explored the following questions: (1) How much of the genomic and phenotypic diversity in nature is adaptive and processed by natural selection? and (2) what is the origin and evolution of adaptation and speciation processes under spatiotemporal variables and stressful macrogeographic and microgeographic environments? The environmental stresses analyzed are thermal, chemical, climatic, and biotic and were studied in the field and under critical laboratory conditions. Nevo (2001) attempted a holistic, integrative analysis of the organism-environment interaction in the evolutionary adaptation and speciation processes. They found genotypic and phenotypic diversity in all species at the protein and DNA levels. The results indicated that genome-phenome organization is nonrandom, heavily structured, and correlated with abiotic and environmental diversity and stress. Natural selection thus mainly drives biodiversity and ultimately overrides mutation, migration, and stochasticity (Nevo, 2001).

In parasites, many early views on the evolution of parasite genomes seem to be reflected in their widespread diversity. The loss of morphological structures is expected to be paralleled by a loss of genetic diversity (Poulin & Randhawa, 2015). As mentioned previously, the apparent loss of morphological structure seen in many parasite lineages has been compensated for by new structures or adaptations tailored to a parasitic existence (Brooks & McLennan, 1993). Therefore, general morphological evolution should not be viewed blindly as a signature of the corresponding evolutionary changes in the genome. Along the same line, strong convergences seen among parasite taxa at the phenotypic level are not necessarily aligned with corresponding genomic changes seen across taxa. Based on different genetic pathways, analogous traits evolving independently in different lineages can still yield similar phenotypes and allow different lineages to meet on the same adaptive peak (a maximum point on a fitness landscape is a schematic tool used to visualize the relationship between reproductive success and an organism's phenotype determined by its genotype) (Poulin, 2011; Poulin & Randhawa, 2015).

Poulin and Randhawa (2015) investigated some evidence of parallel evolution (phenotypic and genotypic) among unrelated parasite taxa concerning genome reduction or compaction and gene losses or gains. They revealed matching such changes in parasite genomes with the broad phenotypic traits that define the convergence of parasites toward six host exploitation strategies, including parasitoids, parasitic castrators, directly transmitted parasites, trophically transmitted parasites, and vector-transmitted parasites. They concluded that no set of genes could be linked directly to the features of the whole organism, such as using multiple hosts per generation or the ability to manipulate a suit of host phenotypic traits to enhance transmission success (Jackson, 2015; Poulin & Randhawa, 2015). However, the dramatically increasing available parasite genomes and molecular advances may detect more apparent trends in the evolution of parasite genome architecture that can represent true convergent adaptive peaks for deep knowledge in the evolution of parasitism along convergent lines. Future studies of phylogenomic analysis may aid in resolving the problematic issues in phylogenetic systematics regarding homoplasious characters and adaptive plasticity, particularly in parasitic flatworms (O'Grady, 1985).

4.12 Genes in Populations

As mentioned in Chap. 3, host-parasite interactions are among the best settings for adaptation in spatially and temporally heterogeneous environments. The effects of parasitism are likely to be highly variable in time and among host populations (Gandon & Michalakis, 2002). Spatially heterogeneous environments may lead to the adaptation of populations to their local environmental conditions. Local adaptation occurs when the mean fitness is higher in its habitat than in a remote habitat. Gandon and Michalakis (2002) analyzed a host-parasite metapopulation model and studied the effects of several factors, such as population size, mutation rates, and migration rates for both host and parasites, on the emergence of local adaptation. Population size, mutation, and migration rates may act on local adaptation through their effects on the evolutionary potential of each species, promoting local adaptation through higher numbers of mutations or migrations. They noted that shorter parasite generation time does not always favor local parasite adaptation when there is limited genetic variability. Local adaptation may not be favored as there is an increase in the strength of selection and a decrease in the parasite's capacity to adapt to an evolving host.

Genes in a population can be transferred from one generation to another, making it possible to trace the lines of descent back through the generations. In the simplest scenario, genes are inherited through only one sex. Each gene is copied from one of the genes in the previous generation, indicated by the line shown in Fig. 4.5. All these lines of descent eventually converge on a single copy of the gene if we go back

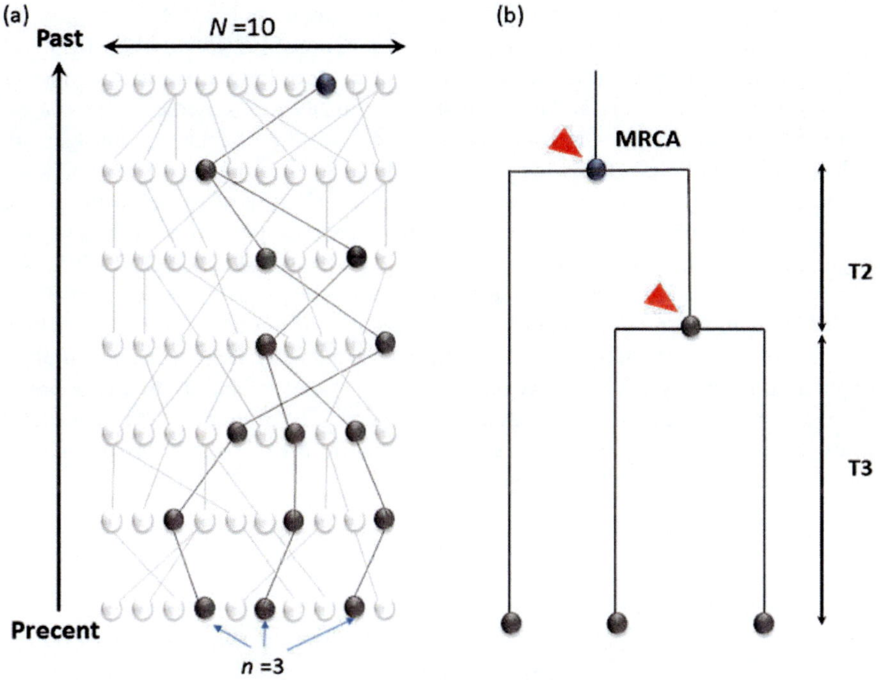

Fig. 4.5 Principle of the coalescent theory. (**a**) The complete genealogy for a population of ten individual alleles (*N* individuals used for a haploid model of diploid populations having 2 *N* individuals) is indicated. Black lines trace the ancestries of three sampled lineages back to a single common ancestor. The black and gray dots represent the individual alleles in the genealogical tree. (**b**) Subgenealoy for the three sampled lineages. The red points indicate the coalescent events of these three lineages. Times between coalescent events are indicated as T3 and T2 from the present to the most recent common ancestor (MRCA)

enough in time. The process is called the coalescence process, which is a model that assumes neither recombination nor natural selection occurs, resulting in no gene flow or population structure. With no recombination or natural selection occurring, each variant is equally likely to have been passed from one generation to the next. Along with a random process in coalescence events, the model looks backward, merging alleles into a single ancestral copy. Under this model, the expected time between successive coalescence events increases almost exponentially in time.

In the population, measuring genetic variation at the level of the DNA sequence, mainly using mitochondrial DNA (mtDNA) as the molecular genetic marker, may differ from sequence to sequence. Observing genetic variation is more useful when presented as gene genealogies, which describe the phylogenetic relationships of the sequence sampled from a population. The coalescent theory is about the gene genealogy, particularly in populations affected only by genetic drift, where changes in population size affect the number of lineages produced without altering the individual's fitness. To explain the coalescence process, we first assume gene sequences are sampled from two individuals in a population subjected to genetic drift. The lineages of these two sequences can be traced back to when they last shared a common ancestor allele or coalesce. If we do this for a larger sample of genes with random mating and no recombination in any past generation, the coalescence of any two lineages is equal. As we continue to go backward, lineages are reduced by one at each coalescence by creating a node on the tree. This backward pruning of lineages gives rise to a bifurcating genealogy and so-called the single ancestral allele from which all the alleles in the recent samples arise—this is known as the most recent common ancestor (MRCA) (Page & Holmes, 2009). The relationship between coalescent theory and genetic drift allows us to estimate further the probabilities of when these alleles last shared a common ancestor (in terms of which generation). The probability of fixation of a neutral allele by genetic drift in a diploid population is simply 1/2 *N*. However, 1/2 *N* is also the probability that two alleles share a common ancestor in the previous generation. The probability that two alleles share an ancestor two generations ago is then (1–1/2 *N*) 1/2 *N*, and the probability that this happened at *G* generations in the past is:

$$f(G) = (1/2N)(1 - 1/2N)^{G-1}$$

which approximates to

$$f(G) = (1/2N)e^{-(G-1)/2N}$$

assuming having neutral alleles from a population that has maintained a constant size throughout its history. This implies that time back to the MRCA is the same as when it takes an allele to fixation—4 *N* generations.

Coalescent events get less likely as we go backward since fewer lineages are left to coalesce, and half of the total coalescence time will occur between the last two lineages. On average, the time until only two lineages are left is 2 *N* generations, but it takes other 2 *N* generations until the last two lineages coalesce as the MRCA. However, the meantime to the MRCA for mitochondrial genes is one-quarter of that for nuclear genes because mtDNA is only transmitted maternally, indicating the haploid state. All the copies of mtDNA found in living organisms must eventually coalesce into a single ancestral mtDNA.

The coalescent process is closely related to classical population genetic theory. Imagine using a stochastic simulation to investigate sample distribution under a model that involves random genetic drift. Using the coalescent method, there is no need to keep track of the entire population. The coalescent method simulates the genealogy of samples by going back in time until the MRCA is found and then adding mutations forward along the branches of the newly generated tree. Although the analysis methods to analyze polymorphisms are borrowed from phylogenetics, the approach and fundamentals that involve phylogenetic trees are very different (Rosenberg & Nordborg, 2002).

Phylogenetic methods to estimate trees were developed to determine the pattern of species descent and are assumed to be treelike. A single sequence from each species of interest is usually analyzed, and the genealogy of the sequences is estimated. The estimated gene tree is then used to conclude relationships between species. Typically the gene tree (phylogeny of alleles or haplotypes for any specified stretch of DNA) is equated with the species tree (the estimated genealogy of genes sampled from different species) (Avise, 1989). Gene trees are the components of population trees or species trees, which typically deal with the frequencies of phylogenetically unordered alleles (Avise, 1989). Each gene tree is a random outcome estimated from different genes to give different trees. Gene and species divergences are likely to be nearly concurrent, and gene trees are likely to be very similar to the species tree. However, due to rapid divergences in adaptive radiation, discordances between gene trees and species trees for phylogenies exist (Rosenberg & Nordborg, 2002; Seehausen, 2004).

Moreover, gene trees and species trees often disagree with gene exchanges, particularly between closely related species (Mallet et al., 2016; Rosenberg, 2002). In these cases, it is helpful to infer that the species tree uses more than one individual per species and more than a single locus. For gene trees, haplotype diversity in mtDNA provided the first extensive empirical data suitable for estimating gene trees on a microevolutionary (intraspecific) scale (Avise, 2012; Brower et al., 1996).

4.13 Mitochondrial Genome

Recently, genes in mitochondria and mitochondrial (mt) genomes have typically been used for studies in population genetics. Mitochondrial genomes provide genetic information located in the cytoplasm of the cell of eukaryotes, which cooperates interactively with the nuclear genome (the majority of the genetic information of a cell). In vertebrates, mtDNA is present in multiple copies, usually 1000–10,000 copies per cell, and programs the expression of 12–13 proteins of the adenosine triphosphate (ATP)-synthesizing respiratory-chain enzyme complexes (Saraste, 1999). The remaining mt components are encoded by nuclear genes and are imported into the mitochondrion by a specific transport system. The mt genome is usually clonal and rarely undergoes recombination. Mitochondria in sperms rarely enter or persist in the zygote, and thus, mtDNA is generally inherited maternally. Typically, animal mitochondrial genomes or mitogenomes possess small, circular mt molecules ranging from about 14–40 kb (Boore, 1999; Wolstenholme, 1992). In metazoans, the majority of mt genomes contain 37 genes consisting of two ribosomal RNAs, a set of 22 tRNAs, and a group of 13 protein-encoding genes such as cytochrome c oxidase subunits 1, 2, and 3 (*cox*1,2,3); NADH dehydrogenase subunits 1, 2, 3, 4 L, 4, 5, and 6 (*nad*1,2,3,4 L,4,5,6); cytochrome b (*cytb*); and adenosine triphosphatase subunits 6 and 8 (*atp*6,8) (Boore, 1999; Saccone et al., 1999). The mt genome is compact with many genes overlapping, but most genes adjoin one another or are separated by only a short intergenic region.

Moreover, the mitochondrion contains at least one relatively sizable noncoding region that harbors the promoter for transcription initiation (Shadel & Clayton, 1997). Codon usage and other features of the mt genome differ from those in the nuclear genome, reflecting the eubacterial ancestry of the mitochondrion. The translation initiation codons of protein-encoding genes are, in most cases, ATG that codes for the amino acid methionine and are consistent with the universal start codon of nuclear genes. In many invertebrate taxa, the initiation codon for translation can start with ATT, ATA, or GTG (Boore, 1999; Wolstenholme, 1992).

Le et al. (2002) summarized the mt genome of parasitic helminths (Le et al., 2002). The first completed mt genome from a parasitic helminth was *Ascaris suum* (Okimoto et al., 1992). Nematode mitogenomes are similar to those of other metazoa but have a few unusual features, including high variation in the conservation of gene order across major branches and the occasional presence of multiple chromosomes. The nematode mitogenome is usually a circular single-molecule ranging in size from 12 to 22 kb containing around 36 genes and around 12 protein-coding genes (*cox1-cox3*, *cytb, nad1-nad6*, *nad4L*, *atp6*, and rarely *atp8*), two ribosomal RNAs (large rRNA and small rRNA; *lr*RNA and *sr*RNA), and 22 tRNA genes. The *atp8* gene is usually absent in nematode mitogenomes, although it appears in *Trichinella* spp. and *Trichuris* spp. (Hawash et al., 2015; Kern et al., 2020; Lavrov & Brown, 2001; Rivero Fernández et al., 2021). Nematode mitogenomes usually show a strong nucleotide compositional bias toward A and T. This composition bias in nematodes is due to a high tendency to use T-rich codons (having more than two Ts in a triplet) and unequal usage of synonymous codons with a bias against C-rich codons in the third codon position in protein-coding genes (Mitreva et al., 2006).

Recently, 200 nematode mitogenomes have been sequenced and are available in GenBank (Appendix Table 4.1). Kern et al. (2020) emphasized that the variable gene order of the nematode mitochondrial genome should be considered when designing studies and selecting analytical methods for comparison (Kern et al., 2020). The mitochondrial genomes of flatworms are similar to those of other metazoans. Similar to parasitic nematodes, the mitogenomes of flatworms do not have the *atp8* gene. Recently, many complete mitogenomes of flatworms have been published, with 76 species of trematodes and 71 species of cestodes listed in Appendix Table 4.2. Differences in gene order were also present between members of the same genus, such as African and Asian schistosomes.

Moreover, the considerable difference in gene order is also reflected in the marked difference between *S. mansoni* and *S. japonicum* (Le et al., 2002). It is now well established that the mitochondria of parasitic flatworms and all rhabditophorans use AAA to specify Asn instead of Lys used in the universal code. Other differences include AGA and AGG to specify for Ser (Arg in the universal code) and TGA for Trp ("stop" codon in the universal code). Furthermore, TAG and TAA act as stop codons for parasitic flatworms (Korkmaz et al., 2014). Interestingly, TAG and TAA also specify the stop codon in Echinodermata, apparently due to convergent evolution. GTG is sometimes used for the start codon, aside from ATG. Like nematodes, the data from mitogenomes of flatworms are valuable for studies on species identification, phylogenies, and biography (Le et al., 2002).

Analyses of mitochondrial genes, including gene content, arrangement, composition, and variation, have revealed unique features among the helminths compared with other metazoans. These mitochondrial gene analyses provide essential data regarding functional and comparative mitochondrial genomics, molecular taxonomy, population genetics, molecular systematics, and evolutionary history. Moreover, mitogenome data for parasitic helminths are essential for diagnosis, epidemiology, and ecology of infections. Mitogenome data offer a source of markers for the systematics and population genetics of socioeconomically important parasitic helminths of humans and other animals (Jia et al., 2012).

4.14 Conclusion

Molecular evolution is the scientific discipline that is important for the molecular systematics of parasitic helminths to reconstruct the phylogenetic relationships and population genetics of organisms. In this chapter, we have provided the principle of molecular evolution to understand forces in the evolutionary process, which act upon the genome makeup of populations. The principles of genes and gene structure, types of mutation occurring in the genomes, and the evolutionary process leading to genetic variations were mentioned. Parasitic adaptation under environmental change was described in terms of the effect of environmental change on parasite genomes from individuals to populations and, finally, interspecies relationships. Finally, to develop further research, including molecular systematics and molecular identification, a summary of the information available for the mitochondrial genome of parasitic helminths was discussed.

Appendix

Table 4.1 List of Nematoda mitochondrial genomes available in NCBI database

Order	Family	Genus	Species	Number of sequences (complete mitochondrial genome)	Number of sequences (partial mitochondrial genome)
Trichocephalida	Trichuridae	*Trichuris*	*Trichuris trichiura*	5	0
Trichocephalida	Trichuridae	*Trichuris*	*Trichuris ovis*	1	0
Trichocephalida	Trichuridae	*Trichuris*	*Trichuris discolor*	1	0
Trichocephalida	Trichuridae	*Trichuris*	*Trichuris suis*	3	0
Trichocephalida	Trichuridae	*Trichuris*	*Trichuris muris*	2	0
Trichocephalida	Trichuridae	*Trichuris*	*Trichuris* sp. ETH232	1	0
Trichocephalida	Trichuridae	*Trichuris*	*Trichuris* sp. LO613	1	0
Trichocephalida	Trichuridae	*Trichuris*	*Trichuris* sp. TMM5	1	0
Trichocephalida	Trichuridae	*Trichuris*	*Trichuris* sp. TTB1	1	0
Trichocephalida	Trichuridae	*Trichuris*	*Trichuris* sp. TTB2	1	0
Trichocephalida	Trichuridae	*Trichuris*	*Trichuris* sp. GHL-2013	1	0

(continued)

Order	Family	Genus	Species	Number of sequences (complete mitochondrial genome)	Number of sequences (partial mitochondrial genome)
Trichocephalida	Trichuridae	*Trichinella*	*Trichinella spiralis*	1	0
Trichocephalida	Trichuridae	*Trichinella*	*Trichinella pseudospiralis*	4	0
Trichocephalida	Trichuridae	*Trichinella*	*Trichinella patagoniensis*	0	2
Trichocephalida	Trichuridae	*Trichinella*	*Trichinella zimbabwensis*	1	0
Trichocephalida	Trichuridae	*Trichinella*	*Trichinella papuae*	1	0
Trichocephalida	Trichuridae	*Trichinella*	*Trichinella nelsoni*	1	0
Trichocephalida	Trichuridae	*Trichinella*	*Trichinella nativa*	2	0
Trichocephalida	Trichuridae	*Trichinella*	*Trichinella murrelli*	1	0
Trichocephalida	Trichuridae	*Trichinella*	*Trichinella britovi*	1	0
Trichocephalida	Trichuridae	*Trichinella*	*Trichinella* sp. T6	1	0
Trichocephalida	Trichuridae	*Trichinella*	*Trichinella* sp. T9	1	0
Trichocephalida	Capillariidae	*Eucoleus*	*Eucoleus annulatus*	1	0
Trichocephalida	Capillariidae	*Pseudocapillaria*	*Pseudocapillaria tomentosa*	1	0
Trichocephalida	Capillariidae	*Capillaria*	*Capillaria* sp. cat-2018	1	0
Dorylaimida	Longidoridae	*Xiphinema*	*Xiphinema pachtaicum*	1	0
Dorylaimida	Longidoridae	*Xiphinema*	*Xiphinema americanum*	1	0
Dorylaimida	Longidoridae	*Xiphinema*	*Xiphinema rivesi*	1	0
Dorylaimida	Longidoridae	*Paralongidorus*	*Paralongidorus litoralis*	1	0
Dorylaimida	Longidoridae	*Longidorus*	*Longidorus vineacola*	1	0
Mermithida	Mermithidae	*Thaumamermis*	*Thaumamermis cosgrovei*	2	0
Mermithida	Mermithidae	*Romanomermis*	*Romanomermis culicivorax*	1	0
Mermithida	Mermithidae	*Romanomermis*	*Romanomermis nielseni*	1	0
Mermithida	Mermithidae	*Romanomermis*	*Romanomermis iyengari*	1	0
Mermithida	Mermithidae	*Agamermis*	*Agamermis* sp. BH-2006	1	0
Mermithida	Mermithidae	*Hexamermis*	*Hexamermis agrotis*	1	0
Mermithida	Mermithidae	*Strelkovimermis*	*Strelkovimermis spiculatus*	2	0
Plectida	Plectidae	*Plectus*	*Plectus aquatilis*	1	0
Plectida	Plectidae	*Plectus*	*Plectus acuminatus*	1	0
Oxyurida	Oxyuridae	*Enterobius*	*Enterobius vermicularis*	2	0
Oxyurida	Oxyuridae	*Oxyuris*	*Oxyuris equi*	1	0
Oxyurida	Oxyuridae	*Wellcomia*	*Wellcomia siamensis*	1	0
Oxyurida	Oxyuridae	*Wellcomia*	*Wellcomia compar*	1	0
Oxyurida	Oxyuridae	*Syphacia*	*Syphacia obvelata*	1	0
Oxyurida	Oxyuridae	*Passalurus*	*Passalurus ambiguus*	1	0
Oxyurida	Oxyuridae	*Aspiculuris*	*Aspiculuris tetraptera*	1	0
Oxyurida	Oxyuridae	*Hammerschmidtiella*	*Hammerschmidtiella* sp. ZengetLiu-2016	1	0
Ascaridida	Ascarididae	*Ortleppascaris*	*Ortleppascaris sinensis*	1	0
Ascaridida	Ascarididae	*Ascaris*	*Ascaris ovis*	2	0
Ascaridida	Ascarididae	*Ascaris*	*Ascaris suum*	2	0
Ascaridida	Ascarididae	*Ascaris*	*Ascaris lumbricoides*	2	1
Ascaridida	Ascarididae	*Ascaris*	*Ascaris* sp.	1	0
Ascaridida	Ascarididae	*Ascaris*	*Ascaris* sp. chimpanzee-cA	1	0
Ascaridida	Ascarididae	*Ascaris*	*Ascaris* sp. gibbon-gA	1	0
Ascaridida	Ascarididae	*Toxascaris*	*Toxascaris leonina*	3	0
Ascaridida	Ascarididae	*Parascaris*	*Parascaris univalens*	1	0

(continued)

Order	Family	Genus	Species	Number of sequences (complete mitochondrial genome)	Number of sequences (partial mitochondrial genome)
Ascaridida	Ascarididae	*Parascaris*	*Parascaris equorum*	1	1
Ascaridida	Ascarididae	*Baylisascaris*	*Baylisascaris procyonis*	1	0
Ascaridida	Ascarididae	*Baylisascaris*	*Baylisascaris schroederi*	1	0
Ascaridida	Ascarididae	*Baylisascaris*	*Baylisascaris ailuri*	1	0
Ascaridida	Ascarididae	*Baylisascaris*	*Baylisascaris transfuga*	1	0
Ascaridida	Ascaridiidae	*Ascaridia*	*Ascaridia columbae*	1	0
Ascaridida	Ascaridiidae	*Ascaridia*	*Ascaridia galli*	1	0
Ascaridida	Ascaridiidae	*Ascaridia*	*Ascaridia* sp. GHL-2013	1	0
Ascaridida	Anisakidae	*Anisakis*	*Anisakis simplex*	18	0
Ascaridida	Anisakidae	*Anisakis*	*Anisakis pegreffii*	1	0
Ascaridida	Anisakidae	*Contracaecum*	*Contracaecum osculatum*	1	0
Ascaridida	Anisakidae	*Contracaecum*	*Contracaecum rudolphii*	1	0
Ascaridida	Anisakidae	*Contracaecum*	*Contracaecum ogmorhini* sensu lato	3	0
Ascaridida	Anisakidae	*Pseudoterranova*	*Pseudoterranova azarasi*	1	0
Ascaridida	Anisakidae	*Pseudoterranova*	*Pseudoterranova bulbosa*	1	0
Ascaridida	Anisakidae	*Pseudoterranova*	*Pseudoterranova cattani*	1	0
Ascaridida	Anisakidae	*Pseudoterranova*	*Pseudoterranova* decipiens sensu lato	1	0
Ascaridida	Anisakidae	*Pseudoterranova*	*Pseudoterranova krabbei*	1	0
Ascaridida	Anisakidae	*Ophidascaris*	*Ophidascaris baylisi*	1	0
Ascaridida	Anisakidae	*Ophidascaris*	*Ophidascaris* sp. CYZ-2020	1	0
Ascaridida	Toxocaridae	*Toxocara*	*Toxocara canis*	3	0
Ascaridida	Toxocaridae	*Toxocara*	*Toxocara cati*	1	0
Ascaridida	Toxocaridae	*Toxocara*	*Toxocara malaysiensis*	1	0
Ascaridida	Heterakidae	*Heterakis*	*Heterakis dispar*	1	0
Ascaridida	Heterakidae	*Heterakis*	*Heterakis beramporia*	1	0
Ascaridida	Heterakidae	*Heterakis*	*Heterakis gallinarum*	1	0
Ascaridida	Cucullanidae	*Cucullanus*	*Cucullanus robustus*	1	0
Ascaridida	Quimperiidae	*Pingus*	*Pingus sinensis*	1	0
Spirurida	Onchocercidae	*Brugia*	*Brugia malayi*	2	0
Spirurida	Onchocercidae	*Brugia*	*Brugia pahangi*	1	0
Spirurida	Onchocercidae	*Wuchereria*	*Wuchereria bancrofti*	5	0
Spirurida	Onchocercidae	*Onchocerca*	*Onchocerca lupi*	1	0
Spirurida	Onchocercidae	*Onchocerca*	*Onchocerca ochengi*	4	0
Spirurida	Onchocercidae	*Onchocerca*	*Onchocerca volvulus*	2	0
Spirurida	Onchocercidae	*Onchocerca*	*Onchocerca flexuosa*	2	0
Spirurida	Onchocercidae	*Mansonella*	*Mansonella perstans*	1	0
Spirurida	Onchocercidae	*Dirofilaria*	*Dirofilaria repens*	4	0
Spirurida	Onchocercidae	*Loa*	*Loa loa*	1	0
Spirurida	Onchocercidae	*Acanthocheilonema*	*Acanthocheilonema viteae*	1	0
Spirurida	Onchocercidae	*Chandlerella*	*Chandlerella quiscali*	1	0
Spirurida	Onchocercidae	*Dirofilaria*	*Dirofilaria* sp. 'hongkongensis'	1	0
Spirurida	Setariidae	*Setaria*	*Setaria labiatopapillosa*	1	0
Spirurida	Setariidae	*Setaria*	*Setaria digitata*	2	0
Spirurida	Gongylonematidae	*Gongylonema*	*Gongylonema pulchrum*	2	0
Spirurida	Camallanidae	*Camallanus*	*Camallanus cotti*	1	0

(continued)

Order	Family	Genus	Species	Number of sequences (complete mitochondrial genome)	Number of sequences (partial mitochondrial genome)
Spirurida	Thelaziidae	*Thelazia*	*Thelazia callipaeda*	1	0
Spirurida	Thelaziidae	*Spirocerca*	*Spirocerca lupi*	2	0
Spirurida	Gnathostomatidae	*Gnathostoma*	*Gnathostoma nipponicum*	1	0
Spirurida	Gnathostomatidae	*Gnathostoma*	*Gnathostoma doloresi*	2	0
Spirurida	Gnathostomatidae	*Gnathostoma*	*Gnathostoma spinigerum*	1	0
Spirurida	Physalopteridae	*Heliconema*	*Heliconema longissimum*	1	0
Spirurida	Philometridae	*Philometroides*	*Philometroides sanguineus*	1	0
Spirurida	Rhigonematidae	*Rhigonema*	*Rhigonema thysanophora*	1	0
Spirurida	Dracunculidae	*Dracunculus medinensis*	*Dracunculus medinensis*	1	0
Rhabditida	Strongyloididae	*Strongyloides*	*Strongyloides ratti*	1	0
Rhabditida	Strongyloididae	*Strongyloides*	*Strongyloides stercoralis*	6	0
Rhabditida	Strongyloididae	*Strongyloides*	*Strongyloides venezuelensis*	1	0
Rhabditida	Strongyloididae	*Strongyloides*	*Strongyloides papillosus*	1	0
Rhabditida	Steinernematidae	*Steinernema*	*Steinernema carpocapsae*	4	0
Rhabditida	Steinernematidae	*Steinernema*	*Steinernema abbasi*	1	0
Rhabditida	Alloionematidae	*Rhabditophanes*	*Rhabditophanes* sp. KR3021 mitochondrial molecule B DNA	1	0
Rhabditida	Alloionematidae	*Rhabditophanes*	*Rhabditophanes* sp. KR3021 mitochondrial molecule A DNA	1	0
Rhabditida	Cephalobidae	*Acrobeloides*	*Acrobeloides varius*	1	0
Rhabditida	Diplogasteridae	*Allodiplogaster*	*Allodiplogaster sudhausi*	1	0
Rhabditida	Neodiplogasteridae	*Pristionchus*	*Pristionchus pacificus*	1	0
Rhabditida	Meloidogynidae	*Meloidogyne*	*Meloidogyne javanica*	1	0
Rhabditida	Meloidogynidae	*Meloidogyne*	*Meloidogyne enterolobii*	1	0
Rhabditida	Meloidogynidae	*Meloidogyne*	*Meloidogyne arenaria*	1	0
Rhabditida	Meloidogynidae	*Meloidogyne*	*Meloidogyne incognita*	1	0
Rhabditida	Meloidogynidae	*Meloidogyne*	*Meloidogyne chitwoodi*	1	0
Rhabditida	Meloidogynidae	*Meloidogyne*	*Meloidogyne graminicola*	1	0
Rhabditida	Meloidogynidae	*Meloidogyne*	*Meloidogyne exigua*	1	0
Rhabditida	Meloidogynidae	*Meloidogyne*	*Meloidogyne oryzae*	1	0
Rhabditida	Meloidogynidae	*Meloidogyne*	*Meloidogyne arenaria*	3	0
Rhabditida	Heteroderidae	*Globodera*	*Globodera ellingtonae*	2	0
Rhabditida	Pratylenchidae	*Pratylenchus*	*Pratylenchus vulnus*	1	0
Rhabditida	Aphelenchoididae	*Aphelenchoides*	*Aphelenchoides besseyi*	1	0
Rhabditida	Aphelenchoididae	*Bursaphelenchus xylophilus*	*Bursaphelenchus xylophilus*	4	0
Rhabditida	Aphelenchoididae	*Bursaphelenchus xylophilus*	*Bursaphelenchus mucronatus*	1	0
Rhabditida	Rhabditidae	*Caenorhabditis*	*Caenorhabditis remanei*	1	0
Rhabditida	Rhabditidae	*Caenorhabditis*	*Caenorhabditis castelli*	1	0
Rhabditida	Rhabditidae	*Caenorhabditis*	*Caenorhabditis macrosperma*	1	0
Rhabditida	Rhabditidae	*Caenorhabditis*	*Caenorhabditis nouraguensis*	1	0
Rhabditida	Rhabditidae	*Caenorhabditis*	*Caenorhabditis wallacei*	1	0

(continued)

Order	Family	Genus	Species	Number of sequences (complete mitochondrial genome)	Number of sequences (partial mitochondrial genome)
Rhabditida	Rhabditidae	*Caenorhabditis*	*Caenorhabditis doughertyi*	1	0
Rhabditida	Rhabditidae	*Caenorhabditis*	*Caenorhabditis virilis*	1	0
Rhabditida	Rhabditidae	*Caenorhabditis*	*Caenorhabditis angaria*	1	0
Rhabditida	Rhabditidae	*Caenorhabditis*	*Caenorhabditis plicata*	1	0
Rhabditida	Rhabditidae	*Caenorhabditis*	*Caenorhabditis brenneri*	1	0
Rhabditida	Rhabditidae	*Caenorhabditis*	*Caenorhabditis nigoni*	1	0
Rhabditida	Rhabditidae	*Caenorhabditis*	*Caenorhabditis tropicalis*	1	0
Rhabditida	Rhabditidae	*Caenorhabditis*	*Caenorhabditis remanei*	2	0
Rhabditida	Rhabditidae	*Caenorhabditis*	*Caenorhabditis tribulationis*	1	0
Rhabditida	Rhabditidae	*Caenorhabditis*	*Caenorhabditis afra*	1	0
Rhabditida	Rhabditidae	*Caenorhabditis*	*Caenorhabditis* sp.	1	0
Rhabditida	Rhabditidae	*Caenorhabditis*	*Caenorhabditis* sp. 38 MB-2015	1	0
Rhabditida	Rhabditidae	*Litoditis*	*Litoditis* aff. marina PmIV	1	0
Rhabditida	Rhabditidae	*Litoditis*	*Litoditis* aff. marina PmIII	1	0
Rhabditida	Rhabditidae	*Litoditis*	*Litoditis* aff. marina PmII	1	0
Rhabditida	Rhabditidae	*Litoditis*	*Litoditis* aff. marina PmI	1	0
Rhabditida	Rhabditidae	*Litoditis*	*Litoditis* aff. marina PmI	1	0
Rhabditida	Rhabditidae	*Litoditis*	*Litoditis* aff. marina PmII	1	0
Rhabditida	Rhabditidae	*Litoditis*	*Litoditis* aff. marina PmIII	1	0
Rhabditida	Rhabditidae	*Litoditis*	*Litoditis* aff. marina PmIV	1	0
Rhabditida	Rhabditidae	*Oscheius*	*Oscheius tipulae*	1	0
Rhabditida	Tetrameridae	*Crassicauda*	*Crassicauda* sp. Ningbo-2019	1	0
Rhabditida	Cephalobidae	*Acrobeloides*	*Acrobeloides varius*	1	0
Rhabditida	Hoplolaimidae	*Rotylenchulus*	*Rotylenchulus reniformis*	1	0
Strongylida	Haemonchidae	*Teladorsagia*	*Teladorsagia circumcincta*	3	0
Strongylida	Haemonchidae	*Haemonchus*	*Haemonchus contortus*	2	0
Strongylida	Haemonchidae	*Haemonchus*	*Haemonchus placei*	1	0
Strongylida	Haemonchidae	*Mecistocirrus*	*Mecistocirrus digitatus*	1	0
Strongylida	Haemonchidae	*Marshallagia*	*Marshallagia marshalli*	1	0
Strongylida	Cooperiidae	*Cooperia*	*Cooperia oncophora*	3	0
Strongylida	Heligmonellidae	*Nippostrongylus*	*Nippostrongylus brasiliensis*	2	0
Strongylida	Heligmonellidae	*Trichostrongylus*	*Trichostrongylus axei*	1	0
Strongylida	Heligmonellidae	*Trichostrongylus*	*Trichostrongylus vitrinus*	1	0
Strongylida	Dictyocaulidae	*Dictyocaulus*	*Dictyocaulus eckerti*	1	0
Strongylida	Dictyocaulidae	*Dictyocaulus*	*Dictyocaulus viviparus*	2	0
Strongylida	Angiostrongylidae	*Angiostrongylus*	*Angiostrongylus cantonensis*	8	0
Strongylida	Angiostrongylidae	*Angiostrongylus*	*Angiostrongylus malaysiensis*	1	0
Strongylida	Angiostrongylidae	*Angiostrongylus*	*Angiostrongylus costaricensis*	3	0
Strongylida	Angiostrongylidae	*Angiostrongylus*	*Angiostrongylus vasorum*	1	0

(continued)

Order	Family	Genus	Species	Number of sequences (complete mitochondrial genome)	Number of sequences (partial mitochondrial genome)
Strongylida	Angiostrongylidae	*Angiostrongylus*	*Angiostrongylus mackerrasae*	1	0
Strongylida	Angiostrongylidae	*Aelurostrongylus*	*Aelurostrongylus abstrusus*	1	0
Strongylida	Metastrongylidae	*Metastrongylus*	*Metastrongylus salmi*	1	0
Strongylida	Metastrongylidae	*Metastrongylus*	*Metastrongylus pudendotectus*	1	0
Strongylida	Filaroididae	*Parafilaroides*	*Parafilaroides normani*	1	0
Strongylida	Filaroididae	*Protostrongylidae*	*Protostrongylus rufescens*	1	0
Strongylida	Strongylidae	*Coronocyclus*	*Coronocyclus labiatus*	1	0
Strongylida	Strongylidae	*Cylicodontophorus*	*Cylicodontophorus bicoronatus*	1	0
Strongylida	Strongylidae	*Cylicocyclus*	*Cylicocyclus ashworthi*	1	0
Strongylida	Strongylidae	*Cylicocyclus*	*Cylicocyclus nassatus*	1	0
Strongylida	Strongylidae	*Cylicocyclus*	*Cylicocyclus insignis*	1	0
Strongylida	Strongylidae	*Cylicocyclus*	*Cylicocyclus auriculatus*	2	0
Strongylida	Strongylidae	*Cyathostomum*	*Cyathostomum pateratum*	1	0
Strongylida	Strongylidae	*Cyathostomum*	*Cyathostomum catinatum*	1	0
Strongylida	Strongylidae	*Cylicostephanus*	*Cylicostephanus minutus*	1	0
Strongylida	Strongylidae	*Poteriostomum*	*Poteriostomum imparidentatum*	1	0
Strongylida	Strongylidae	*Triodontophorus*	*Triodontophorus nipponicus*	1	0
Strongylida	Strongylidae	*Triodontophorus*	*Triodontophorus serratus*	1	0
Strongylida	Strongylidae	*Triodontophorus*	*Triodontophorus brevicauda*	1	0
Strongylida	Strongylidae	*Strongylus*	*Strongylus vulgaris*	2	0
Strongylida	Strongylidae	*Strongylus*	*Strongylus equinus*	1	0
Strongylida	Heterorhabditidae	*Heterorhabditis*	*Heterorhabditis bacteriophora*	2	0
Strongylida	Heterorhabditidae	*Heterorhabditis*	*Heterorhabditis indica*	1	0
Strongylida	Chabertiidae	*Chabertia*	*Chabertia ovina*	2	0
Strongylida	Chabertiidae	*Chabertia*	*Chabertia erschowi*	1	0
Strongylida	Chabertiidae	*Oesophagostomum*	*Oesophagostomum dentatum*	3	0
Strongylida	Chabertiidae	*Oesophagostomum*	*Oesophagostomum columbianum*	1	0
Strongylida	Chabertiidae	*Oesophagostomum*	*Oesophagostomum asperum*	1	0
Strongylida	Chabertiidae	*Oesophagostomum*	*Oesophagostomum quadrispinulatum*	1	0
Strongylida	Chabertiidae	*Oesophagostomoides*	*Oesophagostomoides giltneri*	1	0
Strongylida	Chabertiidae	*Oesophagostomoides*	*Oesophagostomoides longispicularis*	1	0
Strongylida	Chabertiidae	*Oesophagostomoides*	*Oesophagostomoides stirtoni*	1	0
Strongylida	Chabertiidae	*Macropicola*	*Macropicola ocydromi*	1	0
Strongylida	Chabertiidae	*Hypodontus*	*Hypodontus macropi*	3	0
Strongylida	Chabertiidae	*Phascolostrongylus*	*Phascolostrongylus turleyi*	1	0
Strongylida	Chabertiidae	*Phascolostrongylus*	*Paramacropostrongylus iugalis*	2	0

(continued)

Order	Family	Genus	Species	Number of sequences (complete mitochondrial genome)	Number of sequences (partial mitochondrial genome)
Strongylida	Chabertiidae	*Phascolostrongylus*	*Paramacropostrongylus typicus*	1	0
Strongylida	Chabertiidae	*Torquenema*	*Torquenema toraliforme*	1	0
Strongylida	Syngamidae	*Syngamus*	*Syngamus trachea*	1	0
Strongylida	Ancylostomatidae	*Ancylostoma*	*Ancylostoma caninum*	3	0
Strongylida	Ancylostomatidae	*Ancylostoma*	*Ancylostoma ceylanicum*	2	0
Strongylida	Ancylostomatidae	*Ancylostoma*	*Ancylostoma tubaeforme*	1	0
Strongylida	Ancylostomatidae	*Ancylostoma*	*Ancylostoma duodenale*	2	0
Strongylida	Ancylostomatidae	*Bunostomum*	*Bunostomum phlebotomum*	3	0
Strongylida	Ancylostomatidae	*Bunostomum*	*Bunostomum trigonocephalum*	1	0
Strongylida	Ancylostomatidae	*Necator*	*Necator americanus*	2	0
Strongylida	Ancylostomatidae	*Uncinaria*	*Uncinaria sanguinis*	2	0
Strongylida	Stephanuridae	*Stephanurus*	*Stephanurus dentatus*	1	0
Strongylida	Diaphanocephalidae	*Kalicephalus*	*Kalicephalus indicus*	1	0
Strongylida	Molineidae	*Nematodirus*	*Nematodirus oiratianus*	1	0
Strongylida	Molineidae	*Nematodirus*	*Nematodirus spathiger*	1	0

Table 4.2 List of Platyhelminthes mitochondrial genomes available in NCBI database

Subclass	Order	Family	Genus	Species	Number of sequences (complete)	Number of sequences (partial)
Digenea	Plagiorchiida	Opisthorchiidae	*Clonorchis*	*Clonorchis sinensis*	4	184
Digenea	Plagiorchiida	Opisthorchiidae	*Opisthorchis*	*Opisthorchis viverrini*	0	1
Digenea	Plagiorchiida	Opisthorchiidae	*Opisthorchis*	*Opisthorchis felineus*	1	0
Digenea	Plagiorchiida	Opisthorchiidae	*Opisthorchis*	*Opisthorchis sudarikovi*	0	1
Digenea	Plagiorchiida	Opisthorchiidae	*Amphimerus*	*Amphimerus* sp.	1	0
Digenea	Plagiorchiida	Opisthorchiidae	*Metorchis*	*Metorchis orientalis*	1	0
Digenea	Plagiorchiida	Heterophyidae	*Haplorchis*	*Haplorchis taichui*	2	0
Digenea	Plagiorchiida	Heterophyidae	*Metagonimus*	*Metagonimus yokogawai*	1	0
Digenea	Plagiorchiida	Troglotrematidae	*Paragonimus*	*Paragonimus ohirai*	1	5
Digenea	Plagiorchiida	Troglotrematidae	*Paragonimus*	*Paragonimus westermani*	5	0
Digenea	Plagiorchiida	Troglotrematidae	*Paragonimus*	*Paragonimus westermani* complex sp. type 1	1	0
Digenea	Plagiorchiida	Troglotrematidae	*Paragonimus*	*Paragonimus heterotremus*	1	1
Digenea	Plagiorchiida	Troglotrematidae	*Paragonimus*	*Paragonimus kellicotti*	0	1
Digenea	Plagiorchiida	Plagiorchiidae	*Glypthelmins*	*Glypthelmins quieta*	1	0
Digenea	Plagiorchiida	Plagiorchiidae	*Glypthelmins*	*Glypthelmins* sp. LW2G	1	0
Digenea	Plagiorchiida	Haematoloechidae	*Haematoloechus*	*Haematoloechus* sp. CW13H	1	0
Digenea	Plagiorchiida	Orientocreadiidae	*Orientocreadium*	*Orientocreadium* sp. HS	1	0
Digenea	Plagiorchiida	Echinostomatidae	*Echinostoma*	*Echinostoma revolutum*	2	0
Digenea	Plagiorchiida	Echinostomatidae	*Echinostoma*	*Echinostoma miyagawai*	2	0
Digenea	Plagiorchiida	Echinostomatidae	*Echinostoma*	*Echinostoma paraensei*	1	0
Digenea	Plagiorchiida	Echinostomatidae	*Echinostoma*	*Echinostoma hortense*	1	0
Digenea	Plagiorchiida	Echinostomatidae	*Echinostoma*	*Echinostoma* sp. JM-2019	1	0
Digenea	Plagiorchiida	Echinochasmidae	*Echinochasmus*	*Echinochasmus japonicus*	1	0
Digenea	Plagiorchiida	Echinostomatidae	*Hypoderaeum*	*Hypoderaeum* sp. Hubei-2014	1	0
Digenea	Plagiorchiida	Echinostomatidae	*Artyfechinostomum*	*Artyfechinostomum sufrartyfex*	1	0
Digenea	Plagiorchiida	Fasciolidae	*Fasciola*	*Fasciola hepatica*	3	0
Digenea	Plagiorchiida	Fasciolidae	*Fasciola*	*Fasciola gigantica*	2	0
Digenea	Plagiorchiida	Fasciolidae	*Fasciola*	*Fasciola jacksoni*	1	0
Digenea	Plagiorchiida	Fasciolidae	*Fasciola*	*Fasciola* sp. GHL-2014	1	0
Digenea	Plagiorchiida	Fasciolidae	*Fascioloides*	*Fascioloides magna*	1	0
Digenea	Plagiorchiida	Fasciolidae	*Fasciolopsis*	*Fasciolopsis buski*	1	0
Digenea	Plagiorchiida	Eucotylidae	*Tamerlania*	*Tamerlania zarudnyi*	1	0

(continued)

Table 4.2 (continued)

Subclass	Order	Family	Genus	Species	Number of sequences (complete)	Number of sequences (partial)
Digenea	Plagiorchiida	Eucotylidae	*Tanaisia*	*Tanaisia* sp. SS-2020	1	0
Digenea	Plagiorchiida	Eucotylidae	*Acanthoparyphium*	*Acanthoparyphium* sp. WAK-2018	1	0
Digenea	Plagiorchiida	Dicrocoeliidae	*Eurytrema*	*Eurytrema pancreaticum*	1	0
Digenea	Plagiorchiida	Dicrocoeliidae	*Dicrocoelium*	*Dicrocoelium dendriticum*	1	0
Digenea	Plagiorchiida	Dicrocoeliidae	*Dicrocoelium*	*Dicrocoelium chinensis*	1	0
Digenea	Plagiorchiida	Dicrocoeliidae	*Lyperosomum*	*Lyperosomum longicauda*	1	0
Digenea	Plagiorchiida	Paramphistomidae	*Paramphistomum*	*Paramphistomum cervi*	2	0
Digenea	Plagiorchiida	Paramphistomidae	*Explanatum*	*Explanatum explanatum*	1	0
Digenea	Plagiorchiida	Paramphistomidae	*Calicophoron*	*Calicophoron microbothrioides*	1	0
Digenea	Plagiorchiida	Paramphistomidae	*Gastrothylax*	*Gastrothylax crumenifer*	1	0
Digenea	Plagiorchiida	Gastrothylacidae	*Fischoederius*	*Fischoederius elongatus*	2	0
Digenea	Plagiorchiida	Gastrothylacidae	*Fischoederius*	*Fischoederius cobboldi*	1	0
Digenea	Plagiorchiida	Gastrothylacidae	*Fischoederius*	*Fischoederius* sp. 1 RG-2021	1	0
Digenea	Plagiorchiida	Gastrodiscidae	*Homalogaster*	*Homalogaster paloniae*	1	0
Digenea	Plagiorchiida	Diplodiscidae	*Diplodiscus*	*Diplodiscus nigromaculati*	1	0
Digenea	Plagiorchiida	Cyclocoelidae	*Uvitellina*	*Uvitellina* sp. SSS-2019	1	0
Digenea	Plagiorchiida	Cyclocoelidae	*Tracheophilus*	*Tracheophilus cymbius*	1	0
Digenea	Plagiorchiida	Plagiorchiidae	*Plagiorchis*	*Plagiorchis maculosus*	1	0
Digenea	Plagiorchiida	Prosthogonimidae	*Prosthogonimus*	*Prosthogonimus cuneatus*	1	0
Digenea	Plagiorchiida	Prosthogonimidae	*Prosthogonimus*	*Prosthogonimus pellucidus*	1	0
Digenea	Plagiorchiida	Brachycladiidae	*Brachycladium*	*Brachycladium goliath*	1	0
Digenea	Plagiorchiida	Notocotylidae	*Ogmocotyle*	*Ogmocotyle sikae*	1	0
Digenea	Plagiorchiida	Notocotylidae	*Ogmocotyle*	*Ogmocotyle* sp. JM-2015	1	0
Digenea	Plagiorchiida	Notocotylidae	*Notocotylus*	*Notocotylus intestinalis*	1	0
Digenea	Plagiorchiida	Notocotylidae	*Azygiidae*	*Azygia hwangtsiyui*	1	0
Digenea	Plagiorchiida	Clinostomidae	*Clinostomum*	*Clinostomum complanatum*	2	0
Digenea	Diplostomida	Schistosomatidae	*Schistosoma*	*Schistosoma japonicum*	154	0
Digenea	Diplostomida	Schistosomatidae	*Schistosoma*	*Schistosoma haematobium*	7	5
Digenea	Diplostomida	Schistosomatidae	*Schistosoma*	*Schistosoma spindale*	2	0
Digenea	Diplostomida	Schistosomatidae	*Schistosoma*	*Schistosoma indicum*	1	0
Digenea	Diplostomida	Schistosomatidae	*Schistosoma*	*Schistosoma mekongi*	1	0
Digenea	Diplostomida	Schistosomatidae	*Schistosoma*	*Schistosoma margrebowiei*	1	0

(continued)

Table 4.2 (continued)

Subclass	Order	Family	Genus	Species	Number of sequences (complete)	Number of sequences (partial)
Digenea	Diplostomida	Schistosomatidae	*Schistosoma*	*Schistosoma curassoni*	1	0
Digenea	Diplostomida	Schistosomatidae	*Schistosoma*	*Schistosoma mansoni*	1	0
Digenea	Diplostomida	Schistosomatidae	*Schistosoma*	*Schistosoma bovis*	1	0
Digenea	Diplostomida	Schistosomatidae	*Schistosoma*	*Schistosoma guineensis*	1	0
Digenea	Diplostomida	Schistosomatidae	*Orientobilharzia*	*Orientobilharzia turkestanicum*	1	0
Digenea	Diplostomida	Schistosomatidae	*Trichobilharzia*	*Trichobilharzia szidati*	2	0
Digenea	Diplostomida	Schistosomatidae	*Trichobilharzia*	*Trichobilharzia regenti*	1	0
Digenea	Diplostomida	Brachylaimidae	*Postharmostomum*	*Postharmostomum commutatum*	1	0
Digenea	Diplostomida	Brachylaimidae	*Diplostomum*	*Diplostomum ardeae*	1	0
Digenea	Diplostomida	Strigeidae	*Apharyngostrigea*	*Apharyngostrigea pipientis*	1	0
Digenea	Diplostomida	Strigeidae	*Cotylurus*	*Cotylurus marcogliesei*	1	0
Digenea	Diplostomida	Strigeidae	*Cardiocephaloides*	*Cardiocephaloides medioconiger*	1	0
Digenea	Diplostomida	Cyathocotylidae	*Cyathocotyle*	*Cyathocotyle prussica*	1	0
Digenea	Diplostomida	Diplostomidae	*Hysteromorpha*	*Hysteromorpha triloba*	1	0
Digenea	Diplostomida	Diplostomidae	*Posthodiplostomum*	*Posthodiplostomum centrarchi*	1	0
Digenea	Diplostomida	Diplostomidae	*Tylodelphys*	*Tylodelphys immer*	1	0
Digenea	Diplostomida	Diplostomidae	*Alaria*	*Alaria americana*	1	0
Eucestoda	Cyclophyllidea	Taeniidae	*Taenia*	*Taenia hydatigena*	28	1
Eucestoda	Cyclophyllidea	Taeniidae	*Taenia*	*Taenia asiatica*	2	0
Eucestoda	Cyclophyllidea	Taeniidae	*Taenia*	*Taenia solium*	2	1
Eucestoda	Cyclophyllidea	Taeniidae	*Taenia*	*Taenia multiceps*	3	0
Eucestoda	Cyclophyllidea	Taeniidae	*Taenia*	*Taenia pisiformis*	2	0
Eucestoda	Cyclophyllidea	Taeniidae	*Taenia*	*Taenia crassiceps*	1	0
Eucestoda	Cyclophyllidea	Taeniidae	*Taenia*	*Taenia ovis*	1	0
Eucestoda	Cyclophyllidea	Taeniidae	*Taenia*	*Taenia madoquae*	1	0
Eucestoda	Cyclophyllidea	Taeniidae	*Taenia*	*Taenia laticollis*	1	0
Eucestoda	Cyclophyllidea	Taeniidae	*Taenia*	*Taenia twitchelli*	1	0
Eucestoda	Cyclophyllidea	Taeniidae	*Taenia*	*Taenia serialis*	1	0
Eucestoda	Cyclophyllidea	Taeniidae	*Taenia*	*Taenia saginata*	1	0
Eucestoda	Cyclophyllidea	Taeniidae	*Taenia*	*Taenia martis*	1	0
Eucestoda	Cyclophyllidea	Taeniidae	*Taenia*	*Taenia crocutae*	1	0
Eucestoda	Cyclophyllidea	Taeniidae	*Taenia*	*Taenia regis*	1	0

(continued)

Table 4.2 (continued)

Subclass	Order	Family	Genus	Species	Number of sequences (complete)	Number of sequences (partial)
Eucestoda	Cyclophyllidea	Taeniidae	*Taenia*	*Taenia arctos*	1	0
Eucestoda	Cyclophyllidea	Taeniidae	*Taenia*	*Taenia caixuepengi*	1	0
Eucestoda	Cyclophyllidea	Taeniidae	*Taenia*	*Taenia tianguangfui*	1	0
Eucestoda	Cyclophyllidea	Taeniidae	*Taenia*	*Taenia* sp. MZH:127001	1	0
Eucestoda	Cyclophyllidea	Taeniidae	*Taenia*	*Taenia* sp. MPM:20922	1	0
Eucestoda	Cyclophyllidea	Taeniidae	*Taenia*	*Taenia* sp. MZH:127052	1	0
Eucestoda	Cyclophyllidea	Taeniidae	*Versteria*	*Versteria mustelae*	2	0
Eucestoda	Cyclophyllidea	Taeniidae	*Hydatigera*	*Hydatigera taeniaeformis*	2	0
Eucestoda	Cyclophyllidea	Taeniidae	*Hydatigera*	*Hydatigera kamiyai*	1	0
Eucestoda	Cyclophyllidea	Taeniidae	*Hydatigera*	*Hydatigera parva*	1	0
Eucestoda	Cyclophyllidea	Taeniidae	*Hydatigera*	*Hydatigera krepkogorski*	1	0
Eucestoda	Cyclophyllidea	Taeniidae	*Hydatigera*	*Hydatigera* sp. HCFr	1	0
Eucestoda	Cyclophyllidea	Taeniidae	*Echinococcus*	*Echinococcus granulosus* sensu lato genotype G7	68	0
Eucestoda	Cyclophyllidea	Taeniidae	*Echinococcus*	*Echinococcus granulosus* sensu lato genotype G6	28	0
Eucestoda	Cyclophyllidea	Taeniidae	*Echinococcus*	*Echinococcus canadensis*	15	0
Eucestoda	Cyclophyllidea	Taeniidae	*Echinococcus*	*Echinococcus granulosus*	2	339
Eucestoda	Cyclophyllidea	Taeniidae	*Echinococcus*	*Echinococcus equinus*	1	1
Eucestoda	Cyclophyllidea	Taeniidae	*Echinococcus*	*Echinococcus multilocularis*	1	0
Eucestoda	Cyclophyllidea	Taeniidae	*Echinococcus*	*Echinococcus vogeli*	2	0
Eucestoda	Cyclophyllidea	Taeniidae	*Echinococcus*	*Echinococcus shiquicus*	2	0
Eucestoda	Cyclophyllidea	Taeniidae	*Echinococcus*	*Echinococcus ortleppi*	2	0
Eucestoda	Cyclophyllidea	Taeniidae	*Echinococcus*	*Echinococcus felidis*	1	0
Eucestoda	Cyclophyllidea	Taeniidae	*Echinococcus*	*Echinococcus oligarthrus*	2	0
Eucestoda	Cyclophyllidea	Mesocestoididae	*Mesocestoides*	*Mesocestoides vogae*	1	0
Eucestoda	Cyclophyllidea	Mesocestoididae	*Mesocestoides*	*Mesocestoides corti*	1	0
Eucestoda	Cyclophyllidea	Hymenolepididae	*Hymenolepis*	*Hymenolepsis nana*	3	0
Eucestoda	Cyclophyllidea	Hymenolepididae	*Hymenolepis*	*Hymenolepis diminuta*	3	0
Eucestoda	Cyclophyllidea	Hymenolepididae	*Hymenolepis*	*Hymenolepis microstoma*	3	0
Eucestoda	Cyclophyllidea	Hymenolepididae	*Pseudanoplocephala*	*Pseudanoplocephala crawfordi*	1	0
Eucestoda	Cyclophyllidea	Hymenolepididae	*Cloacotaenia*	*Cloacotaenia megalops*	1	0
Eucestoda	Cyclophyllidea	Anoplocephalidae	*Moniezia*	*Moniezia benedeni*	2	0
Eucestoda	Cyclophyllidea	Anoplocephalidae	*Moniezia*	*Moniezia expansa*	2	0

(continued)

Table 4.2 (continued)

Subclass	Order	Family	Genus	Species	Number of sequences (complete)	Number of sequences (partial)
Eucestoda	Cyclophyllidea	Anoplocephalidae	*Moniezia*	*Moniezia* sp. MN-2015	1	0
Eucestoda	Cyclophyllidea	Anoplocephalidae	*Anoplocephala*	*Anoplocephala magna*	1	0
Eucestoda	Cyclophyllidea	Anoplocephalidae	*Anoplocephala*	*Anoplocephala perfoliata*	1	0
Eucestoda	Cyclophyllidea	Paruterinidae	*Paruterina*	*Paruterina candelabraria*	1	0
Eucestoda	Cyclophyllidea	Paruterinidae	*Cladotaenia*	*Cladotaenia vulturi*	1	0
Eucestoda	Cyclophyllidea	Dipylidiidae	*Diphylidium*	*Dipylidium caninum*	3	0
Eucestoda	Bothriocephalidea	Bothriocephalidae	*Senga*	*Senga ophiocephalina*	1	0
Eucestoda	Proteocephalidea	Proteocephalidae	*Gangesia*	*Gangesia oligonchis*	1	0
Eucestoda	Caryophyllidea	Capingentidae	*Breviscolex*	*Breviscolex orientalis*	1	0
Eucestoda	Caryophyllidea	Lytocestidae	*Atractolytocestus*	*Atractolytocestus huronensis*	1	0
Eucestoda	Caryophyllidea	Lytocestidae	*Khawia*	*Khawia sinensis*	2	0
Eucestoda	Caryophyllidea	Caryophyllaeidae	*Parabreviscolex*	*Parabreviscolex niepini*	1	0
Eucestoda	Caryophyllidea	Caryophyllaeidae	*Caryophyllaeus*	*Caryophyllaeus brachycollis*	1	0
Eucestoda	Phyllobothriidea	Phyllobothriidae	*Anindobothrium*	*Anindobothrium anacolum*	3	0
Eucestoda	Phyllobothriidea	Phyllobothriidae	*Anindobothrium*	*Anindobothrium lisae*	1	0
Eucestoda	Phyllobothriidea	Phyllobothriidae	*Caulobothrium*	*Caulobothrium* sp. MZUSP 7990	1	0
Eucestoda	Phyllobothriidea	Phyllobothriidae	*Caulobothrium*	*Caulobothrium* sp. MZUSP 8009	1	0
Eucestoda	Phyllobothriidea	Phyllobothriidae	*Caulobothrium*	*Caulobothrium* sp. MZUSP 7989	1	0
Eucestoda	Phyllobothriidea	Phyllobothriidae	*Caulobothrium*	*Caulobothrium* sp. LRP 10376	1	0
Eucestoda	Phyllobothriidea	Phyllobothriidae	*Scyphophyllidium*	*Scyphophyllidium guariticus*	2	0
Eucestoda	Diphyllobothriidea	Diphyllobothriidae	*Spirometra*	*Spirometra erinaceieuropaei*	8	0
Eucestoda	Diphyllobothriidea	Diphyllobothriidae	*Spirometra*	*Spirometra decipiens*	2	0
Eucestoda	Diphyllobothriidea	Diphyllobothriidae	*Spirometra*	*Spirometra theileri*	1	0
Eucestoda	Diphyllobothriidea	Diphyllobothriidae	*Spirometra*	*Spirometra mansoni*	1	0
Eucestoda	Diphyllobothriidea	Diphyllobothriidae	*Dibothriocephalus*	*Dibothriocephalus latus*	3	0
Eucestoda	Diphyllobothriidea	Diphyllobothriidae	*Dibothriocephalus*	*Dibothriocephalus nihonkaiensis*	3	0
Eucestoda	Diphyllobothriidea	Diphyllobothriidae	*Dibothriocephalus*	*Dibothriocephalus dendriticus*	1	0
Eucestoda	Diphyllobothriidea	Diphyllobothriidae	*Diphyllobothrium*	*Diphyllobothrium stemmacephalum*	1	0
Eucestoda	Diphyllobothriidea	Diphyllobothriidae	*Diphyllobothrium*	*Diphyllobothrium cordatum*	1	0
Eucestoda	Diphyllobothriidea	Diphyllobothriidae	*Diphyllobothrium*	*Diphyllobothrium schistochilos*	1	0
Eucestoda	Diphyllobothriidea	Diphyllobothriidae	*Diplogonoporus*	*Diplogonoporus grandis*	1	0
Eucestoda	Diphyllobothriidea	Diphyllobothriidae	*Diplogonoporus*	*Diplogonoporus balaenoptera*	1	0

(continued)

Table 4.2 (continued)

Subclass	Order	Family	Genus	Species	Number of sequences (complete)	Number of sequences (partial)
Eucestoda	Diphyllobothriidea	Diphyllobothriidae	*Schistocephalus*	*Schistocephalus solidus*	1	0
Eucestoda	Diphyllobothriidea	Diphyllobothriidae	*Schistocephalus*	*Schistocephalus pungitii*	1	0
Eucestoda	Diphyllobothriidea	Diphyllobothriidae	*Ligula*	*Ligula intestinalis*	3	0
Eucestoda	Diphyllobothriidea	Diphyllobothriidae	*Cephalochlamys*	*Cephalochlamys namaquensis*	2	0
Eucestoda	Diphyllobothriidea	Diphyllobothriidae	*Duthiersia*	*Duthiersia expansa*	1	0
Eucestoda	Diphyllobothriidea	Diphyllobothriidae	*Bothridium*	*Bothridium pithonis*	1	0
Eucestoda	Diphyllobothriidea	Diphyllobothriidae	*Adenocephalus*	*Adenocephalus pacificus*	1	0
Eucestoda	Diphyllobothriidea	Diphyllobothriidae	*Digramma*	*Digramma interrupta*	1	0
Eucestoda	Tetraphyllidea	Serendipidae	*Duplicibothrium*	*Duplicibothrium* sp. MZUSP 8039	1	0
Eucestoda	Tetraphyllidea	Onchobothriidae	*Potamotrygonocestus*	*Potamotrygonocestus* sp. MZUSP 7996	1	0
Eucestoda	Tetraphyllidea	Onchobothriidae	*Acanthobothrium*	*Acanthobothrium* sp. MZUSP 7998	1	0
Eucestoda	Tetraphyllidea	Onchobothriidae	*Acanthobothrium*	*Acanthobothrium* sp. MZUSP 8035	1	0
Eucestoda	Tetraphyllidea	Onchobothriidae	*Acanthobothrium*	*Acanthobothrium* sp. MZUSP 8034	1	0
Eucestoda	Tetraphyllidea	Onchobothriidae	*Acanthobothrium*	*Acanthobothrium* sp. MZUSP 8033	1	0
Eucestoda	Tetraphyllidea	Serendipidae	*Serendip*	*Serendip* sp. MZUSP 7987	1	0
Eucestoda	Rhinebothriidea	Rhinebothriidae	*Rhinebothrium*	*Rhinebothrium fulbrighti*	1	0
Eucestoda	Rhinebothriidea	Rhinebothriidae	*Rhinebothrium*	*Rhinebothrium cf. copianullum* DJM-2021	1	0
Eucestoda	Rhinebothriidea	Rhinebothriidae	*Rhinebothrium*	*Rhinebothrium taeniuri*	2	0
Eucestoda	Rhinebothriidea	Rhinebothriidae	*Rhinebothrium*	*Rhinebothrium flexile*	2	0
Eucestoda	Rhinebothriidea	Rhinebothriidae	*Rhinebothrium*	*Rhinebothrium megacanthophallus*	1	0
Eucestoda	Rhinebothriidea	Rhinebothriidae	*Rhinebothrium*	*Rhinebothrium reydai*	1	0
Eucestoda	Rhinebothriidea	Rhinebothriidae	*Rhinebothrium*	*Rhinebothrium* sp. MZUSP 8032	1	0
Eucestoda	Rhinebothriidea	Rhinebothriidae	*Rhinebothrium*	*Rhinebothrium* sp. MZUSP 8031	1	0
Eucestoda	Rhinebothriidea	Rhinebothriidae	*Rhinebothrium*	*Rhinebothrium* sp. MZUSP 8030	1	0
Eucestoda	Rhinebothriidea	Rhinebothriidae	*Rhinebothrium*	*Rhinebothrium* sp. MZUSP 8025	1	0
Eucestoda	Rhinebothriidea	Rhinebothriidae	*Rhinebothrium*	*Rhinebothrium* sp. MZUSP 8024	1	0
Eucestoda	Rhinebothriidea	Rhinebothriidae	*Rhinebothrium*	*Rhinebothrium* sp. MZUSP 8023	1	0
Eucestoda	Rhinebothriidea	Rhinebothriidae	*Rhinebothrium*	*Rhinebothrium* sp. MZUSP 8022	1	0
Eucestoda	Rhinebothriidea	Rhinebothriidae	*Rhinebothrium*	*Rhinebothrium* sp. 1 voucher LRP 10400	1	0
Eucestoda	Rhinebothriidea	Rhinebothriidae	*Rhinebothrium*	*Rhinebothrium* sp. 1 voucher LRP 10401	1	0
Eucestoda	Rhinebothriidea	Rhinebothriidae	*Rhinebothrium*	*Rhinebothrium* sp. 1 voucher LRP 10402	1	0
Eucestoda	Rhinebothriidea	Rhinebothriidae	*Rhinebothrium*	*Rhinebothrium* sp. 1 voucher LRP 10403	1	0

(continued)

Table 4.2 (continued)

Subclass	Order	Family	Genus	Species	Number of sequences (complete)	Number of sequences (partial)
Eucestoda	Rhinebothriidea	Rhinebothriidae	*Rhinebothrium*	*Rhinebothrium* sp. 1 voucher LRP 10404	1	0
Eucestoda	Rhinebothriidea	Rhinebothriidae	*Rhinebothrium*	*Rhinebothrium* sp. MZUSP 8018	1	0
Eucestoda	Rhinebothriidea	Rhinebothriidae	*Rhinebothrium*	*Rhinebothrium* sp. LRP 10405	1	0
Eucestoda	Rhinebothriidea	Rhinebothriidae	*Rhinebothrium*	*Rhinebothrium* sp. LRP 10395	1	0
Eucestoda	Rhinebothriidea	Rhinebothriidae	*Rhinebothrium*	*Rhinebothrium* sp. LRP 10396	1	0
Eucestoda	Rhinebothriidea	Rhinebothriidae	*Rhinebothrium*	*Rhinebothrium* sp. LRP 10407	1	0
Eucestoda	Rhinebothriidea	Rhinebothriidae	*Rhinebothrium*	*Rhinebothrium* sp. 3 DJM-2021	1	0
Eucestoda	Rhinebothriidea	Rhinebothriidae	*Rhinebothrium*	*Rhinebothrium* sp. 1 voucher LRP 10397	1	0
Eucestoda	Rhinebothriidea	Rhinebothriidae	*Rhinebothrium*	*Rhinebothrium* sp. 1 voucher LRP 10398	1	0
Eucestoda	Rhinebothriidea	Rhinebothriidae	*Rhinebothrium*	*Rhinebothrium* sp. 1 voucher LRP 10399	1	0
Eucestoda	Rhinebothriidea	Rhinebothriidae	*Rhinebothrium*	*Rhinebothrium* sp. LRP 10392	1	0
Eucestoda	Rhinebothriidea	Rhinebothriidae	*Rhinebothroides*	*Rhinebothroides* sp. MZUSP 8029	1	0
Eucestoda	Rhinebothriidea	Rhinebothriidae	*Rhinebothroides*	*Rhinebothroides* sp. MZUSP 8028	1	0
Eucestoda	Rhinebothriidea	Rhinebothriidae	*Rhinebothroides*	*Rhinebothroides* sp. MZUSP 8027	1	0
Eucestoda	Rhinebothriidea	Rhinebothriidae	*Rhinebothroides*	*Rhinebothroides* sp. MZUSP 8026	1	0
Eucestoda	Rhinebothriidea	Rhinebothriidae	*Rhinebothroides*	*Rhinebothroides* sp. MZUSP 8021	1	0
Eucestoda	Rhinebothriidea	Rhinebothriidae	*Rhinebothroides*	*Rhinebothroides* sp. MZUSP 8020	1	0
Eucestoda	Rhinebothriidea	Rhinebothriidae	*Rhinebothroides*	*Rhinebothroides* sp. MZUSP 8013	1	0
Eucestoda	Rhinebothriidea	Rhinebothriidae	*Rhinebothroides*	*Rhinebothroides* sp. MZUSP 8014	1	0
Eucestoda	Rhinebothriidea	Rhinebothriidae	*Rhinebothroides*	*Rhinebothroides* sp. MZUSP 8015	1	0
Eucestoda	Rhinebothriidea	Rhinebothriidae	*Rhinebothroides*	*Rhinebothroides* sp. MZUSP 8016	1	0
Eucestoda	Rhinebothriidea	Rhinebothriidae	*Rhinebothroides*	*Rhinebothroides* sp. MZUSP 8017	1	0
Eucestoda	Rhinebothriidea	Rhinebothriidae	*Rhinebothroides*	*Rhinebothroides* sp. MZUSP 8019	1	0
Eucestoda	Rhinebothriidea	Rhinebothriidae	*Rhodobothrium*	*Rhodobothrium cf. paucitesticulare* DJM-2021	2	0
Eucestoda	Rhinebothriidea	Rhinebothriidae	*Rhabdotobothrium*	*Rhabdotobothrium anterophallum*	1	0
Eucestoda	Rhinebothriidea	Rhinebothriidae	*Echeneibothrium*	*Echeneibothrium williamsi*	1	0
Eucestoda	Rhinebothriidea	Rhinebothriidae	*Echeneibothrium*	*Echeneibothrium multiloculatum*	1	0
Eucestoda	Rhinebothriidea	Rhinebothriidae	*Spongiobothrium*	*Spongiobothrium variabile*	1	0
Eucestoda	Rhinebothriidea	Rhinebothriidae	*Spongiobothrium*	*Spongiobothrium* sp. MZUSP 7999	1	0
Eucestoda	Rhinebothriidea	Rhinebothriidae	*Anthocephalum*	*Anthocephalum cf. caira* DJM-2021	1	0
Eucestoda	Rhinebothriidea	Rhinebothriidae	*Anthocephalum*	*Anthocephalum* sp. LRP 10375	1	0
Eucestoda	Rhinebothriidea	Rhinebothriidae	*Anthocephalum*	*Anthocephalum* sp. LRP 10372	1	0

(continued)

Table 4.2 (continued)

Subclass	Order	Family	Genus	Species	Number of sequences (complete)	Number of sequences (partial)
Eucestoda	Rhinebothriidea	Rhinebothriidae	*Anthocephalum*	*Anthocephalum* sp. LRP 10373	1	0
Eucestoda	Rhinebothriidea	Rhinebothriidae	*Anthocephalum*	*Anthocephalum* sp. LRP 10374	1	0
Eucestoda	Rhinebothriidea	Rhinebothriidae	*Scalithrium*	*Scalithrium magniphallum*	2	0
Eucestoda	Rhinebothriidea	Rhinebothriidae	*Scalithrium*	*Scalithrium* sp. LRP 10408	1	0
Eucestoda	Rhinebothriidea	Rhinebothriidae	*Scalithrium*	*Scalithrium* sp. LRP 10409	1	0
Eucestoda	Rhinebothriidea	Rhinebothriidae	*Scalithrium*	*Scalithrium* sp. LRP 10410	1	0
Eucestoda	Rhinebothriidea	Rhinebothriidae	*Scalithrium*	*Scalithrium* sp. LRP 10411	1	0
Eucestoda	Rhinebothriidea	Rhinebothriidae	*Divaricobothrium*	*Divaricobothrium* sp. LRP 10377	1	0
Eucestoda	Rhinebothriidea	Rhinebothriidae	*Alveobothrium*	*Alveobothrium grabatum*	3	0
Eucestoda	Rhinebothriidea	Rhinebothriidae	*Escherbothrium*	*Escherbothrium* sp. LRP 10380	1	0
Eucestoda	Rhinebothriidea	Rhinebothriidae	*Escherbothrium*	*Escherbothrium* sp. LRP 10382	1	0
Eucestoda	Spathebothriidea	Acrobothriidae	*Didymobothrium*	*Didymobothrium rudolphii*	2	0
Eucestoda	Haplobothriidea	Haplobothriidae	*Haplobothrium*	*Haplobothrium globuliforme*	2	0
Eucestoda	Trypanorhyncha	Eutetrarhynchidae	*Dollfusiella*	*Dollfusiella* sp. MZUSP 7992	1	0

References

Abascal-Palacios, G., Ramsay, E. P., Beuron, F., Morris, E., & Vannini, A. (2018). Structural basis of RNA polymerase III transcription initiation. *Nature, 553*(7688), 301–306.

Anisimova, M. (2012). Parametric models of codon evolution. In *Codon evolution: Mechanisms and models* (pp. 12–33).

Arenas, M. (2015). Trends in substitution models of molecular evolution. *Frontiers in Genetics, 6*, 319.

Avise, J. C. (1989). Gene trees and organismal histories: A phylogenetic approach to population biology. *Evolution, 43*(6), 1192–1208.

Avise, J. C. (2012). *Molecular markers, natural history and evolution*. Springer Science & Business Media.

Baer, C. F., Miyamoto, M. M., & Denver, D. R. (2007). Mutation rate variation in multicellular eukaryotes: Causes and consequences. *Nature Reviews Genetics, 8*(8), 619–631.

Bernardi, G. (1993). The vertebrate genome: Isochores and evolution. *Molecular Biology and Evolution, 10*(1), 186–204.

Bernardi, G., Hughes, S., & Mouchiroud, D. (1997). The major compositional transitions in the vertebrate genome. *Journal of Molecular Evolution, 44*(1), S44–S51.

Boore, J. L. (1999). Animal mitochondrial genomes. *Nucleic Acids Research, 27*(8), 1767–1780.

Bowen, N. J., & Jordan, I. K. (2002). Transposable elements and the evolution of eukaryotic complexity. *Current Issues in Molecular Biology, 4*(3), 65–76.

Brooks, D. R., & McLennan, D. A. (1993). Comparative study of adaptive radiations with an example using parasitic flatworms (Platyhelminthes: Cercomeria). *The American Naturalist, 142*(5), 755–778.

Brower, A., DeSalle, R., & Vogler, A. (1996). Gene trees, species trees, and systematics: A cladistic perspective. *Annual Review of Ecology and Systematics, 27*(1), 423–450.

Buckley, T. R., Simon, C., & Chambers, G. K. (2001). Exploring among-site rate variation models in a maximum likelihood framework using empirical data: Effects of model assumptions on estimates of topology, branch lengths, and bootstrap support. *Systematic Biology, 50*(1), 67–86.

Cankara, F. (2020). Prediction of the effects of single amino acid variations on protein functionality with structural and annotation centric modeling [Thesis (M.S.) – Graduate School of Informatics. Bioinformatics.]. Middle East Technical University. https://open.metu.edu.tr/bitstream/handle/11511/45389/index.pdf

Charlier, J., van der Voort, M., Kenyon, F., Skuce, P., & Vercruysse, J. (2014). Chasing helminths and their economic impact on farmed ruminants. *Trends in Parasitology, 30*(7), 361–367.

Cheverud, J. M. (1984). Quantitative genetics and developmental constraints on evolution by selection. *Journal of Theoretical Biology, 110*(2), 155–171.

Consortium, I. H. G. (2019). Comparative genomics of the major parasitic worms. *Nature Genetics, 51*(1), 163.

Ermolaev, M. D. (2001). Synonymous codon usage in bacteria. *Current Issues in Molecular Biology, 3*(4), 91–97.

Erwin, D. H. (2000). Macroevolution is more than repeated rounds of microevolution. *Evolution & Development, 2*(2), 78–84.

Folguera, G., & Lombardi, O. (2012). The relationship between microevolution and macroevolution, and the structure of the extended synthesis. *History and Philosophy of the Life Sciences, 34*, 539–559.

Gandon, S., & Michalakis, Y. (2002). Local adaptation, evolutionary potential and host–parasite coevolution: Interactions between migration, mutation, population size and generation time. *Journal of Evolutionary Biology, 15*(3), 451–462.

Gatto, L., Catanzaro, D., & Milinkovitch, M. C. (2006). Assessing the applicability of the GTR nucleotide substitution model through simulations. *Evolutionary Bioinformatics, 2*, 117693430600200020.

Gilleard, J. S. (2006). Understanding anthelmintic resistance: The need for genomics and genetics. *International Journal for Parasitology, 36*(12), 1227–1239.

Gillespie, J. H. (1991). *The causes of molecular evolution*. Oxford University Press.

Goldman, N., & Yang, Z. (1994). A codon-based model of nucleotide substitution for protein-coding DNA sequences. *Molecular Biology and Evolution, 11*(5), 725–736.

Haaland, T. R., Wright, J., & Ratikainen, I. I. (2019). Bet-hedging across generations can affect the evolution of variance-sensitive strategies within generations. *Proceedings of the Royal Society B: Biological Sciences, 286*(1916), 20192070. https://doi.org/10.1098/rspb.2019.2070

Halabi, K., Karin, E. L., Guéguen, L., & Mayrose, I. (2021). A codon model for associating phenotypic traits with altered selective patterns of sequence evolution. *Systematic Biology, 70*(3), 608–622.

Hall, B. G. (1990). Spontaneous point mutations that occur more often when advantageous than when neutral. *Genetics, 126*(1), 5–16.

Hasegawa, M., Kishino, H., & Yano, T. A. (1985). Dating of the human-ape splitting by a molecular clock of mitochondrial DNA. *Journal of Molecular Evolution, 22*(2), 160–174.

Hawash, M. B., Andersen, L. O., Gasser, R. B., Stensvold, C. R., & Nejsum, P. (2015). Mitochondrial genome analyses suggest multiple *Trichuris* species in humans, baboons, and pigs from different geographical regions. *PLoS Neglected Tropical Diseases, 9*(9), e0004059.

Heininger, K. (2015). Duality of stochasticity and natural selection: A cybernetic evolution theory. *WebmedCentral Ecology, 6*, WMC004796.

Higgs, P. G., & Attwood, T. K. (2005). *Bioinformatics and molecular evolution*. Blackwell.

Jackson, A. P. (2015). The evolution of parasite genomes and the origins of parasitism. *Parasitology, 142*(S1), S1–S5.

Jayaswal, V., Ababneh, F., Jermiin, L. S., & Robinson, J. (2011). Reducing model complexity of the general Markov model of evolution. *Molecular Biology and Evolution, 28*(11), 3045–3059.

Jia, W., Yan, H., Ni, X., Lou, Z., Li, H., Cao, P., & Cai, X. (2012). Advances in the study of helminth mitochondrial genomes and their associated applications. *Chinese Science Bulletin, 57*(1), 54–67.

Jones, J. T., Haegeman, A., Danchin, E. G., Gaur, H. S., Helder, J., Jones, M. G., Kikuchi, T., Manzanilla-López, R., Palomares-Rius, J. E., & Wesemael, W. M. (2013). Top 10 plant-parasitic nematodes in molecular plant pathology. *Molecular Plant Pathology, 14*(9), 946–961.

Jukes, T. H., & Cantor, C. R. (1969). Evolution of protein molecules. In *Mammalian protein metabolism* (Vol. 3, pp. 21–132).

Keane, T. M., Creevey, C. J., Pentony, M. M., Naughton, T. J., & Mclnerney, J. O. (2006). Assessment of methods for amino acid matrix selection and their use on empirical data shows that ad hoc assumptions for choice of matrix are not justified. *BMC Evolutionary Biology, 6*(1), 1–17.

Kern, E., Kim, T., & Park, J.-K. (2020). The mitochondrial genome in nematode phylogenetics. *Frontiers in Ecology and Evolution, 8*, 250.

Kimura, M. (1980). A simple method for estimating evolutionary rates of base substitutions through comparative studies of nucleotide sequences. *Journal of Molecular Evolution, 16*(2), 111–120.

Korkmaz, G., Holm, M., Wiens, T., & Sanyal, S. (2014). Comprehensive analysis of stop codon usage in bacteria and its correlation with release factor abundance. *The Journal of Biological Chemistry, 289*(44), 30334–30342. https://doi.org/10.1074/jbc.M114.606632

Kosiol, C., Holmes, I., & Goldman, N. (2007). An empirical codon model for protein sequence evolution. *Molecular Biology and Evolution, 24*(7), 1464–1479.

Kryazhimskiy, S., & Plotkin, J. B. (2008). The population genetics of dN/dS. *PLoS Genetics, 4*(12), e1000304.

Lamolle, G., Fontenla, S., Rijo, G., Tort, J. F., & Smircich, P. (2019). Compositional analysis of flatworm genomes shows strong codon usage biases across all classes. *Frontiers in Genetics, 10*, 771.

Lavrov, D. V., & Brown, W. M. (2001). *Trichinella spiralis* mtDNA: A nematode mitochondrial genome that encodes a putative ATP8 and normally structured tRNAs and has a gene arrangement relatable to those of coelomate metazoans. *Genetics, 157*(2), 621–637.

Lawrie, D. S., Petrov, D. A., & Messer, P. W. (2011). Faster than neutral evolution of constrained sequences: The complex interplay of mutational biases and weak selection. *Genome Biology and Evolution, 3*, 383–395.

Le, T. H., Blair, D., & McManus, D. P. (2002). Mitochondrial genomes of parasitic flatworms. *Trends in Parasitology, 18*(5), 206–213.

Lemmon, A. R., & Moriarty, E. C. (2004). The importance of proper model assumption in Bayesian phylogenetics. *Systematic Biology*, 265–277.

Li, W.-H., & Graur, D. (1999). *Fundamentals of molecular evolution* (2nd ed.). Sinauer Associates.

Liberles, D. A., Teichmann, S. A., Bahar, I., Bastolla, U., Bloom, J., Bornberg-Bauer, E., Colwell, L. J., De Koning, A. J., Dokholyan, N. V., & Echave, J. (2012). The interface of protein structure, protein biophysics, and molecular evolution. *Protein Science, 21*(6), 769–785.

Liò, P., & Goldman, N. (1998). Models of molecular evolution and phylogeny. *Genome Research, 8*(12), 1233–1244.

Lopez, P., & Philippe, H. (2001). Composition strand asymmetries in prokaryotic genomes: Mutational bias and biased gene orientation. *Comptes Rendus de l'Academie des Sciences. Serie III, Sciences de la Vie, 324*(3), 201–208.

Mallet, J., Besansky, N., & Hahn, M. W. (2016). How reticulated are species? *BioEssays, 38*(2), 140–149.

McVean, G. A., & Charlesworth, B. (1999). A population genetic model for the evolution of synonymous codon usage: Patterns and predictions. *Genetics Research, 74*(2), 145–158.

Minin, V., Abdo, Z., Joyce, P., & Sullivan, J. (2003). Performance-based selection of likelihood models for phylogeny estimation. *Systematic Biology, 52*(5), 674–683.

Mitreva, M., Wendl, M. C., Martin, J., Wylie, T., Yin, Y., Larson, A., Parkinson, J., Waterston, R. H., & McCarter, J. P. (2006). Codon usage patterns in Nematoda: Analysis based on over 25 million codons in thirty-two species. *Genome Biology, 7*(8), 1–19.

Muse, S. V., & Gaut, B. S. (1994). A likelihood approach for comparing synonymous and non-synonymous nucleotide substitution rates, with application to the chloroplast genome. *Molecular Biology and Evolution, 11*(5), 715–724.

Nabiyouni, M., Prakash, A., & Fedorov, A. (2013). Vertebrate codon bias indicates a highly GC-rich ancestral genome. *Gene, 519*(1), 113–119.

Nevo, E. (2001). Evolution of genome–phenome diversity under environmental stress. *Proceedings of the National Academy of Sciences, 98*(11), 6233–6240.

O'Grady, R. T. (1985). Ontogenetic sequences and the phylogenetics of parasitic flatworm life cycles. *Cladistics, 1*(2), 159–170.

Okimoto, R., Macfarlane, J., Clary, D., & Wolstenholme, D. (1992). The mitochondrial genomes of two nematodes, *Caenorhabditis elegans* and *Ascaris suum*. *Genetics, 130*(3), 471–498.

Page, R. D., & Holmes, E. C. (2009). *Molecular evolution: A phylogenetic approach*. Wiley.

Palidwor, G. A., Perkins, T. J., & Xia, X. (2010). A general model of codon bias due to GC mutational bias. *PLoS One, 5*(10), e13431.

Pellissier, L. (2015). Stability and the competition-dispersal trade-off as drivers of speciation and biodiversity gradients. *Frontiers in Ecology and Evolution, 3*, 52.

Piskurek, O., & Jackson, D. J. (2012). Transposable elements: From DNA parasites to architects of metazoan evolution. *Genes, 3*(3), 409–422.

Posada, D., & Crandall, K. A. (2001). Selecting the best-fit model of nucleotide substitution. *Systematic Biology, 50*(4), 580–601.

Poulin, R. (2011). The many roads to parasitism: A tale of convergence. *Advances in Parasitology, 74*, 1–40.

Poulin, R., & Randhawa, H. S. (2015). Evolution of parasitism along convergent lines: From ecology to genomics. *Parasitology, 142*(S1), S6–S15.

Qian, W., He, X., Chan, E., Xu, H., & Zhang, J. (2011). Measuring the evolutionary rate of protein–protein interaction. *Proceedings of the National Academy of Sciences, 108*(21), 8725–8730.

Qin, H., Wu, W. B., Comeron, J. M., Kreitman, M., & Li, W.-H. (2004). Intragenic spatial patterns of codon usage bias in prokaryotic and eukaryotic genomes. *Genetics, 168*(4), 2245–2260.

Rivero Fernández, J., Callejón Fernández, R., & Cutillas Barrios, C. (2021). Complete mitochondrial genome of *Trichuris trichiura* from *Macaca sylvanus* and *Papio papio*. *Life, 11*(2), 1–15.

Rosenberg, N. A. (2002). The probability of topological concordance of gene trees and species trees. *Theoretical Population Biology, 61*(2), 225–247.

Rosenberg, N. A., & Nordborg, M. (2002). Genealogical trees, coalescent theory and the analysis of genetic polymorphisms. *Nature Reviews Genetics, 3*(5), 380–390.

Saccone, C., De Giorgi, C., Gissi, C., Pesole, G., & Reyes, A. (1999). Evolutionary genomics in Metazoa: The mitochondrial DNA as a model system. *Gene, 238*(1), 195–209.

Salim, H. M., & Cavalcanti, A. R. (2008). Factors influencing codon usage bias in genomes. *Journal of the Brazilian Chemical Society, 19*(2), 257–262.

Saraste, M. (1999). Oxidative phosphorylation at the fin de siecle. *Science, 283*(5407), 1488–1493.

Seehausen, O. (2004). Hybridization and adaptive radiation. *Trends in Ecology & Evolution, 19*(4), 198–207.

Seo, T.-K., & Kishino, H. (2008). Synonymous substitutions substantially improve evolutionary inference from highly diverged proteins. *Systematic Biology, 57*(3), 367–377.

Shadel, G. S., & Clayton, D. A. (1997). Mitochondrial DNA maintenance in vertebrates. *Annual Review of Biochemistry, 66*(1), 409–435.

Shapiro, B., Rambaut, A., & Drummond, A. J. (2006). Choosing appropriate substitution models for the phylogenetic analysis of protein-coding sequences. *Molecular Biology and Evolution, 23*(1), 7–9.

Sharp, P. M., Emery, L. R., & Zeng, K. (2010). Forces that influence the evolution of codon bias. *Philosophical Transactions of the Royal Society, B: Biological Sciences, 365*(1544), 1203–1212.

Shoemaker, J., & Fitch, W. (1989). Evidence from nuclear sequences that invariable sites should be considered when sequence divergence is calculated. *Molecular Biology and Evolution, 6*(3), 270–289.

Simons, A. M. (2002). The continuity of microevolution and macroevolution. *Journal of Evolutionary Biology, 15*(5), 688–701.

Simpson, G. (1944). *Tempo and mode in evolution*. Columbia University Press.

Sumner, J. G., Jarvis, P. D., Fernández-Sánchez, J., Ferñandez-Sánchez, J., Kaine, B. T., Woodhams, M. D., & Holland, B. R. (2012). Is the general time-reversible model bad for molecular phylogenetics? *Systematic Biology, 61*(6), 1069–1074.

Tavaré, S. (1986). Some probabilistic and statistical problems in the analysis of DNA sequences. *Lectures on Mathematics in the Life Sciences, 17*(2), 57–86.

Thomas, M. C., Macias, F., Alonso, C., & López, M. C. (2010). The biology and evolution of transposable elements in parasites. *Trends in Parasitology, 26*(7), 350–362.

Vicario, S., Moriyama, E. N., & Powell, J. R. (2007). Codon usage in twelve species of *Drosophila*. *BMC Evolutionary Biology, 7*(1), 1–17.

Wit, J., & Gilleard, J. S. (2017). Resequencing helminth genomes for population and genetic studies. *Trends in Parasitology, 33*(5), 388–399.

Wolstenholme, D. R. (1992). Animal mitochondrial DNA: Structure and evolution. *International Review of Cytology, 141*, 173–216.

Xia, X., Perkins, T. J., & Palidwor, G. A. (2010). A general model of codon bias due to GC mutational bias. *PLoS One, 5*(10), e13431.

Yang, Z. (1994). Maximum likelihood phylogenetic estimation from DNA sequences with variable rates over sites: Approximate methods. *Journal of Molecular Evolution, 39*(3), 306–314.

Yang, Z., & Bielawski, J. P. (2000). Statistical methods for detecting molecular adaptation. *Trends in Ecology & Evolution, 15*(12), 496–503.

Chapter 5
Challenges of Species Identification for Parasitic Helminths

Abstract Species are the fundamental units of systematics, biodiversity, and ecological and evolutionary studies, but their delimitation has been neglected methodologically. Species are not discrete but ever-evolving groups of populations that are more or less connected by gene flow. As mentioned in Chap. 4, natural selection and genetic drift act on heritable variations over time, eventually causing reproductively isolated entities in geographical isolation. In systematics, we generally have two major goals: (1) to discover and describe species and (2) to resolve the phylogenetic relationships of the species studied. Almost all articles published on systematics have been focused on reconstructing phylogenies. Species delimitation is how species boundaries are determined and new species are discovered. Species delimitation has emerged as a critical topic in modern systematics. Recently, DNA methods have been used to study the systematics and population genetics of parasitic helminths, leading to the chance discovery of many genetically distinct but morphologically very similar species. In this chapter, we provided essential knowledge to deal with the challenges of species delimitation. Species concepts, which are the criteria tied to species delimitation and the principle of speciation, particularly parasitic speciation, were mentioned. Additionally, we discussed the pitfalls and problematic issues for species identification based on morphological characteristics, focusing on cryptic species, species hybridization, homoplasy under convergent evolution, and phenotypic plasticity under adaptive evolution.

5.1 Species Delimitation

Not only is the extinction rate of species on the planet due to anthropogenic activity challenging to estimate precisely, but also millions of species have yet to be described. These factors complicate the determination of species boundaries. Therefore, species delimitation and identification is a central activity pivotal for biodiversity conservation. Species delimitation is also important for providing essential

U. Thaenkham et al., *Molecular Systematics of Parasitic Helminths*,
https://doi.org/10.1007/978-981-19-1786-8_5

information on host and parasite diversity, reflecting environmental and ecological changes (Sites Jr & Marshall, 2003; Wiens, 2007).

Previously, species delimitation was based on morphological apparent-fixed differences, codified by Davis and Nixon (1992). This traditional approach is called population aggregation analysis (PAA), which systematically compares character state distributions among populations (polymorphic traits). However, PAA is problematic in that (1) the number of species present may be underestimated because morphological characters with fixed states may not be observed unless many characters are sampled and (2) polymorphic traits may overestimate the number of species unless many individuals are sampled (Wiens & Servedio, 2000). Recently, genetic data have become easier and less expensive to use and accumulate. Vilas et al. (2005) suggested that species delimitation can utilize DNA markers to find evidence of reciprocally monophyletic groups in large population samples, to look for evidence that species may be present. The formal testing of the absence of the gene flow hypothesis can be done by collecting more samples and applying multiple DNA markers. Therefore, using an arbitrary genetic distance yardstick is not recommended as a method to delimit species (Vilas et al., 2005).

Carstens et al. (2013) suggested that researchers apply a broad range of species delimitation analyses to their data and verify the confidence in delimitations to ensure congruence in results across methods. They suggested that the inferences drawn from species delimitation studies should be conservative, as it is better to fail for species delimitation than to delimit entities that do not represent actual evolutionary lineages. Species delimitation is closely and directly tied to the species concept, requiring species delimitation as a criterion (Carstens et al., 2013).

5.2 Species Concepts

Species concepts define species as taxa being individuals (Simpson, 1951), and there are many kinds of concepts defined by intention. Each concept has properties that should provide the necessary and sufficient conditions for speciesness. Species concepts are not only concerned with what constitutes a species but also concerned with the process of speciation. In taxonomic and systematic literature, speciation models rely on factors associated with species concepts (Aldhebiani, 2018). These factors focus on genetics and a mathematical expression of population properties and how those properties were maintained or altered to make a decisive contribution in understanding species evolution. Various models and methods are published and categorized (Fig. 5.1) (Mayden, 1997, 1999; Mayden & Wood, 1995). Confusion and disagreements with species concepts have become part of the *species debate*.

To determine a natural concept suited for general application in phylogenetics, Ghiselin (1974) began sorting through species concepts under the paradigm of species as individuals and concluded that concepts such as morphological species concept (MSC) or phenetic species concept treat individual species as classes. Classes herein are defined as organisms' traits rather than individuals having

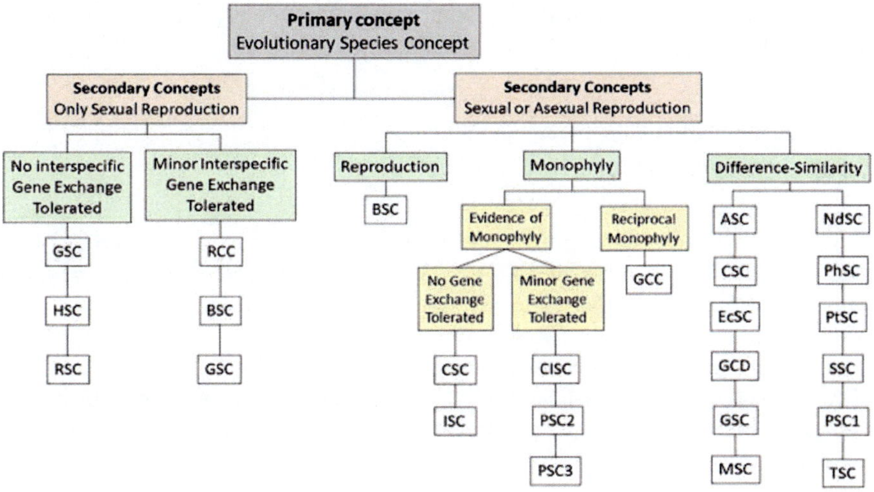

Fig. 5.1 Proposed hierarchy of species concepts. The various proposed species concepts show different levels of consilience (Mayden, 1999). The concept abbreviations are from Mayden (1997): *ASC* Agamospecies Concept, *BSC* Biological Species Concept, *ClSC* Cladistic Species Concept, *CSC* Cohesion Species Concept, *CpSC* Composite Species Concept, *EcSC* Ecological Species Concept, *ESU* Evolutionary Significant Unit, *ESC* Evolutionary Species Concept, *GCC* Genealogical Concordance Concept, *GSC* Genetic Species Concept, *GCD* Genotypic Cluster Definition, *HSC* Hennigian Species Concept, *ISC* Internodal Species Concept, *MSC* Morphological Species Concept, *NdSC* Nondimensional Species Concept, *PhSC* Phenetic Species Concept, *PSC* Phylogenetic Species Concepts, *PSC1* Diagnosable Version, *PSC2* Monophyly Version, *PSC3* Diagnosable/Monophyly Version, *PtSC* Polythetic Species Concept, *RSC* Recognition Species Concept, *RCC* Reproductive Competition Concept, *SSC* Successional Species Concept, *TSC* Taxonomic Species Concept (Mayden, 1999)

properties sufficient for membership in the species category (Ghiselin, 1974). Afterward, Wiley and Mayden (2000) suggested that the species concept for systematists should reflect how these concepts function in systematic and evolutionary theory and should be concerned with operationalism that surfaces from time to time in science (Wiley & Mayden, 2000). They suggested that the philosophy of operationalism is problematic because of the concepts defined by the observations. An operational concept includes a discovery criterion that permits the systematist to apply definite criteria to whether a particular specimen is drawn from a species or simply a variation within a species (Sites Jr & Marshall, 2004). However, Hull (1968) addressed the problems of operationalism in species definition (Hull, 1968). For instance, if there are two operational criteria for species recognition, there are two kinds of species, which means there are two kinds of weight for two kinds of operational criteria, which becomes a heavy load of operationalism.

Nowadays, we have numerous species concepts derived from many sources ranging from the investigative study of particular taxa and character sets to philosophical aptitude and world view to operationalism and nomenclature rules. Several popular species concepts generally used in systematics are explained below (Mayden, 1999).

5.2.1 Biological Species Concept

The biological species concept (BSC) was the first concept and was defined by Mary (Mary, 1942). He defined species as "*groups of actually or potentially interbreeding natural populations which are reproductively isolated from other such groups*." The BSC is a non-phylogenetic species concept because it is based on the potential interbreeding process with no references of ancestry (Bisby & Coddington, 1995; Mayr, 1982). This concept has two main disadvantages: (1) it is inapplicable to asexual organisms, and (2) it is impractical for allopatric populations (geographically isolated) (Cronquist, 1978; Mayr, 2015).

5.2.2 Morphological Species Concept

The morphological species concept (MSC) was defined by Cronquist, and it states that species are the smallest groups that are distinctive and distinguishable by average means (Cronquist, 1978). Therefore, species are the smallest natural populations permanently separated by a distinct discontinuity in the series of biotypes (Bisby & Coddington, 1995; Du Rietz, 1930). The MSC states that "*a species is a community, or a number of related communities, whose distinctive morphological characters are, in the opinion of a competent systematist, sufficiently definite to entitle it, or them, to a specific name*" (Regan, 1925). MSC can be applied to sexual and asexual organisms, contrasting with the BSC. Also, the MSC is helpful in instances where fossil records are available. However, the concerning point is that morphological characters are sometimes subjective and dependent on expert opinion for critical characteristics. The concept is also limited by morphologically indistinguishable species that have different genetic lineages (Aldhebiani, 2018).

5.2.3 Ecological Species Concept

The ecological species concept (ESC) was modified from the reproductive species concept originated by Simpson (1961). The ecological species concept is mainly considered in the aspect of ecological competition, where a species is adapted to a particular set of resources (a niche) in the environment (Simpson, 1961). Van Valen (1976) stated, "*A species is a lineage (or a closely related set of lineage) which occupies an adaptive zone minimally different from that of any other lineage in its range and which evolves separately from all lineages outside its range*" (Van Valen, 1976).

Colinvaux (1986) also emphasized the ecological species concept and stated, "*A species is a number of related populations the members of which compete more with their own kind than with members of other species.*" An ecological niche creates discrete zones, which creates discrete phenetic clusters that are recognized as species. Phenetic clusters are recognized as species because ecological and evolutionary processes controlling the division of resources produce those clusters (O'Neill et al., 1986). However, this species concept is limited by requiring the life histories of individual species members, which is not always true. It also has a problem similar to the morphological species concept of splitting divergent forms into new species based on phenetic clusters. It is not always significant to determine the degree to which two or more entities compete ecologically. Niches tend to be assumed, and it is challenging to define boundaries completely. Many taxa exploit overlapping resources or can suddenly switch if a resource becomes scarce. Moreover, similarities in morphological characters, particularly in cryptic species, make it difficult to completely define what constitutes a species. The advantage of the ecological species concept is that it considers the role the environment plays in controlling morphological development.

5.2.4 Evolutionary Species Concept

Simpson (1961) initially proposed the evolutionary species concept (ESC), which was modified by Wiley (1978, 1981). This concept states that "*a species is a single lineage of ancestral descendant populations of organisms that maintains its identity from other such lineages and has its evolutionary tendencies and historical fate*" (Wiley, 1978). This concept has been relatively accepted compared with other species concepts in the phylogenetic era. The evolutionary species concept has been used in studies of speciation where biologically comparable units are necessary for analysis (Brooks et al., 1991; McLennan & Brooks, 1991, 2002; Wiley, 1981) and is perhaps the most self-consistent of the lineage-based concepts. However, these very same characteristics rob this concept of its empirical content, making some consider it "highly nonoperational" (Panchen, 1992).

5.2.5 Cohesion Species Concept

Templeton (1989) states that species are defined as "*a species in the most inclusive group of organisms having the potential for genetic and/or demographic exchangeability.*" The cohesion species concept (CSC) determines that the most inclusive population of individuals has the potential for phenotypic cohesion through intrinsic cohesion mechanisms (Templeton, 1989). This species concept is similar to the evolutionary species concept, where the cohesion species concept emphasizes mechanisms promoting cohesion, while the evolutionary species concept emphasizes the "*manifestation of cohesion over evolutionary time*" (Endler, 1989; Wiley & Mayden, 2000).

5.2.6 *Phylogenetic Species Concept*

The phylogenetic species concept (PSC) was defined by Cracraft (1983) and can solve issues presented by the BSC (Cracraft, 1983, 1987; Rosen, 1978, 1979). A phylogenetic species is a basal cluster of organisms that is diagnosably distinct from other such clusters. Cracraft (1983) stated, "*A species is the smallest diagnosable cluster of individual organisms within which there is a parental pattern of ancestry and descent.*" A phylogenetic species is advocated to use a grouping criterion of monophyly in a cladistic sense and a ranking (pluralistic) criterion in producing and maintaining lineages in a particular case. This concept emphasizes the most general aspect of taxonomic diversification (differentiation), in which some differentiation results in reproductive isolation and some does not. Therefore, the phylogenetic species concept can reflect an underlying unity that the BSC cannot (Mishler & Brandon, 1987). A relationship hypothesis expressed as a branching diagram has no limitation on the composition of individual species as taxa.

5.3 Speciation

Speciation is the evolutionary process through which populations evolve to become distinct species, producing morphologically distinct species (Mayden, 1999). There are three recognized basic modes of speciation: allopatric speciation, sympatric speciation, and parapatric speciation. Firstly, allopatric speciation involves the physical subdivision of the ancestral gene pool into more descendant gene pools. Several criteria can be used to identify the occurrence of allopatric speciation. Coyne and Orr (2004) codified and summarized the various types of allopatric speciation: (1) species borders correlated with existing geographic or climatic barriers, (2) allopatry of young sister species, (3) geographic congruence between species borders among species of different clades, and (4) absence of sister species in areas where geographic isolation is unlikely. Secondly, sympatric speciation is the formation of new species in the absence of either a geographical or a physical barrier that isolates the population of the ancestral speciation. Sympatric speciation occurs within the range of an ancestral species where geography does not play a causal role in species formation (Coyne & Orr, 2004; Wiley & Lieberman, 2011). Reticulate and divergent sympatric speciation are the two broad classes identified. Reticulate sympatric speciation results in the origin of new species between two existing species via a hybridization event, creating allopolyploid and recombinant species.

Conversely, divergent sympatric speciation results in the origin of new species in response to some mechanisms such as disruptive selection, sexual selection, and autopolyploidy. Lastly, parapatric speciation occurs when there is partial speciation of the gene pool of an ancestral species, which results in a narrow primary zone of hybridization because of selection against hybrids within a hybrid zone. This can create a genetic sink, resulting in genetic divergence along the steep selection gradient (Endler, 1977). Parapatric speciation occurs under conditions of spatial adjacency between two initial species. Under this condition, gene flow and mixing between individuals of those two species may be allowed and engender speciation. These homogenizing forces (gene flow and mixing) may be restricted to the extent that genetic drift or local adaptation is occurring (Dieckmann et al., 2004). Dieckmann et al. (2004) suggested that most speciation processes may initially be parapatric, but they can differ significantly in the level of possible gene flow and amount of ecological contact between the incipient species.

5.4 Parasite Speciation

As mentioned above, the speciation process is fundamental to explaining the diversity of life. Parasites are ideal models for studying speciation processes as they have a high potential for diversification and specialization. The rapidly changing environment of the parasite (within a host) and their breeding site renders a higher chance for diversification for parasites than free-living organisms (De Meeûs et al., 1998; Huyse et al., 2005). However, fundamental studies of parasite speciation are scarce, and most studies were of human parasites and domestic animals (Huyse et al., 2005; Morand, 2015). Studies of naturally undistributed populations are required to understand speciation. Huyse et al. (2005) reviewed the speciation of parasites by focusing on parasite life histories and parasite population dynamics and their influence on population genetics. Coevolutionary interactions between hosts and parasites can drive divergence across time and space. They showed that parasite population ecology and population genetics are tightly linked to each other. They also argued that the structure of parasite populations correlates with (1) host mobility, (2) parasite reproduction mode, (3) parasite life cycle complexity, (4) parasite infrapopulation size, and (5) host specificity. These factors vary from one parasite species to another. Therefore, a comparative approach with a phylogenetic perspective is critical to unraveling the various processes that drive parasite diversification (Huyse et al., 2005).

The proposal that parasite phylogeny reflects host phylogeny (Klassen, 1992) was suggested on the basis that parasite speciation often occurs via cospeciation (Clayton et al., 2004; Page, 2003). However, the mechanisms of parasite adaptation following cospeciation events were rarely investigated (Morand et al., 2008). Host-parasite cospeciation and host-switch events are also the major hypotheses of parasite speciation. Host-switch events have been exemplified in several host-parasite systems (Desdevises et al., 2002), and host shifts can produce congruent host and parasite phylogenies (De Vienne et al., 2007). Speciation following host switch is also broadly observed in fish monogeneans (Desdevises et al., 2002; Huyse & Volckaert, 2005; Mendlova et al., 2012). Moreover, parasites, particularly helminths, typically have evidence for radiation of congeneric species (Inglis, 1971; Poulin, 1999; Schad, 1963).

Allopatric parasite speciation was suggested to follow the geographical isolation of their hosts as a plausible mechanism of cospeciation (Weckstein, 2004). In contrast, sympatric speciation was suggested to be relatively difficult to achieve because of genetic recombination occurring when incipient species interbreed. Genetic recombination breaks the correlation between coadapted genes necessary for species formation (Coyne, 2007). The cost of hybridization is producing offspring with an intermediate phenotype of intermediate fitness (Kondrashov & Shpak, 1998). Several studies suggested that sympatric speciation occurs under very restrictive conditions in natural conditions (Coyne & Price, 2000; Gourbiere, 2004; Jiggins, 2006). Coyne (2007) suggested that sympatric speciation requires several conditions consisting of (1) species must be sympatric, (2) sympatric species must show substantial reproductive isolation, (3) sympatric taxa must be sister species, and (4) biographic and evolutionary history of the group must make the existence of an allopatric phase very unlikely.

Brunner and Eizaguirre (2016) reviewed the effects of environmental changes (either rapid or gradual) on the modification of host immune response, parasite virulence, and specificity of interactions between host and parasite (Brunner & Eizaguirre, 2016). As mentioned in Chap. 3, environmental change can alter host-parasite evolution trajectories and potentially result in host and parasite speciation. Fluctuating environmental conditions could play a key role in host speciation. However, phenotypic variations (plasticity) can result, especially under fluctuating environmental conditions (Schaum & Collins, 2014).

5.5 Problems and Pitfalls for Species Identification by Morphology

Manter (1969) stated, "*Parasitologists have more chances to make mistakes than most other biologists, and they probably do so.*" These mistakes refer to chances for errors mainly in morphological identification as morphological variations of helminthic parasites are a big challenge in species identification. The pitfall for species identification based on morphological characteristics can be affected by evolutionary forces, for example, climate change affecting host and parasite speciation, as mentioned in Chaps. 3 and 4. The problematic issues for morphological identification have been discussed, particularly in the cryptic species groups, species hybridization, homoplasy under convergent evolution, and morphological plasticity under adaptive evolution. The evidence of phenotypic plasticity can promote population divergence by facilitating phenotypic diversification and eventually genetic divergence. Genetic divergence arises gradually and relatively slowly as it relies on random mutations according to genetic assimilation, contributing to the cryptic genetic variation of populations (Schneider & Meyer, 2017). Cryptic genetic variation is not commonly seen but is expressed under new environmental conditions. Gibson and Reed (2008) stated that the measurable component of normal variation is just the tip of an iceberg of genetic possibilities hidden below the visible surface. Moreover, species boundaries are frequently challenging to delimit under the speciation process based on morphology only. Aside from morphological characters, integrative taxonomy is required, and this includes additional sources of information such as molecular data, biogeography, behavior, and ecology (Gibson & Reed, 2008).

5.6 Cryptic Species

Cryptic species are a common phenomenon for metazoan taxa, which can be found in all sorts of habitats and biogeographic zones (Bickford et al., 2007; Pfenninger & Schwenk, 2007). Groups characterized by poor dispersal abilities are especially prone to cryptic speciation. The cryptic species concept is widely used in modern biodiversity studies and implies that morphologically indistinguishable species can be recognized only by molecular data (Bickford et al., 2007; Korshunova et al., 2019). Cryptic species have become popular and supplanted the term sibling species, commonly used in previous cases with difficult-to-distinguish species (Knowlton, 1993). In fact, the taxonomic challenge posed by cryptic species has been recognized for nearly 300 years, but morphological similarity could not be confirmed clearly by their morphological characters. However, with the advent of relatively inexpensive and rapid DNA sequencing, biologists now have a new tool that helps detect species, especially those with uncertain status (Criscione & Blouin, 2004; Xiao et al., 2006).

5.6.1 Cryptic Species of Parasitic Helminths

Carlson et al. (2020) estimated the number of cryptic species and showed unequal distribution among trematodes, cestodes, nematodes, and acanthocephalans. The overall richness for cryptic species was estimated by complying with meta-analysis using big data sources for biodiversity research, such as the London Natural History Museum (NHM)'s host-parasite associations database and the U.S. National Parasite Collection (USNPC). The analysis suggested an average of 2.6 cryptic species per species of acanthocephalan (22,404 species), 2.4 per species of cestode (80,747 species), 1.2 per species of nematode (63,457 species), and 3.1 species per digenean species (181,474 species) (Carlson et al., 2020). The number of cryptic species for trematodes is relatively higher than that of other helminth taxa, which supports the cryptic species observation by Poulin (2011). Reports of cryptic species in helminthic parasites are dramatically increasing because genetic methods have become a standard practice for parasite identification and are used deliberately to prospect for cryptic diversity (Blouin, 2002; de León & Poulin, 2018; Locke et al., 2010; Razo-Mendivil et al., 2015).

Sithithaworn et al. (2015) questioned the significance of the helminth species complex (a group of closely related species that separately evolve metapopulation lineages, including cryptic or sibling species). The outstanding potential problems from the medical point of view consist of the following: (1) Are there differences in virulence within species complex? (2) Does molecular diagnostic work for all species? (3) Is the susceptibility to anthelmintic drugs the same? (4) Has drug resistance developed in some species but not in others? (5) Which hosts must be considered in control programs? (6) What levels of morbidity and mortality are associated with each cryptic species? (Sithithaworn et al., 2015).

Published works for delimiting helminth species generally use neutral molecular genetic markers. Recently, numerous species complexes have been revealed. These include trematodes such as *Opisthorchis viverrini* sensu lato (s.l.), four species complexes of *Paragonimus* spp. (Yoshida et al., 2019), members of the Echinostomatidae (Georgieva et al., 2014), *Schistosoma japonicum* species complex (Van Herwerden et al., 1998), cestodes such as *Echinococcus granulosus* s.l. (Nakao et al., 2013), cryptic species of *Didymobothrium rudolphii* (Marques et al., 2007), nematodes such as *Anisakis simplex* s.l. complex (Mattiucci et al., 2014), *Pseudoterranova decipiens* complex (Zhu et al., 2002), *Angiostrongylus cantonensis* cryptic lineage (Dusitsittipon et al., 2017), and cryptic species of cyathostomins in horses (Louro et al., 2021).

Cryptic species in parasitic helminths are considered scarce compared to the estimated numbers that should be found (Carlson et al., 2020). It should be noted that cryptic species could be present and mixed with species that are assumed to be the same (collected either from the same or different host species). The correct identification of a species is critical to basic and applied research and has implications beyond morphological taxonomy. For example, any biological work done before performing species delimitation without confirming genetic information might be unreliable unless the genetic material is still available, allowing for definitive identification. In a medical aspect, if two or more species are confused under a single name, the data collected may well refer to a mixture of different species or the wrong species, leading to the choice of false diagnostic and treatment options. Therefore, recognizing the helminth species complex, particularly parasitic ones, should be prioritized (Sithithaworn et al., 2015).

5.7 Species Hybridization

Species can be defined as diagnosably distinct populations, reproductively isolated, cohesive, or exclusive groups of organisms. Intrinsic barriers to gene exchange maintain boundaries between species in sympatry; these boundaries may not be uniform in space, time, or across the genome. Harrison and Larson (2014) explored the nature of species boundaries that remain differentiated in the face of potential hybridization and introgression (introducing single genes or chromosomal regions from one species into that of another through repeated backcrossing). They suggested that species boundaries are semipermeable, with permeability (gene exchange) being a function of the genome region, which commonly occurs in a hybrid zone (Harrison & Larson, 2014). Reticulate genetic processes such as allele, gene, or genome introgression can profoundly impact population's and species' ecological/evolutionary dynamics (Detwiler & Criscione, 2010). Reticulate evolution points out the lack of independence between two evolutionary lineages affecting species' survival, fitness, and speciation rates. It can also happen between lineages that have been separated only for a short time, for example, through hybrid speciation in a species complex. Hybridization between species or diverged populations could result in the transfer of adaptive traits, promote divergence via reinforcement when hybrids are less fit than parentals, and lead to homogenization across the interbreeding populations' genomes or promote rapid adaptive diversification via the formation of hybrid species (Barton, 2001; Grant & Grant, 2019; Seehausen, 2004).

Most cases of natural hybridization are challenging to detect by meiotic mismatch because they occur between closely related species by having the same number of chromosomes as the parent species (Seehausen, 2004). Morphology alone cannot detect hybridization, and molecular genetic techniques will have to be used. Recently, discoveries have increased the number of hybrid species in both plants and metazoans, suggesting that there could be many more species of hybrid origin than what was previously estimated (Calvani & Slapeta, 2021; Demarais et al., 1992; Dowling & Secor, 1997; Oey et al., 2019; Whitton et al., 1997). Recent theoretical models and experimental results suggest that the induction of polyploidy as a means of instantaneous reproductive isolation is not required for hybrid speciation (the origin of a new species through the formation of hybrids between existing species), but that hybrid species can rapidly become reproductively isolated by other mechanisms. An example of another mechanism is transgressive segregation (Seehausen, 2004). Through modeling of genome evolution between two allopatrically evolved lineages, it was demonstrated that transgressive segregation through hybridization could facilitate adaptive radiation. Transgressive segregation is most potent when novel vacant ecological niches are highly dissimilar, with a small phenotypic effect size of mutations, and moderate genetic differentiation between parental lineages (Kagawa & Takimoto, 2018). Particularly under changing climate conditions, hybridization and introgression can play an important role in diversification by enhancing species' responses to environmental changes (Hamilton & Miller, 2016).

5.7.1 Consequences of Hybridization in Parasites

Recently, only slight attention has been given to the role that hybridization and introgression have played in the evolution of parasites compared to other organisms like vertebrates and insects. Although some evidence of hybridization and introgression has been reported, particularly for parasitic helminths, strong evidence exists for current hybridization in *Schistosoma*, *Ascaris*, and *Angiostrongylus* (Dusitsittipon et al., 2017; Easton et al., 2020; Webster et al., 2013). Coinfection of parasites in the same host may have an enormous impact resulting from synergistic or antagonistic interactions between parasites (Nowak & May, 1994). Several studies suggested that mixed-species coinfection of parasites influenced parasite establishment, growth, maturation, reproductive success, and drug efficacy (Koukounari et al., 2010; Norton et al., 2008; Webster et al., 2008). Detwiler and Criscione (2010) suggested that coinfections allow heterospecific mate pairing between species or between lineages, resulting in parthenogenesis, introgression, and whole-genome admixture through hybridization.

Recent molecular developments have revealed fertile hybridization and introgression in plants, metazoa, and humans (Arnold, 2004; Baack & Rieseberg, 2007; Boon et al., 2017; Easton et al., 2020; Racimo et al., 2015). For human parasitic helminths causing neglected tropical diseases (NTDs), reticulate evolution has been reported and studied in species such as *Schistosoma*, *Fasciola*, *Ascaris*, *Trichinella*, *Anisakis,* and *Angiostrongylus* (Bello et al., 2021; Criscione et al., 2007; Dunams-Morel et al., 2012; Dusitsittipon et al., 2017; Le et al., 2008; Roca-Geronès et al., 2021; Webster et al., 2013). King et al. (2015) reviewed the consequences of hybridization in parasites and concluded that it is an emerging public health concern with implications for infectious disease biology and evolution interface (King et al., 2015). However, more studies are required to gain more information on the introgression and hybridization of parasitic helminths.

Recent evidence, mainly from nonparasitic animals and plants, suggests that hybridization can have a major evolutionary impact (Arnold, 2004). In contrast, hybridization can result in lower average fitness due to F1 sterility and inviability, caused by negative epistatic interaction and the disruption of beneficial gene complexes (Coyne & Orr, 2004; Dobzhansky, 1937; Edmands, 2002. Thus, a hybrid offspring may be less fit than its purebred offspring counterpart from either parental line. Nonetheless, hybridization produces a whole range of potentially adaptive phenotypes ranging from one parent to the other, broadening the selection surface (King et al., 2015).

Hybridization has already been shown to promote geographic range expansion of populations and adaptation to a new environment in nonparasitic taxa. New alleles or allelic combinations that happen to be beneficial in the new environment are available immediately, contrasting to nonhybrid populations whose adaptation relies on alleles brought in via immigration or novel mutation requiring an extended period of time. For parasites, King et al. (2015) weighed potential drivers of hybridization (human migration, transportation of crops and livestock, climate change, landscape development, mass drug administration) and their consequences in the levels of importance and potentially risky hybridization effects in eukaryotic parasites. These drivers and consequences are (1) rapid parasite evolution generating novel and extreme infection phenotypes and new sites of infection, (2) increase in host range and potential of host switches, (3) increase in transmission potential, (4) increase in parasite evolutionary potential with a consequence for host-parasite coevolution, which leads to being higher virulence, and (5) the breakdown of host-specific adaptations and an altered response to drug therapy, which may become resistant (King et al., 2015).

As we mentioned above, hybridization in parasites can occur by reticulate evolution. Various molecular tools and genetic markers have been applied. For example, Steinauer et al. (2008) studied the introgressive hybridization of human and rodent schistosomes, which aimed to empirically examine the dynamics of hybridization between wild populations of those two schistosome species in Kenya and to determine if there is evidence of introgression by examining the presence or absence of various categories of hybrids. This study used the 16S–12S rRNA genes, ribosomal internal transcribed spacer region 1, and microsatellites to examine F1 hybrids and descendants. This study firstly explored the extent of hybridization of *Schistosoma mansoni* and *S. rodhaini* in a natural hybrid zone in Western Kenya. Individuals of hybrid ancestry were relatively common, although recent hybrids (F1 generation) were rare. Introgression has occurred from the direction of *S. rodhaini* to *S. mansoni*. They also suggested that the introgression detected using neutral markers has theoretically no fitness consequences (Steinauer et al., 2008). However, this finding suggests that other regions in these two species' genomes may be shared, playing an important role in transmission, disease processes, and host specificity. Oey et al. (2019) supported the interspecific hybridization between *S. bovis* and *S. haematobium* by whole-genome sequences. Genome comparison identified large continuous parts of the genome, showing almost 100% identity between these two parasites. The results suggested that interspecific hybridization is unlikely to result from genome conservation and provides further evidence of natural interspecific hybridization between *S. bovis* and *S. haematobium*, particularly in sub-Saharan Africa (Oey et al., 2019).

Additionally, Oey et al. (2019) suggested that foreign DNA obtained by interspecific hybridization was maintained in the population through multiple meiosis cycles and that hybrids were sexually reproductive, which may produce viable offspring. These hybridizations occurring in wild populations may lead to increased genetic diversity and accelerate the risk of zoonotic transmission (from animals to humans). Thus, parasite transmission rates, disease dynamics, control interventions, and parasite evolution are implications of hybridization events (Oey et al., 2019). Easton et al. (2020) also identified over 11 million nuclear-encoded SNPs in hybrids of *Ascaris lumbricoides* and *A. suum* through comparative phylogenomic analyses. They suggested that these two closely related species form a genetic complex that is capable of interbreeding, which has undergone a recent expansion worldwide because no barrier exists for anthroponotic transmission of these hybrid worms (Easton et al., 2020). Moreover, as *A. lumbricoides* is a major parasitic helminth in humans, this finding has important public health implications concerning the extensive albendazole use in either human or pig populations. Extensive use of the anthelmintic drug albendazole can lead to resistance in both human and pig *Ascais* populations if cross-species infections are common and able to produce fertile offspring.

Hybridization and introgression can occur not only in dioecious (have separate sexes) helminthic parasites like schistosomes and other nematodes but also in monoecious (hermaphroditic) worms like Platyhelminthes. Calvani and Slapeta (2021) proposed that hybridization and eventual introgression are important drivers of evolutionary change and adaptation of parasites, including *Fasciola* spp. Recently, increased demand for animal-derived protein from *F. hepatica*-endemic countries has led to a growing number of hybridization reports between *F. hepatica* and *F. gigantica* in Southeast Asia, where *F. gigantica* is endemic (Calvani & Slapeta, 2021). F1 hybrids between these two species were detected using nuclear ribosomal ITS sequences obtained from Sanger sequencing, where heterogeneous chromatogram peaks were present (Huang et al., 2004). However, the results can be identical to the hybrids in the case of coinfection by detecting eggs in a fecal sample (Calvani & Slapeta, 2021). Hybridization between *F. hepatica* and *F. gigantica* is increasingly reported in coexisting areas where parasites are in sympatry (Ichikawa-Seki et al., 2017; Le et al., 2008; Nguyen et al., 2012; Sanad & Al-Megrin, 2005). Itagaki et al. (2011) demonstrated the result of interspecific mating between *F. hepatica* and *F. gigantica*, and F1 hybrids exhibited an intermediate body-width to the body-length ratio between the two species (Itagaki et al., 2011). This intermediate appearance was also found in the samples collected from cattle in Thailand (Sumruayphol et al., 2020). However, using single-locus methods for differentiation between *F. hepatica* and *F. gigantica* adults is limited because introgression could be misrepresented after repeated backcrossing between F1 hybrids and pure *F. gigantica* (Fig. 5.2).

Hybridization introgression can be detected using rDNA genetic markers compared to multi-loci and/or whole-genome methods to illustrate two hypothetical *target* regions. Calvani and Slapeta (2021) suggested that identifying introgressed individuals requires pre-existing knowledge of markers susceptible to introgression.

5.8 Homoplasy

Homoplasy is the inverse of homology, where features shared by two organisms are not present in their nearest common ancestor (Wood, 1999). Meyer (1999) characterized three classes of homoplasy, including convergence (superficial similarity arising through independent evolution), parallelism (a feature present in closely related organisms but not present continuously in all in the lineage), and reversals (phenotypes similar to those seen in ancestors in the lineage) (Meyer, 1999; Papini, 2002) (Fig. 5.3).

The traits/features in homoplasy are the recurrences of similarity, which is not the persistence of similarity in evolution. By the phylogenetic content, homoplasy is different from homology by reflecting the presence of a feature in two species that do not share a recent common ancestor (convergence) or the presence of a feature in two species when the feature is not found in their most recent common ancestors but present in basal ancestor (parallelism) (Hall, 2012). The homoplasy character states

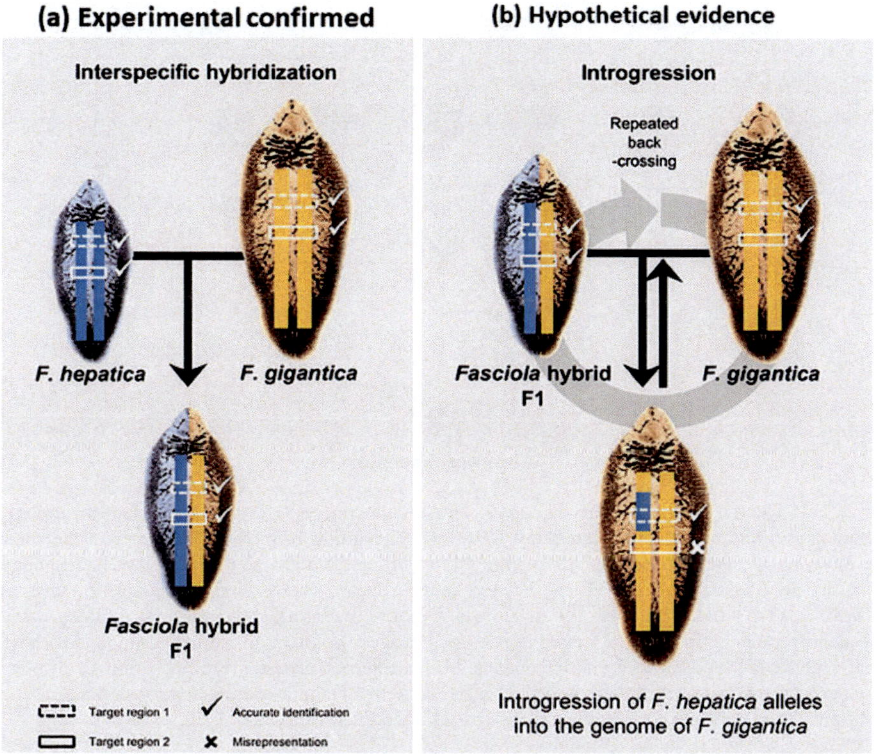

Fig. 5.2 Hybridization and introgression between *Fasciola hepatica* and *F. gigantica*. (**a**) The hybridization between two *Fasciola* species was confirmed by SNP/ single loci analysis. The blue and yellow thick lines in each species represent the nuclear genomes of *F. hepatica* and *F. gigantica*, respectively. Two-color lines indicate *Fasciola* hybrid adult, a theoretical representation of the F1 generation. (**b**) The introgression between the F1 *Fasciola* hybrid and *F. gigantica* can potentially occur (hypothesis evidence). The introgression forms can be detected using rDNA compared to multilocus and/or whole-genome methods. Ticks indicate accurately identified regions, while the crosses indicate misidentification due to using an inappropriate molecular genetic marker that cannot detect the introgression (Calvani & Slapeta, 2021)

have been problematic in taxonomic accuracy until molecular data confirm the species status, particularly for cryptic species that display low phenotypic disparity in relation to divergence time.

In hypothesized cryptic species driven by recent divergence, they are sister taxa or members of a species complex with short divergence times, which are too recent for substantial morphological differences to accumulate (Reidenbach et al., 2012). The accumulation rate of morphological disparity might not differ significantly from older non-cryptic species. Consequently, similarity in morphology might not be expected in very young species. Based on morphological similarity, Struck et al. (2018) suggested that recently diverged taxa showing significantly lower rates of morphological disparity might be constrained by stabilizing selection and

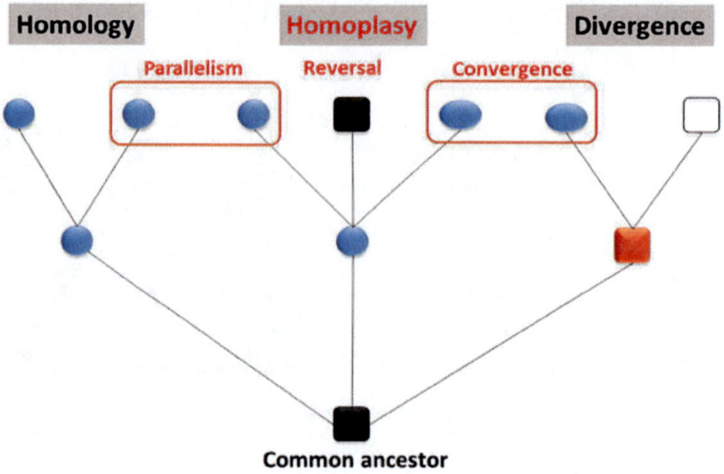

Fig. 5.3 The major evolutionary outcomes. Three main concepts of evolution include homology, homoplasy, and divergence. Homology is similarity resulting from shared ancestry in the same underlying mechanism. Homoplasy is similarity resulting from shared ecological pressures, comprising parallelism (closely related lineages independently evolve similar characters), reversal (a character in a descendent recovers features of a remote ancestor), and convergence (independent similarity traits or features in related species or lineages occurring in similar ecological niches). Divergence is the dissimilarity in descendants from a common ancestor resulting from the different underlying mechanisms or differential selection pressure. The diagram represents similarities and differences in shape and colored geometric forms. (Modified from Papini, 2002)

representing early stasis stages. They also suggested that cryptic species evolved by parallelism, which is not sister taxa but is phylogenetically separated by independently evolving from morphologically similar ancestors. Moreover, cryptic species may be defined by the morphological similarity resulting from the independent evolution of morphologically dissimilar ancestors.

At the early stage of divergence, cryptic and non-cryptic species are expected to show similar rates of morphological differentiation. Cryptic species would begin to converge morphologically at some point in time and eventually have low morphological disparity over time. Nevertheless, convergence as a mechanism for cryptic species is rare. The last evolutionary process involving morphological similarity is stasis. Under stasis, cryptic species are sister taxa or complex members that retain a high degree of morphological similarity over extended periods. Compared with non-cryptic species, the rate of morphological change is substantially reduced (Struck et al., 2018) (Fig. 5.4).

Homoplasy in parasitic helminths has been increasingly reported; for example, Hasegawa and Dewi (2017) explored the morphological similarity between two new species, *Trichuris musseri* and *T. mallomyos*, collected from endemic murines of Indonesia. *Trichuris musseri* is readily distinguished from *T. mallomyos* by having a much smaller body and a large number of nuclei per subdivision of stichosome (Hasegawa & Dewi, 2017). The presence of a similar spicule between these two

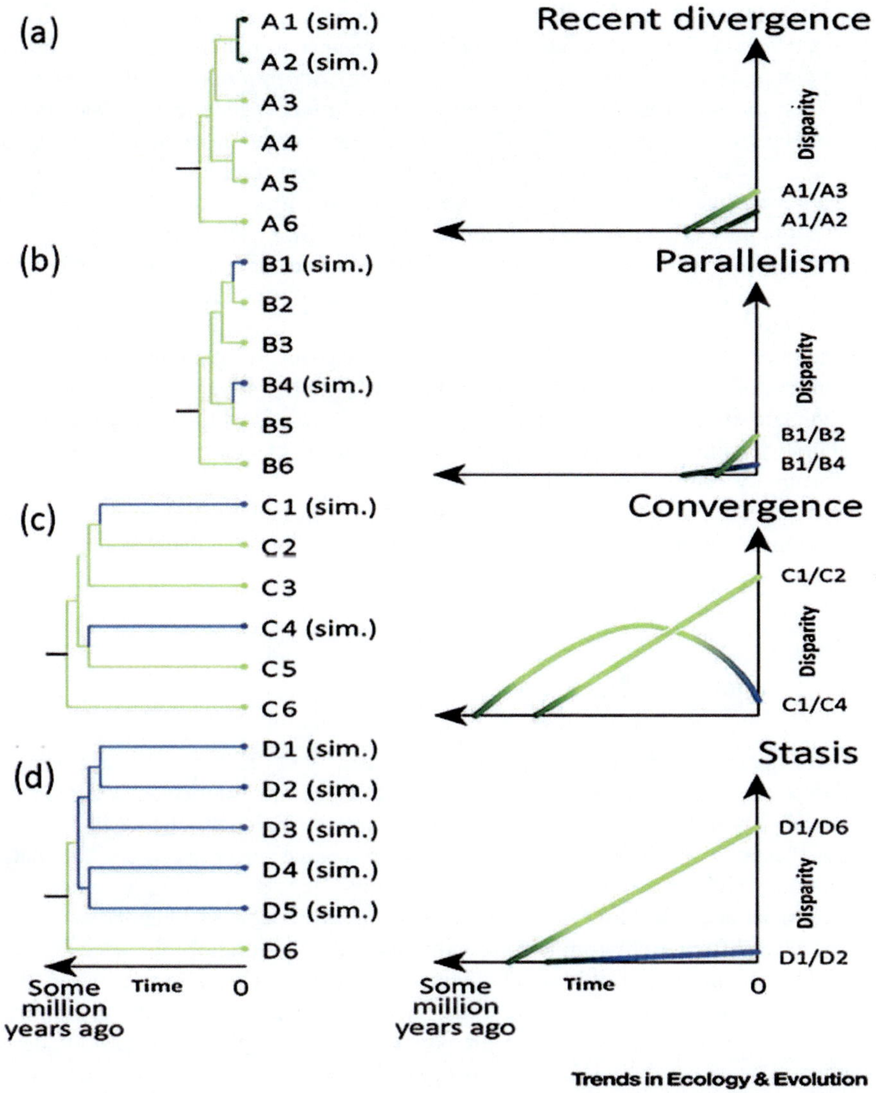

Fig. 5.4 Expected signatures of four evolutionary processes leading to cryptic species. The species with similar (identical) morphotypes are denoted with "sim." The phylogenetic relationships among taxa are shown in the left panel. At the same time, the graphs of the evolution of morphological disparity through an evolutionary time of a pair of cryptic species (for example, A1/A2) and non-cryptic species (for example, A1/A3) are indicated in the right panel. The disparity between the cryptic species progressed into the dark blue-shaded line, while the disparity between non-cryptic species evolved from dark to light green-shaded lines (Struck et al., 2018)

species was suggested because both hosts belong to different tribes and have different habitats. It remains to be elucidated whether morphological similarity is just homoplasy or reflects a close phylogenetic relationship between the parasites. However, molecular information is required to confirm how genetically related these two species are. Kiontke et al. (2007) studied the evolution of nematode vulva development in 51 rhabditid species by analyzing more than 40 characteristics of vulva developments. This study was also performed with concatenated small ribosomal subunit (SSU) and large ribosomal subunit (LSU) rRNA sequences and the largest subunit of RNA polymerase II. Variations and an even larger number of evolutionary changes were surprisingly found, suggesting a high degree of homoplasy (convergence and reversals). This study showed that the evolution of vulva development is strongly biased, and only a few aspects are likely to change in an unconstrained stochastic fashion (Kiontke et al., 2007). In parasitic trematodes, Thaenkham et al. (2010) revealed the phylogenetic relationships of species in the subfamily Haplorchiinae and analyzed the homologs of common morphological characters by comparisons with molecular phylogenies. Phylogenetic analysis and morphological comparisons suggested that the modified ventral sucker of the ventrogenital complex seems to be a homoplasious character (Thaenkham et al., 2010).

5.9 Phenotypic Plasticity

As we mentioned in Chap. 3, phenotypic plasticity is the capacity of a single genotype to exhibit variable phenotypes in different environments (Whitman & Agrawal, 2009). Phenotypic plasticity can facilitate evolutionary change and speciation. Plasticity can increase fitness, develop individual phenotypes, and facilitate evolution, which structures ecological communities. The adaptation process to novel environmental conditions can occur via (1) the expression of phenotypes when exposed to different environmental conditions and (2) evolution via selection for particular phenotypes resulting in the modification of genetic variation in the population. Once it loses environmental sensitivity (plasticity), traits can become fixed or expressed constitutively in the population or through genetic assimilation, leading to genetic polymorphisms (Fox et al., 2019).

Phenotypic plasticity is typical of all living organisms that respond to genes and the environment (Sommer, 2020). The plasticity of phenotypes among life-cycle stages is apparent, particularly among parasitic nematodes. Different life-cycle stages can have different morphological characters, physiologies, lifestyles, etc. Some life-cycle stages in a species can be parasitic, while the rest are free-living, such as *Strongyloides* and *Parastrongyloides*. These two genera, in the free-living phase of the life cycle, have a phenotypic plasticity developmental choice between (1) development as the larval stage (leading to the development of an infective stage) and (2) development of a free-living dioecious adult generation (Viney & Diaz, 2012). Environmental conditions control the developmental choice between these two routes. Sommer et al. (2017) studied phenotypic plasticity using the free-living nematode *Pristionchus pacificus* as a model. This worm has been established as a genetic system with forward and reverse genetic tools, allowing unbiased genetic studies of mouth-form plasticity (Sommer, 2020; Sommer et al., 2017). They showed *P. pacificus* exhibited feeding plasticity allowing for predatory or bacterivorous feeding. Also, they found that developmental switch genes (genetic switch genes) control feeding plasticity, and these genes are under epigenetic control to facilitate morphological novelty and diversity induced by the environment.

5.10 Other Inherent Factors

Potential problems confounding the use of morphology in helminthic parasites to identify species have been mentioned above. However, fixation issues and host-induced variation, particularly body size, can also affect morphological variation without changing the parasite's genetics (Perkins et al., 2011).

Firstly, for helminthic parasite fixation, several factors affect the ability to recover morphological characters, including the source where the helminth was isolated from (freshly caught, kept on ice, or isolated host's organ preserved in ethanol), completeness of specimens removed from the host, preservation method, and specimen relaxation and mounting techniques (Perkins et al., 2011). There are various techniques used to flatten or compress trematodes on slides. For example, the weight of the cover glass can be used to apply slight pressure to the trematodes. The technique can be modified in the case of a relatively large body of parasites like *Fasciola* spp. for geometric morphometric analysis (Sumruayphol et al., 2020). A specimen is fixed between two transparent glass slides containing two 0.4 mm-thick paper pieces to prepare the *Fasciola* specimen. The two slides are gently held using a rubber band and then placed in normal saline to relax the worm body. The specimens are then fixed in 70% ethanol and stored at 4 °C for further staining (Sumruayphol et al., 2020).

Morphological variations can easily occur based on different fixation techniques and the researchers' expertise. Also, as mentioned above, preparation methods are not consistent between groups of parasites as these methods can be modified according to the researcher's preferences. Criscione and Font (2001) suggested that morphological variations prior to fixation, especially in the size and shape of the specimen structures, can also occur due to the host's condition (fresh versus preserved material). Thus, the species description from morphological characteristics may also differ depending on the source quality that the helminth was isolated from (Criscione & Font, 2001).

Secondly, size variation of the helminthic parasites can also be induced by the host's body condition variation. For example, a larger host or better nutritional resources for parasites may allow more growth, causing them to be larger in size (Sánchez et al., 2018). Read and Rothman (1957) observed that the length of *Hymenolepis diminuta* was greater in older rats or those that obtained a better quality of carbohydrates in their diet (Read & Rothman, 1957). In another example, Bruce et al. (1961) observed a large degree of size variation in *Schistosoma mansoni* that infected thirteen experimental animals. They also showed the size variation of *S. mansoni* obtained from different infection sites among the various host species (Bruce et al., 1961). Other factors that can also influence the size of parasites include the total number of parasites per host (parasite intensity), a host's previous exposure to the parasite, and the coexistence of other parasite species within the same host.

5.11 Problems for Species Delimitation with Molecular Analysis

5.11.1 Sufficient Genetic Materials

Regarding the challenges in systematic revisionary work, morphology alone may become the pitfall to taxonomic identification and systematics. Many helminthic parasites are fragile, and it is challenging to isolate quality genomic DNA without DNA degradation, especially after the helminths have been prepared for morphological identification. Additionally, helminth samples are frequently either vouchered as formalin-fixed or mounted specimens, for example, the specimen of formalin-fixed paraffin-embedded cyst tissue samples of *Echinococcus granulosus* (Schneider et al., 2008) and *Taenia* specimens after long-term preservation in formalin (Jeon et al., 2011). If genomic DNA for molecular studies is required, there might be a lower yield of genomic DNA and DNA that is already degraded. To circumvent the challenge of low quality and yield of DNA, researchers have resolved this through a solution that will not distort specimens for morphological studies and yield usable DNA for molecular studies. By using ethanol as a preservation tool, the act between morphological identification and molecular use can be balanced, rendering it a good choice for morphological and molecular research to be studied in parallel, useful for small or large helminthic parasites (Chan et al., 2020, 2021; Sumruayphol et al., 2020). Nevertheless, regardless of the helminth preservation method, the quality of specimens for successful morphological and molecular research depends on how fresh the specimens are before preservation.

5.11.2 Suitable Molecular Markers

In the past decades, considerable evidence for species delimitation has utilized molecular markers for the systematics of helminthic parasites. The most popular gene is the small subunit ribosomal gene (SSU or 18S rRNA gene) that is applied for molecular systematics of a broad spectrum of organisms. In most eukaryotic organisms, the ribosomal RNA genes have two properties that make them suitable molecular markers for molecular systematics. Firstly, they exist in multiple copies in a genome. Bik et al. (2013) measured the repeat copy number for five free-living nematode species (*Caenorhabditis brenneri*, *C. remanei*, *C. briggsae*, *C. japonica*, and *P. pacificus*) and one parasitic nematode (*Brugia malayi*) by mapping the sequences from whole-genome shotgun libraries against reference sequences for their rRNA repeat (Bik et al., 2013). They revealed a sixfold variation in repeat copy number among taxa investigated with levels of intragenomic variation ranging from 56 to 323 copies of the rRNA array. Secondly, rRNA genes contain highly conserved regions, allowing a simple PCR primer to cover the variable regions for phylogenetic information. However, there is a caveat for using the nuclear internal transcribed spacer 1 and 2 regions (ITS1 and ITS2) for nematodes, trematodes, and cestodes. The ambiguity in their alignment because of clusters of tandem repeats contained in the ITS regions renders this region particularly difficult for alignment and subsequent phylogenetic use (Manuel et al., 2021).

Other popular choices as molecular markers for molecular systematics of parasitic helminths are the mitochondrial genes, such as cytochrome b (*cytB*), cytochrome c oxidase subunit I (*cox1*), and 12S and 16S rRNA genes. As these genes evolve at a higher rate than the nuclear rRNA genes, they are advantageous for discriminating against closely related species. Recently, the *cox1* gene has been the preferred marker in DNA barcoding studies across a diversity of parasite taxa (Blasco-Costa et al., 2016; Evans & Paulay, 2012; Steinauer et al., 2007). Nevertheless, the lack of conserved regions of the *cox1* gene to design primers across parasitic helminth taxa limits the availability of universal *cox1* primers for both nematodes and Platyhelminthes. Chan et al. (2021) suggested that the 12S and 16S rRNA genes are the potential molecular markers for molecular systematics and the universal molecular marker for parasitic helminths. More details on utilizing molecular genetic markers for parasitic helminths will be mentioned in Chaps. 10 to 12.

Whole-genome resources have been made available, with increasing studies being conducted now that whole-genome sequencing costs have dramatically decreased. However, genomes of nonmajor medical or veterinary importance are still scarce. They are not widely applied for broad systematic studies among parasites. In the future, whole genomes and next-generation sequencing technologies will gain momentum, and a better taxonomic sampling of parasites will undoubtedly be obtained. New technologies will be helpful to overcome the challenges of insufficient quantity of parasite genetic material, together with the diminishing pool of taxonomic experts for helminth morphological identification.

5.12 Conclusion

Species delimitation is a critical topic, and this chapter addressed the problems of species delimitation in parasitic helminths. The species concepts that are intricately tied to species delimitation were described and defined. Along with examples from studies on parasitic helminths, the pitfalls for morphology-based species identification under the speciation process, such as cryptic species, species hybridization, homoplasy, morphological plasticity, and other inherent factors, were discussed. Also, the problems associated with molecular-based species delimitation, such as defective genetic materials and unsuitable molecular markers, were addressed. Chapter 6 will focus on the molecular genetic markers generally used in parasitic helminth research, such as molecular identification and systematics.

References

Aldhebiani, A. Y. (2018). Species concept and speciation. *Saudi Journal of Biological Sciences, 25*(3), 437–440. https://doi.org/10.1016/j.sjbs.2017.04.013

Arnold, M. L. (2004). Natural hybridization and the evolution of domesticated, pest and disease organisms. *Molecular Ecology, 13*(5), 997–1007. https://doi.org/10.1111/j.1365-294X.2004.02145.x

Baack, E. J., & Rieseberg, L. H. (2007). A genomic view of introgression and hybrid speciation. *Current Opinion in Genetics & Development, 17*(6), 513–518. https://doi.org/10.1016/j.gde.2007.09.001

Barton, N. H. (2001). The role of hybridization in evolution. *Molecular Ecology, 10*(3), 551–568.

Bello, E., Palomba, M., Webb, S. C., Paoletti, M., Cipriani, P., Nascetti, G., & Mattiucci, S. (2021). Investigating the genetic structure of the parasites *Anisakis pegreffii* and *A. berlandi* (Nematoda: Anisakidae) in a sympatric area of the southern Pacific Ocean waters using a multilocus genotyping approach: First evidence of their interspecific hybridization. *Infection, Genetics and Evolution, 92*, 104887. https://doi.org/10.1016/j.meegid.2021.104887

Bickford, D., Lohman, D. J., Sodhi, N. S., Ng, P. K., Meier, R., Winker, K., Ingram, K. K., & Das, I. (2007). Cryptic species as a window on diversity and conservation. *Trends in Ecology & Evolution, 22*(3), 148–155.

Bik, H. M., Fournier, D., Sung, W., Bergeron, R. D., & Thomas, W. K. (2013). Intra-genomic variation in the ribosomal repeats of nematodes. *PLoS One, 8*(10), e78230. https://doi.org/10.1371/journal.pone.0078230

Bisby, F., & Coddington, J. A. (1995). Biodiversity from a taxonomic and evolutionary perspective. In *Global biodiversity assessment* (pp. 27–57).

Blasco-Costa, I., Cutmore, S. C., Miller, T. L., & Nolan, M. J. (2016). Molecular approaches to trematode systematics:'best practice'and implications for future study. *Systematic Parasitology, 93*(3), 295–306.

Blouin, M. S. (2002). Molecular prospecting for cryptic species of nematodes: Mitochondrial DNA versus internal transcribed spacer. *International Journal for Parasitology, 32*(5), 527–531.

Boon, N. A. M., Fannes, W., Rombouts, S., Polman, K., Volckaert, F. A. M., & Huyse, T. (2017). Detecting hybridization in African schistosome species: Does egg morphology complement molecular species identification? *Parasitology, 144*(7), 954–964. https://doi.org/10.1017/s0031182017000087

Brooks, D. R., McLennan, D. A., & McLennan, D. A. (1991). *Phylogeny, ecology, and behavior: A research program in comparative biology*. University of Chicago Press.

Bruce, J. I., Llewellyn, L. M., & Sadun, E. H. (1961). Susceptibility of wild mammals to infection by *Schistosoma mansoni*. *The Journal of Parasitology, 47*(5), 752–756.

Brunner, F. S., & Eizaguirre, C. (2016). Can environmental change affect host/parasite-mediated speciation? *Zoology, 119*(4), 384–394.

Calvani, N. E. D., & Slapeta, J. (2021). *Fasciola* species introgression: Just a fluke or something more? *Trends in Parasitology, 37*(1), 25–34. https://doi.org/10.1016/j.pt.2020.09.008

Carlson, C. J., Dallas, T., Alexander, L., Phelan, A., & Phillips, A. (2020). What would it take to describe the global diversity of parasites? *Proceedings of the Royal Society B, 287*(1939), 2020.1841.

Carstens, B. C., Pelletier, T. A., Reid, N. M., & Satler, J. D. (2013). How to fail at species delimitation. *Molecular Ecology, 22*(17), 4369–4383.

Chan, A. H. E., Chaisiri, K., Morand, S., Saralamba, N., & Thaenkham, U. (2020). Evaluation and utility of mitochondrial ribosomal genes for molecular systematics of parasitic nematodes. *Parasites & Vectors, 13*(1), 364. https://doi.org/10.1186/s13071-020-04242-8

Chan, A. H. E., Chaisiri, K., Saralamba, S., Morand, S., & Thaenkham, U. (2021). Assessing the suitability of mitochondrial and nuclear DNA genetic markers for molecular systematics and species identification of helminths. *Parasites & Vectors, 14*(1), 233. https://doi.org/10.1186/s13071-021-04737-y

Clayton, D. H., Bush, S. E., & Johnson, K. P. (2004). Ecology of congruence: Past meets present. *Systematic Biology, 53*(1), 165–173.

Colinvaux P. (1986) Ecology. John Wiley and Sons, New York, p. 152

Coyne, J. A. (2007). Sympatric speciation. *Current Biology, 17*(18), R787–R788. https://doi.org/10.1016/j.cub.2007.06.056

Coyne, J. A., & Orr, H. A. (2004). *Speciation* (Vol. 37). Sinauer Associates.

Coyne, J. A., & Price, T. D. (2000). Little evidence for sympatric speciation in island birds. *Evolution, 54*(6), 2166–2171.

Cracraft, J. (1983). Species concepts and speciation analysis. In *Current ornithology* (pp. 159–187). Springer.

Cracraft, J. (1987). Species concepts and the ontology of evolution. *Biology and Philosophy, 2*(3), 329–346.

Criscione, C. D., & Blouin, M. S. (2004). Life cycles shape parasite evolution: Comparative population genetics of salmon trematodes. *Evolution, 58*(1), 198–202.

Criscione, C. D., & Font, W. F. (2001). The guest playing host: Colonization of the introduced Mediterranean gecko, *Hemidactylus turcicus*, by helminth parasites in southeastern Louisiana. *The Journal of Parasitology, 87*, 1273–1278.

Criscione, C. D., Anderson, J. D., Sudimack, D., Peng, W., Jha, B., Williams-Blangero, S., & Anderson, T. J. (2007). Disentangling hybridization and host colonization in parasitic roundworms of humans and pigs. *Proceedings of the Biological Sciences, 274*(1626), 2669–2677. https://doi.org/10.1098/rspb.2007.0877

Cronquist, A. (1978). Once again, what is a species? Biosystematics in agriculture. *Beltsville Symposia in Agricultural Research, 2*, 3–20.

Davis, J. I., & Nixon, K. C. (1992). Populations, genetic variation, and the delimitation of phylogenetic species. *Systematic Biology., 41*(4), 421–435. https://doi.org/10.2307/2992584

de León, G. P.-P., & Poulin, R. (2018). An updated look at the uneven distribution of cryptic diversity among parasitic helminths. *Journal of Helminthology, 92*(2), 197–202.

De Meeûs, T., Michalakis, Y., & Renaud, F. (1998). Santa Rosalia revisited: Or why are there so many kinds of parasites in 'The garden of earthly Delights'? *Parasitology Today, 14*(1), 10–13.

De Vienne, D., Giraud, T., & Shykoff, J. (2007). When can host shifts produce congruent host and parasite phylogenies? A simulation approach. *Journal of Evolutionary Biology, 20*(4), 1428–1438.

Demarais, B. D., Dowling, T. E., Douglas, M. E., Minckley, W., & Marsh, P. C. (1992). Origin of *Gila seminuda* (Teleostei: Cyprinidae) through introgressive hybridization: Implications for evolution and conservation. *Proceedings of the National Academy of Sciences, 89*(7), 2747–2751.

Desdevises, Y., Morand, S., Jousson, O., & Legendre, P. (2002). Coevolution between *Lamellodiscus* (Monogenea: Diplectanidae) and Sparidae (Teleostei): The study of a complex host-parasite system. *Evolution, 56*(12), 2459–2471.

Detwiler, J. T., & Criscione, C. D. (2010). An infectious topic in reticulate evolution: Introgression and hybridization in animal parasites. *Genes (Basel), 1*(1), 102–123. https://doi.org/10.3390/genes1010102

Dieckmann, U., Doebeli, M., Metz, J. A., & Tautz, D. (2004). *Adaptive speciation*. Cambridge University Press.

Dobzhansky, T. (1937). Genetic nature of species differences. *The American Naturalist, 71*(735), 404–420. https://doi.org/10.1086/280726

Dowling, T. E., & Secor, C. L. (1997). The role of hybridization and introgression in the diversification of animals. *Annual Review of Ecology and Systematics, 28*(1), 593–619.

Du Rietz, G. E. (1930). *The fundamental units of biological taxonomy*. Svensk Botaniska Foreningen Uppsala.

Dunams-Morel, D. B., Reichard, M. V., Torretti, L., Zarlenga, D. S., & Rosenthal, B. M. (2012). Discernible but limited introgression has occurred where *Trichinella nativa* and the T6 genotype occur in sympatry. *Infection, Genetics and Evolution, 12*(3), 530–538. https://doi.org/10.1016/j.meegid.2012.01.004

Dusitsittipon, S., Criscione, C. D., Morand, S., Komalamisra, C., & Thaenkham, U. (2017). Cryptic lineage diversity in the zoonotic pathogen *Angiostrongylus cantonensis*. *Molecular Phylogenetics and Evolution, 107*, 404–414. https://doi.org/10.1016/j.ympev.2016.12.002

Easton, A., Gao, S., Lawton, S. P., Bennuru, S., Khan, A., Dahlstrom, E., Oliveira, R. G., Kepha, S., Porcella, S. F., & Webster, J. (2020). Molecular evidence of hybridization between pig and human *Ascaris* indicates an interbred species complex infecting humans. *eLife, 9*, e61562.

Edmands, S. (2002). Does parental divergence predict reproductive compatibility. *Trends in Ecology & Evolution, 17*, 520–527.

Endler, J. (1977). *Geographic variation, speciation, and clines*. Princeton University Press.

Endler, J. A. (1989). In D. Otte & J. Endler (Eds.), *Conceptual and other problems in speciation*. Sinauer Associates.

Evans, N., & Paulay, G. (2012). DNA barcoding methods for invertebrates. In *DNA barcodes* (pp. 47–77). Springer.

Fox, R. J., Donelson, J. M., Schunter, C., Ravasi, T., & Gaitan-Espitia, J. D. (2019). Beyond buying time: The role of plasticity in phenotypic adaptation to rapid environmental change. *Philosophical Transactions of the Royal Society of London. Series B, Biological Sciences, 374*(1768), 20180174. https://doi.org/10.1098/rstb.2018.0174

Georgieva, S., Faltýnková, A., Brown, R., Blasco-Costa, I., Soldánová, M., Sitko, J., Scholz, T., & Kostadinova, A. (2014). *Echinostoma'revolutum'*(Digenea: Echinostomatidae) species complex revisited: Species delimitation based on novel molecular and morphological data gathered in Europe. *Parasites & Vectors, 7*(1), 1–18.

Ghiselin, M. T. (1974). A radical solution to the species problem. *Systematic Biology, 23*(4), 536–544.

Gibson, G., & Reed, L. K. (2008). Cryptic genetic variation. *Current Biology, 18*(21), R989–R990. https://doi.org/10.1016/j.cub.2008.08.011

Gourbiere, S. (2004). How do natural and sexual selection contribute to sympatric speciation? *Journal of Evolutionary Biology, 17*(6), 1297–1309.

Grant, P. R., & Grant, B. R. (2019). Hybridization increases population variation during adaptive radiation. *Proceedings of the National Academy of Sciences, 116*(46), 23216–23224.

Hall, B. K. (2012). Parallelism, deep homology, and evo-devo. *Evolution & Development, 14*(1), 29–33.

Hamilton, J. A., & Miller, J. M. (2016). Adaptive introgression as a resource for management and genetic conservation in a changing climate. *Conservation Biology, 30*(1), 33–41.

Harrison, R. G., & Larson, E. L. (2014). Hybridization, introgression, and the nature of species boundaries. *The Journal of Heredity, 105*(Suppl 1), 795–809. https://doi.org/10.1093/jhered/esu033

Hasegawa, H., & Dewi, K. (2017). Two new species of *Trichuris* (Nematoda: Trichuridae) collected from endemic murines of Indonesia. *Zootaxa, 4254*(1), 127–135.

Huang, W. Y., He, B., Wang, C. R., & Zhu, X. Q. (2004). Characterisation of *Fasciola* species from mainland China by ITS-2 ribosomal DNA sequence. *Veterinary Parasitology, 120*(1–2), 75–83. https://doi.org/10.1016/j.vetpar.2003.12.006

Hull, D. L. (1968). The operational imperative: Sense and nonsense in operationism. *Systematic Biology, 17*(4), 438–457.

Huyse, T., & Volckaert, F. A. (2005). Comparing host and parasite phylogenies: *Gyrodactylus* flatworms jumping from goby to goby. *Systematic Biology, 54*(5), 710–718.

Huyse, T., Poulin, R., & Theron, A. (2005). Speciation in parasites: A population genetics approach. *Trends in Parasitology, 21*(10), 469–475.

Ichikawa-Seki, M., Peng, M., Hayashi, K., Shoriki, T., Mohanta, U. K., Shibahara, T., & Itagaki, T. (2017). Nuclear and mitochondrial DNA analysis reveals that hybridization between *Fasciola hepatica* and *Fasciola gigantica* occurred in China. *Parasitology, 144*(2), 206–213.

Inglis, W. G. (1971). Speciation in parasitic nematodes. *Advances in Parasitology, 9*, 185–223.

Itagaki, T., Ichinomiya, M., Fukuda, K., Fusyuku, S., & Carmona, C. (2011). Hybridization experiments indicate incomplete reproductive isolating mechanism between *Fasciola hepatica* and *Fasciola gigantica*. *Parasitology, 138*(10), 1278–1284.

Jeon, H.-K., Kim, K.-H., & Eom, K. S. (2011). Molecular identification of *Taenia* specimens after long-term preservation in formalin. *Parasitology International, 60*(2), 203–205.

Jiggins, C. D. (2006). Sympatric speciation: Why the controversy? *Current Biology, 16*(9), R333–R334.

Kagawa, K., & Takimoto, G. (2018). Hybridization can promote adaptive radiation by means of transgressive segregation. *Ecology Letters, 21*(2), 264–274.

King, K. C., Stelkens, R. B., Webster, J. P., Smith, D. F., & Brockhurst, M. A. (2015). Hybridization in parasites: Consequences for adaptive evolution, pathogenesis, and public health in a changing world. *PLoS Pathogens, 11*(9), e1005098. https://doi.org/10.1371/journal.ppat.1005098

Kiontke, K., Barrière, A., Kolotuev, I., Podbilewicz, B., Sommer, R., Fitch, D. H., & Félix, M.-A. (2007). Trends, stasis, and drift in the evolution of nematode vulva development. *Current Biology, 17*(22), 1925–1937.

Klassen, G. J. (1992). Coevolution: A history of the macroevolutionary approach to studying host-parasite associations. *The Journal of Parasitology, 78*, 573–587.

Knowlton, N. (1993). Sibling species in the sea. *Annual Review of Ecology and Systematics, 24*(1), 189–216.

Kondrashov, A. S., & Shpak, M. (1998). On the origin of species by means of assortative mating. *Proceedings of the Royal Society of London. Series B: Biological Sciences, 265*(1412), 2273–2278.

Korshunova, T., Picton, B., Furfaro, G., Mariottini, P., Pontes, M., Prkić, J., Fletcher, K., Malmberg, K., Lundin, K., & Martynov, A. (2019). Multilevel fine-scale diversity challenges the 'cryptic species' concept. *Scientific Reports, 9*(1), 1–23.

Koukounari, A., Donnelly, C. A., Sacko, M., Keita, A. D., Landouré, A., Dembelé, R., Bosqu-é-Oliva, E., Gabrielli, A. F., Gouvras, A., Traoré, M., Fenwick, A., & Webster, J. P. (2010). The impact of single versus mixed schistosome species infections on liver, spleen and bladder morbidity within Malian children pre- and post-praziquantel treatment. *BMC Infectious Diseases, 10*, 227. https://doi.org/10.1186/1471-2334-10-227

Le, T. H., De, N. V., Agatsuma, T., Thi Nguyen, T. G., Nguyen, Q. D., McManus, D. P., & Blair, D. (2008). Human fascioliasis and the presence of hybrid/introgressed forms of *Fasciola hepatica* and *Fasciola gigantica* in Vietnam. *International Journal for Parasitology, 38*(6), 725–730. https://doi.org/10.1016/j.ijpara.2007.10.003

Locke, S. A., Daniel McLaughlin, J., & Marcogliese, D. J. (2010). DNA barcodes show cryptic diversity and a potential physiological basis for host specificity among Diplostomoidea (Platyhelminthes: Digenea) parasitizing freshwater fishes in the St. Lawrence River, Canada. *Molecular Ecology, 19*(13), 2813–2827. https://doi.org/10.1111/j.1365-294X.2010.04713.x

Louro, M., Kuzmina, T. A., Bredtmann, C. M., Diekmann, I., de Carvalho, L. M. M., von Samson-Himmelstjerna, G., & Krucken, J. (2021). Genetic variability, cryptic species and phylogenetic relationship of six cyathostomin species based on mitochondrial and nuclear sequences. *Scientific Reports, 11*(1), 8245. https://doi.org/10.1038/s41598-021-87500-8

Manuel, M., Ramanujam, K., & Ajjampur, S. S. R. (2021). Molecular tools for diagnosis and surveillance of soil-transmitted helminths in endemic areas. *Parasitologia, 1*(3), 105–118. https://doi.org/10.3390/parasitologia1030012

Marques, J., Santos, M., Gibson, D., Cabral, H., & Olson, P. (2007). Cryptic species of *Didymobothrium rudolphii* (Cestoda: Spathebothriidea) from the sand sole, Solea lascaris, off the Portuguese coast, with an analysis of their molecules, morphology, ultrastructure and phylogeny. *Parasitology, 134*(7), 1057–1072.

Mary, E. (1942). *Systematics and the origin of species*. Columbia University Press.

Mattiucci, S., Cipriani, P., Webb, S. C., Paoletti, M., Marcer, F., Bellisario, B., Gibson, D. I., & Nascetti, G. (2014). Genetic and morphological approaches distinguish the three sibling species of the *Anisakis simplex* species complex, with a species designation as *Anisakis berlandi* n. sp. for *a. simplex* sp. C (Nematoda: Anisakidae). *The Journal of Parasitology, 100*(2), 199–214.

Mayden, R. L. (1997). *A hierarchy of species concepts: The denouement in the saga of the species problem*. Chapman & Hall.

Mayden, R. L. (1999). Consilience and a hierarchy of species concepts: Advances toward closure on the species puzzle. *Journal of Nematology, 31*(2), 95–116.

Mayden, R. L., & Wood, R. M. (1995). *Systematics, species concepts, and the evolutionarily significant unit in biodiversity and conservation biology* (Vol. 17). American Fisheries Society.

Mayr, E. (1982). *The growth of biological thought: Diversity, evolution, and inheritance*. Harvard University Press.

Mayr, E. (2015). *Principles of systematic zoology*. Scientific Publishers.

McLennan, D. A., & Brooks, D. R. (1991). Parasites and sexual selection: A macroevolutionary perspective. *The Quarterly Review of Biology, 66*(3), 255–286.

McLennan, D. A., & Brooks, D. R. (2002). Complex histories of speciation and dispersal in communities: A re-analysis of some Australian bird data using BPA. *Journal of Biogeography, 29*(8), 1055–1066.

Mendlova, M., Desdevises, Y., Civáňová, K., Pariselle, A., & Šimková, A. (2012). Monogeneans of west African cichlid fish: Evolution and cophylogenetic interactions. *PLoS One, 7*(5), e37268.

Meyer, A. (1999). Homology and homoplasy: The retention of genetic programmes. In *Homology* (pp. 141–157).

Mishler, B. D., & Brandon, R. N. (1987). Individuality, pluralism, and the phylogenetic species concept. *Biology and Philosophy, 2*(4), 397–414.

Morand, S. (2015). (Macro-) evolutionary ecology of parasite diversity: From determinants of parasite species richness to host diversification. *International Journal for Parasitology: Parasites and Wildlife, 4*(1), 80–87.

Morand, S., Vetešníková Šimková, A., & Gourbiere, S. (2008). *Beyond the paradigms of cospeciation and host-switch: Is sympatric speciation an important mode of speciation for parasites?* (pp. 125–132).

Nakao, M., Lavikainen, A., Yanagida, T., & Ito, A. (2013). Phylogenetic systematics of the genus *Echinococcus* (Cestoda: Taeniidae). *International Journal for Parasitology, 43*(12–13), 1017–1029.

Nguyen, S., Amer, S., Ichikawa, M., Itagaki, T., Fukuda, Y., & Nakai, Y. (2012). Molecular identification of *Fasciola* spp.(Digenea: Platyhelminthes) in cattle from Vietnam. *Parasite: Journal de la Société Française de Parasitologie, 19*(1), 85.

Norton, A., Webster, J., Kane, R., & Rollinson, D. (2008). Inter-specific parasite competition: Mixed infections of *Schistosoma mansoni* and *S. rodhaini* in the definitive host. *Parasitology, 135*(4), 473–484.

Nowak, M., & May, R. (1994). Superinfection and the evolution of parasite virulence. *Proceedings of the Royal Society of London. Series B: Biological Sciences, 255*(1342), 81–89.

Oey, H., Zakrzewski, M., Gravermann, K., Young, N. D., Korhonen, P. K., Gobert, G. N., Nawaratna, S., Hasan, S., Martínez, D. M., & You, H. (2019). Whole-genome sequence of the bovine blood fluke *Schistosoma bovis* supports interspecific hybridization with *S. haematobium*. *PLoS Pathogens, 15*(1), e1007513.

O'Neill, R. V., Deangelis, D. L., Waide, J. B., Allen, T. F., & Allen, G. E. (1986). *A hierarchical concept of ecosystems*. Princeton University Press.

Page, R. D. (2003). *Tangled trees: Phylogeny, cospeciation, and coevolution*. University of Chicago Press.

Panchen, A. L. (1992). *Classification, evolution, and the nature of biology*. Cambridge University Press.

Papini, M. R. (2002). Pattern and process in the evolution of learning. *Psychological Review, 109*(1), 186–201. https://doi.org/10.1037/0033-295x.109.1.186

Perkins, S. L., Martinsen, E. S., & Falk, B. G. (2011). Do molecules matter more than morphology? Promises and pitfalls in parasites. *Parasitology, 138*(13), 1664–1674. https://doi.org/10.1017/S0031182011000679

Pfenninger, M., & Schwenk, K. (2007). Cryptic animal species are homogeneously distributed among taxa and biogeographical regions. *BMC Evolutionary Biology, 7*(1), 1–6.

Poulin, R. (1999). The functional importance of parasites in animal communities: Many roles at many levels? *International Journal for Parasitology, 29*(6), 903–914.

Poulin, R. (2011). Uneven distribution of cryptic diversity among higher taxa of parasitic worms. *Biology Letters, 7*(2), 241–244. https://doi.org/10.1098/rsbl.2010.0640

Racimo, F., Sankararaman, S., Nielsen, R., & Huerta-Sánchez, E. (2015). Evidence for archaic adaptive introgression in humans. *Nature Reviews. Genetics, 16*(6), 359–371. https://doi.org/10.1038/nrg3936

Razo-Mendivil, U., Rosas-Valdez, R., Rubio-Godoy, M., & De León, G. P.-P. (2015). The use of mitochondrial and nuclear sequences in prospecting for cryptic species in *Tabascotrema verai* (Digenea: Cryptogonimidae), a parasite of *Petenia splendida* (Cichlidae) in middle America. *Parasitology International, 64*(2), 173–181.

Read, C., & Rothman, A. (1957). The role of carbohydrates in the biology of cestodes. I: The effect of dietary carbohydrate quality on the size of *Hymenolepis diminuta*. *Experimental Parasitology, 6*(1), 1–7.

Regan, C. T. (1925). Organic evolution. *Nature, 116*(2915), 398–401.

Reidenbach, K. R., Neafsey, D. E., Costantini, C., Sagnon, N. F., Simard, F., Ragland, G. J., Egan, S. P., Feder, J. L., Muskavitch, M. A. T., & Besansky, N. J. (2012). Patterns of genomic differentiation between ecologically differentiated M and S forms of *Anopheles gambiae* in west and Central Africa. *Genome Biology and Evolution, 4*(12), 1202–1212. https://doi.org/10.1093/gbe/evs095

Roca-Geronès, X., Alcover, M. M., Godínez-González, C., Montoliu, I., & Fisa, R. (2021). Hybrid genotype of *Anisakis simplex* (s.s.) and *A. pegreffii* identified in third- and fourth-stage larvae from sympatric and allopatric Spanish marine waters. *Animals, 11*(8), 2458. Retrieved from https://www.mdpi.com/2076-2615/11/8/2458

Rosen, D. E. (1978). Vicariant patterns and historical explanation in biogeography. *Systematic Zoology, 27*(2), 159–188.

Rosen, D. E. (1979). Fishes from the uplands and intermontane basins of Guatemala: Revisionary studies and comparative geography. *Bulletin of the AMNH, 162*, 5.

Sanad, M. M., & Al-Megrin, W. A. (2005). Fascioliasis among local and imported sheep in Saudi Arabia: parasitological and serological diagnosis. *Journal of the Egyptian Society of Parasitology, 35*(3 Suppl), 1121–1134.

Sánchez, C. A., Becker, D. J., Teitelbaum, C. S., Barriga, P., Brown, L. M., Majewska, A. A., Hall, R. J., & Altizer, S. (2018). On the relationship between body condition and parasite infection in wildlife: A review and meta-analysis. *Ecology Letters, 21*(12), 1869–1884.

Schad, G. (1963). Niche diversification in a parasitic species flock. *Nature, 198*(4878), 404–406.

Schaum, C. E., & Collins, S. (2014). Plasticity predicts evolution in a marine alga. *Proceedings of the Royal Society B: Biological Sciences, 281*(1793), 20141486.

Schneider, R. F., & Meyer, A. (2017). How plasticity, genetic assimilation and cryptic genetic variation may contribute to adaptive radiations. *Molecular Ecology, 26*(1), 330–350. https://doi.org/10.1111/mec.13880

Schneider, R., Gollackner, B., Edel, B., Schmid, K., Wrba, F., Tucek, G., Walochnik, J., & Auer, H. (2008). Development of a new PCR protocol for the detection of species and genotypes (strains) of *Echinococcus* in formalin-fixed, paraffin-embedded tissues. *International Journal for Parasitology, 38*(8–9), 1065–1071. https://doi.org/10.1016/j.ijpara.2007.11.008

Seehausen, O. (2004). Hybridization and adaptive radiation. *Trends in Ecology & Evolution, 19*(4), 198–207.

Simpson, G. G. (1951). The species concept. *Evolution, 5*(4), 285–298.

Simpson, G. G. (1961). *Principles of animal taxonomy*. Columbia University Press.

Sites, J., Jr., & Marshall, J. (2003). Delimiting species: A renaissance issue in systematic biology. *Trends in Ecology & Evolution, 18*(9), 462–470.

Sites, J. W., Jr., & Marshall, J. C. (2004). Operational criteria for delimiting species. *Annual Review of Ecology, Evolution, and Systematics, 35*, 199–227.

Sithithaworn, P., Petney, T. N., & Andrews, R. H. (2015). *What significance do helminths species-complexes have for the prevention, diagnosis and treatment of human infections?* Oxford University Press.

Sommer, R. J. (2020). Phenotypic plasticity: From theory and genetics to current and future challenges. *Genetics, 215*(1), 1–13. https://doi.org/10.1534/genetics.120.303163

Sommer, R. J., Dardiry, M., Lenuzzi, M., Namdeo, S., Renahan, T., Sieriebriennikov, B., & Werner, M. S. (2017). The genetics of phenotypic plasticity in nematode feeding structures. *Open Biology, 7*(3), 160332. https://doi.org/10.1098/rsob.160332

Steinauer, M. L., Nickol, B. B., & Orti, G. (2007). Cryptic speciation and patterns of phenotypic variation of a highly variable acanthocephalan parasite. *Molecular Ecology, 16*(19), 4097–4109. https://doi.org/10.1111/j.1365-294X.2007.03462.x

Steinauer, M. L., Hanelt, B., Mwangi, I. N., Maina, G. M., Agola, L. E., Kinuthia, J. M., Mutuku, M. W., Mungai, B. N., Wilson, W. D., Mkoji, G. M., & Loker, E. S. (2008). Introgressive hybridization of human and rodent schistosome parasites in western Kenya. *Molecular Ecology, 17*(23), 5062–5074. https://doi.org/10.1111/j.1365-294X.2008.03957.x

Struck, T. H., Feder, J. L., Bendiksby, M., Birkeland, S., Cerca, J., Gusarov, V. I., Kistenich, S., Larsson, K.-H., Liow, L. H., Nowak, M. D., Stedje, B., Bachmann, L., & Dimitrov, D. (2018). Finding evolutionary processes hidden in cryptic species. *Trends in Ecology & Evolution, 33*(3), 153–163. https://doi.org/10.1016/j.tree.2017.11.007

Sumruayphol, S., Siribat, P., Dujardin, J. P., Dujardin, S., Komalamisra, C., & Thaenkham, U. (2020). *Fasciola gigantica, F. hepatica* and *Fascio*la intermediate forms: Geometric morphometrics and an artificial neural network to help morphological identification. *PeerJ, 8*, e8597. https://doi.org/10.7717/peerj.8597

Templeton, A. R. (1989). The meaning of species and speciation: A genetic perspective. In *The units of evolution: Essays on the nature of species, 1992* (pp. 159–183).

Thaenkham, U., Dekumyoy, P., Komalamisra, C., Sato, M., & Waikagul, J. (2010). Systematics of the subfamily Haplorchiinae (Trematoda: Heterphyidae), based on nuclear ribosomal DNA genes and ITS2 region. *Parasitology International, 59*(3), 460–465.

Van Herwerden, L., Blair, D., & Agatsuma, T. (1998). Intra-and inter-specific variation in nuclear ribosomal internal transcribed spacer 1 of the *Schistosoma japonicum* species complex. *Parasitology, 116*(4), 311–317.

Van Valen, L. (1976). Ecological species, multispecies, and oaks. *Taxon, 25*, 233–239.

Vilas, R., Criscione, C., & Blouin, M. (2005). A comparison between mitochondrial DNA and the ribosomal internal transcribed regions in prospecting for cryptic species of platyhelminth parasites. *Parasitology, 131*(6), 839–846.

Viney, M., & Diaz, A. (2012). Phenotypic plasticity in nematodes: Evolutionary and ecological significance. *Worm, 1*, 98–106.

Webster, J. P., Gower, C. M., & Norton, A. J. (2008). Evolutionary concepts in predicting and evaluating the impact of mass chemotherapy schistosomiasis control programmes on parasites and their hosts. *Evolutionary Applications, 1*(1), 66–83. https://doi.org/10.1111/j.1752-4571.2007.00012.x

Webster, B. L., Diaw, O. T., Seye, M. M., Webster, J. P., & Rollinson, D. (2013). Introgressive hybridization of *Schistosoma haematobium* group species in Senegal: Species barrier break down between ruminant and human schistosomes. *PLoS Neglected Tropical Diseases, 7*(4), e2110.

Weckstein, J. D. (2004). Biogeography explains cophylogenetic patterns in toucan chewing lice. *Systematic Biology, 53*(1), 154–164.

Whitman, D. W., & Agrawal, A. A. (2009). What is phenotypic plasticity and why is it important. In *Phenotypic plasticity of insects: Mechanisms and consequences* (pp. 1–63).

Whitton, J., Wolf, D., Arias, D., Snow, A., & Rieseberg, L. (1997). The persistence of cultivar alleles in wild populations of sunflowers five generations after hybridization. *Theoretical and Applied Genetics, 95*(1–2), 33–40.

Wiens, J. J. (2007). Species delimitation: New approaches for discovering diversity. *Systematic Biology, 56*(6), 875–878.

Wiens, J. J., & Servedio, M. R. (2000). Species delimitation in systematics: Inferring diagnostic differences between species. *Proceedings of the Royal Society of London. Series B: Biological Sciences, 267*(1444), 631–636.

Wiley, E. O. (1978). The evolutionary species concept reconsidered. *Systematic Zoology, 27*(1), 17–26.

Wiley, E. (1981). *Phylogenetic systematics*. Wiley.

Wiley, E. O., & Lieberman, B. S. (2011). *Phylogenetics: Theory and practice of phylogenetic systematics*. Wiley.

Wiley, E. O., & Mayden, R. L. (2000). A defense of the evolutionary species concept. In *Species concepts and phylogenetic theory. A debate* (pp. 198–208). Columbia University Press.

Wood, B. (1999). *Homoplasy: Foe and friend?* Wiley Online Library.

Xiao, N., Qiu, J., Nakao, M., Li, T., Yang, W., Chen, X., Schantz, P. M., Craig, P. S., & Ito, A. (2006). *Echinococcus shiquicus*, a new species from the Qinghai–Tibet plateau region of China: Discovery and epidemiological implications. *Parasitology International, 55*, S233–S236.

Yoshida, A., Doanh, P. N., & Maruyama, H. (2019). *Paragonimus* and paragonimiasis in Asia: An update. *Acta Tropica, 199*, 105074. https://doi.org/10.1016/j.actatropica.2019.105074

Zhu, X., D'amelio, S., Palm, H., Paggi, L., George-Nascimento, M., & Gasser, R. (2002). SSCP-based identification of members within the *Pseudoterranova decipiens* complex (Nematoda: Ascaridoidea: Anisakidae) using genetic markers in the internal transcribed spacers of ribosomal DNA. *Parasitology, 124*(6), 615–623.

Chapter 6
Molecular Genetic Markers: General Use in Parasitic Helminth Researches

Abstract A molecular genetic marker, or a gene, is a segment of the DNA located within the genome, in nuclear or mitochondrial regions. As mentioned in Chap. 5, advancements in molecular biological tools and technologies such as PCR and DNA sequencing in the past decades opened possibilities for utilizing DNA sequences in the study of biology. Inadvertently, the increasing use of DNA data facilitated the development and use of numerous molecular genetic markers for species identification and phylogenetic inferences. However, not all molecular genetic markers are suitable for phylogenetic inferences. Various factors such as the differences in the evolutionary rates of each marker, the different selective forces leading to different degrees of sequence conservation, and inherent differences between groups of organisms make it challenging to make sweeping generalizations about the utility of each molecular genetic marker. Thus, this chapter focused on the properties of popular genetic markers used for molecular based studies, as it is crucial to know their properties, which can then be appropriately chosen to deduce and infer true phylogenetic relationships.

6.1 General Properties of Nuclear and Mitochondrial Molecular Genetic Markers

In eukaryotes, molecular genetic markers can be present in nuclear or mitochondrial DNA. Different properties of the molecular genetic markers result due to their different evolutionary rates and function in the cell. This section will review the general properties of nuclear and mitochondrial DNA markers that are frequently used in molecular studies of parasitic helminths.

U. Thaenkham et al., *Molecular Systematics of Parasitic Helminths*,
https://doi.org/10.1007/978-981-19-1786-8_6

6.1.1 Nuclear DNA: Ribosomal RNA Genes and Spacer Regions

The nuclear DNA in eukaryotic cells has high copy numbers of up to 5000, and the ribosomal RNA genes (rRNA) and spacer regions are organized in tandem, with repeating units. Each repeating unit consists of coding genes and spacer regions (Hwang & Kim, 1999; McManus & Bowles, 1996; Sonnenberg et al., 2007). As depicted in Fig. 6.1, the coding genes contain the small nuclear subunit (18S rRNA), the large subunit (28S rRNA), and 5.8S rRNA. These genes are, in turn, separated from each other by the spacer regions – internal transcribed spacer 1 (ITS1) and the internal transcribed spacer 2 (ITS2). The 28S rRNA gene of one repeating unit is separated from the 18S rRNA gene of the following repeating unit by the external transcribed spacer (ETS) and the non-transcribed spacer (NTS). The ETS and NTS are collectively known as the intergenic spacer region (IGS). The rRNA genes and spacer regions evolve in concert rather than independently, resulting in a tendency to be homologous within an individual (Hillis & Dixon, 1991; Hwang & Kim, 1999; McManus & Bowles, 1996; Nolan & Cribb, 2005; Sonnenberg et al., 2007). Concerted evolution also results in a single individual being representative of a population, as the process acts to fix variants within reproductive units (McManus & Bowles, 1996; Nolan & Cribb, 2005).

18S, 28S, and 5.8S rRNA Genes

The 18S and 28S rRNAs encode for the ribosomal subunits, and the subunits are then responsible for the translation of mRNA into polypeptide chains, and the polypeptide chains are subsequently folded into proteins. As the 18S and 28S rRNA genes encode for proteins essential for cellular function, possessing a high rate of evolution will affect the success of the ribosome subunit construction and subsequently affect the downstream translation process. Thus, they are highly conserved, and this property enables them to be successfully used for molecular systematics of many groups of organisms (Hwang & Kim, 1999; Patwardhan et al., 2014; Sonnenberg et al., 2007). The 18S rRNA genetic marker is useful for

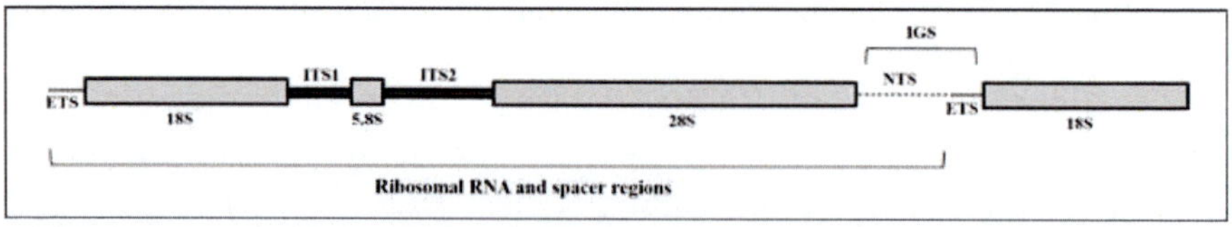

Fig. 6.1 Schematic representation of the nuclear ribosomal RNA and spacer regions. Each repeat consists of the 18S, 28S, and 5.8S ribosomal RNA genes and the internal transcribed spacer 1 (ITS1), internal transcribed spacer 2 (ITS2), and the intergenic spacer regions (IGS). The intergenic spacer regions contain both the non-transcribed spacer (NTS) and external transcribed spacer (ETS)

reconstructing deep phylogenetic branches of higher taxonomic categories ranging from phylum, class, and order (Hwang & Kim, 1999). The 28S rRNA gene, on the other hand, shows slightly more sequence variation as compared to the 18S rRNA gene (Hillis & Dixon, 1991; Hwang & Kim, 1999; McManus & Bowles, 1996; Nolan & Cribb, 2005; Sonnenberg et al., 2007; Waikagul & Thaenkham, 2014). The 28S rRNA gene contains many divergent domains useful for phylogenetic inferences. As a result, the 28S rRNA gene has been applied to slightly lower taxonomic categories, such as the order or family level (Hwang & Kim, 1999). The 5.8S gene is the shortest in the rRNA repeat and has a similar degree of sequence variation to the 18S rRNA gene (Hillis & Dixon, 1991). However, it is too short to contain sufficient phylogenetic information for analysis and is thus not widely used. In addition to the primary DNA sequences being used as molecular genetic markers, the highly conserved secondary structures of the rRNA genes have also served as useful markers for phylogenetic studies.

Internal Transcribed Spacer Regions

The ITS regions are noncoding and nonfunctional regions, and they evolve much faster relative to the 18S and 28S rRNA genes (Choudhary et al., 2015; Grover & Sharma, 2016; Hwang & Kim, 1999; Nolan & Cribb, 2005). The faster rate of evolution is because any substitutions that occur in the ITS regions will not show lethal effects on the organism, in contrast to the nuclear rRNA genes. The higher sequence variation enables the ITS regions to be used as molecular genetic markers to resolve phylogenetic issues in lower taxonomic categories, between genera and species (Choudhary et al., 2015; Hwang & Kim, 1999). The amount of variation between ITS1 and ITS2 and the length of repeat regions vary for each organism and taxonomic category group. Like the rRNA genes, the ITS regions are relatively conserved within the same species due to concerted evolution in play.

6.1.2 Nuclear DNA: Protein-Coding Genes

The nuclear protein-coding genes are the genomic regions that encode for various proteins with different functions in a cell, expressed under DNA transcription and translation (Li & Graur, 2000). There are thousands of protein-coding genes in the eukaryotic nuclear genome, and they present a rich source of molecular genetic markers (Wild & Maddison, 2008). The evolutionary rate and sequence variability of each gene are dependent on the function of the gene. For example, housekeeping genes are required to maintain basal cellular functions that are essential for a cell's existence and generally conserved (Eisenberg & Levanon, 2013; Hillis & Dixon, 1991). Due to a variety of nuclear protein-coding genes to choose from and their moderately conserved nature, they have been used as molecular genetic markers for molecular identification and the detection of hybridization events (Knapp et al.,

2011; Shoriki et al., 2016). Examples of nuclear protein-coding genes that have been used in helminths include the elongation factor 1-alpha (*Ef-1α*) and heat shock proteins (HSP), which will be discussed further in Sect. 6.3.

6.1.3 Mitochondrial DNA Markers

Mitochondrial DNA (mtDNA) is a small, closed circular DNA molecule, ranging from 14 to 40 kilobases in size (Jia et al., 2012; Le et al., 2000). The mitochondria have their own organelle-specific DNA replication, transcription, and translation systems, and the mtDNA evolves independently of the nuclear genome (McManus & Bowles, 1996). It is present in multiple copies of up to 100,000 copies per cell. Compared to nuclear DNA, the mtDNA does not undergo recombination, does not contain introns, contains very few noncoding regions, and is mostly maternally inherited (Grover & Sharma, 2016; Le et al., 2000).

The metazoan mtDNA generally contains 36 or 37 genes, and among them are 2 rRNAs, 22 transfer RNAs (tRNA), 12 or 13 subunits of multimeric proteins of the inner mitochondrial membrane, and a control region. These multimeric proteins of the inner mitochondrial membrane are vital components in the respiratory chain enzyme complexes for oxidative phosphorylation for the generation of adenosine triphosphate (ATP) (Grechko, 2002; Jia et al., 2012; Le et al., 2000, 2002). The subunits of the multimeric proteins are cytochrome *c* oxidase subunits 1 to 3 (*COI* to *COIII*), ATP synthase 6 and 8 (*ATP6* and *ATP8*), nicotinamide adenine dinucleotide 1 to 6 (*ND1* to *ND6*) and 4 L, and cytochrome B (*cytB*). In helminths, except for *Trichuris* and *Trichinella*, *ATP8* is absent in the mitochondrial genome (Le et al., 2000, 2002). Figure 6.2 depicts a representation of a typical mitochondrial genome of helminths with 36 genes.

Generally, mtDNA has a faster evolution rate than nuclear DNA (Grechko, 2002; Le et al., 2000; McManus & Bowles, 1996). In vertebrates, the rate is estimated to be approximately 5 to 10 times faster than nuclear DNA (McManus & Bowles, 1996). The faster rate of evolution of mtDNA makes them useful and powerful molecular genetic markers in resolving species-level phylogenies. However, the rate of sequence variation varies among groups of organisms and among molecular genetic markers.

Ribosomal RNA Genes

The two mitochondrial rRNA genes (12S rRNA and 16S) encode for the mitochondrial small and large ribosomal subunits, respectively. Together with other cytosolic proteins imported into the mitochondria, the small and large subunits are assembled to form the mitochondrial ribosomes (Grover & Sharma, 2016; Sharma et al., 2003; Yang et al., 2014). These ribosomes are essential for translating the mitochondrial mRNA into proteins, like the function of the nuclear 18S and 28S rRNA genes. The

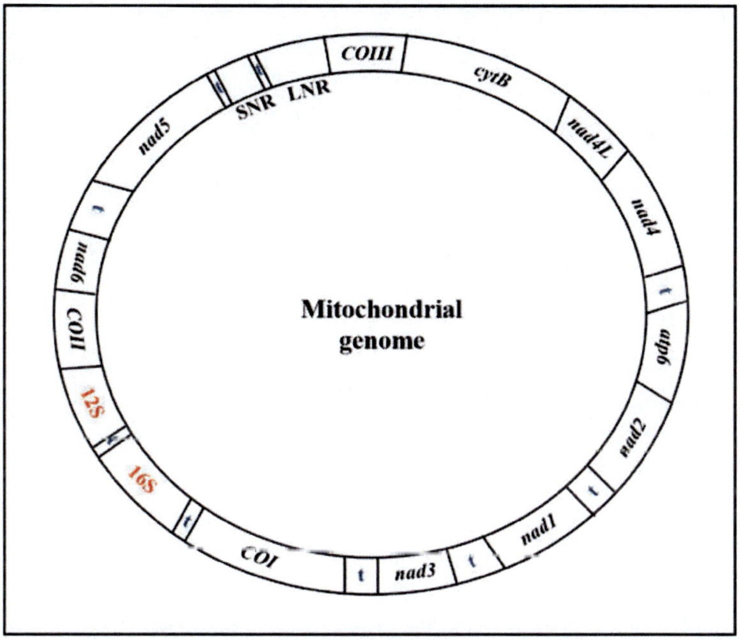

Fig. 6.2 Schematic representation of the mitochondrial genome of a typical helminth. The circular mitochondrial DNA of *Fasciola hepatica* is represented here, with 36 genes, consisting of 12 protein-coding genes, two ribosomal RNA genes (12S and 16S rRNA), 22 transfer RNA genes, the short noncoding regions (SNR), and the long noncoding region (LNR) (Le et al., 2001). The gene order and arrangement might differ, depending on the species

12S and 16S rRNA genes occupy approximately 1/16 and 1/10 of the entire mitochondrial genome (Sharma et al., 2003). Among the mitochondrial genes, the tRNA and rRNA genes are the most conserved due to their important function in the translation process in the mitochondria (Le et al., 2000). Possessing a high rate of evolution could result in detrimental consequences for the cell and is thus not ideal. As a result of this property, the mitochondrial rRNA genes are like the nuclear rRNA genes (although the mtDNA possesses more sequence variation than the nuclear DNA). The 12S rRNA gene is more conserved than the 16S rRNA gene and has been used to illustrate the molecular phylogeny of higher taxonomic categories such as phylum or class. Contrarily, the slightly more variable 16S rRNA gene is helpful for family-level phylogenies (Hwang & Kim, 1999).

Protein-Coding Genes

The mitochondrial protein-coding genes encode various proteins in the mitochondria's respiratory chain enzyme complexes to synthesize ATP (Le et al., 2000). The protein-coding genes contain more sequence variation than the 12S and 16S rRNA

Table 6.1 The general properties of each group of molecular genetic markers

Location	Molecular genetic marker	Function	Properties	Possible suitable taxonomic categories
Nuclear DNA	rRNA	Encodes for the nuclear small and large ribosomal subunit for downstream translation of mRNA into proteins	Highly conserved, slower rate of evolution compared to nuclear ITS regions and protein-coding genes	High
	ITS regions	Noncoding	Varying length of repeat regions, more variation compared to nuclear rRNA genes	Low
	Protein coding	Encodes for various proteins for responsible for different cellular functions	More variable compared to nuclear rRNA genes	Mid
Mt DNA	rRNA	Encodes for the mitochondrial small and large ribosomal subunit for downstream translation of mRNA into proteins	More variable than nuclear DNA molecular genetic markers, more conserved than mitochondrial protein-coding genes	High to mid
	Protein-coding	Encodes for mitochondrial proteins for respiratory process for the generation of ATP	More variable than mitochondrial rRNA genes	Mid to low

Compiled from (Grover & Sharma, 2016; Hwang & Kim, 1999; Nolan & Cribb, 2005; Patwardhan et al., 2014)

genes (Jia et al., 2012; Le et al., 2000). As a result of higher levels of sequence variation, they are useful molecular genetic markers to examine phylogenetic relationships, especially at lower taxonomic categories of genus, species, and population levels (Hwang & Kim, 1999; Le et al., 2000). Moreover, the DNA sequences of protein-coding genes can be translated into amino acids, resulting in more conserved sequences, and thus they can also be applied for phylogenetic studies at the family level. Of the 12 or 13 protein-coding genes, sequence variation between them varies. Generally, the *COI* gene, which encodes for the main subunit of the cytochrome *c* oxidase complex, possesses the smallest sequence variation among all the mitochondrial protein-coding genes (Hwang & Kim, 1999). In addition, the *COI* gene is also the most widely used molecular genetic marker for molecular-based species identification and is currently the DNA barcode of choice for most metazoans (Hebert et al., 2003; Kress, 2017).

Table 6.1 summarizes the general properties of the molecular genetic markers discussed, and some of the important criteria for selecting an appropriate molecular genetic marker will be discussed in the next section.

6.2 Important Criteria for the Selection of Molecular Genetic Markers

The selection of an appropriate molecular genetic marker is crucial for accurate inferences and interpretations about phylogenetic relationships (Grover & Sharma, 2016). Not all molecular genetic markers are suitable, given the varying properties of each molecular genetic marker, as mentioned in the above section. To select an appropriate molecular genetic marker, some essential guidelines are provided below:

First, the choice of utilizing molecular genetic markers is highly dependent on the objective of the research and the type of application (for molecular systematics or molecular identification) (Grover & Sharma, 2016; Patwardhan et al., 2014). In Chap. 1, we defined molecular systematics as the molecular study of the taxonomic classification of organisms and the understanding of their evolutionary relationships, whereas molecular identification is concerned with the accurate identification of species. Overlap between the two applications exists; however, it is essential to differentiate between both, as the different properties of each molecular genetic marker result in different utilities for the resulting application. Higher to mid taxonomic categories such as kingdom, phylum, class, order, and family are usually looked at for molecular systematics to study the classification of groups of organisms. In contrast, mid to lower taxonomic categories such as families, genuses, species, and populations are analyzed for molecular identification.

Second, the level of sequence variation of the molecular genetic marker should be appropriate to accurately determine phylogenetic relationships for the relevant taxonomic category of interest (McManus & Bowles, 1996; Patwardhan et al., 2014). For example, a molecular genetic marker that is too slow to evolve (more conserved) might not be suitable to infer phylogenetic relationships between species, as low levels of sequence variation between species can result. In this case, a molecular genetic marker with higher sequence variation levels with more phylogenetically informative sites for species differentiation is appropriate. Contrarily, a more conserved molecular genetic marker might be helpful in inferences of deep phylogenetic relationships at a higher taxonomic category, whereas a molecular genetic marker with a faster rate of evolution could result in too much sequence variation, obscuring the true phylogeny due to convergences (Grechko, 2002; McManus & Bowles, 1996). Moreover, a molecular genetic marker that is too fast evolving may reach a state of saturation due to nucleotide substitutions, potentially leading to inaccurate phylogenetic inferences (Grechko, 2002). A balance in the level of sequence variation will also be needed to enable primers to successfully bind to the DNA target for amplification through PCR, especially for studies involving species with a broad taxonomic range (Grover & Sharma, 2016; McManus & Bowles, 1996).

Third, sequence alignment should not be complicated. Sequence alignment can be complicated with multiple insertions and deletions (indels) in the sequences, resulting in many gaps being inserted during the alignment process (Patwardhan et al., 2014). The presence of indels can also lead to different sequence lengths for each species, complicating analysis, especially when species from different higher taxonomic categories are aligned together.

Last, reference sequences of the target molecular genetic marker of the organisms of interest studied should ideally be available in public databases. As both molecular systematics and molecular identification rely on the accurate grouping of the unknown DNA sequences with other sequences, it is important to have reference sequences to the target organism as a basis for comparison. DNA sequences on their own, without any reference, are not as informative as to when reference sequences are available for phylogenetic comparison and confirmation of the selected species in question.

In summary, the four critical criteria for selecting a suitable molecular genetic marker are the type of study that the molecular genetic marker will be applied to, the level of sequence variation of the molecular genetic marker, the ease of sequence alignment, and the availability of reference sequences of the molecular genetic marker for the organism studied. Table 6.2 summarizes the essential criteria for the molecular genetic markers that should be considered when deciding which can be used for molecular systematics or molecular identification studies.

6.3 Utilities and Limitations of Nuclear and Mitochondrial Molecular Genetic Markers for Parasitic Helminths

From the previous sections, we have seen how the functions and properties of the different molecular genetic markers differ, which influences the selection of suitable molecular genetic markers in molecular studies in terms of molecular systematics and molecular identification. In this section, we will discuss the nuclear and mitochondrial molecular genetic markers (based on the four criteria previously discussed) and their use, with some molecular studies conducted for parasitic helminths as examples. Although specific examples will be provided along the way, the list of examples is far from complete but will serve as references to bring out the intended point.

6.3.1 Nuclear DNA Markers

The nuclear DNA genetic markers discussed here include the 18S and 28 rRNA genes, ITS1 and ITS2 regions, and the nuclear protein-coding genes.

Table 6.2 Summary of the essential criteria to be considered when selecting an appropriate molecular genetic marker

	Sequence variation	Sequence alignment	Reference sequences	Other considerations
Nuclear rRNA	*Highly conserved*: – Low discriminatory power for species discrimination – Primers can be designed to cover a broad species range – Not affected by nucleotide substitution saturation	– Presence of indels can affect sequence alignment – Different regions used for amplification and sequencing	– Partial sequences usually available	
Nuclear spacer regions	*Variable*: – Good resolution for species discrimination – Affected by nucleotide substitution saturation – Concerted evolution resulting in low intraspecific variation – Difficult to design primers for broad species range	– Presence of indels – Different lengths		
Nuclear protein coding	*Moderately conserved (more variable than nuclear rRNA, but more conserved than spacers and mtDNA)* – Primers can be designed to cover a broad species range	– Presence of indels	– Dependent on group of organisms studied	– Might be paralogous genes – Might be difficult to amplify due to single-copy genes
Mt rRNA	*Moderately conserved (more variable than nuclear DNA, but more conserved than Mt protein-coding):* – Sufficient resolution for species discrimination – Primers can be designed to cover a broad species range – Not affected by nucleotide substitution saturation	– Lack of indels	– Full-length sequences from whole mt genomes	– Potential presence of NUMTs
Mt protein-coding	*Variable*: – Good resolution for species discrimination – Intraspecific variation and thus can be applied at population level – Difficult to design primers for broad species range	– Lack of indels	– Full-length sequences from whole mt genomes	– Potential presence of NUMTs

(continued)

Table 6.2 (continued)

	Sequence variation	Sequence alignment	Reference sequences	Other considerations
Whole mt genome	Good resolution for species discrimination and phylogenetic relationships due to large size	– Lack of indels	– Full-length	– Potential presence of NUMTs – Advanced molecular technologies resulting in higher cost

Nuclear rRNA (18S and 28S)

The nuclear rRNA genes are highly conserved relative to the other nuclear and mitochondrial DNA markers. This property enables the molecular genetic markers to help study phylogenetic relationships in higher taxonomic categories. Both nuclear rRNA genes have been used in molecular systematics to provide a phylogenetic framework for classifying nematodes, trematodes, and cestodes. The first comprehensive study on nematode molecular systematics using 53 18S rRNA sequences by Blaxter et al. identified five major clades within the phylum Nematoda (Blaxter et al., 1998). The successful use of the 18S rRNA gene for nematode classification prompted researchers to expand the range of species further to include marine and free-living nematodes (Holterman et al., 2006; Meldal et al., 2007). Both the 18S and 28S rRNA genes were also used for trematode and cestode molecular systematics, with 163 and 28 taxa sampled for trematodes and cestodes, respectively (Olson et al., 2003; Waeschenbach & Littlewood, 2017; Waeschenbach et al., 2007).

Moreover, the use of both nuclear rRNA genes provided increased phylogenetic resolution for both groups of helminths. These studies on the molecular systematics of nematodes, trematodes, and cestodes provide evidence of the robustness and suitability of the nuclear rRNA genes for molecular systematics of helminths at higher taxonomic categories. The molecular systematics for each group of helminths will be discussed in further detail in Chaps. 10 to 12. A critical factor that could have significantly contributed to the successful use of the nuclear rRNA genes is due to the conserved sequences. The conserved sequences also meant that suitable primer binding sites for primer binding across many species are not difficult to find (Nolan & Cribb, 2005). With just a "universal" primer set, a broad range of taxa can be sampled, supporting the use of the nuclear rRNA molecular genetic markers for molecular systematics studies.

Contrarily, the amount of sequence variation due to the slow evolutionary rate of the nuclear rRNA genes might render them unsuitable for molecular identification at the species level. Previous studies have demonstrated that the 18S rRNA genes had low sequence variation using the 18S rRNA gene, in contrast to the nuclear ITS regions and mtDNA molecular genetic markers with the cestodes *Diphyllobothrium dentricum* and *D. ditremum* (Wicht et al., 2010). Among the liver flukes of

Opisthorchis viverrini, *O. felineus,* and *Clonorchis sinensis*, Kang et al. (2008) also showed no sequence variation when the 18S rRNA gene was used (Kang et al., 2008). No sequence variation was also observed between the nematodes *Trichuris muris* and *T. arvicolae*, while other species within *Trichuris* were able to be discriminated, although low sequence variation was found between them, with 95 to 100% similarity (Callejon et al., 2013). The low sequence variation observed when utilizing the 18S rRNA gene can complicate molecular identification, especially among morphologically similar species.

Another example is between the morphologically similar *Haemonchus contortus* and *Haemonchus placei*, which infects ruminants. However, the 18S rRNA sequences were identical between these two morphologically similar species (Patwardhan et al., 2014). These examples demonstrate that the 18S rRNA gene possesses low discriminatory power to successfully differentiate helminths at lower taxonomic levels, especially for closely related species. Moreover, the presence of phenotypic plasticity and cryptic species (explained in Chap. 5) can hinder accurate species identification, especially for helminths. Difficulties in morphological identification, coupled with the low sequence variation when using the 18S rRNA gene, might cause this molecular genetic marker to be unsuitable for molecular identification of some parasitic helminths.

The 28S rRNA gene, as compared to the 18S rRNA gene, possesses slightly more sequence variation with many divergent domains. The minute intestinal flukes of *Metagonimus yokogawai*, *M. takahashii*, and *M. miyatai* were successfully differentiated using the 28S rRNA gene (Lee et al., 2004). A comparison between the 28S rRNA gene and the mitochondrial *COI* gene to differentiate species within *Hymenolepis* was performed, and the results showed that both molecular genetic markers were able to successfully differentiate between the species (Rezabkova et al., 2019). However, as the 28S rRNA gene spans a total of 12 domains, it might be challenging to find available partial reference sequences in the nucleotide database for the specific domain of interest. The multiple domains can complicate the alignment process during multiple sequence alignment with other reference sequences, as usually only the partial length of the 28S rRNA gene is amplified and sequenced. Despite possessing more sequence variation than the 18S rRNA gene, other challenges with sequence alignment and reference sequence availability should be considered when utilizing the 28S rRNA gene for molecular identification of parasitic helminths.

Nuclear Spacers (ITS1 and ITS2)

The nuclear spacer regions possess higher sequence variation when compared to the nuclear rRNA genes. The higher level of sequence variation can enable the nuclear spacer regions to be useful molecular genetic markers for lower taxonomic categories, especially between species. In helminths, numerous studies have demonstrated the effectiveness of using both ITS1 and ITS2 in molecular identification. Among the soil-transmitted helminths, successfully discriminating the morphologically indistinguishable eggs of hookworms *Necator americanus* and *Ancylostoma duodenale* was achieved using species-specific ITS primers (Yong et al., 2007). Similar success was achieved with the morphologically similar eggs of *O. viverrini, C. sinensis,* and *Haplorchis taichui*, using the ITS2 region (Traub et al., 2009). For trematodes, the ITS2 region is more frequently used than ITS1 as it contains more variable sites and does not contain a repeat element to complicate sequence alignment (Choudhary et al., 2015; Nolan & Cribb, 2005; Waikagul & Thaenkham, 2014). Aside from molecular identification, the ITS regions have also been used in disease diagnosis and the quantification of helminths. TaqMan quantitative PCR was developed to detect *Anisakis* in fisheries products and applied to assess the prevalence and intensity of intestinal parasites in humans (Seesao et al., 2017). These examples provide ample evidence of the robustness of ITS regions for the molecular identification of helminths.

Despite the successful use of the ITS regions for species discrimination, low levels of intraspecific variations were observed among populations from different geographic localities. The process of concerted evolution tends to fix variants within reproductive units; thus, species variation at the population level is low. Low levels of intraspecific variation were observed in different geographical isolates of *C. sinensis* using ITS1, and a reliable geographical phylogeny could not be constructed (Tatonova et al., 2012). Thus, the ITS regions might not possess sufficient sequence variation for population-level studies.

Notwithstanding the robustness of the ITS regions for molecular identification, the ITS regions might not be appropriate for molecular systematics where higher taxonomic categories are concerned. Firstly, the variable sequences limit the success of primer design to covering a broad range of species. When targeting the ITS regions, species-specific primers are usually designed to conduct a multiplex PCR for the simultaneous detection of multiple species. The prospect of having "universal" primers, especially for molecular systematics, is thus hindered by this limitation. Secondly, differences in the lengths of the ITS sequences between species have been observed, and these differences in sequence lengths are due to the varying lengths of the repeat regions among taxa (Nolan & Cribb, 2005). The varying lengths of the repeat regions complicate sequence alignment, especially when sampling taxa across taxonomic levels. Studies have shown that the ITS regions are inappropriate for phylogenetic comparisons between distantly related taxa from a broad taxonomic hierarchy (Le et al., 2017, 2020; Thaenkham et al., 2012). Moreover, the fast-evolving ITS regions may reach a state of saturation due to nucleotide substitutions and are therefore not useful for phylogenetic inferences at higher taxonomic categories.

Nuclear Protein-Coding Genes

The nuclear protein-coding genes are moderately conserved and can also be helpful molecular genetic markers. A group of genes that have been used for molecular studies on helminths include the heat shock protein (HSPs). They are a gene family that is highly conserved and are found in both prokaryotes and eukaryotes, with key roles in response to stress conditions (Asea & Kaur, 2017). Both HSP70 and HSP90 have been used as molecular genetic markers for nematode phylogenies. Skantar and Carta (2004) tested the phylogenetic utility of HSP90 for nematodes and compared it with the nuclear rRNA genes, concluding that the tree topology of HSP90 was consistent with the nuclear rRNA genes (Skantar & Carta, 2004). The eukaryotic translation elongation factor 1 α (*EF-1α*) promoter is another highly conserved gene encoding used for phylogenetic inferences of helminths. The phylogenetic relationships of cestodes were inferred using both the *EF-1α* and the 18S rRNA genes, and the markers showed the ability to determine the molecular systematics of cestodes (Olson & Caira, 1999). The *EF-1α* gene was also used together with three other nuclear genes (transforming growth factor-beta receptor kinase, thioredoxin peroxidase, and calreticulin) to reconstruct the phylogeny of *Echinococcus* (Saarma et al., 2009). Other nuclear protein-coding molecular genetic markers that have been used include the RNA polymerase II subunit and DNA polymerase delta, among others. The conserved nature of these nuclear protein-coding genes enables them to be useful molecular genetic markers for investigating deep evolutionary relationships.

On the flip side, as there are many nuclear protein-coding genes, finding a suitable molecular genetic marker, together with sufficient reference sequences, might be a challenge. Moreover, some of the genes might be paralogous due to the process of gene duplication (Skantar & Carta, 2004). Relationship inferences among organisms using these paralogous genes will not be due to having a common ancestor and will distort true phylogenetic relationships.

6.3.2 Mitochondrial DNA Markers

The mtDNA molecular genetic markers possess higher sequence variation than the nuclear DNA molecular genetic markers due to their faster rate of evolution. This property enabled them to be popular molecular genetic markers, especially at the species and population level. In the subsequent paragraphs, we will first discuss the general utilities and limitations of mtDNA molecular genetic markers, on the whole, followed by the specific molecular genetic markers for the 12S and 16S rRNA genes, the protein-coding genes of *COI*, *cytB*, and *ND1*, and whole mitochondrial genomes.

The availability of whole mitochondrial genomes in reference databases can provide a comprehensive reference for phylogenetic analysis and reference comparisons (Jia et al., 2012). With the advances in next-generation sequencing techniques, many whole mitochondrial genomes are available and are advantageous for molecular systematics and molecular identification studies using mtDNA molecular genetic markers. Full-length gene sequences from the mitochondrial rRNA and protein-coding genes can be mined from the complete mitochondrial genome of interest, serving as an advantage over nuclear DNA molecular genetic markers, where only partial sequences are usually available. In addition, the lack of indels in the mtDNA (in contrast to the nuclear DNA) enables multiple sequence alignment to be conducted with ease, providing an additional advantage for molecular studies.

However, a drawback of utilizing mtDNA molecular genetic markers is the presence of pseudogenes. Pseudogenes, also known as nuclear insertions of mitochondrial sequences (NUMTs), are mitochondrial DNA sequences transferred into the nuclear genome (Bensasson et al., 2003; Rubinoff & Holland, 2005). Pseudogenes can commonly affect mtDNA-based phylogenetic analysis, resulting in incorrect phylogenetic inferences. Several studies utilizing mtDNA molecular genetic markers have shown the presence of NUMTs, leading to incorrect phylogenetic placements, and care will therefore have to be taken when making inferences with mtDNA molecular genetic markers (Dusitsittipon et al., 2018).

Mitochondrial rRNA (12S and 16S)

Among the mitochondrial molecular genetic markers, the 12S and 16S rRNA genes are the most conserved (Le et al., 2000). This more conserved property has enabled the mitochondrial rRNA genes to illustrate the molecular phylogeny for high to mid taxonomic categories. The 16S rRNA gene and the nuclear rRNA genes have been used in conjunction to provide increased resolution across 16 orders within Cestoda (Waeschenbach et al., 2012). Moreover, the 12S rRNA gene was used to infer phylogenetic relationships between the order Cyclophyllidea (Cestoda) and the phylogenetic tree corresponding to the systematic classification for this order (von Nickisch-Rosenegk et al., 1999). Both the 12S and 16S rRNA genes were also evaluated for the molecular systematics of parasitic nematodes, supporting their use for molecular systematics for helminths (Chan et al., 2020b). The more conserved property of the mitochondrial rRNA genes enabled the design of primers to cover a wide range of species, in contrast to the mitochondrial protein-coding genes, rendering them useful molecular genetic markers for molecular systematics.

Compared to the nuclear 18S and 28S rRNA genes, the mitochondrial rRNA molecular genetic markers possess higher levels of sequence variation. Mitochondrial DNA generally evolves faster than nuclear DNA, and thus the mitochondrial 12S and 16S rRNA genes are more variable than the 18S and 28S rRNA genes. With this property in mind, the mitochondrial rRNA genes have also been used for molecular identification studies. Yang et al. (2014) developed a primer set targeting the 12S and 16S rRNA genes for the species identification of a wide variety of animals such as mammals, fishes, birds, and reptiles (Yang et al., 2014). In addition, the 16S rRNA gene has been proposed as an alternative molecular genetic marker for amphibian DNA barcoding (instead of the *COI* gene) (Vences et al., 2005). For molecular identification of helminths, *Toxocara canis* and *Toxocara vitulorum* were successfully differentiated with the 12S rRNA gene (Wickramasinghe et al., 2009). The 16S rRNA gene was used for species identification in *Trichuris* (Callejon et al., 2009). In addition, the closely related *Angiostrongylus cantonensis* and *Angiostrongylus malaysiensis* were successfully differentiated with both the 12S and 16S rRNA genes (Chan et al., 2020a). These studies demonstrate the ability of the mitochondrial rRNA genes to provide effective molecular identification among helminths.

Mitochondrial Protein-Coding Genes

The mitochondrial protein-coding genes possess a higher level of sequence variation as compared to the mitochondrial rRNA genes. Due to this trait, they are helpful molecular genetic markers to examine phylogenetic relationships, especially at lower taxonomic categories of genus, species, and population levels.

A prominent molecular genetic marker is the *COI* gene, which was proposed as a universal molecular genetic marker for DNA barcoding and species identification of animals by Herbert and co-workers (Hebert et al., 2003). Studies encompassing many groups of taxa have shown the success of utilizing the *COI* gene for species identification, including arthropods, birds, fishes, mammals, and amphibians (Trivedi et al., 2020). Helminths are no exception. Examples include successful inter- and intraspecies discrimination among three species of *Diphyllobothrium*, species discrimination among *Trichuris*, and discrimination of the closely related *O. lobatus* from *O. viverrini* (Thaenkham et al., 2011; Wicht et al., 2010). Moreover, the vast utility of the *COI* gene enabled many *COI* sequences from many species to be deposited in the reference database. The probability of matching an unknown sequence is thus increased, enabling easy species identification of unknowns. The *ND1* gene has been used for molecular identification, especially for cestodes. Bowles and McManus (1993) used this molecular genetic marker to compare species and strains within *Echinococcus*, revealing the usefulness of the *ND1* gene for inter- and intraspecific variations (Bowles & McManus, 1993). *ND1* was also used among nine species of *Taenia*, and the results showed that the *ND1* gene had higher levels of interspecies variation as compared to the *COI* gene (Gasser et al., 1999).

On the other hand, the *cytB* gene has been used for the molecular identification of nematodes. An example is the species discrimination of the closely related *Angiostrongylus cantonensis* and *Angiostrongylus malaysiensis* (Yong et al., 2015). Like the *COI* and *ND1* genetic markers, the *cytB* genetic marker was also successfully used in population genetic studies for *A. cantonensis* (Dusitsittipon et al., 2015). Undoubtedly, the mitochondrial protein-coding genes are useful molecular genetic markers for molecular identification and population-level studies.

However, despite the successes with molecular identification, the high sequence variation of the mitochondrial protein-coding genes also presents some challenges, especially for helminths. Although universal primers have been developed and utilized for the *COI* gene among other groups of organisms, low PCR amplification success was observed for helminths. Moszczynska et al. (2009) illustrated the difficulties in designing and amplifying the *COI* gene among Digenea, despite good resolution for species identification (Moszczynska et al., 2009). The high sequence variation also hindered the use of these markers for molecular systematics, where too much sequence variation can reflect biasness in true ancestry (Patwardhan et al., 2014).

6.3.3 *Whole Mitochondria Genomes*

Next-generation sequencing technologies enabled the sequencing of whole mitochondrial genomes, and they are another source of molecular genetic markers for helminth molecular studies. The first completed mitochondrial genome from a parasitic helminth was *Ascaris suum*, and thereafter, more than 100 mitochondrial genomes of helminths have been sequenced thus far (Jia et al., 2012; Le et al., 2000). The whole mitochondrial genome of helminths typically contains 37 genes (Fig. 6.2) and is generally adenosine-thymine (AT) rich. The size and gene arrangement can differ, depending on the group of helminths (Jia et al., 2012). The size ranges from 12.6 to 26.2 kilobases for nematodes, 13.3 to 14.2 kilobases for cestodes, and 13.9 to 16.9 kilobases for trematodes. The gene arrangement within cestodes is nearly identical, whereas the arrangement for trematodes and nematodes is more variable.

With the faster evolution rate for the mtDNA compared to the nuclear DNA, the mitochondrial genome is also a useful molecular genetic marker for molecular phylogenetic studies. Many phylogenetic studies have been performed using whole mitochondrial genomes for each group of parasitic helminths, showing the robustness of mitochondrial genomes for elucidating and clarifying phylogenetic relationships among helminths (Kim et al., 2015; Sun et al., 2014). With the varying substitution rates of each gene in the mitochondrial genome, the mitochondrial genome can provide resolution for a broad range of phylogenetic depths, from shallow divergences in the lower taxonomic categories to deep divergences at the higher taxonomic categories (Kern et al., 2020).

As compared to amplicon sequencing of a short segment of the DNA using a specific molecular genetic marker, sequencing of whole mitochondria genomes requires the use of next-generation sequencing technology, long-PCR amplification, and laboratory capabilities for this technique. The prospect of utilizing whole mitochondrial genomes as a molecular genetic marker is thus subjected to the budget allocated for the research.

6.4 Conclusion

We saw how the functions of the molecular genetic markers are linked to their resultant properties, which affects the utility and limitation of the respective molecular genetic marker, together with various examples from parasitic helminths in the above sections. Selecting an appropriate molecular genetic marker for the appropriate application is the first step to conducting convincing molecular studies to produce reliable inferences and results. In the next chapter, we will review the types of molecular biology techniques that have been used, in addition to general considerations and guidelines for primer design and PCR amplification of the target molecular genetic marker.

References

Asea, A. A., & Kaur, P. (2017). *Heat shock proteins in veterinary medicine and sciences* (1st ed.). Springer.

Bensasson, D., Feldman, M. W., & Petrov, D. A. (2003). Rates of DNA duplication and mitochondrial DNA insertion in the human genome. *Journal of Molecular Evolution, 57*(3), 343–354. https://doi.org/10.1007/s00239-003-2485-7

Blaxter, M. L., De Ley, P., Garey, J. R., Liu, L. X., Scheldeman, P., Vierstraete, A., Vanfleteren, J. R., Mackey, L. Y., Dorris, M., Frisse, L. M., Vida, J. T., & Thomas, W. K. (1998). A molecular evolutionary framework for the phylum Nematoda. *Nature, 392*(6671), 71–75. https://doi.org/10.1038/32160

Bowles, J., & McManus, D. P. (1993). NADH dehydrogenase 1 gene sequences compared for species and strains of the genus *Echinococcus*. *International Journal for Parasitology, 23*(7), 969–972. https://doi.org/10.1016/0020-7519(93)90065-7

Callejon, R., de Rojas, M., Ariza, C., Ubeda, J. M., Guevara, D. C., & Cutillas, C. (2009). Cytochrome oxidase subunit 1 and mitochondrial 16S rDNA sequences of *Trichuris skrjabini* (Tricocephalida: Trichuridae). *Parasitology Research, 104*(3), 715–716. https://doi.org/10.1007/s00436-008-1280-1

Callejon, R., Nadler, S., De Rojas, M., Zurita, A., Petrasova, J., & Cutillas, C. (2013). Molecular characterization and phylogeny of whipworm nematodes inferred from DNA sequences of *cox1* mtDNA and 18S rDNA. *Parasitology Research, 112*(11), 3933–3949. https://doi.org/10.1007/s00436-013-3584-z

Chan, A. H. E., Chaisiri, K., Dusitsittipon, S., Jakkul, W., Charoennitiwat, V., Komalamisra, C., & Thaenkham, U. (2020a). Mitochondrial ribosomal genes as novel molecular genetic markers for discrimination of closely related species in the *Angiostrongylus cantonensis* lineage. *Acta Tropica, 211*, 105645. https://doi.org/10.1016/j.actatropica.2020.105645

Chan, A. H. E., Chaisiri, K., Morand, S., Saralamba, N., & Thaenkham, U. (2020b). Evaluation and utility of mitochondrial ribosomal genes for molecular systematics of parasitic nematodes. *Parasites & Vectors, 13*(1), 364. https://doi.org/10.1186/s13071-020-04242-8

Choudhary, K., Verma, A. K., Swaroop, S., & Agrawal, N. (2015). A review on the molecular characterization of digenean parasites using molecular markers with special reference to ITS region. *Helminthologia, 52*(3), 167–187. https://doi.org/10.1515/helmin-2015-0031

Dusitsittipon, S., Thaenkham, U., Watthanakulpanich, D., Adisakwattana, P., & Komalamisra, C. (2015). Genetic differences in the rat lungworm, *Angiostrongylus cantonensis* (Nematoda: Angiostrongylidae), Thailand. *Journal of Helminthology, 89*(5), 545–551. https://doi.org/10.1017/s0022149x14000388

Dusitsittipon, S., Criscione, C. D., Morand, S., Komalamisra, C., & Thaenkham, U. (2018). Hurdles in the evolutionary epidemiology of *Angiostrongylus cantonensis*: Pseudogenes, incongruence between taxonomy and DNA sequence variants, and cryptic lineages. *Evolutionary Applications, 11*(8), 1257–1269. https://doi.org/10.1111/eva.12621

Eisenberg, E., & Levanon, E. Y. (2013). Human housekeeping genes, revisited. *Trends in Genetics, 29*(10), 569–574. https://doi.org/10.1016/j.tig.2013.05.010

Gasser, R. B., Zhu, X., & McManus, D. P. (1999). NADH dehydrogenase subunit 1 and cytochrome *c* oxidase subunit I sequences compared for members of the genus *Taenia* (Cestoda). *International Journal for Parasitology, 29*(12), 1965–1970. https://doi.org/10.1016/s0020-7519(99)00153-8

Grechko, V. V. (2002). Molecular DNA markers in phylogeny and systematics. *Russian Journal of Genetics, 38*(8), 851–868. https://doi.org/10.1023/A:1016890509443

Grover, A., & Sharma, P. C. (2016). Development and use of molecular markers: Past and present. *Critical Reviews in Biotechnology, 36*(2), 290–302. https://doi.org/10.3109/07388551.2014.959891

Hebert, P. D., Cywinska, A., Ball, S. L., & deWaard, J. R. (2003). Biological identifications through DNA barcodes. *Proceedings of the Biological Sciences, 270*(1512), 313–321. https://doi.org/10.1098/rspb.2002.2218

Hillis, D. M., & Dixon, M. T. (1991). Ribosomal DNA: Molecular evolution and phylogenetic inference. *The Quarterly Review of Biology, 66*(4), 411–453. https://doi.org/10.1086/417338

Holterman, M., van der Wurff, A., van den Elsen, S., van Megen, H., Bongers, T., Holovachov, O., Bakker, J., & Helder, J. (2006). Phylum-wide analysis of SSU rDNA reveals deep phylogenetic relationships among nematodes and accelerated evolution toward crown clades. *Molecular Biology and Evolution, 23*(9), 1792–1800. https://doi.org/10.1093/molbev/msl044

Hwang, U. W., & Kim, W. (1999). General properties and phylogenetic utilities of nuclear ribosomal DNA and mitochondrial DNA commonly used in molecular systematics. *The Korean Journal of Parasitology, 37*(4), 215–228. https://doi.org/10.3347/kjp.1999.37.4.215

Jia, W., Yan, H., Ni, X., Lou, Z., Li, H., Cao, P., & Cai, X. (2012). Advances in the study of helminth mitochondrial genomes and their associated applications. *Chinese Science Bulletin, 57*(1), 54–67. https://doi.org/10.1007/s11434-011-4748-9

Kang, S., Sultana, T., Loktev, V. B., Wongratanacheewin, S., Sohn, W. M., Eom, K. S., & Park, J. K. (2008). Molecular identification and phylogenetic analysis of nuclear rDNA sequences among three opisthorchid liver fluke species (Opisthorchiidae: Trematoda). *Parasitology International, 57*(2), 191–197. https://doi.org/10.1016/j.parint.2007.12.007

Kern, E. M. A., Kim, T., & Park, J.-K. (2020). The mitochondrial genome in nematode phylogenetics. *Frontiers in Ecology and Evolution, 8*, 250. https://doi.org/10.3389/fevo.2020.00250

Kim, J., Lee, S.-H., Gazi, M., Kim, T., Jung, D., Chun, J.-Y., Kim, S., Seo, T.-K., Park, C., Baldwin, J. G., Nadler, S. A., & Park, J.-K. (2015). Mitochondrial genomes advance phylogenetic hypotheses for Tylenchina (Nematoda: Chromadorea). *Zoologica Scripta, 44*(4), 446–462. https://doi.org/10.1111/zsc.12112

Knapp, J., Nakao, M., Yanagida, T., Okamoto, M., Saarma, U., Lavikainen, A., & Ito, A. (2011). Phylogenetic relationships within *Echinococcus* and *Taenia* tapeworms (Cestoda: Taeniidae): An inference from nuclear protein-coding genes. *Molecular Phylogenetics and Evolution, 61*(3), 628–638. https://doi.org/10.1016/j.ympev.2011.07.022

Kress, W. J. (2017). Plant DNA barcodes: Applications today and in the future. *Journal of Systematics and Evolution, 55*(4), 291–307. https://doi.org/10.1111/jse.12254

Le, T. H., Blair, D., & McManus, D. P. (2000). Mitochondrial genomes of human helminths and their use as markers in population genetics and phylogeny. *Acta Tropica, 77*(3), 243–256. https://doi.org/10.1016/s0001-706x(00)00157-1

Le, T. H., Blair, D., & McManus, D. P. (2001). Complete DNA sequence and gene organization of the mitochondrial genome of the liver fluke, *Fasciola hepatica* L. (Platyhelminthes; Trematoda). *Parasitology, 123*(Pt 6), 609–621. https://doi.org/10.1017/s0031182001008733

Le, T. H., Blair, D., & McManus, D. P. (2002). Mitochondrial genomes of parasitic flatworms. *Trends in Parasitology, 18*(5), 206–213. https://doi.org/10.1016/s1471-4922(02)02252-3

Le, T. H., Nguyen, K. T., Nguyen, N. T., Doan, H. T., Dung, D. T., & Blair, D. (2017). The ribosomal transcription units of *Haplorchis pumilio* and *H. taichui* and the use of 28S rDNA sequences for phylogenetic identification of common heterophyids in Vietnam. *Parasites & Vectors, 10*(1), 17. https://doi.org/10.1186/s13071-017-1968-0

Le, T. H., Pham, K. L. T., Doan, H. T. T., Le, T. K. X., Nguyen, K. T., & Lawton, S. P. (2020). Description and phylogenetic analyses of ribosomal transcription units from species of Fasciolidae (Platyhelminthes: Digenea). *Journal of Helminthology, 94*, e136. https://doi.org/10.1017/S0022149X20000164

Lee, S. U., Huh, S., Sohn, W. M., & Chai, J. Y. (2004). Sequence comparisons of 28S ribosomal DNA and mitochondrial cytochrome c oxidase subunit I of *Metagonimus yokogawai, M. takahashii* and *M. miyatai*. *The Korean Journal of Parasitology, 42*(3), 129–135. https://doi.org/10.3347/kjp.2004.42.3.129

Li, W. H., & Graur, D. (2000). *Fundamentals of molecular evolution*. Sinauer Associates.

McManus, D. P., & Bowles, J. (1996). Molecular genetic approaches to parasite identification: Their value in diagnostic parasitology and systematics. *International Journal for Parasitology, 26*(7), 687–704. https://doi.org/10.1016/0020-7519(96)82612-9

Meldal, B. H., Debenham, N. J., De Ley, P., De Ley, I. T., Vanfleteren, J. R., Vierstraete, A. R., Bert, W., Borgonie, G., Moens, T., Tyler, P. A., Austen, M. C., Blaxter, M. L., Rogers, A. D., & Lambshead, P. J. (2007). An improved molecular phylogeny of the Nematoda with special emphasis on marine taxa. *Molecular Phylogenetics and Evolution, 42*(3), 622–636. https://doi.org/10.1016/j.ympev.2006.08.025

Moszczynska, A., Locke, S. A., McLaughlin, J. D., Marcogliese, D. J., & Crease, T. J. (2009). Development of primers for the mitochondrial cytochrome *c* oxidase I gene in digenetic trematodes (Platyhelminthes) illustrates the challenge of barcoding parasitic helminths. *Molecular Ecology Resources, 9 Suppl s1*, 75–82. https://doi.org/10.1111/j.1755-0998.2009.02634.x

Nolan, M. J., & Cribb, T. H. (2005). The use and implications of ribosomal DNA sequencing for the discrimination of digenean species. *Advances in Parasitology, 60*, 101–163. https://doi.org/10.1016/s0065-308x(05)60002-4

Olson, P. D., & Caira, J. N. (1999). Evolution of the major lineages of tapeworms (Platyhelminthes: Cestoidea) inferred from 18S ribosomal DNA and elongation factor-1alpha. *The Journal of Parasitology, 85*(6), 1134–1159.

Olson, P. D., Cribb, T. H., Tkach, V. V., Bray, R. A., & Littlewood, D. T. J. (2003). Phylogeny and classification of the Digenea (Platyhelminthes: Trematoda). *International Journal for Parasitology, 33*(7), 733–755. https://doi.org/10.1016/s0020-7519(03)00049-3

Patwardhan, A., Ray, S., & Roy, A. (2014). Molecular markers in phylogenetic studies- a review. *Journal of Phylogenetics & Evolutionary Biology, 2*(2), 131.

Rezabkova, L., Brabec, J., Jirku, M., Dellerba, M., Kuchta, R., Modry, D., Parker, W., & Jirku Pomajbikova, K. (2019). Genetic diversity of the potentially therapeutic tapeworm *Hymenolepis diminuta* (Cestoda: Cyclophyllidea). *Parasitology International, 71*, 121–125. https://doi.org/10.1016/j.parint.2019.04.009

Rubinoff, D., & Holland, B. S. (2005). Between two extremes: Mitochondrial DNA is neither the panacea nor the nemesis of phylogenetic and taxonomic inference. *Systematic Biology, 54*(6), 952–961. https://doi.org/10.1080/10635150500234674

Saarma, U., Jogisalu, I., Moks, E., Varcasia, A., Lavikainen, A., Oksanen, A., Simsek, S., Andresiuk, V., Denegri, G., Gonzalez, L. M., Ferrer, E., Garate, T., Rinaldi, L., & Maravilla, P. (2009). A novel phylogeny for the genus *Echinococcus*, based on nuclear data, challenges relationships based on mitochondrial evidence. *Parasitology, 136*(3), 317–328. https://doi.org/10.1017/S0031182008005453

Seesao, Y., Gay, M., Merlin, S., Viscogliosi, E., Aliouat-Denis, C. M., & Audebert, C. (2017). A review of methods for nematode identification. *Journal of Microbiological Methods, 138*, 37–49. https://doi.org/10.1016/j.mimet.2016.05.030

Sharma, M. R., Koc, E. C., Datta, P. P., Booth, T. M., Spremulli, L. L., & Agrawal, R. K. (2003). Structure of the mammalian mitochondrial ribosome reveals an expanded functional role for its component proteins. *Cell, 115*(1), 97–108. https://doi.org/10.1016/s0092-8674(03)00762-1

Shoriki, T., Ichikawa-Seki, M., Suganuma, K., Naito, I., Hayashi, K., Nakao, M., Aita, J., Mohanta, U. K., Inoue, N., Murakami, K., & Itagaki, T. (2016). Novel methods for the molecular discrimination of *Fasciola* spp. on the basis of nuclear protein-coding genes. *Parasitology International, 65*(3), 180–183. https://doi.org/10.1016/j.parint.2015.12.002

Skantar, A. M., & Carta, L. K. (2004). Molecular characterization and phylogenetic evaluation of the *hsp90* gene from selected nematodes. *Journal of Nematology, 36*(4), 466–480.

Sonnenberg, R., Nolte, A. W., & Tautz, D. (2007). An evaluation of LSU rDNA D1-D2 sequences for their use in species identification. *Frontiers in Zoology, 4*, 6. https://doi.org/10.1186/1742-9994-4-6

Sun, L., Zhuo, K., Lin, B., Wang, H., & Liao, J. (2014). The complete mitochondrial genome of *Meloidogyne graminicola* (Tylenchina): A unique gene arrangement and its phylogenetic implications. *PLoS One, 9*(6), e98558. https://doi.org/10.1371/journal.pone.0098558

Tatonova, Y. V., Chelomina, G. N., & Besprosvannykh, V. V. (2012). Genetic diversity of nuclear ITS1-5.8S-ITS2 rDNA sequence in *Clonorchis sinensis* Cobbold, 1875 (Trematoda: Opisthorchidae) from the Russian Far East. *Parasitology International, 61*(4), 664–674. https://doi.org/10.1016/j.parint.2012.07.005

Thaenkham, U., Nuamtanong, S., Vonghachack, Y., Yoonuan, T., Sanguankiat, S., Dekumyoy, P., Prommasack, B., Kobayashi, J., & Waikagul, J. (2011). Discovery of *Opisthorchis lobatus* (Trematoda: Opisthorchiidae): A new record of small liver flukes in the greater Mekong sub-region. *The Journal of Parasitology, 97*(6), 1152–1158. https://doi.org/10.1645/GE-2764.1

Thaenkham, U., Blair, D., Nawa, Y., & Waikagul, J. (2012). Families Opisthorchiidae and Heterophyidae: Are they distinct? *Parasitology International, 61*(1), 90–93. https://doi.org/10.1016/j.parint.2011.06.004

Traub, R. J., Macaranas, J., Mungthin, M., Leelayoova, S., Cribb, T., Murrell, K. D., & Thompson, R. C. (2009). A new PCR-based approach indicates the range of *Clonorchis sinensis* now extends to Central Thailand. *PLoS Neglected Tropical Diseases, 3*(1), e367. https://doi.org/10.1371/journal.pntd.0000367

Trivedi, S., Rehman, H., Saggu, S., Panneerselvam, C., & Ghosh, S. K. (2020). *DNA barcoding and molecular phylogeny* (2nd ed.). Springer.

Vences, M., Thomas, M., van der Meijden, A., Chiari, Y., & Vieites, D. R. (2005). Comparative performance of the 16S rRNA gene in DNA barcoding of amphibians. *Frontiers in Zoology, 2*(1), 5. https://doi.org/10.1186/1742-9994-2-5

von Nickisch-Rosenegk, M., Lucius, R., & Loos-Frank, B. (1999). Contributions to the phylogeny of the Cyclophyllidea (Cestoda) inferred from mitochondrial 12S rDNA. *Journal of Molecular Evolution, 48*(5), 586–596. https://doi.org/10.1007/PL00006501

Waeschenbach, A., & Littlewood, D. T. J. (2017). *A molecular framework for the Cestoda* (pp. 431–451). University of Kansas.

Waeschenbach, A., Webster, B. L., Bray, R. A., & Littlewood, D. T. (2007). Added resolution among ordinal level relationships of tapeworms (Platyhelminthes: Cestoda) with complete small and large subunit nuclear ribosomal RNA genes. *Molecular Phylogenetics and Evolution, 45*(1), 311–325. https://doi.org/10.1016/j.ympev.2007.03.019

Waeschenbach, A., Webster, B. L., & Littlewood, D. T. (2012). Adding resolution to ordinal level relationships of tapeworms (Platyhelminthes: Cestoda) with large fragments of mtDNA. *Molecular Phylogenetics and Evolution, 63*(3), 834–847. https://doi.org/10.1016/j.ympev.2012.02.020

Waikagul, J., & Thaenkham, U. (2014). *Approaches to the research on the systematics of fish-borne trematodes*. Academic Press.

Wicht, B., Ruggeri-Bernardi, N., Yanagida, T., Nakao, M., Peduzzi, R., & Ito, A. (2010). Inter- and intra-specific characterization of tapeworms of the genus *Diphyllobothrium* (Cestoda: Diphyllobothriidea) from Switzerland, using nuclear and mitochondrial DNA targets. *Parasitology International, 59*(1), 35–39. https://doi.org/10.1016/j.parint.2009.09.002

Wickramasinghe, S., Yatawara, L., Rajapakse, R. P., & Agatsuma, T. (2009). *Toxocara canis* and *Toxocara vitulorum*: Molecular characterization, discrimination, and phylogenetic analysis based on mitochondrial (ATP synthase subunit 6 and 12S) and nuclear ribosomal (ITS-2 and 28S) genes. *Parasitology Research, 104*(6), 1425–1430. https://doi.org/10.1007/s00436-009-1345-9

Wild, A. L., & Maddison, D. R. (2008). Evaluating nuclear protein-coding genes for phylogenetic utility in beetles. *Molecular Phylogenetics and Evolution, 48*(3), 877–891. https://doi.org/10.1016/j.ympev.2008.05.023

Yang, L., Tan, Z., Wang, D., Xue, L., Guan, M. X., Huang, T., & Li, R. (2014). Species identification through mitochondrial rRNA genetic analysis. *Scientific Reports, 4*, 4089. https://doi.org/10.1038/srep04089

Yong, T. S., Lee, J. H., Sim, S., Lee, J., Min, D. Y., Chai, J. Y., Eom, K. S., Sohn, W. M., Lee, S. H., & Rim, H. J. (2007). Differential diagnosis of *Trichostrongylus* and hookworm eggs via PCR using ITS-1 sequence. *The Korean Journal of Parasitology, 45*(1), 69–74. https://doi.org/10.3347/kjp.2007.45.1.69

Yong, H. S., Eamsobhana, P., Song, S. L., Prasartvit, A., & Lim, P. E. (2015). Molecular phylogeography of *Angiostrongylus cantonensis* (Nematoda: Angiostrongylidae) and genetic relationships with congeners using cytochrome b gene marker. *Acta Tropica, 148*, 66–71. https://doi.org/10.1016/j.actatropica.2015.04.020

Chapter 7
PCR and DNA Sequencing: Guidelines for PCR, Primer Design, and Sequencing for Molecular Systematics and Identification

Abstract From Chap. 6, we have seen the properties and utilities of the various classes of genetic markers that have been used for molecular systematics and the identification of parasitic helminths. How can these genetic markers be used for subsequent analysis and interpreted for the relevant research or application? This chapter discussed the principles behind some of the techniques for polymerase chain reaction (PCR) and DNA sequencing and presented practical steps for PCR and post-PCR analysis. The molecular techniques discussed in this chapter are not exhaustive but cover the main methods used for parasitic helminth molecular-based studies. Chapter 8 will then link the DNA sequences obtained after post-PCR analysis to the generation of phylogenetic trees, connecting theory to interpretation and practical applications.

The field of molecular biology is an area of biology concerned with the roles, functions, and interactions of nucleic acids and proteins that carry out molecular processes essential for cellular function (Vitale, 2017). Molecular processes such as DNA replication, transcription, and translation serve as the cornerstone of molecular biology (Fry, 2016; Vitale, 2017). Since the publication and description of the structure of DNA by James Watson and Francis Crick in 1953, research advances in molecular biology accelerated and led to the invention of various molecular techniques. These techniques include identifying, isolating, and manipulating the molecular components in cells such as DNA, RNA, and proteins. PCR and DNA sequencing are examples of these (Fry, 2016). PCR and DNA sequencing applications are extensive and link together with other biology fields, including medicine, molecular genetics, biochemistry, etc. Progress in molecular biology techniques has drastically facilitated the molecular diagnostics and detection of diseases, forensics, biodiversity, the discovery of new species, and our understanding of the intricate tapestry of relationships among groups of organisms. These techniques have also greatly advanced helminth molecular systematics and identification, speeding up the process of helminth identification and discovery, detection of helminthic diseases, and investigation of host-parasite interactions. Figure 7.1 provides a brief overview of the procedures involved in molecular studies, linking organisms to phylogeny.

U. Thaenkham et al., *Molecular Systematics of Parasitic Helminths*,
https://doi.org/10.1007/978-981-19-1786-8_7

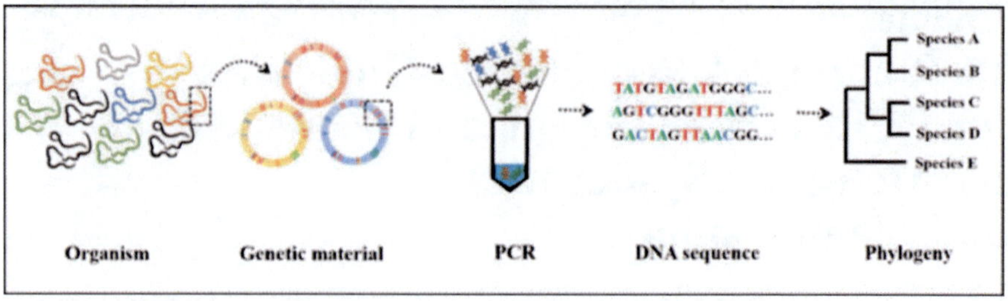

Fig. 7.1 Overview of the steps involved in molecular studies

7.1 Polymerase Chain Reaction (PCR)

PCR is based on the principle of using DNA polymerase to replicate a specific target of DNA sequence through in vitro means, resulting in the synthesis and amplification of new DNA copies. The origins of PCR can be traced back to the 1980s, when Kary Mullis invented this idea, thus earning him the Nobel Prize in Chemistry in 1993 (Bartlett & Stirling, 2003; Kadri, 2019). Exponential amplification of DNA can be achieved in vitro, generating tens of billions of copies of a particular DNA fragment (Bartlett & Stirling, 2003; Kadri, 2019). The isolation of a heat-resistant enzyme from the bacterium *Thermophilus aquatics* that was discovered in a thermal spring also contributed to the success of PCR (Caetano-Anollés, 2013). These discoveries resulted in PCR becoming one of the most widely used techniques, and this idea has been the cornerstone for many areas of research, not only in molecular biology.

7.1.1 Principles of PCR

Simply put, PCR is the use of enzymes to amplify a specific region of the DNA sequence to generate many copies of the target DNA. Essential components for PCR include a thermo-resistant enzyme, also commonly known as DNA polymerase, the deoxynucleotides (dNTPs), which are the building blocks of DNA (adenine, thymine, cytosine, guanine), oligonucleotides (primers) that can bind to the targeted DNA sequence, the DNA template, and a thermal cycler (Caetano-Anollés, 2013).

PCR is performed in three phases: denaturation, annealing, and elongation (Bartlett & Stirling, 2003; Hoy, 2013; Kadri, 2019). Figure 7.2 depicts the stages in the PCR process. First, the double-stranded DNA is separated into single strands in the denaturation phase by increasing temperature to 90 °C to 95 °C. This increase in temperature causes the hydrogen bonds between the complementary nucleotide bases to break, resulting in the denaturation of the double-stranded DNA into single-strand DNA (Kadri, 2019). In the annealing phase, the temperature is decreased, allowing the forward and reverse primers to anneal to the single-strand DNA

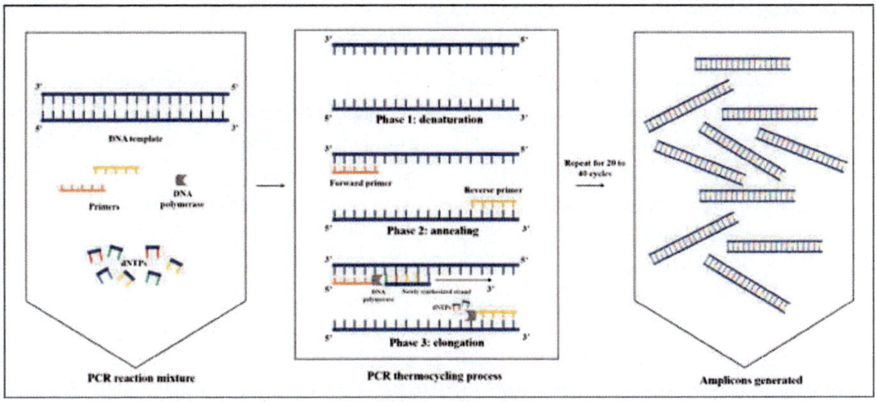

Fig. 7.2 Components and the PCR process conducted via thermocycling

template (Kadri, 2019). The decrease in temperature allows the reformation of hydrogen bonds so that the primers can bind to the single-stranded DNA template. As the primers are significantly shorter in length than the DNA template, the primers can bind more easily to the DNA template than the reformation by the longer strand of the DNA template (preventing the formation of a double-stranded DNA again). The range of temperatures that allows successful primer annealing to the single-stranded DNA template is known as the annealing temperature (Caetano-Anollés, 2013). Once the primers have annealed to the DNA template, the elongation step takes place. During the elongation phase, DNA polymerase binds to the single-strand DNA and catalyzes DNA replication using dNTPs present in the in vitro reaction mixture (Caetano-Anollés, 2013; Kadri, 2019). The synthesis of a new strand that is complementary to the DNA template results. The elongation phase takes place at 72 °C as it is the optimal temperature for the enzymatic activity of DNA polymerase. The three phases are then repeated for approximately 20 to 40 cycles, with each cycle using the synthesized DNA from the previous cycles. A final elongation phase occurs at the end of the 20 to 40 cycles, ensuring that complete elongation occurs and eliminating the chance for partial amplicons to remain in the reaction mixture (Bartlett & Stirling, 2003).

7.1.2 Types of PCR Methods Used for Molecular Systematics and Identification

Various types of PCR methods exist, with slight modifications to the basic principle. We will briefly go through some of the PCR methods that are used most widely.

Conventional Single and Multiplex PCR

Conventional PCR, comprising both single and multiplex PCR, is generally used for molecular identification to detect species of interest. It requires gel electrophoresis to visualize the amplified DNA once amplification of the targeted DNA is completed in a thermal cycler (Bartlett & Stirling, 2003). Gel electrophoresis enables the amplicons to be separated based on their size and then subsequently visualized when the gel is viewed under ultraviolet (UV) light. An appropriately sized band indicates a successful amplification of the target DNA. For single PCR, one pair of primers is used in a single PCR reaction, targeting one region in the DNA.

An example is the molecular identification of human hookworms *Necator americanus* and *Anyclostoma duodenale* from human infections targeting the mitochondrial cytochrome *c* oxidase subunit I gene (*COI*) (Chang et al., 2020). Molecular identification of hookworms was achieved after DNA sequencing of the partial *COI* amplicon. Multiplex PCR uses more than one pair of primers in a single PCR reaction to target more than one region in the DNA. Multiplex PCR generates amplicons of varying sizes (depending on each primer pair) and is then visualized via gel electrophoresis. Multiplex PCR allows for the simultaneous detection of more than one region of the targeted DNA within a single PCR reaction. For example, a conventional multiplex PCR was developed to simultaneously detect the soil-transmitted helminths *Ascaris lumbricoides*, *N. americanus*, and *Strongyloides stercoralis*. Molecular identification of the three species was achieved with the differently sized amplicon sizes of each species after gel electrophoresis (Sanprasert et al., 2019).

Nested PCR

Generally, nested PCR is used for molecular identification of a specific species of interest with increased sensitivity and specificity. It is a modification of the conventional PCR to amplify a more specific DNA target, and it involves two rounds of PCR. The first pair of primers binds and amplifies the DNA target in the first round. In the second round of PCR, PCR products from the first round serve as the DNA template for a second primer pair. The second pair of primers used in the second round of PCR are designed to bind to a region within the first round of PCR products, functioning like "inner" primers. The second round of PCR serves to amplify a more specific DNA target. For example, nested PCR was used for molecular identification of *Schistosoma japonicum* from goats and buffalos and the detection of *S. stercoralis* in human fecal samples (Sharifdini et al., 2015; Zhang et al., 2017). Sensitivity is enhanced when utilizing the nested PCR strategy as low amounts of DNA (from the first round of PCR) can be detected due to the second round of amplification. Additionally, nested PCR can minimize nonspecific products in a PCR reaction. The second round of PCR with the second pair of primers reduces the nonspecific and spurious PCR products, resulting in a more specific method than the conventional PCR.

PCR-Restriction Fragment Length Polymorphism

The PCR-restriction fragment length polymorphism (PCR-RFLP) is another technique commonly used for molecular identification of the species of interest. This technique utilizes restriction enzymes to cut specific sites in the amplified DNA sequence to generate species-specific band patterns for gel electrophoresis visualization (Rasmussen, 2012). It serves as a simple and helpful technique for species identification, taking advantage of the variations in DNA sequences between species. The main steps involved in PCR-RFLP are PCR amplification of the gene target, followed by restriction enzyme digestion to generate fragments of different sizes, and gel electrophoresis to visualize band patterns. PCR-RFLP was applied to the molecular differentiation of six species of minute intestinal flukes (*Haplorchis taichui*, *H. pumilio*, *H. yokogawai*, *Procerovum varium*, *Stellantchasmus falcatus*, and *Centrocestus formosanus*) (Thaenkham et al., 2011). The six species were successfully differentiated using the nuclear 28S rRNA gene based on different band patterns after digestion using the restriction enzyme *MboII*. The hookworms *Ancylostoma caninum*, *A. ceylanicum*, and *A. braziliense* were also successfully differentiated from dog fecal samples through PCR-RFLP using the ITS region (Traub et al., 2004).

Quantitative Real-Time PCR (qPCR)

Real-time PCR or quantitative PCR is based on the concept that amplification can be visualized in real-time without imagining the PCR amplicons through gel electrophoresis (Lenstra, 1995). Based on a standard curve, the amount of DNA in a sample can be quantified and visualized in real-time through fluorescence through qPCR. This technique is commonly used for molecular identification and diagnosis of helminthic infection from various sample types. Using fluorescent dyes to monitor the accumulation of PCR amplicons, the amplification of the DNA target can be visualized in real time. Monitoring amplification can be done in real time due to the thermal cycler measuring the change in fluorescence emission from the fluorescent DNA binding dyes. Two types of fluorescence methods are used for the monitoring of amplification. The first is the use of a fluorescent dye that binds nonspecifically to the double-stranded DNA as it is generated during PCR. A fluorescent signal is emitted upon intercalating with the newly synthesized DNA. An example is the SYBR® Green dye. As the fluorescent dye binds nonspecifically to the DNA, a melt curve analysis is required to ensure that the target of interest is genuinely detected. A SYBR® green qPCR assay was developed and subsequently used for molecular identification of *Angiostrongylus cantonensis* and *A. malaysiensis* in snail intermediate hosts and human cerebrospinal fluid samples using the mitochondrial cytochrome b gene (*cytB*) (Jakkul et al., 2021; Watthanakulpanich et al., 2021).

The second method uses target-specific fluorescent hydrolysis probes added to the PCR mixture. An example is the TaqMan® probes, which are specific sequences designed to bind downstream of the PCR primers; thus, they are target-specific. As DNA is synthesized and extended during PCR, the probe is cleaved, and fluorescence is detected. The second method using specific probes ensures that the fluorescence signal obtained is from the target gene of interest compared to the first method. However, a drawback of using specific probes is in finding suitable probe binding sites downstream of the primer binding sites. (Bisht & Panda, 2014; Lenstra, 1995; Life Technologies, 2012; Löfström et al., 2015; Viljoen et al., 2005b). A TaqMan multiplex qPCR assay was developed for species identification of *Trichinella spiralis*, *T. nativa*, *T. murrelli*, and *Trichinella* T6 in human and animal infected tissues using the internal transcribed spacer (ITS) regions (Almeida et al., 2018).

7.2 DNA Sequencing

DNA sequencing, which determines the precise order of nucleotides within a DNA molecule, has significantly revolutionized the world of research and biological sciences (Bisht & Panda, 2014; Graham & Hill, 2001). In 1977, Maxam and Gilbert "birthed" the first generation of DNA sequencing, where the DNA sequences were determined through chemical degradation and radio-labeled nucleotides (Bisht & Panda, 2014; Shendure et al., 2017). Sanger sequencing was developed by Frederick Sanger, where it utilizes the concept of chain termination (Shendure et al., 2017). To date, Sanger sequencing is by far the most common DNA sequencing technique used. With technical advances and automation, the concept of Sanger sequencing was refined, giving rise to second-generation sequencing. These techniques are different from first-generation sequencing in that they do not depend on radio-labeled or fluorescent-labeled oligonucleotides to determine the DNA sequence. In addition, second-generation sequencing has a high-throughput capability compared to first-generation sequencing (McGinn & Gut, 2013). Examples of second-generation sequencing techniques include Pyrosequencing, Ion Torrent, and Illumina. Recent advances in technology have brought about a new era for DNA sequencing, giving rise to third-generation sequencing techniques (McGinn & Gut, 2013). These advances in DNA sequencing will inadvertently accelerate biological sciences research in the near future.

The two molecular biology techniques, DNA sequencing and PCR, have been the bedrock in widening our understanding of organisms, including molecular systematics and molecular identification studies. An overview of the three generations of DNA sequencing will be briefly explained in the next section.

7.2.1 *Types of DNA Sequencing Methods*

Maxam-Gilbert

The Maxam-Gilbert sequencing method uses a two-step catalytic process involving chemicals to attack purines and pyrimidines selectively. The process results in creating a single-stranded DNA substrate with a radioactive label on the 5′ end of the DNA (Bisht & Panda, 2014; Heather & Chain, 2016). The radio-labeled substrate undergoes four separated cleavage reactions, whereby each reaction creates a population of labeled nucleotides that are known. The DNA fragments are resolved by polyacrylamide gel electrophoresis, where the bands can be separated by size and visualized (Bisht & Panda, 2014).

Sanger

Sanger sequencing involves the concept of chain termination, where the incorporation of a dideoxynucleotide (ddNTP) during the process of elongation results in the termination of the elongation process (Heather & Chain, 2016; Munshi, 2012). ddNTPs is another form of ribose, where the hydroxyl (−OH) group is not present at the 3′ carbon of the sugar. As a result of the lack of the −OH group, the phosphodiester bond between one nucleotide and the next is unable to form. The cycle repeats, and all possible single-stranded DNA molecules that are complementary to the DNA template are formed. Like in Maxam-Gilbert sequencing, the DNA fragments are resolved by polyacrylamide gel electrophoresis. Automation of Sanger sequencing was subsequently developed, and this advancement allowed for the labeling of the ddNTPs by incorporating a fluorescent dye which can be read through capillary tube electrophoresis (Bisht & Panda, 2014).

Pyrosequencing and Ion Torrent

Pyrosequencing and Ion Torrent are second-generation sequencing methods, which utilize the sequencing by synthesis principle. Pyrosequencing relies on the detection of pyrophosphate release when a nucleotide gets incorporated (McGinn & Gut, 2013). The enzyme adenosine-5′-triphosphate sulfurylase (ATP sulfurylase) converts pyrophosphate into adenosine triphosphate (ATP), which is then used as the substrate for luciferase. The luciferase enzyme is responsible for producing light, which can then be detected with an ion-sensitive field-effect transistor (McGinn & Gut, 2013). The amount of light produced is proportional to the amount of pyrophosphate released and the incorporation of the nucleotides. Likewise, the Ion Torrent sequencing method is based on detecting hydrogen ions (instead of light) released when a nucleotide is incorporated into the single strand of DNA (Ansorge, 2009).

Illumina

The Illumina platform, which is another second-generation sequencing method, also utilizes the sequencing by synthesis principle, with specific adaptors binding to the end of the DNA fragment. A "bridge amplification" is performed, whereby both ends of the single-stranded DNA are attached to the surface and form a bridge. The single-stranded DNA strand is extended by DNA polymerase and becomes double-stranded DNA. The double-stranded DNA bridge is then denatured, resulting in two copies of single-stranded DNA templates. The process of bridge amplification repeats and the sequencing primer allows for the determination of sequences through the unique wavelengths of each of the four DNA bases (Ansorge, 2009; Bisht & Panda, 2014; Wong et al., 2019).

Third-Generation Nanopore

The third-generation sequencing methods build on the technological advances of the second-generation techniques and utilize the nanopore approach (McGinn & Gut, 2013). Examples include the single-molecule real-time sequencing (SMRT) approach and the Oxford nanopore sequencers. Pacific Biosciences developed the SMRT platform, enabling the visualization of single fluorophore molecules close to the bottom of the arrays of microfabricated nanostructures (Rhoads & Au, 2015). Oxford developed the MinIon platform, a small mobile phone-sized USB device that can directly analyze the DNA strand when the molecule is drawn through a tiny pore suspended in a membrane (Jain et al., 2016). The MinIon offers a unique, convenient, and portable platform for DNA sequencing to be performed almost anywhere.

7.3 From Theory to Practice: Practical Steps for Use

This section will present some practical steps required for PCR and post-PCR analysis to enable practical use and to put the methods discussed previously into context for research that can be applied for helminth molecular systematics and identification.

7.3.1 Data Mining from Nucleotide Database

Nucleotide sequences of the specific genetic marker of the target organism can be obtained from reference databases that are freely available online (Posada, 2009; Selzer et al., 2018). The National Center for Biotechnology Information (NCBI) GenBank is an example of a reference database and contains all publicly available

Table 7.1 Examples of nucleotide databases and primer design tools available online

Nucleotide database	
Name of database	Website URL
National Center for Biotechnology Information (NCBI)	http://www.ncbi.nlm.nih.gov
DNA Data Bank of Japan (DDBJ)	https://www.ddbj.nig.ac.jp/index-e.html
European Bioinformatics Institute (EBI)	https://www.ebi.ac.uk/
European Nucleotide Archive (ENA)	https://www.ebi.ac.uk/ena/browser/home
Online tools for primer properties	
Name of online tool	Website URL
Oligo Calc (Kibbe, 2007)	http://biotools.nubic.northwestern.edu/OligoCalc.html
NetPrimer	http://www.premierbiosoft.com/index.html
DNA mate (Panjkovich et al., 2005)	http://melolab.org/dnaMATE/
Online tools for primer design	
Name of online tool	Website URL
Primer-BLAST (Ye et al., 2012)	https://www.ncbi.nlm.nih.gov/tools/primer-blast/
Primer3Plus (Untergasser et al., 2007)	http://www.bioinformatics.nl/cgi-bin/primer3plus/primer3plus.cgi
GeneFisher (Giegerich et al., 1996)	https://bibiserv.cebitec.uni-bielefeld.de/genefisher2

nucleotide sequences (http://www.ncbi.nlm.nih.gov). Examples of other reference databases are provided in Table 7.1 below.

To start, nucleotide sequences can be obtained by keying in the scientific name of the organism and the name of the genetic marker in the search box in the NCBI database. A list of the available sequences will be generated, and users can select the sequences they want. It is crucial to screen through the available sequences and only select the appropriate or reliable sequences for the required work. The types of available information for a sequence record include the taxonomic classification of the organism, the organism and collection details, authors and attribution details, the publication status of the manuscript (if any), as well as the sequencing technology that was used. The respective sequence can be obtained from the sequence record and saved in a notepad file in the FASTA format. A FASTA format is text-based and used for representing nucleotide sequences using single-letter codes (Posada, 2009). It begins with a description line, followed by lines of nucleotide sequence data. The description line begins with a ">" symbol.

Once all the required sequences have been mined from the NCBI database, we are ready to move on to primer design and in silico PCR.

7.3.2 Primer Design and In Silico PCR

Designing good primers is a crucial step in ensuring the success of the PCR reaction; thus, here are some important properties of a good primer (Álvarez-Fernández, 2013; Basu, 2015; Dieffenbach et al., 1993). They should be:

1. 18 to 24 nucleotides in length for maximum efficiency.
2. G + C content of 40 to 60% with a G or C at the 3′ end for improved stability of the primer binding.
3. Absence of hairpin formation and self-complementary binding for increased amplification yield.
4. A melting temperature (T_m) of between 55 °C to 65 °C and the T_m between the primer pairs should be within 5 °C of each other. Having too much of a difference between the T_m can lead to loss of specificity for the primer with the lower T_m and the primer with the higher T_m has a greater chance of mispriming. On the other hand, too high a T_m can lead to no primer binding at all. To calculate T_m, a general formula based on Suggs et al. (1981) is

$$T_m = 2x(\text{number of A} + \text{number of } T) + 4x(\text{number of G} + \text{number of C}).$$

Alternatively, the T_m and other properties of the designed primers can be checked using online tools as provided in Table 7.1.

With the FASTA file containing the list of sequences of the target amplicon, perform multiple sequence alignment using sequence alignment software. More in-depth information about multiple sequence alignment will be discussed in Chap. 8. Once the sequences have been aligned, look for conserved sites among all the sequences where the forward and reverse primers are able to bind. Do take note to do a reverse complement for the sequences for the reverse primer. Users should also ensure that there are sufficient sequence variations between the two primer flanking sites for species discrimination. The choice of primer flanking sites will dictate how long the amplicon will be. Once primers have been designed, primer quality can be checked for the various properties mentioned in the paragraph above. Alternatively, many online tools have been made available to aid in designing primers (Table 7.1). However, these tools do not often include the capability to design primers across many target sequences. It is, therefore, always good to be able to design primers for the selected target organisms manually.

In addition to checking the properties of the primer, in silico analysis on primer specificity and in silico PCR can be performed to ensure that the primers are able to bind and produce amplicons successfully. Using Primer-BLAST (http://www.ncbi.nlm.nih.gov/tools.primer-blast), the specificity of the designed primers can be checked against all the sequences (or user-specified) in the NCBI database (Ye et al., 2012). Stringency requirements and primer parameters for the BLAST search can also be adjusted to suit users' preferences. In silico PCR can be performed using the software FastPCR to check for the specific binding of the primers to the desired sequences of the target organism (Kalendar et al., 2017). In FastPCR, the FASTA file containing the target sequences will have to be uploaded with the designed primer sequences. An output showing the primer binding sites and percentage of success with the respective amplicon sequences and predicted annealing temperature would be given. Users can also adjust the stringency of the primer binding or the number of mismatches allowed at the 3′ end of the primer.

After the newly designed primers are appropriate for the selected application, downstream PCR amplification can be carried out with the designed primers on the target organism.

7.3.3 PCR Amplification

General Setup and Conditions

The general workflow for conducting PCR is the preparation of master mix, followed by the addition of DNA template, setting up the thermal-cycling conditions and allowing the reactions to run, and then post-PCR processing (Bartlett & Stirling, 2003). For the setup prior to performing PCR amplification, it is important to assign separate workbenches in the laboratory to minimize the risk of contamination and ensure accurate results. A common strategy is to physically separate the different parts of the laboratory processes and arrange them in a unidirectional workflow. For example, the preparation of the PCR master mix should be performed on a separate bench or within a PCR cabinet (clean-hood) separated from the area where the DNA template will be added. The post-PCR processing area should also be separated from the area where the PCR master mix will be prepared.

The components of a PCR reaction include the PCR master-mix buffer, DNA polymerase, dNTPs, magnesium chloride ($MgCl_2$), primers, and DNA template. Traditionally, each of the components needs to be added individually for PCR. However, the PCR master-mix buffer comes in premixed with DNA polymerase, dNTPs, $MgCl_2$, proprietary additives, and a color dye (Bio-Rad Laboratories, n.d.). Only the primers, DNA template, and water (to make up to the final volume) must be added to the PCR reaction mix. The final concentration for each primer usually ranges from 0.05 μm to 1.0 μm (Bartlett & Stirling, 2003; Viljoen et al., 2005a). The final volume for a single PCR generally ranges from 20 to 50 μL, depending on the purpose and the recommended conditions by the manufacturer of the PCR master-mix buffer. To start, the PCR master mix (all of the components except the DNA template) will be mixed together in a 0.5 mL PCR tube and performed in the PCR hood or the "clean" area. Once the master mix has been prepared, addition of the DNA template can be performed. As a rule of thumb, a positive and negative control is recommended to be included for each PCR run. The positive control is to ensure that the PCR is successful, whereas the negative control is to ensure that no contamination occurred.

As mentioned in Sect. 7.1.1, PCR is divided into three phases: denaturation, annealing, and elongation. An initial denaturation step takes place, where the thermal-cycling program is set at 94 °C to 95 °C for 2 to 5 min to allow for the dissociation of the double-stranded DNA to single-stranded DNA. After the initial denaturation step, 30 to 35 cycles of denaturation, annealing, and elongation take place for amplification. Thirty to 40 cycles are sufficient to ensure sufficient amplification of the target amplicon (Bartlett & Stirling, 2003; Viljoen et al., 2005a). For primer annealing, the annealing temperature is usually 5 °C lower than the calculated T_m of the primers, ranging anywhere from 45 °C to 70 °C; 30 s to 1 min is sufficient time to allow the primers to anneal to the template DNA. The elongation occurs at the temperature for maximum DNA polymerase activity (usually 72 °C). The time for elongation is dependent on the expected size of the PCR amplicon, where approximately 1 min is needed per kilobase (Bartlett & Stirling, 2003). A final extension step is performed to ensure that all amplicons are fully extended.

Optimization and Troubleshooting

To ensure that the PCR conditions are optimal, there is a need to optimize the conditions prior to the actual experiment. Some considerations for PCR optimization are primer concentration, template DNA quality and concentration, and the annealing temperature. Firstly, the concentration of primers will influence PCR amplification specificity and efficiency. High primer concentrations can result in nonspecific binding and the formation of primer dimers, whereas low primer concentrations may affect PCR efficiency (Viljoen et al., 2005a). The forward and reverse primers should also be ideally mixed at equal ratios. Secondly, the quality and concentration of the DNA template affect the success of PCR amplification. The quality of the DNA template can be checked using gel electrophoresis or a photo spectrometer prior to carrying out the PCR. Lastly, the annealing temperature should be optimized as incorrect temperatures might either lead to no amplicons or nonspecific amplicons. A high annealing temperature increases amplification specificity but might result in reduced PCR products, whereas a lower annealing temperature might lead to nonspecific amplification. Ideally, a gradient PCR should be performed with a range of temperatures to determine the optimal annealing temperature for maximum efficiency.

For troubleshooting, we list some common issues encountered when performing PCR amplification in Table 7.2 (Lorenz, 2012; Thermo Fisher Scientific, n.d.; Viljoen et al., 2005a). Do note that the reasons and solutions are non-exhaustive, and the solutions might not bring about total success in a PCR reaction, as the success is highly dependent on the individual PCR protocol, reagents, and DNA.

Table 7.2 Examples of issues encountered during PCR and possible solutions

Issues	Possible reasons	Possible solutions
No bands	Too few cycles	Increase the number of cycles
	Too high annealing temperature	Reduce the annealing temperature
	High G/C content of DNA template	Increase the annealing temperature
	Low quality or low concentration of DNA template	Use another DNA template or add more DNA into the PCR mix
	PCR inhibitors	Dilute the DNA template
	Too low primer concentration	Increase primer concentration
	Not optimal primer design	Redesign primers
Multiple nonspecific bands/primer dimers	Too many cycles	Reduce the number of cycles
	Too long extension time	Reduce the extension time (1 min per kilobase)
	Too low annealing temperature	Increase the annealing temperature
	Too high $MgCl_2$ concentration	Reduce the $MgCl_2$ concentration
	Too high primer concentration	Reduce the primer concentration
Negative control with positive band	Contamination of reagents or contamination during the process	Repeat the PCR with a freshly prepared batch of reagents

7.3.4 Post-PCR Processing

After successfully performing a PCR, the PCR amplicons can be visualized via gel electrophoresis to measure the success of the PCR (Bartlett & Stirling, 2003). Through gel electrophoresis, the amplicons can be separated based on their size or molecular weight with an electric current, with the rate of migration proportional to the size of the amplicon (Isbir et al., 2013; Scitable, n.d.). The larger the amplicon is, the slower it will migrate, rendering its position higher up relative to the smaller amplicon. The success of the PCR is indicated by an appropriately sized band on the gel.

To perform gel electrophoresis, an agarose gel will have to be prepared, containing agarose powder and either the Tris-borate-EDTA or Tris-acetate-EDTA buffer (the same buffer will have to be used in the gel electrophoresis run). Agarose consists of repeated L- and D-galactose subunits, where the agarose polymers associate non-covalently and form a network of bundles during gelation. This network of bundles (matrix size) determines the ability of the gel to perform the separation process (Lee et al., 2012; Voytas, 2000). Generally, the higher the agarose concentration, the smaller the matrix size. The concentration of agarose gel to prepare depends on the size of the PCR amplicons. A 0.8% to 1% agarose gel is typically prepared for an amplicon size of 400 base pairs (bp) to 1000 bp, whereas a higher percentage of gel is prepared for amplicons that are less than 400 bp (Thermo Fisher Scientific, 2012). The voltage of the gel electrophoresis run can also be adjusted according to how separated the different sized amplicons will be. A lower voltage is used when the amplicons need to be well separated.

Typically, in a gel electrophoresis run, 1 to 5 μL of the PCR products are loaded into each well, together with the DNA ladder. If the PCR master-mix buffer does not contain a color dye, a loading dye will have to be mixed and loaded with the PCR products before loading the PCR products into the wells. The loading dye enables the physical visualization of the loaded PCR samples and facilitates the sinking of the products into the well (Voytas, 2000). An example of a loading dye is bromophenol blue. Visualization of the amplicons under a UV transilluminator after the gel electrophoresis run is performed by staining the gel with dyes. Staining of the gel can be performed after gel electrophoresis or by the direct addition of the dye when preparing the agarose gel. Traditionally, ethidium bromide (EtBr) is used, where it works by intercalating with the DNA molecule. However, EtBr is not frequently used nowadays as it is a potential carcinogen. Other dyes with low toxicity, such as SYBR green and SYBR safe, are used in its place.

A final step for the post-PCR processing includes the purification of the PCR amplicons prior to sending them for sequencing. This step is necessary to remove unincorporated primers, enzymes, leftover dNTPs, and other components present in the PCR master mix, which could interfere with the downstream sequencing reactions (New England Biolabs Inc, n.d.). Purification of PCR amplicons can be performed using commercially available kits and spin columns.

7.4 Conclusion

The applications of PCR and DNA sequencing are vast, and they are undoubtedly valuable tools that can be applied in molecular-based studies for parasitic helminths. In this chapter, we have reviewed the principles behind PCR and DNA sequencing and provided examples of some of the techniques commonly used in molecular systematics and identification studies. Practical steps, guidelines, considerations, and troubleshooting tips have been provided later in this chapter to aid users in applying molecular-based studies.

In Chap. 8, we will move on to look at some of the theory behind sequence alignments and phylogenetic analysis, as well as provide some guidelines for practical applications, which can then enable the completion of the picture—by linking organisms to DNA sequences and finally to phylogenetic trees and interpretation.

References

Almeida, M. D., Bishop, H., Nascimento, F. S., Mathison, B., Bradbury, R. S., & Silva, A. D. (2018). Multiplex TaqMan qPCR assay for specific identification of encapsulated *Trichinella* species prevalent in North America. *Memórias do Instituto Oswaldo Cruz, 113*(11), e180305. https://doi.org/10.1590/0074-02760180305

Álvarez-Fernández, R. (2013). Explanatory chapter: PCR primer design. *Methods in Enzymology, 529*, 1–21. https://doi.org/10.1016/b978-0-12-418687-3.00001-x
Ansorge, W. J. (2009). Next-generation DNA sequencing techniques. *New Biotechnology, 25*(4), 195–203. https://doi.org/10.1016/j.nbt.2008.12.009
Bartlett, J. M. S., & Stirling, D. (2003). *PCR protocols* (2nd ed.). Humana Press.
Basu, C. (2015). *PCR primer design* (2nd ed.). Humana Press.
Bio-Rad Laboratories. (n.d.). *PCR Master Mix*. Retrieved from https://www.bio-rad.com/featured/en/pcr-master-mix.html#:~:text=PCR%20Master%20Mixes%20and%20Supermixes,a%20buffer%20optimized%20for%20PCR
Bisht, S. S., & Panda, A. K. (2014). DNA sequencing: Methods and applications. In I. Ravi, M. Baunthiyal, & J. Saxena (Eds.), *Advances in biotechnology* (pp. 11–23). https://doi.org/10.1007/978-81-322-1554-7_2
Caetano-Anollés, D. (2013). Polymerase Chain reaction. In *Brenner's encyclopedia of genetics* (pp. 392–395). https://doi.org/10.1016/b978-0-12-374984-0.01186-4
Chang, T., Jung, B. K., Sohn, W. M., Hong, S., Shin, H., Ryoo, S., Lee, J., Lee, K. H., Khieu, V., Huy, R., & Chai, J. Y. (2020). Morphological and molecular diagnosis of *Necator americanus* and *Ancylostoma ceylanicum* recovered from villagers in northern Cambodia. *The Korean Journal of Parasitology, 58*(6), 619–625. https://doi.org/10.3347/kjp.2020.58.6.619
Dieffenbach, C. W., Lowe, T. M., & Dveksler, G. S. (1993). General concepts for PCR primer design. *PCR Methods and Applications, 3*(3), S30–S37. https://doi.org/10.1101/gr.3.3.s30
Fry, M. (2016). Introduction. In *Landmark experiments in molecular biology* (pp. 1–12). https://doi.org/10.1016/b978-0-12-802074-6.00001-1
Giegerich, R., Meyer, F., & Schleiermacher, C. (1996). GeneFisher—software support for the detection of postulated genes. *Proceedings of the International Conference on Intelligent Systems for Molecular Biology, 4*, 68–77.
Graham, C. A., & Hill, A. J. M. (2001). *DNA sequencing protocols* (2nd ed.). Humana Press.
Heather, J. M., & Chain, B. (2016). The sequence of sequencers: The history of sequencing DNA. *Genomics, 107*(1), 1–8. https://doi.org/10.1016/j.ygeno.2015.11.003
Hoy, M. A. (2013). DNA amplification by the polymerase chain reaction. In *Insect molecular genetics* (pp. 307–372). https://doi.org/10.1016/b978-0-12-415874-0.00008-1
Isbir, T., Kirac, D., Demircan, B., & Dalan, B. (2013). Gel electrophoresis. In S. Maloy & K. Hughes (Eds.), *Brenner's encyclopedia of genetics* (2nd ed., pp. 165–167). Academic Press. https://doi.org/10.1016/B978-0-12-374984-0.00580-5
Jain, M., Olsen, H. E., Paten, B., & Akeson, M. (2016). The Oxford Nanopore MinION: Delivery of nanopore sequencing to the genomics community. *Genome Biology, 17*(1), 239. https://doi.org/10.1186/s13059-016-1103-0
Jakkul, W., Chaisiri, K., Saralamba, N., Limpanont, Y., Dusitsittipon, S., Charoennitiwat, V., Chan, A. H. E., & Thaenkham, U. (2021). Newly developed SYBR green-based quantitative real-time PCRs revealed coinfection evidence of *Angiostrongylus cantonensis* and *A. malaysiensis* in *Achatina fulica* existing in Bangkok metropolitan, Thailand. *Food and Waterborne Parasitology, 23*, e00119. https://doi.org/10.1016/j.fawpar.2021.e00119
Kadri, K. (2019). Polymerase Chain reaction (PCR): Principle and applications. In M. Nagpal, O. Boldura, C. Baltă, & S. Enany (Eds.), *Synthetic biology- new interdisciplinary science*. IntechOpen.
Kalendar, R., Khassenov, B., Ramankulov, Y., Samuilova, O., & Ivanov, K. I. (2017). FastPCR: An in silico tool for fast primer and probe design and advanced sequence analysis. *Genomics, 109*(3–4), 312–319. https://doi.org/10.1016/j.ygeno.2017.05.005
Kibbe, W. A. (2007). OligoCalc: an online oligonucleotide properties calculator. *Nucleic Acids Research, 35*(Web Server issue), W43–W46. https://doi.org/10.1093/nar/gkm234
Lee, P. Y., Costumbrado, J., Hsu, C. Y., & Kim, Y. H. (2012). Agarose gel electrophoresis for the separation of DNA fragments. *Journal of Visualized Experiments, 62*, 3923. https://doi.org/10.3791/3923

Lenstra, J. A. (1995). The applications of the polymerase chain reaction in the life sciences. *Cellular and Molecular Biology (Noisy-le-Grand, France), 41*(5), 603–614.

Life Technologies. (2012). *Real-time PCR handbook* (p. 70) Retrieved from https://www.lifetechnologies.com/qpcr

Löfström, C., Josefsen, M. H., Hansen, T., Søndergaard, M. S. R., & Hoorfar, J. (2015). Fluorescence-based real-time quantitative polymerase chain reaction (qPCR) technologies for high throughput screening of pathogens. In *High throughput screening for food safety assessment* (pp. 219–248). https://doi.org/10.1016/b978-0-85709-801-6.00009-5

Lorenz, T. C. (2012). Polymerase chain reaction: Basic protocol plus troubleshooting and optimization strategies. *Journal of Visualized Experiments, 63*, e3998. https://doi.org/10.3791/3998

McGinn, S., & Gut, I. G. (2013). DNA sequencing—Spanning the generations. *New Biotechnology, 30*(4), 366–372. https://doi.org/10.1016/j.nbt.2012.11.012

Munshi, A. (2012). *DNA sequencing—Methods and applications*. InTech Open. Retrieved from https://www.intechopen.com/books/1722

New England Biolabs Inc. (n.d.). *PCR & reaction cleanup*. Retrieved from https://www.neb.com/applications/dna-amplification-pcr-and-qpcr/pcr-and-reaction-cleanup

Panjkovich, A., Norambuena, T., & Melo, F. (2005). dnaMATE: a consensus melting temperature prediction server for short DNA sequences. *Nucleic Acids Research, 33*(Web Server issue), W570–W572. https://doi.org/10.1093/nar/gki379

Posada, D. (2009). *Bioinformatics for DNA sequence analysis* (1st ed.). Humana Press.

Rasmussen, H. B. (2012). Restriction fragment length polymorphism analysis of PCR-amplified fragments (PCR-RFLP) and gel electrophoresis—Valuable tool for genotyping and genetic fingerprinting. In S. Magdeldin (Ed.), *Gel Electrophoresis–Principles and Basics*. IntechOpen.

Rhoads, A., & Au, K. F. (2015). PacBio sequencing and its applications. *Genomics, Proteomics & Bioinformatics, 13*(5), 278–289. https://doi.org/10.1016/j.gpb.2015.08.002

Sanprasert, V., Kerdkaew, R., Srirungruang, S., Charuchaibovorn, S., Phadungsaksawasdi, K., & Nuchprayoon, S. (2019). Development of conventional multiplex PCR: a rapid technique for simultaneous detection of soil-transmitted helminths. *Pathogens (Basel, Switzerland), 8*(3), 152. https://doi.org/10.3390/pathogens8030152

Scitable. (n.d.). *Gel electrophoresis*. Retrieved from https://www.nature.com/scitable/definition/gel-electrophoresis-286/

Selzer, P. M., Marhöfer, R. J., & Koch, O. (2018). *Applied bioinformatics—an introduction*. Springer.

Sharifdini, M., Mirhendi, H., Ashrafi, K., Hosseini, M., Mohebali, M., Khodadadi, H., & Kia, E. B. (2015). Comparison of nested polymerase chain reaction and real-time polymerase chain reaction with parasitological methods for detection of *Strongyloides stercoralis* in human fecal samples. *The American Journal of Tropical Medicine and Hygiene, 93*(6), 1285–1291. https://doi.org/10.4269/ajtmh.15-0309

Shendure, J., Balasubramanian, S., Church, G. M., Gilbert, W., Rogers, J., Schloss, J. A., & Waterston, R. H. (2017). DNA sequencing at 40: past, present and future. *Nature, 550*(7676), 345–353. https://doi.org/10.1038/nature24286

Suggs, S. V., Hirose, T., Myake, T., Kawashima, E. H., Johnson, M. J., & Wallace, R. B. (1981). Developmental Biology Using Purified Genes. In D. Brown (Ed.), *Proceedings of the 1981 ICN-UCLA Symposia on Developmental Biology Using Purified Genes, Keystone, Col., March 15–20, 1981: Also: tenth UCLA Symposia*. Academic. Retrieved from https://books.google.co.th/books?id=Pt93wgEACAAJ

Thaenkham, U., Phuphisut, O., Pakdee, W., Homsuwan, N., Sa-nguankiat, S., Waikagul, J., Nawa, Y., & Dung do, T. (2011). Rapid and simple identification of human pathogenic heterophyid intestinal fluke metacercariae by PCR-RFLP. *Parasitology International, 60*(4), 503–506. https://doi.org/10.1016/j.parint.2011.09.004

Thermo Fisher Scientific. (2012). *General recommendations for DNA electrophoresis*. Retrieved from https://assets.thermofisher.com/TFS-Assets/LSG/manuals/MAN0012614_Gen_Recommend_DNA_Electrophoresis_UG.pdf

Thermo Fisher Scientific. (n.d.). *PCR Troubleshooting Guide*. Retrieved from https://www.thermofisher.com/th/en/home/life-science/cloning/cloning-learning-center/invitrogen-school-of-molecular-biology/pcr-education/pcr-reagents-enzymes/pcr-troubleshooting.html

Traub, R. J., Robertson, I. D., Irwin, P., Mencke, N., & Thompson, R. C. A. (2004). Application of a species-specific PCR-RFLP to identify *Ancylostoma* eggs directly from canine faeces. *Veterinary Parasitology, 123*(3), 245–255. https://doi.org/10.1016/j.vetpar.2004.05.026

Untergasser, A., Nijveen, H., Rao, X., Bisseling, T., Geurts, R., & Leunissen, J. A. (2007). Primer3Plus, an enhanced web interface to Primer3. *Nucleic Acids Research, 35*(Web Server issue), W71–W74. https://doi.org/10.1093/nar/gkm306

Viljoen, G. J., Nel, L. H., & Crowther, J. R. (2005a). *Molecular diagnostic PCR handbook* (1st ed.). Springer.

Viljoen, G. J., Nel, L. H., & Crowther, J. R. (2005b). Specific types of PCR assays. In *Molecular diagnostic PCR handbook* (pp. 74–100). Springer. https://doi.org/10.1007/1-4020-3404-0_5

Vitale, I. (2017). Molecular biology. In *Reference module in life sciences*. https://doi.org/10.1016/b978-0-12-809633-8.12381-7

Voytas, D. (2000). Agarose gel electrophoresis. *Current Protocols in Molecular Biology, 51*(1), 2.5A.1–2.5A.9. https://doi.org/10.1002/0471142727.mb0205as51

Watthanakulpanich, D., Jakkul, W., Chanapromma, C., Ketboonlue, T., Dekumyoy, P., Lv, Z., Chan, A. H. E., Thaenkham, U., & Chaisiri, K. (2021). Co-occurrence of *Angiostrongylus malaysiensis* and *Angiostrongylus cantonensis* DNA in cerebrospinal fluid: Evidence from human eosinophilic meningitis after ingestion of raw snail dish in Thailand. *Food and Waterborne Parasitology, 24*, e00128. https://doi.org/10.1016/j.fawpar.2021.e00128

Wong, K.-C., Zhang, J., Yan, S., Li, X., Lin, Q., Kwong, S., & Liang, C. (2019). DNA sequencing technologies. *ACM Computing Surveys, 52*(5), 1–30. https://doi.org/10.1145/3340286

Ye, J., Coulouris, G., Zaretskaya, I., Cutcutache, I., Rozen, S., & Madden, T. L. (2012). Primer-BLAST: A tool to design target-specific primers for polymerase chain reaction. *BMC Bioinformatics, 13*(1), 134. https://doi.org/10.1186/1471-2105-13-134

Zhang, X., He, C. C., Liu, J. M., Li, H., Lu, K., Fu, Z. Q., Zhu, C. G., Liu, Y. P., Tong, L. B., Zhou, D. B., Zha, L., Hong, Y., Jin, Y. M., & Lin, J. J. (2017). Nested-PCR assay for detection of *Schistosoma japonicum* infection in domestic animals. *Infectious Diseases of Poverty, 6*(1), 86. https://doi.org/10.1186/s40249-017-0298-y

Chapter 8
DNA Sequence Alignment and Phylogenetic Inferences: Guidelines for Analysis and the Selection of Appropriate Methods for Molecular Systematics

Abstract Changes that occur in the genome (or DNA) are pieces of genetic information containing valuable information to reveal the evolutionary history of organisms. As mentioned in the previous chapter, the elucidation of DNA sequences is a major advancement in the field of molecular biology. These advancements allowed sequences to be compared between different organisms, looking at similarities and differences. Through looking at molecular level changes, the reconstruction of phylogenies based on these sequences can be performed. Molecular phylogenies provide insights into the evolutionary relationships among organisms and a closer look at the processes behind species diversity. In addition, molecular phylogenies are also beneficial in the taxonomical classification and identification of organisms. Regarding Chap. 7, we reviewed the principles behind molecular-based techniques like PCR and DNA sequencing, together with examples of each method popular among researchers. In this chapter, we looked at the types of mutations that can occur in the DNA and the terminologies and main concepts of DNA sequence alignment. The principles behind phylogenetic inferences were introduced and the explanation of various models and methods. Finally, guidelines for analysis and practical steps for sequence alignment and the reconstruction of phylogenetic trees were discussed to complete the link between physical specimens, DNA sequences, and phylogenetic trees.

8.1 Types of Mutations

Mutations are changes that spontaneously occur in the DNA sequence, giving rise to variation and playing an important role in evolution (Carlin, 2011; University of California, 2021). If these mutations get inherited and passed on to the next generation, they serve as a basis for evolutionary changes. Generally, mutations can be classified based on the type of change caused by the mutation event—substitution mutations, insertions, and deletions (Li & Graur, 2000). Aside from DNA mutations, chromosomal mutations are another type of mutation that can also occur. Examples of mutations that can happen at the chromosomal level include recombination,

U. Thaenkham et al., *Molecular Systematics of Parasitic Helminths*,
https://doi.org/10.1007/978-981-19-1786-8_8

duplication, deletion and insertions, translocation, and inversion. A large majority of chromosomal mutations occur during the process of meiosis, often leading to diseases that are visible (phenotypic). In Chap. 4, types of mutations that can arise in both the chromosomal and nucleotide levels were mentioned. For our purpose in this chapter, we will focus solely on mutations that occur in the DNA sequence, including substitution mutations and insertions/deletions, affecting cellular processes downstream.

8.1.1 *Substitutions*

Nucleotide substitution, also known as point mutations, replaces one nucleotide with another (Li & Graur, 2000). Substitution mutations are further divided into two categories—transitions and transversions. Transitions occur when a purine (adenine/guanine) is substituted for another purine or when a pyrimidine (cytosine/thymine) is substituted for another pyrimidine. Transversions occur when a purine is substituted for pyrimidine and vice versa. In protein-coding regions, synonymous and non-synonymous mutations can also occur. As mentioned in Chap. 4, mRNA translation into proteins occurs with a set of "rules" called the genetic code and each codon codes for a particular amino acid. Thus, any mutations in the protein-coding regions can affect the amino acid (product of translation) being coded. A synonymous mutation occurs when the amino acid does not change from the originally specified one. This type of mutation is also known as silent mutation. Conversely, a change in amino acid being coded for due to substitution mutations is known as a non-synonymous mutation. A change in amino acid could have altering effects on the protein produced if the amino acid is altered from the original that was previously encoded, and a missense mutation results. Suppose a stop codon results from the substitution mutation and prematurely ends the translation process; a nonsense mutation results in the production of a truncated protein.

8.1.2 *Insertions and Deletions*

Another type of mutation that can occur in the DNA sequence is insertions and deletions (Page & Holmes, 1998). Insertions and deletions are collectively known as indels. The length of indel ranges from a few nucleotides to longer stretches, depending on the error type that led to the formation. Indels are more likely to occur in non-coding DNA; however, deleterious effects could arise if they appear in coding regions. For example, if the presence of indels changes the DNA sequence such that the original triplet codon frame is altered (due either to insertions or deletions that disrupt the triplet frame), the resulting sequence downstream will be read, and the amino acids being coded will be read in the wrong frame. This type of mutation is also known as frameshift mutation due to the shifting of the codon frame

(Page & Holmes, 1998). Consequently, the protein that results from a frameshift mutation can lead to harmful effects downstream.

Mutations are random changes that occur in the DNA sequence and are essential to evolution. They are a source of genetic variation that can allow us to measure genetic changes and differences between and within species (Carlin, 2011). The study of evolution will not be possible without these random changes in the DNA sequences that occur in the genome. Differences in the DNA sequence of organisms can be observed through DNA sequence alignment and the interpretation of relationships between organisms through phylogenetic analysis. These two areas will be further discussed in the subsequent sections.

8.2 DNA Sequence Alignment

A key aspect linking the DNA sequences obtained to construct a phylogenetic tree for studies on molecular systematics and the eventual interpretation of the data is DNA sequence alignment. Sequence alignment is an important step prior to phylogenetics, as it allows for the analysis of similarities and differences between the set of biological sequences. The information obtained after sequence alignment will offer important clues to the biological relationships between and within the species where these biological sequences were obtained.

8.2.1 Basics

First, we will introduce some terminologies and concepts that are essential in understanding sequence alignments. Generally, DNA sequence alignment analysis aims to look for similarities or differences between the set of sequences studied and find conserved regions between them (Lemey et al., 2009). As mentioned in Chap. 3, all organisms evolved from a common ancestor, and the various mechanisms of obtaining these genetic variations eventually led to the diversity of species we see today. Based on the assumption that the aligned DNA sequences are from a homologous region, we assume that the sequences have diverged from a common ancestral state. We can then infer the degree of similarity between the set of sequences studied by looking at the conserved and not conserved regions of the DNA sequence alignment.

How can the similarities between the set of sequences be obtained? The process involves the insertion of gaps (−) into the sequences so that the resulting sequences all have the same length. From an evolutionary standpoint, these gaps represent insertions and deletions within the genome that are hypothesized to have occurred during evolution (Chatzou et al., 2016). Theoretically, any number of gaps can be introduced into the sequences to align them successfully. Having gap penalties will restrict the number of gaps introduced during the alignment process to circumvent this. Simply put, they are penalties that are subtracted from the alignment score and include the gap opening penalty and gap extension penalty (Lemey et al., 2009; Selzer et al., 2018). A gap opening penalty occurs whenever a gap opening is introduced in the alignment, whereas a gap extension penalty occurs when a gap is added after the gap opening.

Sequence alignment can be grouped into either global or local alignments (Kumar & Filipski, 2007; Posada, 2009). The sequences are aligned throughout their entire length for global alignments from beginning to end. This method is appropriate for a targeted segment of genes and short stretches of genomic sequences. On the other hand, local alignments find local matches along the stretch of sequences. They are applicable to sequences that differ in lengths and are applied for longer sequences to account for large insertions and deletions in the sequence.

8.2.2 Pairwise Sequence Alignment

Sequence similarity and comparison between two sequences can be obtained through pairwise sequence alignment. Methods have been utilized for comparison, including the dot-plot matrix and dynamic programming.

Dot-Plot Matrix

Gibbs and McIntyre first published the dot-plot matrix method in 1970, and it is performed by creating a matrix to compare two sequences (Gibbs & McIntyre, 1970; Lemey et al., 2009). Simply put, the rows of the matrix (x-axis) are the nucleotides of the first sequence, and the columns of the matrix (y-axis) are the nucleotides of the second sequence. The sequences are compared by placing a dot in the matrix if the nucleotide of the first sequence matches the nucleotide of the second sequence. Eventually, the region of high nucleotide similarity is shown as the longest diagonal line across the matrix. Other shorter diagonal lines might be generated, and these are the "noise" regions that are not informative. Using the dot-plot matrix can generate a simple visualization of the measure of similarity between two DNA sequences.

Dynamic Programming

The best alignment is found in the dynamic programming method by scoring all the possible pairs of aligned nucleotides and penalizing gaps through a scoring scheme (Keith, 2017). Similar to the dot-plot matrix, a matrix is constructed based on the sequences to be compared. In addition, a scoring scheme is introduced where it allows for the scoring of matches and mismatches of nucleotides and gaps to define the quality of the alignment, whereby the more optimal alignment has the higher score. Examples of the dynamic programming method implemented include

Needleman and Wunsch for global alignments, Smith-Waterman for local alignments, and word methods utilized by BLAST (Li & Graur, 2000).

8.2.3 *Multiple Sequence Alignment*

Multiple sequence alignment (MSA) extends pairwise sequence alignment that involves more than two DNA sequences (Selzer et al., 2018). Similar to pairwise alignment, multiple sequences can be aligned together by matching the nucleotides through the insertion of gaps. The difference is that manual visualization of the sequences is nearly impossible for multiple sequences, and performing alignment is required. Computer software, therefore, utilizes different methods to perform multiple sequence alignment; we will discuss some standard methods and widely used software in the following paragraphs.

Progressive

The progressive alignment method was developed by Paulien Hogeweg and Ben Hesper, where it utilizes the dynamic programming method to build the MSA (Hogeweg & Hesper, 1984; Lemey et al., 2009). An example of a freely available MSA software that utilizes the progressive method is the Clustal suite, which includes Clustal W, Clustal X, and Clustal Omega (Larkin et al., 2007; Sievers et al., 2011). First, a heuristic guide tree using neighbor-joining (NJ) or unweighted pair group method with arithmetic means (UPGMA) is created based on the pairwise sequence comparisons. Thereafter, the sequences are progressively added based on their sequence similarity from the guide tree. This method is computationally faster than other MSA methods and can align sequences of different lengths and varying similarities. However, due to the greediness of the progressive alignment method, alignment errors such as the incorrect introduction of gaps in the early alignment stages cannot be corrected. The wrong gaps will be propagated into the subsequent and final MSA.

Consistency

The consistency alignment method builds on the progressive method and aims to reduce the initial alignment errors from progressive alignment (Katoh, 2021; Lemey et al., 2009). Intermediate sequence information improves pairwise comparison quality and searches for the alignment that maximizes the agreement with the set of pairwise input sequences. However, this method is more computationally demanding as compared to the progressive alignment method. An example is the T-Coffee software, where Clustal and L-align perform both global and local alignments, respectively (Katoh, 2021; Keith, 2017; Notredame et al., 2000).

Iterative

Another method that aims to reduce the errors from the progressive alignment method is the iterative method. The iterative method tries to work towards an optimal alignment with a more accurate distance measure than the progressive methods to assess the relatedness of two sequences (Keith, 2017; Lemey et al., 2009). MUltiple Sequence Comparison by Log-Expectation (MUSCLE) and Multiple Alignments using Fast Fourier Transform (MAFFT) are examples of MSA software that uses the iterative method (Edgar, 2004; Katoh et al., 2002). A guide tree using UPGMA is first determined based on pairwise sequence similarity, and following the progressive alignment method, the sequences are aligned following the guide tree branching order. Thereafter, improvements to the alignment are made through iteration cycles until an optimal alignment is achieved. In the MUSCLE software, improvements are performed using a tree-dependent restricted partitioning method, whereas MAFFT uses a fast Fourier transformation to rapidly detect homologous segments yielding a higher overall accuracy of alignments.

Simply put, DNA sequence alignment is an important step prior to phylogenetic analysis, and various methods and programs have been utilized to look at the similarities and differences between sequences. Having adequate knowledge on the concepts of DNA sequence alignment can aid in deciding which method is the most suitable for a given set of data, in which the appropriately aligned sequences can be used for downstream phylogenetic analysis and inferences.

8.3 Phylogenetic Inferences

The goal of phylogeny is to determine the possible evolutionary relationships among organisms as accurately as possible, as insights into the chronicle of events along evolutionary lineages and divergence time between organisms (Baldauf, 2003; Li & Graur, 2000; Yang & Rannala, 2012). The study of phylogeny can explain relationships among species, the history of lineages, and the epidemiological dynamics of pathogens. Phylogenetic inferences can be performed using molecules (DNA sequences or amino acids), which are the building blocks of life, as well as using morphological characters. However, for our purpose in this chapter, we will only focus on phylogenetics using molecular characters.

8.3.1 Basics

Parts of a Phylogenetic Tree

To begin, phylogenetic relationships among organisms are visualized with a phylogenetic tree. In simple terms, a phylogenetic tree is a diagram depicting various

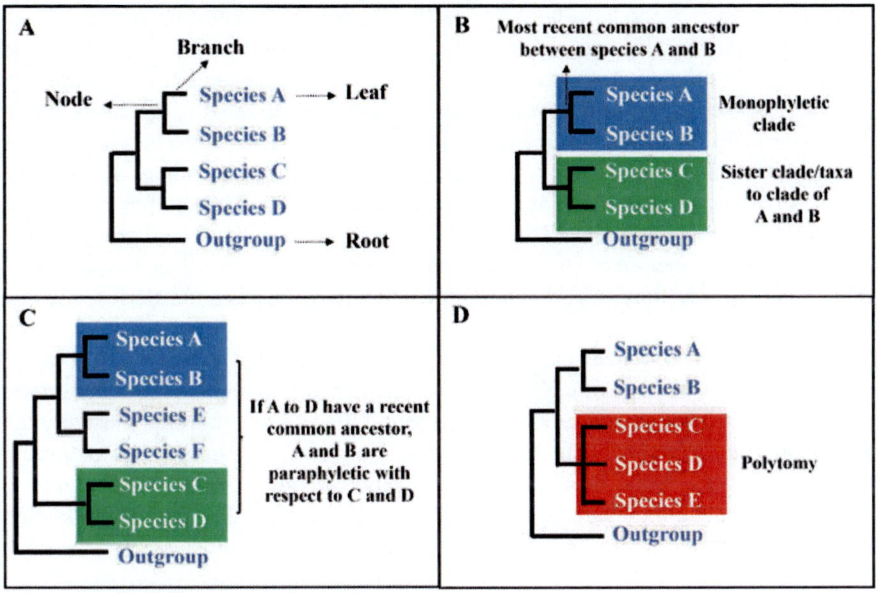

Fig. 8.1 Phylogenetic tree terminologies

genetic lineages and their relationships with respect to each other (Ajawatanawong, 2017; Shaik et al., 2019). As shown in Fig. 8.1a, a typical phylogenetic tree comprises leaves, branches, nodes, and a root. The leaves represent the operational taxonomic units (OTUs) or individual species/populations studied, while the branches that bifurcate to the leaves represent the divergence of one OTU from another. The nodes connect the branches, and an internal node corresponds to the most recent common ancestor of the branches arising from it (Li & Graur, 2000; Shaik et al., 2019). The root of the phylogenetic tree mainly serves to give the phylogenetic tree direction, and it usually is the taxa most distantly related (outgroup) to all the taxa being studied (ingroup).

Tree Terminologies

Now that we are familiar with the various parts of a phylogenetic tree, we can move on to various terminologies that are useful for inferring phylogenetic relationships. First, the general shape and branching order of a phylogenetic tree can also be termed as the topology of a tree. The branch lengths usually depict the number of evolutionary distances between nodes, and the branching order indicates how related the organisms are to each other (Garamszegi, 2014).

Second, relationships between groups of organisms are inferred using a common ancestor. We can determine whether the taxa in question share a most recent common ancestor or whether they are distantly related through the branching order. If taxa in question have been derived from a common ancestor, a group of taxa including common ancestor is defined as a clade. A monophyletic clade is a clade consisting of all the descendants of the most recent common ancestor and implying that they are the most similar to each other (as compared to other taxa that are not in the same clade) (Baldauf, 2003; Baum, 2008; Li & Graur, 2000). The taxa that are most closely related to the monophyletic clade (aside from members of the same clade) is defined as sister taxa, and they share another immediate common ancestor (Subbotin et al., 2013) (Fig. 8.1b). On the other hand, a paraphyletic group occurs when not all the members share the same common ancestor (Li & Graur, 2000) (Fig. 8.1c). A polytomy occurs when more than two branches are derived from a node (Fig. 8.1d).

Third, phylogenetic trees can either be rooted or unrooted. As mentioned in the paragraph above, the root of a phylogenetic tree provides a direction of evolutionary relationships, termed the outgroup (Garamszegi, 2014). The outgroup is usually the most distant taxa out of all the taxa being studied. In the absence of an outgroup, phylogenetic trees are not rooted. Unrooted phylogenetic trees will not be able to provide information on ancestor-descendant relationships, in contrast to rooted phylogenetic trees (Garamszegi, 2014).

8.3.2 Genetic Distances and Nucleotide Substitution Models

An important aspect of phylogenetic inferences is the measurement of the degree of relatedness between species or groups of organisms, also known as genetic distances (Lemey et al., 2009). Genetic distances are usually expressed as a percentage, and the closer the value is to 100%, the more similar they are. The most accessible approach to measuring the degree of relatedness is counting the sites where the sequences differ (p-distance). However, given that this method may underestimate the actual amount of evolutionary change, various models for nucleotide substitutions have been derived to correct the underestimation of the observed nucleotide substitutions. In addition to nucleotide substitution models, nucleotide substitution saturation and homogeneity between sequences are crucial factors to test for as a lack of testing could result in errors during phylogenetic analysis.

Nucleotide Substitution Saturation

With the accumulation of nucleotide substitutions over time, the nucleotide sequences can become saturated for nucleotide substitutions. Substitution saturation occurs due to the accumulation of multiple substitutions per site, resulting in most of the sites being changed before (Lemey et al., 2009; Page & Holmes, 1998). As a result, the nucleotide changes that we observe are most likely due to chance rather than the result of common ancestry. The number of differences between the sequences becomes a less accurate estimate of the number of substitutions due to actual divergence from their common ancestor. Thus, when nucleotide substitution saturation is reached, the sequences are no longer informative about the underlying evolutionary process and are not helpful to make accurate phylogenetic inferences (Xia, 2000). A test for nucleotide substitution saturation should be performed before phylogenetic analysis to ensure that the nucleotide sequences can be useful for phylogenetic inferences.

Homogeneity Between Sequences

As the general assumption is that the sequences being analyzed have evolved with the same process and patterns of nucleotide substitutions and nucleotide frequencies, this assumption could result in errors in phylogenetic inferences if the assumption does not hold (Kumar & Gadagkar, 2001; Rosenberg & Kumar, 2003). Error in phylogenetic inferences results because now unrelated clades with similar nucleotide substitution patterns and frequencies will be grouped in a phylogenetic tree. Thus, there is a need to test for homogeneity between the analyzed sequences by measuring the observed differences in substitution patterns and nucleotide frequencies in the sequences.

Types of Nucleotide Substitution Models

Substitution models are important parameters in phylogenetic inferences as they can better mimic sequence evolution and actual evolutionary processes that are occurring. A variety of substitution models have been developed for nucleotides (DNA substitution and codon substitution) and amino acids. From all possible combinations to mimic the DNA substitution process after considering the different types and frequencies of nucleotide substitutions, more than 1600 different models have been introduced and implemented to refine the accuracy of phylogenetic analysis (Arenas, 2015). In the context of this chapter, DNA substitution models like the Jukes and Cantor; Kimura; Hasegawa, Kishino, and Yano; and general time reversible models will be elaborated on to illustrate the incorporation of various considerations in substitution models (Fig. 8.2).

Jukes and Cantor introduced the Jukes-Cantor (JC) model in 1969, where the model specifies that the frequencies of the four nucleotides are 25% each and any nucleotide has an equal probability of being substituted by another (Erickson, 2010; Lemey et al., 2009). It is the simplest substitution model for sequence evolution as it assumes that substitutions for all four bases are equally likely. However, this assumption does not necessarily hold true. Taking into account that the substitutions for all four bases do not have equal probability, the Kimura 2 parameter (K2P) model adds the assumption that the rates of transition and transversion substitutions are different (transition to transversion ratio) to allow for transition/transversion bias (Lemey et al., 2009). However, the K2P model assumes that the nucleotide base frequencies are equal. Likewise, Felsenstein took into account varying nucleotide

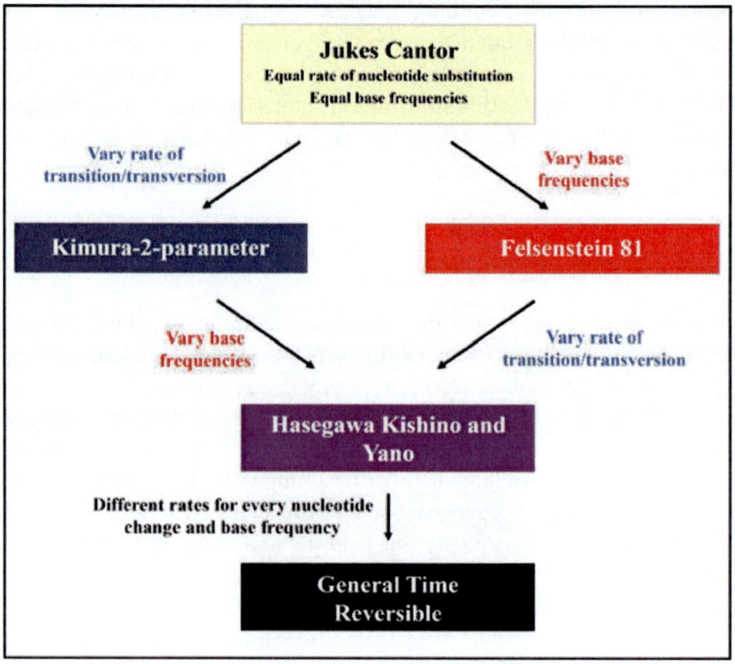

Fig. 8.2 Types of nucleotide substitution models and the parameters incorporated for each model

base frequencies to include in the F81 model (Lemey et al., 2009). To account for unequal nucleotide base frequencies, together with the different rates of nucleotide substitutions, the Hasegawa, Kishino, and Yano (HKY85) model accounts for these two assumptions (Lemey et al., 2009; Page & Holmes, 1998). The most complex model is the general time reversible (GTR) model, where it assumes that each possible substitution has its probability, allowing different rates for every nucleotide change and different nucleotide base frequencies (Page & Holmes, 1998). Additionally, the proportion of invariable sites (+I), the rate of variation across sites (+G), or both can be incorporated into each DNA substitution model (Arenas, 2015).

The selection of the best and most realistic nucleotide substitution model, together with the tests for nucleotide substitution saturation and homogeneity between sequences, should be carried out before constructing a phylogenetic tree to reflect true ancestral relationships. We will discuss more on how to perform these tests in Sect. 8.4.

8.3.3 Types of Phylogenetic Trees

Phylogenetic trees can mainly be divided into two types—distance-based and character-based trees. Distance-based methods rely on calculating genetic distances between each pair of sequences, whereas character-based methods are based on the information available at each homologous site of the sequence (De Bruyn et al., 2014). Methods for phylogenetic inferences can also be categorized into traditional and Bayesian approaches, where examples of traditional approaches include the neighbor-joining, maximum parsimony, and maximum likelihood algorithm, while Bayesian approaches consist of Bayesian inference (Holder & Lewis, 2003). In the subsequent paragraphs, we will go through some of the commonly used phylogenetic trees and their principles.

Unweighted Pair Group Method with Arithmetic Means (UPGMA)

UPGMA is a distance-based method clustering group of sequences that exhibit the smallest genetic distance with each other. This method was initially used to reflect phenotypic similarities among OTUs and assume that evolution rates are approximately constant among the different lineages. A distance matrix is firstly derived from the sequences. Then, an average distance between each pair is calculated, and a rooted tree is built by creating internal nodes for each pair of taxa, starting from the most similar to the least similar (Shaik et al., 2019). An advantage of using UPGMA is its fast computational speed. However, as it is a distance-based method, UPGMA tends to disregard large amounts of potentially evolutionary informative information as the sequences are converted into genetic distances (Subbotin et al., 2013). Moreover, multiple substitutions are not taken into consideration when using UPGMA.

Neighbor-Joining (NJ)

Another distance-based method is the NJ method created by Saitou and Nei (Saitou & Nei, 1987). It sequentially finds neighbors that minimize the total length of the tree and clusters them together. It is one of the most utilized methods for phylogenetic analysis, and this is partly due to the fast computational speed of NJ analysis. Similar to UPGMA, a distance matrix is first derived from the sequences, and the distance matrix represents an estimate of the evolutionary distance between sequences. Starting with a star-like tree, pairs of sequences are then found that minimize the total branch lengths. The process is repeated until the tree is fully resolved (Page & Holmes, 1998; Shaik et al., 2019). The advantage of NJ over UPGMA is that it does not assume that all lineages evolve at the same rate.

Maximum Parsimony (MP)

The MP method is a character-based method that aims at finding the tree that contains the minimum number of changes (most parsimonious tree) (De Bruyn et al., 2014). The most parsimonious tree is the tree that invokes the smallest number of evolutionary changes during sequence divergence. Initially, the MP method was developed for morphological trait analysis but also applied to DNA sequences. For DNA sequences, the parsimony-informative sites are where nucleotide substitutions occurred with a minimum of two-character states (Ajawatanawong, 2017; Page & Holmes, 1998). Firstly, the informative sites between the sequences are identified, and the minimum number of substitutions is calculated for each possible tree. An algorithm using either the branch and bound or heuristic search is performed to search for the most parsimonious tree (Lemey et al., 2009). The advantage of the MP method is its fast computational speed. However, substitution models are not used in MP analysis, which could skew phylogenetic inferences. Also, MP is generally more prone to long-branch attraction, whereby highly divergent sequences might mislead the MP method into joining the branches together, which results in an incorrect phylogeny (a more distantly related sequence is now depicted as closely related to a particular sequence) (Yang & Rannala, 2012).

Maximum Likelihood (ML)

This character-based method was first applied in phylogenetics by Felsenstein. Based on the concept that computes the probability of the data under a given tree with a specified model of nucleotide substitution, the ML method searches for a tree that has the highest likelihood of explaining the data (Subbotin et al., 2013). An initial tree is built from the sequences using a fast optimal method. Then, together with the specified substitution model, the branch lengths of the tree are adjusted to maximize the likelihood for the dataset. The tree with the highest maximum likelihood value with the model fit to the sequences is then displayed as the most likely tree (Page & Holmes, 1998). An advantage of the ML method is that it accounts for an evolutionary model of sequence evolution. Comparing the traditional methods of UPGMA, NJ, MP, and ML, the ML method is the most appealing for phylogenetic estimation. Although slower in computational speed, general models for possible sequence evolution are considered, and the likelihood function is a consistent and powerful basis for assessing tree confidence (Holder & Lewis, 2003). However, the general models for sequence evolution are simpler than the true process underlying actual sequence evolution.

Bayesian Inference (BI)

Following the Bayesian theory, the BI method reconstructs a phylogenetic tree using Bayesian statistics. The BI method aims to search for the tree that maximizes the chance of seeing the tree with the given data and searches for the tree with the highest posterior probability (Ajawatanawong, 2017; Keith, 2017). The main steps of BI include specifying a prior, defining models of sequence evolution, computing the posterior distribution, and checking that the model used is an appropriate description of the data.

In BI, a probability distribution is used to describe the uncertainty of all unknowns, including the model parameters. A prior distribution, $f(\theta)$, which is a function of the unknown parameter θ, is assigned based on the information available about θ before analysis. Bayes' theorem is then used to calculate the posterior distribution of θ along with the likelihood (Nascimento et al., 2017). BI considers models like estimating divergence times without assuming a molecular clock, searching for residues important to natural selection, and detecting recombination. Together with these models and the prior distribution, BI integrates the information to determine the posterior probability. The posterior probability is interpreted as the probability that the model and tree are correct, and this involves a sum of each of and all the trees, with possible combinations of branch lengths and substitution model parameter values (Subbotin et al., 2013). It provides a measure of confidence of the estimates for the tree and makes direct probabilistic statements about the model or θ. Additionally, BI relies on Markov Chain Monte Carlo (MCMC) to approximate probability distributions over a wide variety of contexts. MCMC is a simulation technique for sampling from a probability distribution known up to a normalizing constant. A sample from the posterior distribution is generated, which is used to estimate the mean and the standard deviation of the posterior (Nascimento et al., 2017).

The main difference between BI and traditional approaches is that BI allows for complex models of sequence evolution to be implemented (Holder & Lewis, 2003). In terms of phylogenetic accuracy and computational speed compared to ML, both methods are similar. However, BI is especially useful for simultaneously estimating phylogenetic trees and obtaining uncertainty measurements for each branch by implementing other sequence evolution models (Holder & Lewis, 2003). Table 8.1 shows the comparison between BI and traditional methods of phylogeny.

8.3.4 Measurements of Tree Confidence

Confidence levels are important for drawing sound conclusions regarding the evolutionary relationships among the taxa sampled in the phylogeny. To assess the confidence levels of a phylogenetic tree, statistical methods such as bootstrap and posterior probability are used. Generally, lower values suggest certainties associated with the particular node, and higher values indicate strong support.

Bootstrap

Bootstrap is used to obtain estimates of tree confidence in traditional methods of phylogeny and is commonly used mainly in NJ and ML analysis. Felsenstein (1985) formally proposed bootstrapping as a method for obtaining confidence limits on phylogenies, and the confidence levels for each node are indicated as percentages (bootstrap value) (Soltis & Soltis, 2003). Bootstrapping involves the generation of

Table 8.1 Comparison of traditional and Bayesian methods for phylogenetic inference

	Traditional methods				Bayesian
Type of analysis	*Distance-based*: Rely on calculating genetic distances between each pair of sequences		*Character-based*: Rely on information available at each homologous site of the sequence		*Bayesian statistics*: Rely on Bayes' theorem for the posterior distribution of unknown parameters given the data
Examples	UPGMA Clusters based on sequences that exhibit the smallest genetic distance	NJ Sequentially finds neighbors that minimize the total length of the tree	MP Finds the tree that invokes the minimum number evolutionary changes	Ml Finds the tree that has the highest likelihood of explaining the data under a specified model of nucleotide substitution	BI Uses probability distribution and model parameters to search for the tree with the highest posterior probability
Input data used	Distance matrix	Distance matrix	Parsimony-informative sites	Tree branch lengths and substitution models	Prior distribution and model parameters
Nucleotide substitution model	None	None	None	Incorporated	Incorporated
Tree confidence measurement	Bootstrap	Bootstrap	Bootstrap	Bootstrap	Posterior probability
Disadvantages	Loss of information with conversion to distance matrix		Prone to long-branch attraction	Computationally intensive	Need to specify prior distribution
Advantages	Computationally fast			Captures data using substitution models	Computationally more efficient than ML and can incorporate complex models

Compiled from De Bruyn et al. (2014), Holder and Lewis (2003), Nascimento et al. (2017), Page and Holmes (1998), Subbotin et al. (2013) and Yang and Rannala (2012)

pseudo-replicates of the dataset and resampling with the replacement sites in the original data matrix (Holder & Lewis, 2003). By using the pseudo-replicates (subsamples) of the dataset, trees are then built and scored. Repeated sampling of the dataset is performed (dependent on the number of replicates as specified by the user), where each resulting tree that is accepted is added to the final collection of trees. After a specified number of replicates (as defined by the user), a consensus tree is constructed to summarize the results of all the replicates. The frequency at which a particular grouping of sequences or clades found is derived and recovered is then represented as the bootstrap value (Lemey et al., 2009; Sanderson & Shaffer, 2002).

Posterior Probability

Posterior probability values, on the other hand, are used as a measure of tree confidence levels in BI. In BI, the posterior probability is the probability that the model and tree are correct, accounting for all other models and prior information (Keith, 2017). Using the MCMC method, the probability distribution of the input data is estimated, and finally, an estimate of the probability that the final tree is the true evolutionary tree given the input data and relying on the prior distribution and models specified is generated (Holder & Lewis, 2003). A new tree is proposed with the specified model and condition on the initial tree. The posterior for the proposed tree is calculated, where the proposed tree is either rejected or accepted. If this new posterior value is less than the acceptance probability, the new state is accepted, and the state of the chain is updated. This process is repeated thousands or millions of times where the number of times a single tree is visited during the course of the chain is an approximation of its posterior probability. Similar to bootstrap, the frequency is expressed as a percentage for the confidence level of a node.

8.3.5 *Other Types of Phylogenies*

Besides using DNA sequences for phylogenetic analysis, the secondary structure of the rRNA and the amino acid sequences of protein-coding genes can also be used. These alternative characters can serve as additional information for inferring evolutionary relationships.

Ribosomal RNA Secondary Structure

The rRNA units are non-coding regions in the DNA and collectively contain the internal transcribed spacers, external transcribed spacers, and the ribosomal RNAs (refer to Chap. 6 for more details on the function and properties of the rRNA transcription unit). They can fold into secondary structures from DNA into RNA, where secondary structures form before tertiary structure folding. The folding of the primary nucleotide sequence into a secondary structure involves the interaction of the nucleotides through the formation of hydrogen bonds. The nucleotides pair initiated by Watson-Crick is non-canonical base pairing (Li et al., 2014; Subbotin et al., 2007). Secondary structures such as helices, stem and loops, bulges, and single-stranded portions result, and these structures can provide helpful information about the evolutionary relationships between organisms (Keller et al., 2010; Prasad et al., 2009). Compared to the primary nucleotide sequence, the secondary structure is more conserved despite variations in the primary nucleotide sequence. This is because these structures are important for the function of the cell and are evolutionarily conserved due to their functional constraints (Prasad et al., 2009). Thus, the patterns and sequences of these structures can provide an additional source of information for phylogenetic inferences.

Amino Acid Sequences for Protein-Coding Genes

Amino acid sequences can also serve as an additional source of information for phylogenetic inferences. For protein-coding genes, the translation of mRNA results in the production of polypeptide chains (comprising of amino acids), which then undergo folding processes to form a functional protein. Based on the genetic code, the mRNA sequence is read in triplets (also known as codons), and a corresponding amino acid is coded for each codon (Li & Graur, 2000). As mentioned in Sect. 8.1, non-synonymous or synonymous mutations can result due to nucleotide substitutions. In protein-coding sequences, the third base of a codon may frequently be substituted without affecting the amino acid that the codon codes for, whereas mutations in the first and second base positions of a codon will almost certainly lead to a change in amino acid being coded for (Lemey et al., 2009; Li & Graur, 2000). Thus, mutations in the third position of the codon are more frequent than the first and second positions. DNA sequences can be subjected to frequent mutations over time; however, the amino acid sequences might remain unaltered (depending on which base position the mutation occurs). Comparing to DNA sequences, amino acid sequences are less subjected to changes and could provide additional phylogenetic information, especially among distantly related taxa (Hall, 2005; Margoliash & Fitch, 1971). Also, having information about the translated DNA sequences can be helpful, especially if the protein that the DNA is being coded for is an essential protein needed for cell function.

Clearly, there are different aspects of phylogenetic inferences that we should consider due to the complex nature of nucleotide substitutions and uncertainties of the actual situation that undermines our interpretation of the evolutionary process in organisms. However, the goal of molecular phylogenetics is to determine evolutionary relationships as accurately as possible with the information we have.

8.4 From Theory to Practice: Practical Steps for Use

This section will explore some of the essential practical steps for performing multiple sequence alignment, phylogenetic construction, and finally, phylogenetic inferences. At the end of this section, we will be able to successfully execute, analyze, and interpret a phylogenetic tree, which then completes the picture for molecular-based studies and can be applied for parasitic helminths.

8.4.1 Multiple Sequence Alignment

From where we left off from Chap. 7 (Sect. 7.3), the DNA sequences will be ready for analysis after post-PCR processing and DNA sequencing. The subsequent steps for MSA include checking the DNA sequences, performing and checking MSA, and preparing the aligned sequences for phylogenetic analysis.

Checking of Electropherograms and Creating Contigs

After DNA sequencing, we will first need to check the electropherograms of the sequences. The electropherograms will be in a ".ab1" format and can be viewed in various software. A list of software is provided in Table 8.2. The electropherograms

Table 8.2 List of some freely available software for analysis

Purpose	Software	Website URL
Viewing electropherogram and editing sequences	Chromas	https://technelysium.com.au/wp/chromas/
	BioEdit	https://bioedit.software.informer.com/
Multiple sequence alignment	Clustal X	http://www.clustal.org/clustal2/#Download
	MUSCLE	http://doua.prabi.fr/software/seaview
	T-Coffee	http://tcoffee.crg.cat/
	MAFFT	https://mafft.cbrc.jp/alignment/software/
Substitution saturation	DAMBE	http://dambe.bio.uottawa.ca/Include/software.aspx
Homogeneity	MEGA	https://www.megasoftware.net/
Phylogenetic analysis	MEGA	https://www.megasoftware.net/
	MrBayes	https://nbisweden.github.io/MrBayes/download.html
	PAUP	http://paup.phylosolutions.com/
	PHYLIP	https://evolution.gs.washington.edu/phylip.html
Viewing and editing phylogenetic trees	FigTree	http://tree.bio.ed.ac.uk/software/figtree/
	TreeView	https://treeview-x.en.softonic.com/

should be checked manually by eye, where the highest peak should correspond to their respective nucleotide base. In instances where the electropherograms have messy peaks and we are unable to check the nucleotide bases with accuracy, the sequence should not be used. After checking that the peaks correspond to the nucleotide bases, the beginning and end of the DNA sequence should be trimmed out. The beginning and end of the DNA sequences are usually either low in signal or are messy and should not be used. Once the electropherograms have been checked, we can paste the final trimmed DNA sequences into a notepad file in FASTA format (refer to Chap. 7, Sect. 7.3 for FASTA format). The DNA sequences can be put together in the same Notepad file containing all the DNA sequences you would like to perform MSA.

In instances where both the forward and reverse strands were performed for DNA sequencing, we can use both the forward and reverse sequences to create a contig sequence. A contig sequence is where both strands of DNA are put together and the overlap generates the DNA sequence that we can use for analysis. To generate a contig sequence, we will have to reverse complement the "reverse" sequence and match the newly reverse complement sequence to the "forward" sequence. The overlap portion of both DNA strands can then be used for MSA.

The last step is to perform nucleotide BLAST (https://blast.ncbi.nlm.nih.gov/Blast.cgi) of the DNA sequence on the NCBI server to check the identity or the closest match of the DNA sequence with the sequences that are available in the NCBI database.

MSA and Preparation of Data for Phylogenetic Analysis

With the Notepad file that we have prepared containing the sequences we would like to perform MSA on, we can open and load the sequences on any MSA software from Table 8.2. Do bear in mind that each software utilizes different methods and gap penalties and there is a need to compare the results of each alignment method. The general steps for MSA and preparation of data are similar for each alignment method, but for the purpose of illustration, we will use Clustal X (Larkin et al., 2007) as an example.

After loading the sequences into Clustal X, check that all the sequences have been successfully loaded. Then, we can specify the "output" file formats after performing sequence alignment. Ensure that the "FASTA" format box is checked to view and edit the alignment in BioEdit (Hall, 1999). MSA can be performed on all the selected sequences, and the resulting "output" files will be saved in the same location that the Notepad file was saved. With the aligned FASTA file, we can move on to the next step to look at the alignment and prepare the data for phylogenetic analysis.

To view the alignment after performing MSA, we can open the FASTA file using BioEdit. In BioEdit, we can view the conserved and the variable nucleotides easily as compared to using Clustal X. The alignment should be checked manually by eye to observe the number of nucleotide variations between the species. Doing this step will gauge the amount of sequence variation and allow us to check the DNA sequence of interest with other sequences that we took from the NCBI database. We will notice that gaps are present after alignment, and these gaps can either be removed before phylogenetic analysis or allow the phylogenetic analysis software to remove them. Gaps are not taken into consideration when performing phylogenetic analysis as they are either insertions or deletions in the DNA sequence and are considered "missing" information which could result in inaccurate phylogenetic inference. To manually remove the gaps in BioEdit, we can edit the sequences and delete the gaps in the alignment until there are no more gaps. This FASTA file, after the deletion of gaps, should be saved with another file name.

Test for Nucleotide Substitution Saturation and Homogeneity

To test whether the regions are saturated for nucleotide substitution, we can perform the test in DAMBE (Xia, 2017). With the FASTA file of aligned sequences, we can load the file into DAMBE and perform a test of saturation by Xia et al. To interpret the results, we look at the *Iss* (simple index of substitution) and *Iss.c* values. If the *Iss* value is more than *Iss.c*, the region is saturated for nucleotide saturation and not useful for phylogenetics.

To test for homogeneity between the sequences, we can perform the test in MEGA (Kumar et al., 2018) using the "Disparity Index of Pattern Heterogeneity." A matrix of values will be generated with the disparity index per site value for each sequence pair.

8.4.2 Phylogenetic Analysis

Once the relevant tests and MSA have been performed, we can proceed with phylogenetic tree reconstruction. To do so, we will have to select an appropriate outgroup and a suitable substitution model and finally finish with viewing and editing the resulting phylogenetic tree.

Selection of Outgroup

An appropriate outgroup should be selected as the outgroup will affect the resulting phylogenetic relationships among the ingroup taxa. Prior knowledge of the relationships among the taxa studied will aid in choosing the appropriate outgroup. Additionally, we should also perform an evaluation using our sequences to determine which is the best outgroup. Phylogenetic trees can be reconstructed using different outgroup taxa, and we can then choose the "best" phylogenetic tree. Factors to consider when selecting the "best" phylogenetic tree include the strength of bootstrap or posterior probability support, phylogenetic relationships between the ingroup taxa, and the accuracy of the phylogenetic placement of the ingroup taxa.

Phylogenetic Tree Reconstruction and Genetic Distances

For the reconstruction of phylogenetic trees, we will use the MEGA software as an example, as it is one of the more widely used software and has a user-friendly interface. Moreover, multiple types of phylogenetic trees can be constructed in MEGA. Depending on the purpose of the study and phylogenetic tree reconstruction, we will first need to consider which phylogenetic tree to construct. In the "Phylogeny" icon in MEGA, five types of trees can be reconstructed: UPGMA, NJ, MP, ML, and minimum evolution (ME). Once the type of tree is selected, we will be able to choose the type of data (nucleotide or amino acid sequences) and the type of genetic marker used. Other categories of options include the number of bootstrap replicates to perform, the type of substitution model to use, and how to treat missing data.

Additionally, an appropriate substitution model should be chosen prior to reconstructing ML phylogenetic tree. The ML method considers nucleotide substitutions to search for a tree with the highest likelihood of explaining the data. The time for analysis depends on the type of tree and the number and length of the input sequences.

In MEGA, we can also determine the pairwise genetic distances for the sequences. Genetic distances indicate how much or slight sequence variation there is between pairs of sequences or between groups. This can be performed using the "Distance" icon, where an output of a matrix of values will be produced with the genetic distances between each pair.

Inferring and Aesthetics

The final step will be to view, interpret, and "beautify" the reconstructed phylogenetic tree. The common format for saving tree files is the "Newick" format, which can be used in many tree viewers. In addition, MEGA provides users the option of saving the tree as a PDF or an image file. Phylogenetic trees can be inferred by looking at the relationships between the species and the phylogenetic placement of our species (the sequences obtained after DNA sequencing) with the reference sequences from the NCBI database. Also, bootstrap values (indicated on each node) can provide information on the confidence measurements for each clade.

Adding aesthetics to the reconstructed phylogenetic trees can be performed in either TreeView (Page, 1996) or FigTree (Rambaut, 2010), where colors to the leaves and branches can be added. Other accessories include labeling and changing the fonts and the branching style of the phylogenetic tree.

For more details on the theory behind each software and more detailed instructions, users are advised to refer to respective manuals and "Help" sections available on the website.

8.5 Conclusions

This chapter introduces the concepts behind evolutionary changes that can happen in a DNA sequence, the main types of DNA sequence alignments, and phylogenetic methods for analyzing relationships among organisms. The importance of choosing appropriate methods for both sequence alignments and phylogenetic analysis was discussed, together with considerations and practical steps in phylogenetic tree reconstruction. With this chapter, we have closed the loop from obtaining DNA from an organism to reconstructing a phylogeny that can depict the actual phylogenetic relationships among organisms as accurately as possible. Molecular phylogenetics undoubtedly provides valuable information for understanding the intricate tapestry and relationships among organisms in the "tree" of life.

In Chap. 9 and the subsequent chapters, we will focus on parasitic helminths in detail. Through the molecular systematics and molecular-based studies that have been employed for the three groups of parasitic helminths, we will look at prominent studies that have been performed, aiding in the inference of phylogenetic relationships among parasitic helminths.

References

Ajawatanawong, P. (2017). Molecular phylogenetics: Concepts for a newcomer. *Advances in Biochemical Engineering/Biotechnology, 160*, 185–196. https://doi.org/10.1007/10_2016_49

Arenas, M. (2015). Trends in substitution models of molecular evolution. *Frontiers in Genetics, 6*, 319–319. https://doi.org/10.3389/fgene.2015.00319

Baldauf, S. L. (2003). Phylogeny for the faint of heart: A tutorial. *Trends in Genetics, 19*(6), 345–351. https://doi.org/10.1016/s0168-9525(03)00112-4

Baum, D. (2008). Reading a phylogenetic tree: the meaning of monophyletic groups. *Nature Education, 1*(1), 190.

Carlin, J. (2011). Mutations are the raw materials of evolution. *Nature Education Knowledge, 2*(1), 10. Retrieved from https://www.nature.com/scitable/knowledge/library/mutations-are-the-raw-materials-of-evolution-17395346/

Chatzou, M., Magis, C., Chang, J. M., Kemena, C., Bussotti, G., Erb, I., & Notredame, C. (2016). Multiple sequence alignment modeling: Methods and applications. *Briefings in Bioinformatics, 17*(6), 1009–1023. https://doi.org/10.1093/bib/bbv099

De Bruyn, A., Martin, D. P., & Lefeuvre, P. (2014). Phylogenetic reconstruction methods: An overview. *Methods in Molecular Biology, 1115*, 257–277. https://doi.org/10.1007/978-1-62703-767-9_13

Edgar, R. C. (2004). MUSCLE: A multiple sequence alignment method with reduced time and space complexity. *BMC Bioinformatics, 5*, 113. https://doi.org/10.1186/1471-2105-5-113

Erickson, K. (2010). The jukes-cantor model of molecular evolution. *Primus, 20*(5), 438–445. https://doi.org/10.1080/10511970903487705

Felsenstein, J. (1985). Phylogenies and the comparative method. *The American Naturalist, 125*(1), 1–15.

Garamszegi, L. (2014). *Modern phylogenetic comparative methods and their application in evolutionary biology* (1st ed.). Springer.

Gibbs, A. J., & McIntyre, G. A. (1970). The diagram, a method for comparing sequences. Its use with amino acid and nucleotide sequences. *European Journal of Biochemistry, 16*(1), 1–11. https://doi.org/10.1111/j.1432-1033.1970.tb01046.x

Hall, T. (1999). BioEdit: A user-friendly biological sequence alignment editor and analysis program or windows 95/98/NT. *Nucleic Acids Symposium, 41*, 95–98.

Hall, B. G. (2005). Comparison of the accuracies of several phylogenetic methods using protein and DNA sequences. *Molecular Biology and Evolution, 22*(3), 792–802. https://doi.org/10.1093/molbev/msi066

Hogeweg, P., & Hesper, B. (1984). The alignment of sets of sequences and the construction of phyletic trees: An integrated method. *Journal of Molecular Evolution, 20*(2), 175–186. https://doi.org/10.1007/BF02257378

Holder, M., & Lewis, P. O. (2003). Phylogeny estimation: Traditional and Bayesian approaches. *Nature Reviews. Genetics, 4*(4), 275–284. https://doi.org/10.1038/nrg1044

Katoh, K. (2021). *Multiple sequence alignment*. Humana.

Katoh, K., Misawa, K., Kuma, K., & Miyata, T. (2002). MAFFT: A novel method for rapid multiple sequence alignment based on fast Fourier transform. *Nucleic Acids Research, 30*(14), 3059–3066. https://doi.org/10.1093/nar/gkf436

Keith, J. (2017). *Bioinformatics* (2nd ed.). Humana.

Keller, A., Förster, F., Müller, T., Dandekar, T., Schultz, J., & Wolf, M. (2010). Including RNA secondary structures improves accuracy and robustness in reconstruction of phylogenetic trees. *Biology Direct, 5*, 4. https://doi.org/10.1186/1745-6150-5-4

Kumar, S., & Filipski, A. (2007). Multiple sequence alignment: In pursuit of homologous DNA positions. *Genome Research, 17*(2), 127–135. https://doi.org/10.1101/gr.5232407

Kumar, S., & Gadagkar, S. R. (2001). Disparity index: A simple statistic to measure and test the homogeneity of substitution patterns between molecular sequences. *Genetics, 158*(3), 1321–1327. https://doi.org/10.1093/genetics/158.3.1321

Kumar, S., Stecher, G., Li, M., Knyaz, C., & Tamura, K. (2018). MEGA X: Molecular evolutionary genetics analysis across computing platforms. *Molecular Biology and Evolution, 35*(6), 1547–1549. https://doi.org/10.1093/molbev/msy096

Larkin, M. A., Blackshields, G., Brown, N. P., Chenna, R., McGettigan, P. A., McWilliam, H., Valentin, F., Wallace, I. M., Wilm, A., Lopez, R., Thompson, J. D., Gibson, T. J., & Higgins, D. G. (2007). Clustal W and Clustal X version 2.0. *Bioinformatics, 23*(21), 2947–2948. https://doi.org/10.1093/bioinformatics/btm404

Lemey, P., Salemi, M., & Vandamme, A. (2009). *The phylogenetic handbook* (2nd ed.). Cambridge University Press.

Li, W. H., & Graur, D. (2000). *Fundamentals of molecular evolution*. Sinauer Associates.

Li, H., Zhu, D., Zhang, C., Han, H., & Crandall, K. A. (2014). Characteristics and prediction of RNA structure. *BioMed Research International, 2014*, 690340. https://doi.org/10.1155/2014/690340

Margoliash, E., & Fitch, W. M. (1971). Amino acid sequences and phylogeny. *Taxon, 20*(1), 51–53. https://doi.org/10.2307/1218533

Nascimento, F. F., Reis, M., & d., & Yang, Z. (2017). A biologist's guide to Bayesian phylogenetic analysis. *Nature Ecology & Evolution, 1*(10), 1446–1454. https://doi.org/10.1038/s41559-017-0280-x

Notredame, C., Higgins, D. G., & Heringa, J. (2000). T-coffee: A novel method for fast and accurate multiple sequence alignment. *Journal of Molecular Biology, 302*(1), 205–217. https://doi.org/10.1006/jmbi.2000.4042

Page, R. D. (1996). TreeView: An application to display phylogenetic trees on personal computers. *Computer Applications in the Biosciences, 12*(4), 357–358. https://doi.org/10.1093/bioinformatics/12.4.357

Page, R., & Holmes, E. (1998). *Molecular evolution: A phylogenetic approach*. Blackwell.

Posada, D. (2009). *Bioinformatics for DNA sequence analysis* (1st ed.). Humana.

Prasad, P. K., Tandon, V., Biswal, D. K., Goswami, L. M., & Chatterjee, A. (2009). Phylogenetic reconstruction using secondary structures and sequence motifs of ITS2 rDNA of *Paragonimus westermani* (Kerbert, 1878) Braun, 1899 (Digenea: Paragonimidae) and related species. *BMC Genomics, 10*(Suppl 3), S25. https://doi.org/10.1186/1471-2164-10-S3-S25

Rambaut, A. (2010). *FigTree*. Retrieved from http://tree.bio.ed.ac.uk/software/figtree/

Rosenberg, M. S., & Kumar, S. (2003). Heterogeneity of nucleotide frequencies among evolutionary lineages and phylogenetic inference. *Molecular Biology and Evolution, 20*(4), 610–621. https://doi.org/10.1093/molbev/msg067

Saitou, N., & Nei, M. (1987). The neighbor-joining method: A new method for reconstructing phylogenetic trees. *Molecular Biology and Evolution, 4*(4), 406–425. https://doi.org/10.1093/oxfordjournals.molbev.a040454

Sanderson, M. J., & Shaffer, H. B. (2002). Troubleshooting molecular phylogenetic analyses. *Annual Review of Ecology and Systematics, 33*(1), 49–72. https://doi.org/10.1146/annurev.ecolsys.33.010802.150509

Selzer, P. M., Marhöfer, R. J., & Koch, O. (2018). *Applied bioinformatics—An introduction*. Springer.

Shaik, N., Hakeem, K., Banaganapalli, B., & Elango, R. (2019). *Essentials of bioinformatics* (Vol. 1, 1st ed.). Springer.

Sievers, F., Wilm, A., Dineen, D., Gibson, T. J., Karplus, K., Li, W., Lopez, R., McWilliam, H., Remmert, M., Söding, J., Thompson, J. D., & Higgins, D. G. (2011). Fast, scalable generation of high-quality protein multiple sequence alignments using Clustal omega. *Molecular Systems Biology, 7*, 539. https://doi.org/10.1038/msb.2011.75

Soltis, P. S., & Soltis, D. E. (2003). Applying the bootstrap in phylogeny reconstruction. *Statistical Science, 18*(2), 256–267. Retrieved from http://www.jstor.org/stable/3182855

Subbotin, S. A., Sturhan, D., Vovlas, N., Castillo, P., Tambe, J. T., Moens, M., & Baldwin, J. G. (2007). Application of the secondary structure model of rRNA for phylogeny: D2-D3 expansion segments of the LSU gene of plant-parasitic nematodes from the family Hoplolaimidae Filipjev, 1934. *Molecular Phylogenetics and Evolution, 43*(3), 881–890. https://doi.org/10.1016/j.ympev.2006.09.019

Subbotin, S., Waeyenberge, L., & Moens, M. (2013). Molecular systematics. In R. N. Perry & M. Moens (Eds.), *Plant nematology* (pp. 40–72). CAB International.

University of California, MoP. (2021). *Understanding evolution*. Retrieved from http://evolution.berkeley.edu/

Xia, X. (2000). *Data analysis in molecular biology and evolution* (1st ed.). Springer.

Xia, X. (2017). DAMBE6: New tools for microbial genomics, phylogenetics, and molecular evolution. *The Journal of Heredity, 108*(4), 431–437. https://doi.org/10.1093/jhered/esx033

Yang, Z., & Rannala, B. (2012). Molecular phylogenetics: Principles and practice. *Nature Reviews. Genetics, 13*(5), 303–314. https://doi.org/10.1038/nrg3186

Chapter 9
Parasitic Helminth Sample Preparation for Taxonomic Study

Abstract Classical identification of parasitic helminths is traditionally based on differences of key morphological characteristics, particularly at the parasitic adult stage, using either gross or microscopic platforms. It is still one of the essential tools for taxonomic studies to compare morphological, functional, and ecological aspects, which is suitable for quantitative analyses and population studies, and provides fast results with an affordable budget. However, classical identification based on morphological characters requires appropriate and high-quality sample preparation methods to maintain the best sample condition for further investigations. In addition, classical taxonomists' knowledge, skills, and expertise are required, but we nowadays face a gradually decreasing number of helminthologists proficient in classical taxonomy. In this chapter, concepts and classical protocols for obtaining and preparing adult flatworm and nematode specimens for morphological studies were presented and discussed. The pictorial protocols of selected methods also provided better visualization and clarity for further implementation. Accordingly, proper preservation methods to maintain a high quality of genetic materials of helminth specimens are inevitably necessary to be aware of. The general concepts and recommendations of helminth specimen preservation for subsequent molecular applications were also presented at the end of this chapter.

9.1 Obtaining Helminth Specimens for Taxonomic Study

Parasitological laboratory procedures play a crucial role in diagnosing and confirming taxonomic identification in parasitological research. Identification of parasitic helminths, particularly at a species level, requires observation and discrimination of the essential morphological characteristics, particularly at the adult stage. Appropriate specimen collection techniques and preservation methods are needed to depict the key morphological characteristics for identification successfully. Also, it is rather challenging to identify parasites (even at genus or family level) from improperly preserved specimens.

Figure 9.1 shows the potential routes of parasitic helminth specimen acquisition for systematic and taxonomic studies. Adult specimens may be obtained via either

U. Thaenkham et al., *Molecular Systematics of Parasitic Helminths*,
https://doi.org/10.1007/978-981-19-1786-8_9

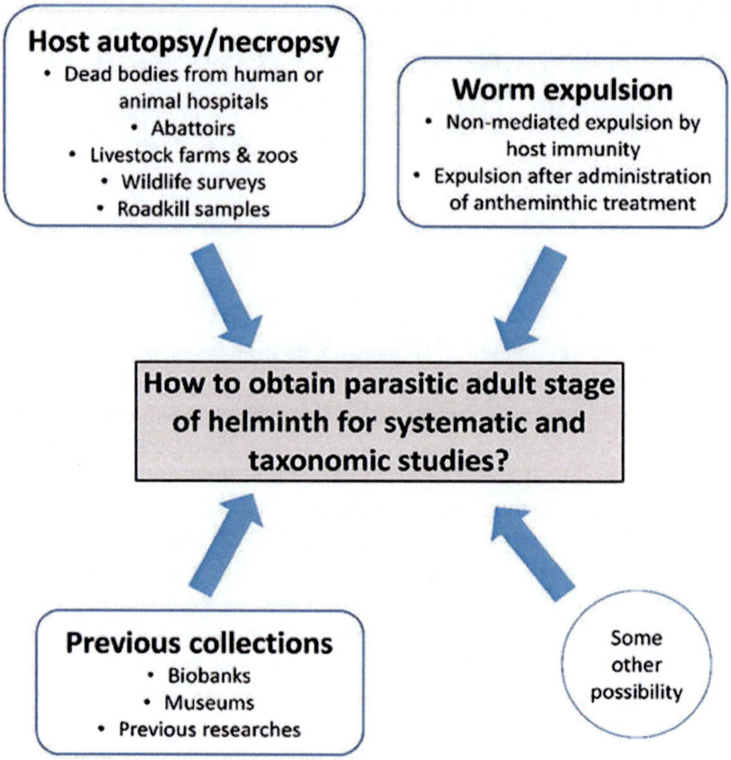

Fig. 9.1 Schematic diagram showing potential parasitic helminth specimen acquisition routes for systematic and taxonomic studies

necropsy of a dead host in the field or from parasite expulsion through feces after administering anthelmintic drugs and laxatives. The latter approach requires a protocol approved by a responsible, ethical committee and should be performed under the supervision and closely monitored by veterinary or medical doctors. Adult specimens can also be obtained from museum collections or hospital patients. In addition, roadkill and other dead animals (particularly wildlife) can be used as materials for parasitological research (Auricchio et al., 2014).

9.1.1 Obtaining Specimens from Host Necropsy and Dissection

For the animal necropsy approach, an ideal method for collecting helminths in fresh and good condition is to examine a host within 4–6 h post-death as most parasitic helminths are still alive in the host (Sepulveda & Kinsella, 2013). Deterioration or morphological changes to the helminth specimens may result if helminth collection is performed on hosts that were dead for more than 24 h or frozen. Deterioration and morphological changes render difficulties for morphological identification. Damage or loss of some important characteristics on the helminth specimen can result, e.g., rostellar hooks of tapeworm, body and collar spines of trematodes and cuticular eruption, decay, or shrinkage in nematodes.

Most helminths are usually recovered from the gastrointestinal tract of hosts, including the esophagus, stomach, and intestines. However, some other organs of the host like the liver, gallbladder, lung, urinary bladder, eyes, and brain should also be examined. In organs from the thoracic cavity (lung, heart, and trachea), filarial nematodes, metastrongylid nematodes, and large trematodes such as *Paragonimus* spp. may present. In the liver, several groups of helminths, including nematodes (e.g., *Capillaria hepatica* and other migratory nematodes larvae of ascarids and strongylids), trematodes (e.g., fasciolids and opisthorchids), and cestodes (e.g., larval cysts of taeniids and other tapeworms), may be found as this organ is also an ideal microhabitat. Some specific nematode taxa, such as thelaziids and filarial nematodes, are usually found in the eyes.

Morphological examination should be performed using a stereomicroscope to maximize the efficacy of specimen observation. For helminth examination in brief, place organs (e.g., gastrointestinal tract, liver, lung, and other tissues) in petri dishes with a small amount of tap water to facilitate dissection. Carefully cut or tear organs with a pair of dissecting needles or small scissors, taking special care to avoid accidentally damaging helminths present in the organs. Medium- to large-sized helminths can be seen easily by naked eyes. It is strongly recommended to search under a stereomicroscope for small- and tiny-sized helminths.

To separate the digestive tract, cut the rectum before the anus, and lift the whole gastrointestinal tract by carefully trimming fibrous tissues and ligaments attaching the visceral organs. Cut the other end (at the esophagus), and place the digestive tract in a petri dish or dissecting tray. If the gastrointestinal tract cannot be examined on the same day, it can be kept at 4 °C for a few days or preserved in 80–85% ethanol at a low temperature. Adding a minimum of 4:1 volume of ethanol and viscera ratio to the digestive tract samples is recommended for better preservation (Herbreteau et al., 2011). Changing the ethanol on the following day is ideal as water would be released from the tissues, diluting the ethanol concentration. Note that ethanol preservation is the least preferred alternative, as helminths should be isolated from the organs as soon as possible if time permits.

To conduct helminth examination of the digestive tract, separate the organs, e.g., esophagus, stomach, small intestine, and large intestine, and place each part in individual petri dishes or trays with a small amount of tap water. Open each organ lengthwise either from the anterior downward or posterior upward, by using small scissors to cut or dissecting needles to tear the organ apart. Carefully scrape the mucosa and examine for helminths that may still attach to the mucosa or are already detached in the content. Regularly change the water if the solution contains too much debris and becomes cloudy, obscuring inspection visibility with naked eyes or under a microscope. In addition, before throwing the content away, it is recommended that each opened part be placed under running water, washed, and sieved (sieve size largely depends on the worm size and stage of interest) (Sepulveda & Kinsella, 2013). Helminth specimens can be picked up using needles, paintbrushes, fine forceps, or droppers.

Helminth abundance and intensity of infection can be quantified by counting the helminths obtained. After examination and collection from the respective organs, separate helminths based on their differences in gross morphology by group and by an individual host; in general, nematode and trematode specimens are rarely damaged during specimen collection. However, cestodes are longer and more fragile and can be easily damaged or broken, particularly at the scolex. Thus, the collection of cestodes should therefore be carried out with caution. If cestode scolex is still attached to the intestinal mucosa, leave the specimen to relax in the water for a few hours before trying to collect the specimen again (Schmidt, 1986).

9.1.2 Obtaining Specimens from Worm Expulsion

Occasionally, parasitic worms (e.g., intestinal helminths) are expelled in the stool due to the host immune response (i.e., type 2 cytokine production), epithelial cell proliferation, and regeneration to defend against helminth colonization (Anthony et al., 2007). Helminths expelled in stool make it possible to obtain samples for further investigation. Some helminths can shed their body parts via the host's stool (e.g., cestode gravid proglottids) as a mechanism for the reproductive distribution into the environment. These cestode segments can be morphologically studied for further taxonomic identification.

In addition, the administration of anthelmintic drugs is another acceptable method of expelling helminths from the body. For example, physicians can collect intestinal trematode and cestode from human patients by administering praziquantel (20–40 mg/kg concentration) followed by a purgative (saturated magnesium sulfate solution) to activate bowel movement facilitating stool sample collection (Crellen et al., 2021; Cruz, 1991; Waikagul et al., 2008). Stool samples collected are then washed repeatedly by sedimentation, and helminths are successfully collected from the cleaned sediments. However, praziquantel still has limitations in terms of its therapeutic efficacy in some parasite species. Although praziquantel is a broad-spectrum anthelmintic targeting several platyhelminthic infections, e.g., schistosomiasis, clonorchiasis, opisthorchiasis, paragonimiasis, heterophyidiasis, echinostomiasis, fasciolopsiasis, taeniasis, and hymenolepiasis (Andrews et al., 1983; Chai, 2013), researches revealed that *Fasciola* spp. infections and other metacestodes (i.e., hydatid cysts and plerocercoid larvae) are refractory to praziquantel (Andrews et al., 1983; Becker et al., 1980; Bylund et al., 1977). Alternatively, other drug compounds such as niclosamide are another option for killing parasitic cestodes (Lateef et al., 2008; Pearson & Hewlett, 1985; Perera et al., 1970) and used for helminth expulsion. For example, to obtain cestode specimens from whole bowel movement samples, 2 g niclosamide (4 tablets of 500 mg) followed by a purgative (60 ml of saturated magnesium sulfate solution) 2 h after the treatment can be administered to patients with taeniasis (Anantaphruti et al., 2010; Maipanich et al., 2011).

In terms of nematode expulsion following deworming, parasitologists can collect strongyle nematodes from horses by using three different anthelmintic drugs, i.e., ivermectin (0.2 mg/kg), pyrantel pamoate (19 mg/kg), and fenbendazole (7.5 mg/kg) (Osterman Lind et al., 2003). Cyathostomin nematodes are expelled in horse fecal samples the first day post-treatment with ivermectin and pyrantel pamoate, whereas fenbendazole-treated horses expelled nematodes the second day post-treatment. In school, in children infected with soil-transmitted helminth, adult specimens of hookworm species, i.e., *Necator americanus* and *Ancylostoma duodenale*, were expelled in bowel movements from day 1 to day 7 post-treatment with mebendazole (100 mg twice daily for 3 days) (Anantaphruti et al., 2002). In another example, following a single oral dose of albendazole (400 mg/kg) treatment in *Ascaris lumbricoides*-infected human patients, helminths were expelled day 1 post-treatment till day 10 when no adult worm was found (Levecke et al., 2018).

9.2 Sample Fixation, Preservation, and Preparation of Adult Worms for Morphological Study

Preservation means using appropriate methods for long-term specimen preservation in the best possible condition for systematic storage and further uses. Ideally, living parasitic helminths newly isolated from hosts are recommended to be washed by shaking in normal saline solution or sodium bicarbonate solution ($NaHCO_3$) to remove debris or mucus on the parasite surface (Berland, 2004). Consequently, the specimens for morphological study should be fixed and preserved in an appropriate manner according to each helminth group. Fixation is defined in the parasitological process as killing and fixing parasite specimens simultaneously by either using temperature (heat or freeze) or chemical agents. Fixatives are chemical agents used to modify tissues so that the form is retained. Fixation also makes the tissue easy to stain (Kenedy, 1979).

Fixatives can be classified into the four main groups: (1) aldehyde group (e.g., formaldehyde and glutaraldehyde), (2) oxidizing agents (e.g., osmium tetroxide and potassium permanganate), (3) alcohol-based fixatives (methyl and ethyl alcohol), and (4) metallic group (e.g., mercuric chloride and picric acid). The first two groups fix tissue by forming cross-link proteins and/or amino acids; alcohol-based fixatives act as protein-denaturing agents, whereas the metallic group can form insoluble metallic precipitation and coagulation of protein structure (Grizzle et al., 2008; Thavarajah et al., 2012). Common fixatives for classical and modern helminthology include ethanol (70% is generally good) and formaldehyde (2–10% formalin), glacial acetic acid (GAA) and Looss' fluid (1 part of glycerol in 9 parts of 70% ethanol), and a combination of fixatives from different groups such as Bouin's fixative (compound fixative containing picric acid, formaldehyde, and acetic acid) and alcohol-formalin-acetic acid solution (AFA). AFA is one of the most commonly used fixatives in helminthology (Auricchio et al., 2014; Berland, 2004; Garvin et al., 1997; Kenedy, 1979; Schmidt, 1986). Helminth specimens fixed in AFA are usually stained well in carmine-based staining, which is one of the most widely used dyes in helminthology today.

9.2.1 *Preparation of Platyhelminth Specimens for Morphological Study*

Live cestode and trematode specimens should be relaxed in cold or ambient water for a couple of hours to overnight, depending on the size of specimens (Fig. 9.2a, b). Avoid placing flatworms in saline solution as it does not allow osmotic activity and prevents subsequent relaxation of the specimens (Gardner et al., 2012). After specimen relaxation, medium- to large-sized flatworms should be subjected to press-fixation. Specimens are flattened between two microscopic glass slides and submerged in a fixative such as AFA or 5–10% formalin for 24–48 h (Fig. 9.2c–i). Then, the unpressed specimens are transferred to 70% ethanol for neutralization and further storage. The pressed worms retain their flattened shape if the specimens are appropriately fixed. Very small-sized flatworms such as minute intestinal flukes or other small-sized species may not require the pressing step. They can be alternatively handled by briefly submerging in warm 10% formalin and transferred to AFA for fixation (Garvin et al., 1997).

After fixation, cestode and trematode specimens should be subjected to the processes of permanent slide preparation, which involve specimen staining, de-colorization, dehydration, clearing, and mounting on slides for analysis of morphological characters and taxonomic identification. For staining, carmine and hematoxylin are commonly used for cestodes and trematodes, producing satisfactory results (Schell, 1970; Schmidt, 1986). Fast green FCF is also occasionally used as a counterstaining dye for better visualization of body tegument and some other internal organs of trematodes (Blankespoor & Ulmer, 1970; Jitra Waikagul & Thaenkham, 2014; Wanas et al., 1995). In addition, staining with lactophenol cotton blue is a simple technique invented to rapidly and clearly differentiate the internal structures of Platyhelminthes and is suitable for wide-scale studies and easily performed in field conditions (Henedi & El-Azazy, 2013). Various herbal extract dyes, e.g., red beet, rose madder, bulb onion, and Persian walnut, have also been applied to differentiate internal organs of helminths as alternatives for safer, more eco-friendly, and inexpensive options compared to carmine dye (Al-Amura et al., 2012; Marhaba & Haniloo, 2018).

Here, we present the procedures of acetic carmine staining and whole mounting for Platyhelminthes as an example. The pictorial protocol is shown in Fig. 9.2. After fixation in AFA or formalin, transfer specimens to 70% ethanol for a couple of hours,

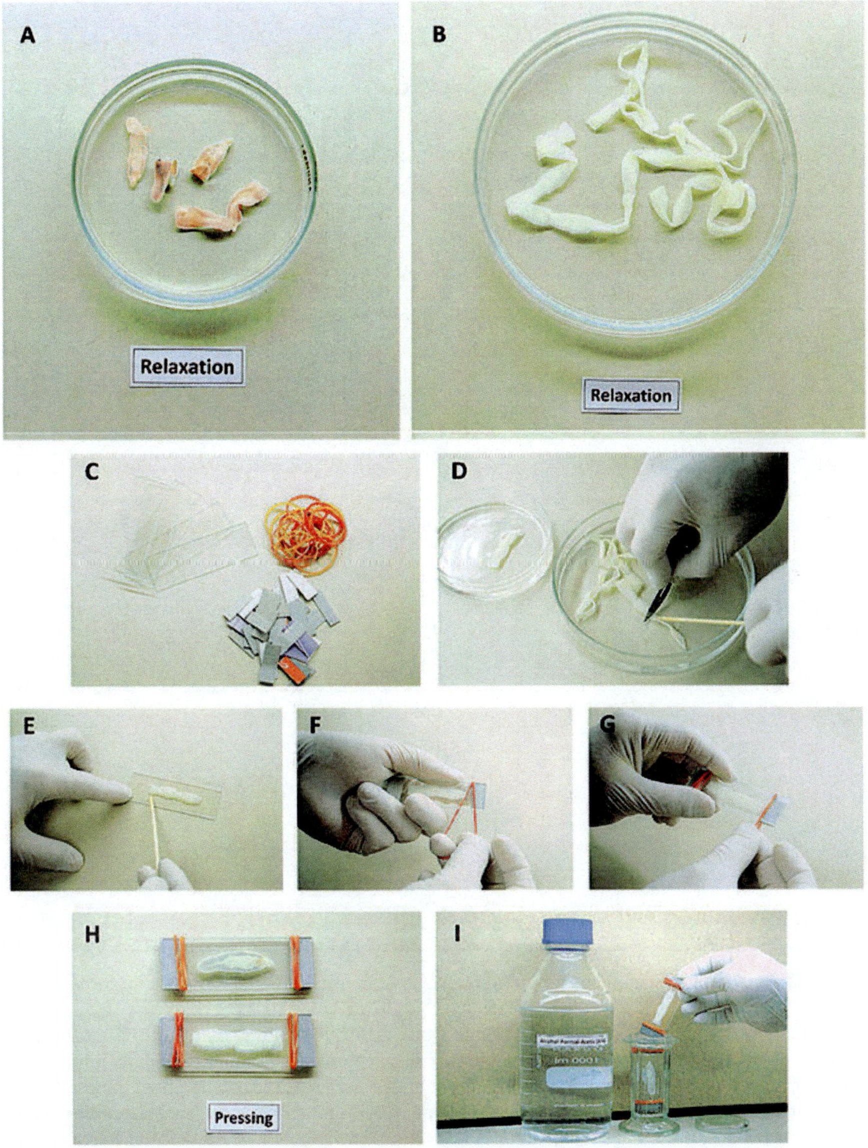

Fig. 9.2 Pictorial protocol for permanent slide preparation of flatworms. Relaxation of flatworm specimens (**a**) trematodes and (**b**) cestodes in distilled water (ambient temperature) overnight. (**c**) A set of simple equipment for specimen pressing and fixing: glass slides, rubber bands, and pieces of paper for inserting between glass slides for cushioning; (**d–h**) specimen preparation in the press-fixing step, flatten worm specimens between two microscopic glass slides, and hold with rubber bands; (**i**) then, submerge the specimens in AFA solution for a couple of days. After fixation, (**j–l**) unpress the slides, remove the worm specimens, and place into 70% alcohol; (**m, n**) stain the

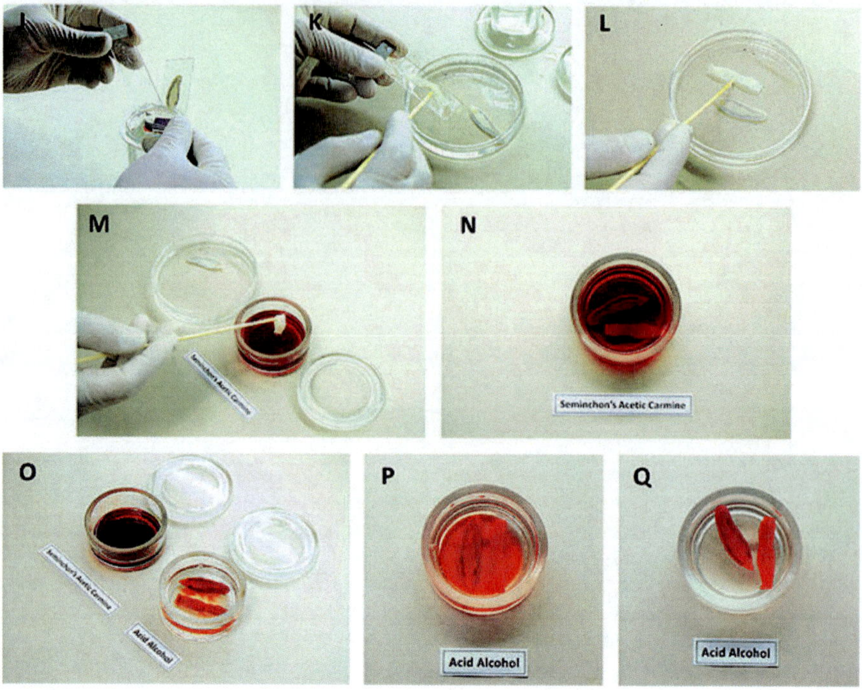

Fig. 9.2 (continued)

Fig. 9.2 (continued) specimens in carmine-based dye (i.e., Semichon's acetic carmine) for 5–10 min, depending on the size of specimens; (**o–q**) de-color or remove excess dye from the specimens by soaking in acid alcohol solution (i.e., 1% hydrochloric acid in 70% ethanol) for 5–10 min or until the specimen color become pale red or pink. After staining and de-coloring, (**r**, **s**) pass the specimens to the dehydration step through the series of alcohol gradient concentrations (gradually from low to high alcohol concentrations) for 15–30 min at each step, depending on the size of specimens; (**t**) soak the specimens in alcohol-xylene solution (1:1 ratio of absolute alcohol and xylene) for about 15–30 min for neutralization; (**u**) then, place the specimens in xylene for clearing step before mounting on a slide. Permanent slide mounting steps: (**v**) apply a proper amount of a mounting medium (i.e., Permount) on a glass slide; (**w**) place the specimen on the mounting media, and adjust the specimen to a good position; (**x**) carefully cover the specimen with a coverslip; (**y**) mounting medium can be added to a side of the coverslip in the case of insufficient amount of solution; (**z**) label important information on left side of the slide, rest the specimens on an aluminum tray, and incubate in a hot air oven at around 60 °C for a couple of days until the mounting medium is properly set

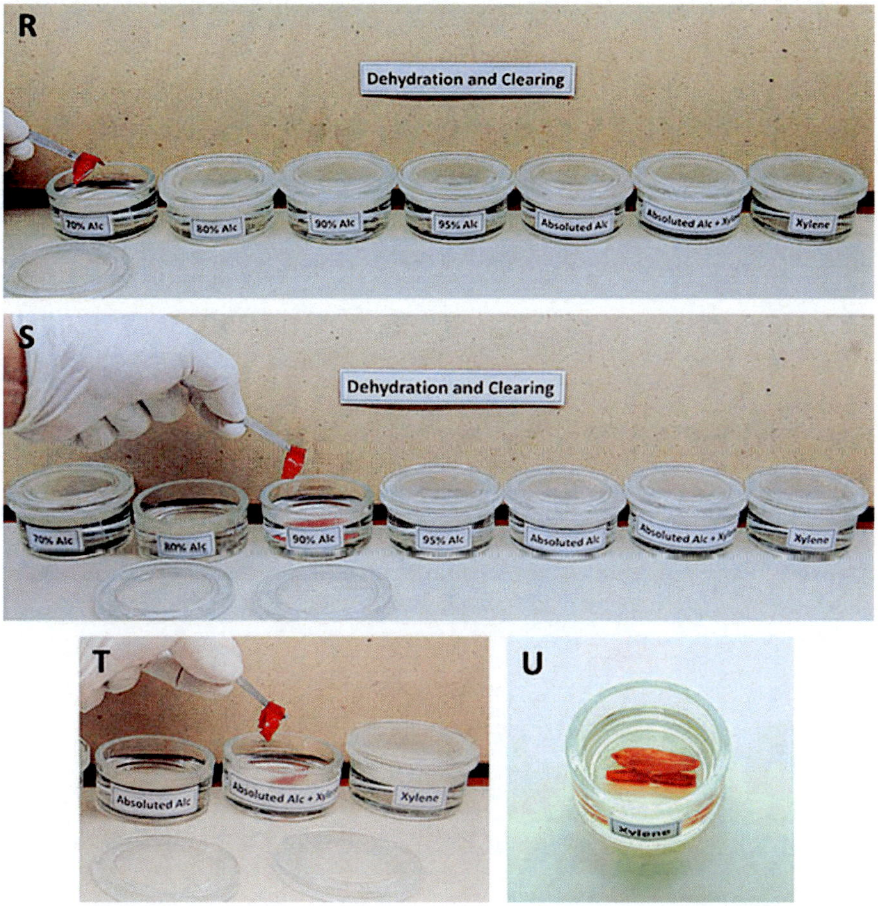

Fig. 9.2 (continued)

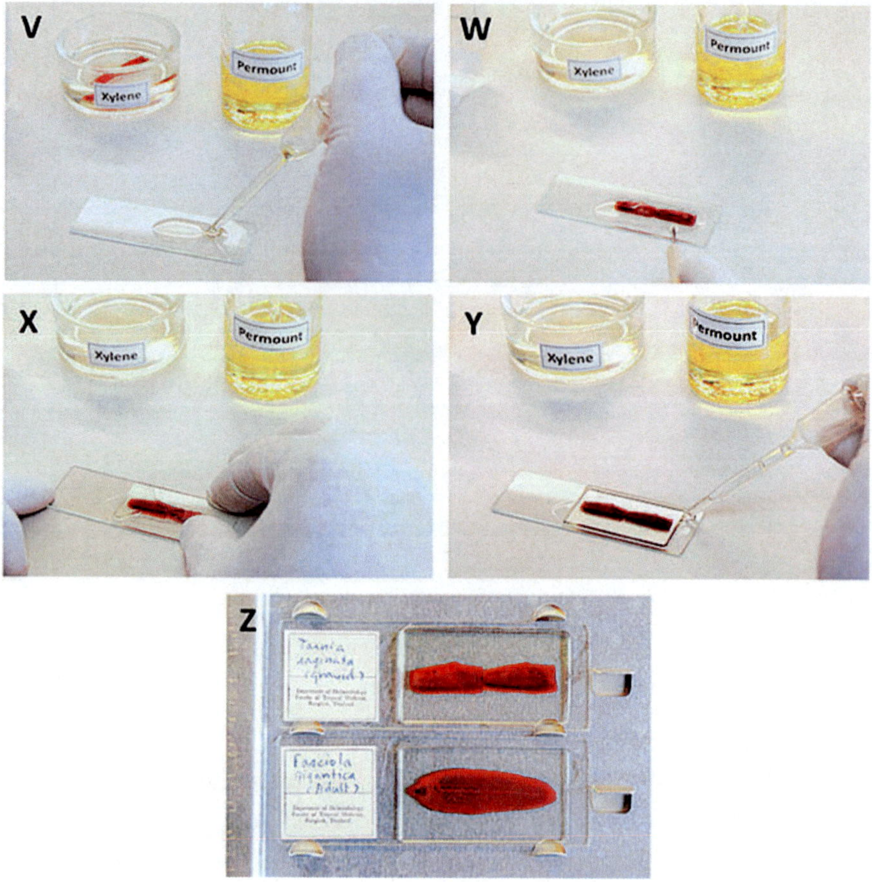

Fig. 9.2 (continued)

and then place the worms in acetic carmine staining solution for 5–60 min, depending on the size of specimens (Fig. 9.2j–m). Consequently, transfer the stained specimens into acid alcohol solution (1% hydrochloric acid in 70% ethanol) for de-colorization to remove excess stains until the internal organs become pale red or pink (Fig. 9.2o–q). Then, the specimens are subjected to dehydration through a series of ethanol gradient concentrations, i.e., 70%, 80%, 90%, 95%, and absolute ethanol, for 15–30 min at each step depending on the size of specimens (Fig. 9.2r, s). It is recommended to soak the specimens in ethanol-xylene solution (1:1 absolute ethanol and xylene) for about 15–30 min for neutralization (Fig. 9.2t). Finally, dehydrated specimens will be placed in xylene for the clearing step and then mounted on glass slides with an adequate resinous mounting medium (Fig. 9.2u–z). The prepared slides should be dried at 60 °C in a hot air oven for a couple of days.

Do not forget to label important information on the slides properly. The specimens are then ready for downstream microscopic examination.

9.2.2 Preparation of Nematode Specimens for Morphological Study

In general, fixing nematodes in cold fixatives such as buffered formalin or ethanol may result in the specimen being coiled up or contracted, while fixing in a hot (70–90 °C) fixative yields a good result, particularly those small-sized and fragile nematodes (Sepulveda & Kinsella, 2013). GAA (glacial acetic acid) is an alternative fixation to satisfactory outcomes. It rapidly penetrates nematode tissue making the worm body swell and softer than other fixatives due to increasing turgidity (Berland, 1961). GAA is accepted as one of the standard nematode fixatives, particularly for medium- to large-sized nematodes (Berland, 2004; Sepulveda & Kinsella, 2013). Glycerin alcohol (e.g., 5% glycerin in 70% ethanol) is another widely used fixative for nematodes, and it is recommended for storage purposes (Kenedy, 1979; Sepulveda & Kinsella, 2013). AFA is also occasionally used to fix nematodes, similar to Platyhelminthes (Auricchio et al., 2014).

Unlike Platyhelminthes, nematodes generally do not require staining. Important structures and internal organs are exhibited well in samples mounted in clearing agents. Lactophenol and glycerin are acceptable and widely used for preserving and clearing processes of nematode preparation (Kenedy, 1979; Seinhorst, 1959; Sepulveda & Kinsella, 2013; Trajan, 1973). Temporary mount (or wet whole mount) of nematodes in lactophenol between a glass slide and coverslip is easy for rapid helminth examination. In this way, the specimen can be adjusted to a better position, aiding better visualization of some key characters under a microscope. In the case of large-sized nematodes, the cephalic and caudal ends may be cut and mounted separately in lactophenol to specifically observe some other details such as the mouthpart and male reproductive organs. After morphological examination of the temporary-mounted specimens, specimens should be removed and briefly washed in acid ethanol solution before storing in 5% glycerin alcohol.

For permanent slide preparation, specimens are mounted in synthetic resinous media such as Canada balsam or Permount. Specimens must be first passed through the dehydration steps, and the clearing agent used in the process, e.g., xylene or toluene, should be mixable with the resin (Kenedy, 1979; May, 1922). Most of these steps are similar to the permanent slide preparation of parasitic flatworms explained in the previous section. This classical mounting method with Canada balsam or Permount requires a considerable amount of work and time and continued attention during the process to ensure a good outcome; otherwise, failure and specimen damage will result. Alternatively, the double glass mounting technique can be applied with nematodes as some small-sized nematodes can be challenging to dehydrate before the mounting step. Accordingly, the double glass mounting technique is efficient for mounting small-sized nematodes without exposing them to the alcohol gradient dehydration steps.

For permanent mount, the pictorial protocol is presented in Fig. 9.3. In brief, nematodes are placed in warm 5% glycerine alcohol at 60 °C for about an hour (Fig. 9.3a, b). During this time, keep watching and be careful not to let the alcohol completely evaporate. Next, place a drop of pre-warm glycerine jelly (see Fig. 9.3c–g for glycerine jelly preparation) on a glass slide, and place the specimen on the solution. Adjust the specimen to its proper position and cover it with 18 × 18 mm coverslip. Leave the slide at room temperature for a few minutes to allow the glycerine jelly to cool (Fig. 9.3h–k). Subsequently, perform double mounting by adding a drop of resinous mountant, e.g., Canada balsam or Permount, on the top of pre-mounted specimen in glycerine jelly. Place a bigger size coverslip, e.g., 22 × 22 mm or 24 × 50 mm, on the specimen to cover the area of the smaller coverslip. The prepared slides should be dried at 60 °C in a hot air oven for 2 days. Do not forget to label important information on the slides properly, and then these are ready for further microscopic examination (Fig. 9.3l–o).

As another option, a novel procedure of using formaldehyde alcohol azocarmine lactophenol (FAAL) as a combined fixative and stain, followed by classic dehydration through alcohol gradient concentrations and permanently mounting in Canada balsam medium, has been proven as a potential method for small-sized nematode preparation (Zahabiun et al., 2015).

9.3 Sample Preservation for Molecular Study

The choice of reagents and methods for helminth preservation depends largely on the purpose and objectives for which the specimen is to be used in the future. Ideally, it is highly recommended that if the number of specimens of interest in a population is numerous, it is good practice to fix and preserve them in various reagents for different downstream purposes. Some reagents may be very effective in maintaining morphological conditions, but in turn, they may adversely affect the quality of genetic material, e.g., formaldehyde preservatives. Formaldehyde mainly induces interstrand cross-links in DNA and cross-links between DNA-protein, causing DNA lesion and damage in general (Anandarajan et al., 2020; Kawanishi et al., 2014). Specimen preservation in alcohol-based solutions is cheap, eco-friendly and versatile and has become universal in biological science (Justine et al., 2012; Lutz et al., 2017). Several protocols and guidelines advocate the use of 70% alcohol as the single universal preservative for routine morphological and molecular studies (Galbreath et al., 2019; Garvin et al., 1997; Justine et al., 2012; Lutz et al., 2017; Sepulveda & Kinsella, 2013). It is well known that higher alcohol concentration (e.g., 95% to absolute alcohol) results in better preservation of genetic materials. However, highly concentrated alcohol makes specimens brittle, deformed, and easily damaged, making it difficult to study morphology.

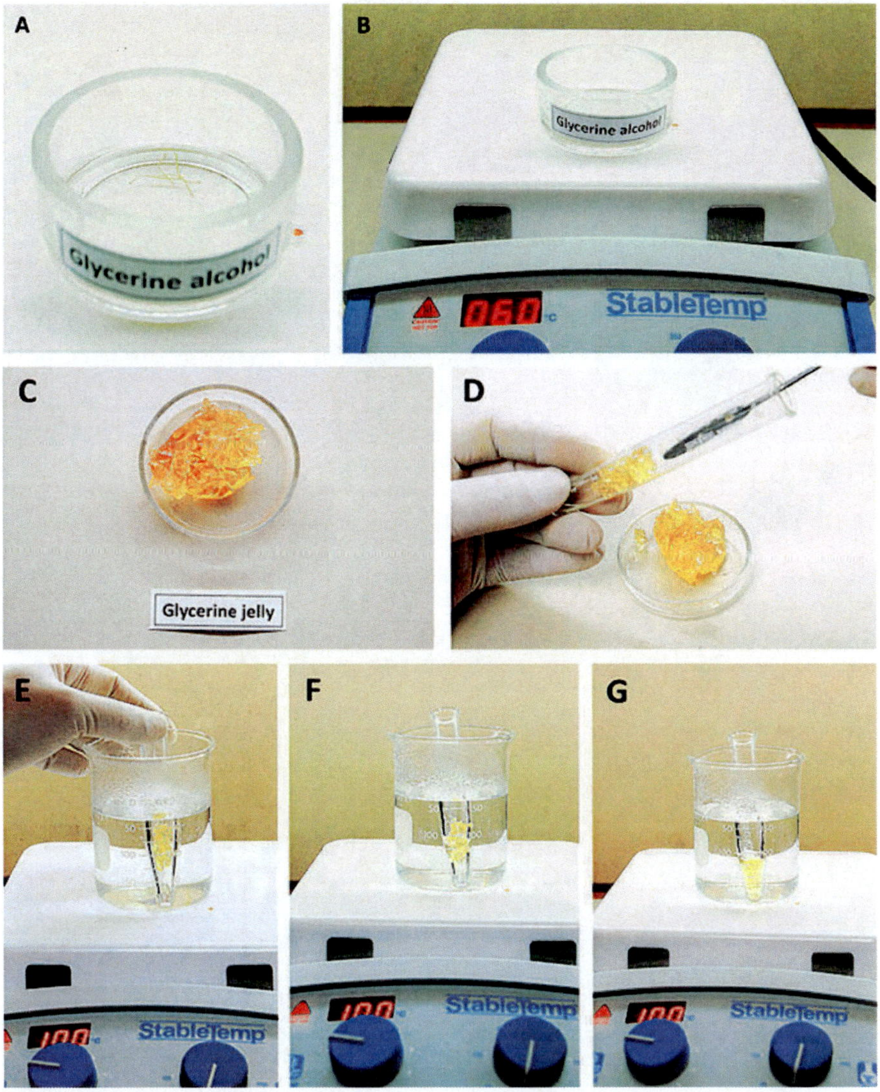

Fig. 9.3 Pictorial protocol for permanent slide preparation of parasitic nematodes using double glass mounting technique. Fixation of nematode specimens in (**a**) glycerin alcohol solution (5% glycerin in 70% alcohol); (**b**) warm the specimens in glycerin alcohol fixative at 60 °C for about an hour. At this step, nematodes will be stretched and dehydrated as the alcohol concentration will be gradually concentrated due to the heat. While waiting for the specimen fixation and preservation in glycerin alcohol, the first mounting medium, i.e., glycerin jelly, can be prepared; (**c**) stocked glycerin jelly kept in a fridge is semisolid; (**d**) aliquot a required amount of glycerin jelly in a test tube; (**e–g**) warm the glycerin jelly on a hot plate until it melts into a liquid. Permanent slide mounting steps: (**h, i**) apply a proper amount of the first mounting medium (i.e., warm glycerin jelly) on a glass slide; (**j**) place the specimen on the mounting media and adjust the specimen to a

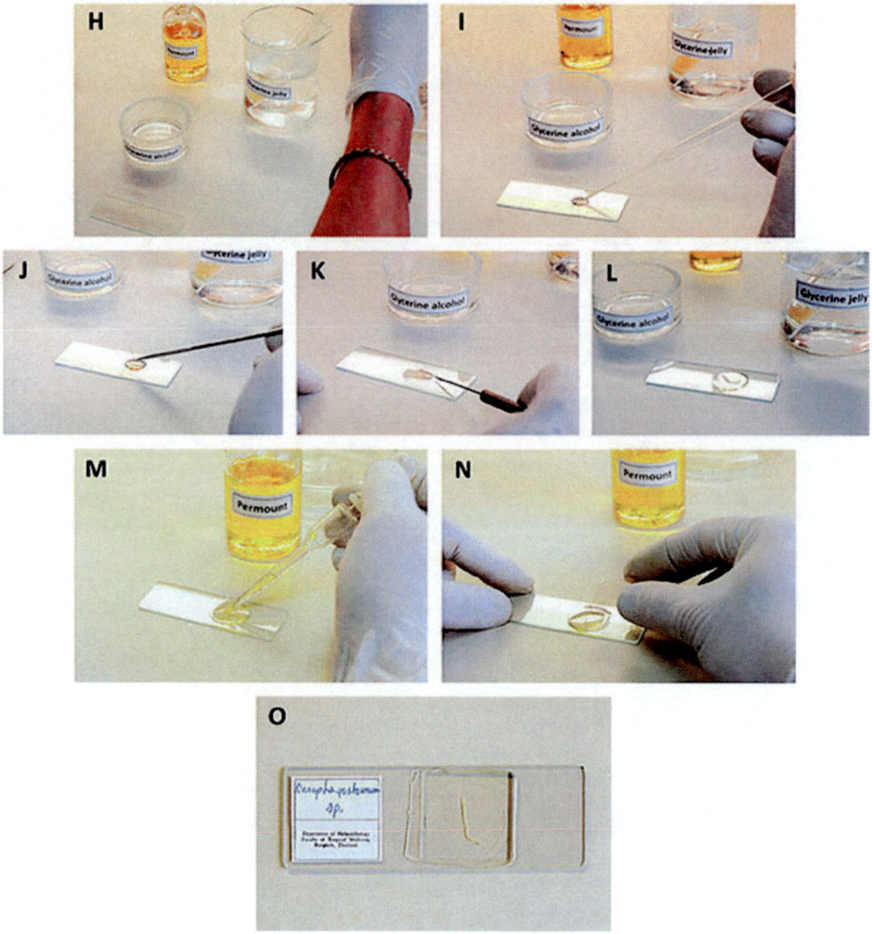

Fig. 9.3 (continued)

Fig. 9.3 (continued) good position; (**k, l**) carefully cover the specimen with a coverslip. Leave the specimen at room temperature briefly until glycerin jelly hardens. This first step is to obtain a temporary slide that the worm specimen can still move. (**m**) Then, perform double mounting by applying the second mounting medium (i.e., Permount) to the top of the pre-mounted parasite in glycerin jelly. (**n**) Place a bigger-sized coverslip on the specimen to cover the area of the smaller coverslip underneath. (**o**) Label important information on the left side of the slide, rest the specimens on a tray, and incubate in a hot air oven at around 60 °C for a couple of days until the mounting medium is appropriately set

Parasitologists have to be aware of the clear trade-off between morphological study and genetic analysis. In terms of storage temperature, genetic materials of alcohol-preserved specimens can survive briefly at room temperature, but likely more degradation occurred over time. Storing alcohol-preserved helminths at −20 °C or colder largely preserves DNA for several years and yields satisfactory results of consequent molecular applications (Galbreath et al., 2019). It is highly recommended that specimens should be permanently stored in the freezer (or at least in a fridge) as soon as they are isolated from hosts. Alternatively, washing newly isolated helminth samples thoroughly in distilled water and then freezing immediately at −80 °C without any preservatives is acceptable as a suitable method of preserving specimens for further use in molecular studies.

Acetone is also known to be highly capable of maintaining genetic materials of several organisms such as insects, zooplankton, and fungi (Fukatsu, 1999; Goetze & Jungbluth, 2013; Punekar et al., 2003). Subsequent molecular and proteomic experiments such as PCR, sodium dodecyl sulfate-polyacrylamide gel electrophoresis (SDS-PAGE), and immunoblotting successfully yielded good results from the samples stored in acetone for years (Bisanti et al., 2009; Fukatsu, 1999). However, it is not commonly used due to its toxic effect on health, and it is unsuitable for plastic containers as it can dissolve plastics.

Recently, scientists have been trying to devise alternative reagents that can preserve specimens for use in a broader range of purposes. A novel preservative reagent containing dimethyl sulfoxide, disodium EDTA, and saturated NaCl known as "DESS" has been invented and tested for its efficacy in the preservation of nematodes which combined suitability for morphological and molecular analyses (Beknazarova et al., 2017; Naem et al., 2010; Yoder et al., 2006).

9.4 Conclusion

Proper preservation techniques of helminths are essential for downstream morphological and molecular analysis. High-quality specimens can be obtained when proper collection and preservation of helminths are performed. This chapter presented protocols for the preservation and preparation of adult platyhelminths and nematodes for morphological identification, along with pictorial protocols for visualization. Additionally, recommendations for helminth preservation for molecular analysis were mentioned, providing a strong foundation for downstream molecular applications like molecular systematics and molecular identification.

References

Al-Amura, M. F., Hassen, Z. A., & Al-Mhanawi, B. H. (2012). Staining technique for helminth parasites by use red beet (*Beta vulgaris* L.) extract. *Basrah Journal of Veterinary Research, 11* (1), 283–292.

Anandarajan, V., Noguchi, C., Oleksak, J., Grothusen, G., Terlecky, D., & Noguchi, E. (2020). Genetic investigation of formaldehyde-induced DNA damage response in *Schizosaccharomyces pombe*. *Current Genetics, 66*(3), 593–605. https://doi.org/10.1007/s00294-020-01057-z

Anantaphruti, M. T., Maipanich, W., Muennoo, C., Pubampen, S., & Sanguankiat, S. (2002). Hookworm infections of school children in southern Thailand. *The Southeast Asian Journal of Tropical Medicine and Public Health, 33*(3), 468–473.

Anantaphruti, M. T., Okamoto, M., Yoonuan, T., Saguankiat, S., Kusolsuk, T., Sato, M., Sato, M. O., Sako, Y., Waikagul, J., & Ito, A. (2010). Molecular and serological survey on taeniasis and cysticercosis in Kanchanaburi Province, Thailand. *Parasitology International, 59*(3), 326–330. https://doi.org/10.1016/j.parint.2010.03.007

Andrews, P., Thomas, H., Pohlke, R., & Seubert, J. (1983). Praziquantel. *Medicinal Research Reviews, 3*(2), 147–200. https://doi.org/10.1002/med.2610030204

Anthony, R. M., Rutitzky, L. I., Urban, J. F. J., Stadecker, M. J., & Gause, W. C. (2007). Protective immune mechanisms in helminth infection. *Nature Reviews. Immunology, 7*(12), 975–987. https://doi.org/10.1038/nri2199

Auricchio, P., Catenacci, L. S., Santos, K. R., & Britto, F. B. (2014). A protocol for the use of roadkill or stranded animals as material for research and teaching. *SITIENTIBUS série Ciências Biológicas, 14*, 1–6. https://doi.org/10.13102/scb237

Becker, B., Mehlhorn, H., Andrews, P., Thomas, H., & Eckert, J. (1980). Light and electron microscopic studies on the effect of praziquantel on *Schistosoma mansoni*, *Dicrocoelium dendriticum*, and *Fasciola hepatica* (Trematoda) in vitro. *Zeitschrift Fur Parasitenkunde (Berlin, Germany), 63*(2), 113–128. https://doi.org/10.1007/BF00927527

Beknazarova, M., Millsteed, S., Robertson, G., Whiley, H., & Ross, K. (2017). Validation of DESS as a DNA preservation method for the detection of *Strongyloides* spp. in canine feces. *International Journal of Environmental Research and Public Health, 14*(6). https://doi.org/10.3390/ijerph14060624

Berland, B. (1961). Use of glacial acetic acid for killing parasitic nematodes for collection purposes. *Nature, 191*(4795), 1320–1321. https://doi.org/10.1038/1911320a0

Berland, B. (1961). Use of glacial acetic acid for killing parasitic nematodes for collection purposes. *Nature, 191*(4795), 1320–1321. https://doi.org/10.1038/1911320a0

Bisanti, M., Ganassi, S., & Mandrioli, M. (2009). Comparative analysis of various fixative solutions on insect preservation for molecular studies. *Entomologia Experimentalis et Applicata, 130*(3), 290–296. https://doi.org/10.1111/j.1570-7458.2008.00821.x

Blankespoor, H. D., & Ulmer, M. J. (1970). Helminths from six species of Iowa bats. *Proceedings of the Iowa Academy of Science, 77*(30), 200–206.

Bylund, G., Bång, B., & Wikgren, K. (1977). Tests with a new compound (Praziquantel) against *Diphyllobothrium latum*. *Journal of Helminthology, 51*(2), 115–119. https://doi.org/10.1017/s0022149x00007343

Chai, J.-Y. (2013). Praziquantel treatment in trematode and cestode infections: An update. *Infection & Chemotherapy, 45*(1), 32–43. https://doi.org/10.3947/ic.2013.45.1.32

Crellen, T., Sithithaworn, P., Pitaksakulrat, O., Khuntikeo, N., Medley, G. F., & Hollingsworth, T. D. (2021). Towards evidence-based control of *Opisthorchis viverrini*. *Trends in Parasitology, 37*(5), 370–380. https://doi.org/10.1016/j.pt.2020.12.007

Cruz, A. C. (1991). Treatment of human taeniasis in the Philippines: A review. *The Southeast Asian Journal of Tropical Medicine and Public Health, 22*(Suppl), 271–274.

Fukatsu, T. (1999). Acetone preservation: A practical technique for molecular analysis. *Molecular Ecology, 8*(11), 1935–1945. https://doi.org/10.1046/j.1365-294x.1999.00795.x

Galbreath, K. E., Hoberg, E. P., Cook, J. A., Armién, B., Bell, K. C., Campbell, M. L., Dunnum, J. L., Dursahinhan, A. T., Eckerlin, R. P., Gardner, S. L., Greiman, S. E., Henttonen, H., Jiménez, F. A., Koehler, A. V. A., Nyamsuren, B., Tkach, V. V., Torres-Pérez, F., Tsvetkova, A., & Hope, A. G. (2019). Building an integrated infrastructure for exploring biodiversity: Field collections and archives of mammals and parasites. *Journal of Mammalogy, 100*(2), 382–393. https://doi.org/10.1093/jmammal/gyz048

Gardner, S. L., Fisher, R. N., & Barry, S. J. (2012). Collecting and preserving parasites during reptile biodiversity survey. *Faculty Publications from the Harold W Manter Laboratory of Parasitology, 738*, 114–121.

Garvin, M. C., Bates, J. M., & Kinsella, J. M. (1997). Field techniques for collecting and preserving helminth parasites from birds, with new geographic and host Records of Parasitic Nematodes from Bolivia. *Ornithological Monographs, 48*, 261–266. http://www.jstor.org/stable/40157537.

Goetze, E., & Jungbluth, M. J. (2013). Acetone preservation for zooplankton molecular studies. *Journal of Plankton Research, 35*(5), 972–981. https://doi.org/10.1093/plankt/fbt035

Grizzle, W. E., Fredenburgh, J. L., & Myers, R. B. (2008). 4 - Fixation of tissues. In J. D. Bancroft & M. Gamble (Eds.), *Theory and practice of histological techniques* (6th ed., pp. 53–74). Churchill Livingstone. https://doi.org/10.1016/B978-0-443-10279-0.50011-7

Henedi, A. A. M., & El-Azazy, O. M. E. (2013). A simple technique for staining of platyhelminths with the lactophenol cotton blue stain. *Journal of the Egyptian Society of Parasitology, 43*(2), 419–423. https://doi.org/10.12816/0006398

Herbreteau, V., Jittapalapong, S., Rerkamnuaychoke, W., Chaval, Y., Cosson, J. F., & Morand, S. (2011). In V. Herbreteau, S. Jittapalapong, W. Rerkamnuaychoke, Y. Chaval, J. F. Cosson, & S. Morand (Eds.), *Protocols for field and laboratory rodent studies* (1st ed.). Kasetsart University Press.

Justine, J.-L., Briand, M. J., & Bray, R. A. (2012). A quick and simple method, usable in the field, for collecting parasites in suitable condition for both morphological and molecular studies. *Parasitology Research, 111*(1), 341–351. https://doi.org/10.1007/s00436-012-2845-6

Kawanishi, M., Matsuda, T., & Yagi, T. (2014). Genotoxicity of formaldehyde: Molecular basis of DNA damage and mutation. *Frontiers in Environmental Science, 2*, 36. https://doi.org/10.3389/fenvs.2014.00036

Kenedy, M. J. (1979). *Basic methods of specimen preparation in parasitology*. IDRC Manuscript Reports.

Lateef, M., Zargar, S. A., Khan, A. R., Nazir, M., & Shoukat, A. (2008). Successful treatment of niclosamide- and praziquantel-resistant beef tapeworm infection with nitazoxanide. *International Journal of Infectious Diseases, 12*(1), 80–82. https://doi.org/10.1016/j.ijid.2007.04.017

Levecke, B., Easton, A. V., Cools, P., Albonico, M., Ame, S., Gilleard, J. S., Keiser, J., Montresor, A., Prichard, R., Vlaminck, J., & Vercruysse, J. (2018). The optimal timing of post-treatment sampling for the assessment of anthelminthic drug efficacy against *Ascaris* infections in humans. *International Journal for Parasitology: Drugs and Drug Resistance, 8*(1), 67–69. https://doi.org/10.1016/j.ijpddr.2017.12.004

Lutz, H. L., Tkach, V. V., & Weckstein, J. D. (2017). Methods for specimen-based studies of avian symbionts. In M. S. Webster (Ed.), *The extended specimen: Emerging frontires in collection-based ornithological research* (1st ed., p. 28). CRC Press.

Maipanich, W., Sato, M., Pubampen, S., Sanguankiat, S., Kusolsuk, T., Thaenkham, U., & Waikagul, J. (2011). Abnormal *Taenia saginata* tapeworms in Thailand. *The Southeast Asian Journal of Tropical Medicine and Public Health, 42*(5), 1065–1071.

Marhaba, Z., & Haniloo, A. (2018). Staining of parasitic helminths by extracts of *Allium cepa*, *Juglans regia*, and *Rubia tinctorum*: An approach to herbal dyes. *Iranian Journal of Parasitology, 13*(2), 293–300.

May, H. G. (1922). Killing, staining and mounting parasitic nematodes. *Transactions of the American Microscopical Society, 41*(2), 103–105.

Naem, S., Pagan, C., & Nadler, S. A. (2010). Structural restoration of nematodes and acanthocephalans fixed in high percentage alcohol using DESS solution and rehydration. *The Journal of Parasitology, 96*(4), 809–811. https://doi.org/10.1645/GE-2402.1

Osterman Lind, E., Eysker, M., Nilsson, O., Uggla, A., & Höglund, J. (2003). Expulsion of small strongyle nematodes (cyathostomin spp) following deworming of horses on a stud farm in Sweden. *Veterinary Parasitology, 115*(4), 289–299. https://doi.org/10.1016/s0304-4017(03)00200-0

Pearson, R. D., & Hewlett, E. L. (1985). Niclosamide therapy for tapeworm infections. *Annals of Internal Medicine, 102*(4), 550–551. https://doi.org/10.7326/0003-4819-102-4-550

Perera, D. R., Western, K. A., & Schultz, M. G. (1970). Niclosamide treatment of cestodiasis. Clinical trials in the United States. *The American Journal of Tropical Medicine and Hygiene, 19*(4), 610–612. https://doi.org/10.4269/ajtmh.1970.19.610

Punekar, N. S., Suresh Kumar, S. V., Jayashri, T. N., & Anuradha, R. (2003). Isolation of genomic DNA from acetone-dried *Aspergillus mycelia*. *Fungal Genetics Newsletter, 50*, 15–16.

Schell, S. C. (Ed.). (1970). *How to know the trematodes*. WM. C. Brown.

Schmidt, G. D. (1986). *CRC handbook of tapeworm identification*. CRC Press.

Seinhorst, J. W. (1959). A rapid method for the transfer of nematodes from fixative to anhydrous glycerin. *Nematologica, 4*(1), 67–69. https://doi.org/10.1163/187529259X00381

Sepulveda, M. S., & Kinsella, J. M. (2013). Helminth collection and identification from wildlife. *Journal of Visualized Experiments, 82*, e51000. https://doi.org/10.3791/51000

Thavarajah, R., Mudimbaimannar, V. K., Elizabeth, J., Rao, U. K., & Ranganathan, K. (2012). Chemical and physical basics of routine formaldehyde fixation. *Journal of Oral and Maxillofacial Pathology, 16*(3), 400–405. https://doi.org/10.4103/0973-029X.102496

Trajan, A. C. (1973). A rapid lactophenol-glycerin method for preserving nematodes. *Proceedings of the Helminthological Society of Washington, 40*(2), 296–297.

Waikagul, J., Jongsuksantigul, P., Rattanawitoon, U., Radomyos, P., Kojima, S., & Takeuchi, T. (2008). Parasitological monitoring of helminth control program in Northern Thailand. *Southeast Asian Journal of Tropical Medicine and Public Health, 39*(6), 1008–1014.

Waikagul, J., & Thaenkham, U. (2014). Chapter 2 - Collection of fish-borne trematodes from the final host. In J. Waikagul & U. Thaenkham (Eds.), *Approaches to research on the systematics of fish-borne trematodes* (pp. 17–37). Academic Press. https://doi.org/10.1016/B978-0-12-407720-1.00002-9

Wanas, M. Q., Rashed, A. A., Al Shareef, A. M., Al Attar, M. N., & Abdalla, H. M. (1995). Morphological and anatomical studies on larval trematode of genus *Brachylaima* (Brachylaimidae) from the land snail, Monacha obstructa. *Journal of the Egyptian Society of Parasitology, 25*(2), 407–415.

Yoder, M., De Ley, I. T., King, I. W., Mundo-Ocampo, M., Mann, J., Blaxter, M., Poiras, L., & Ley, P. D. (2006). DESS: A versatile solution for preserving morphology and extractable DNA of nematodes. *Nematology, 8*(3), 367–376. https://doi.org/10.1163/156854106778493448

Zahabiun, F., Sadjjadi, S. M., & Esfandiari, F. (2015). Development of a double glass mounting method using formaldehyde alcohol Azocarmine Lactophenol (FAAL) and its evaluation for permanent mounting of small nematodes. *Iranian Journal of Parasitology, 10*(4), 617–624.

Chapter 10
Molecular Systematics of Parasitic Nematodes

Abstract The study of molecular systematics, where the concept was first introduced in Chap. 1, enables the elucidation of relationships among groups of organisms using molecular means. Molecular systematics allows us to unravel the evolutionary history embedded in the organism's genetic material by reconstructing phylogenetic trees, which illuminate the driving forces behind speciation and biodiversity. This chapter focused on the molecular systematics of parasitic nematodes that afflict humans and animals, where the phylum Nematoda is the second-largest phylum in the kingdom Animalia. Relationships among the major groups were detailed, as evidenced through molecular phylogenetic studies performed over the decades. A summary of the major findings, together with significant milestones in elucidating parasitic nematode phylogenetic relationships, was discussed. With such vast species diversity, coupled with the complex nature of host-parasite relationships, a robust phylogenetic framework through the study of nematode molecular systematics is essential to make meaningful comparisons across taxa, for classification and the generation of evolutionary hypotheses for the phylum.

10.1 The Molecular Phylogenetic Backbone of Nematodes and Their Classification

Traditional classification and systematics of nematodes are based on sharing key morphological characters at the various taxonomical hierarchy levels. However, this feat is hindered by nematodes' species and morphological diversity, causing difficulties in choosing reliable morphological characters (Perkins et al., 2011; Ricciardi & Ndao, 2015). Various proposals based on morphology have resulted, but none was as prominent as the division proposed by Chitwood (1937) (Chitwood, 1937; Lorenzen, 1994; Maggenti, 1983). The most well-known proposal was by Chitwood (1937), where they proposed dividing the phylum into subclass Adenophorea and Secernentea (Chitwood, 1937). The division was based on Adenophorea being "gland bearers," whereas Secernentea is "secretors." Adenophorea includes aquatic nematodes in the order Enoplida and Chromadorida, along with nematodes that are terrestrial omnivores or plant-feeders in the order Dorylaimia. Secernentea consists

U. Thaenkham et al., *Molecular Systematics of Parasitic Helminths*,
https://doi.org/10.1007/978-981-19-1786-8_10

of most parasitic species in the order Strongylida, Tylenchida, Ascaridida, and Spirurida, along with terrestrial, free-living nematodes belonging to Rhabditida (Anderson, 2009).

A significant milestone for nematode systematics occurred with the accessibility of molecular techniques. Blaxter et al. (1998) proposed a molecular phylogenetic framework for the phylum using the nuclear 18S ribosomal RNA (rRNA) gene sequences for 53 nematode species (Blaxter et al., 1998). Five clades resulted from the analysis, with clades I and II belonging to Adenophorea and clades III, IV, and V belonging to Secernentea. The molecular phylogenetic framework of nematodes revealed by Blaxter et al. (1998) has been the backbone for nematode molecular systematics. Over the years, a rise in the number of phylum-wide molecular analyses has been performed for nematodes using the 18S rRNA gene and genetic markers from the mitochondrial DNA and whole mitochondrial genomes. Another prominent milestone and progress was made by the International Helminth Genomes Consortium, where they published 31 new nematodes and 14 platyhelminth genomes and combined them with 36 other previously published genomes (International Helminth Genomes Consortium, 2019). The resulting phylogeny supports the division of the phylum into the five major clades as proposed by Blaxter et al. (1998). These molecular studies greatly increased the proportion and groups of species sampled and the number of available sequences in molecular databases. Other molecular genetic markers have also been employed, but they were chiefly performed at lower taxonomic levels (either at superfamily or family level) instead of phylum-wide analysis.

The following subsections will focus on the main nematode phylogenies obtained from the nuclear 18S rRNA gene and phylogenies obtained using either mitochondrial genomes or mitochondrial genetic markers. The clades will mostly follow the numberings based on Blaxter et al. (1998) for easy reference. Emphasis on the groups of medical importance to humans and animals will be placed in the context of their phylogenetic relevance and significance.

10.1.1 The Nuclear 18S rRNA Gene Phylogeny

The nuclear 18S rRNA gene phylogeny first proposed by Blaxter et al. (1998) divided the phylum into five clades using 53 sequences (Blaxter et al., 1998). Figure 10.1a shows the 18S rRNA gene phylogeny that was obtained. Clades I and II are in subclass Adenophorea (Blaxter et al., 1998; De Ley, 2006; De Ley & Blaxter, 2002; Dorris et al., 1999), where analysis indicates that Adenophorea might be paraphyletic. Clade I contain the vertebrate-parasitic Trichocephalida, the insect-parasitic Mermithida, the plant-parasitic Dorylaimida, and the free-living Monochida. Clade II contains the plant-parasitic Triplonchida and the free-living Enoplida. Clades III, IV, and V are in the monophyletic subclass Secernentea (Blaxter et al., 1998; De Ley, 2006; De Ley & Blaxter, 2002; Dorris et al., 1999). Clade III contains the vertebrate- and arthropod-parasitic taxa from four major

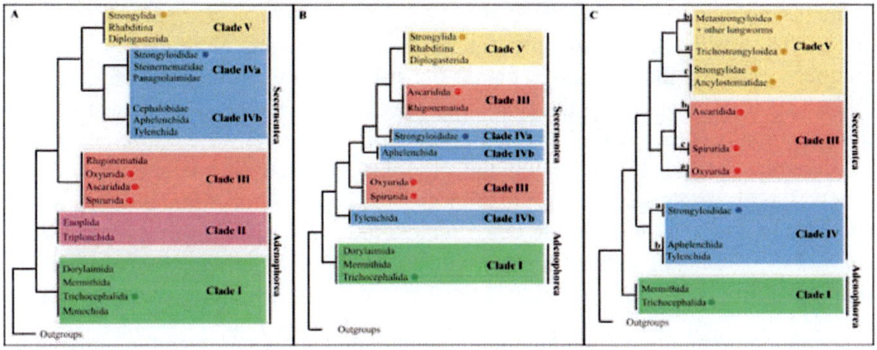

Fig. 10.1 Phylogeny of Nematoda with the (**a**) nuclear 18S rRNA gene, (**b**) mitochondrial genome, and (**c**) whole genome. The 18S rRNA phylogenetic tree was adapted from Blaxter et al. (1998) and De Ley and Blaxter (2002). The mitochondrial genome phylogeny was adapted based on a compilation of studies utilizing the complete mitochondrial genomes (Kang et al., 2009; Kern et al., 2020; Liu et al., 2013; Park et al., 2011; Sultana et al., 2013). The whole-genome phylogeny was adapted from the International Helminth Genomes Consortium (International Helminth Genomes Consortium, 2019) Parasitic nematode groups of medical importance to humans and animals are indicated in colored circles

orders—Ascaridida (large intestinal roundworms), Oxyurida (pinworms), Spirurida (filarial worms), and Rhigonematida (millipede parasites). Clade IV combines plant-parasitic Tylenchida, insect-parasitic *Steinernema*, free-living bacteriovores, and vertebrate-parasitic *Strongyloides*. Clade V contains the vertebrate-parasitic Strongylida, insect-parasitic Diplogasterida, and Heterorhabditis, along with the model organism *Caenorhabditis elegans* and suborder Rhabditina. For nematodes relevant to animals and humans, they are in clades I, III, IV, and V, consisting of orders Trichocephalida, Ascaridida, Oxyurida, Spirurida, and Strongylida, and *Strongyloides*. The proposed phylogeny is largely congruent with the traditional taxonomical classification based on morphology, showing the robustness of the nuclear 18S rRNA gene for phylum-wide analysis for nematodes (Blaxter et al., 1998). Additionally, the phylogeny obtained revealed the vast ecological habitats of nematodes and their capability to parasitize various groups of organisms.

Attempts to further refine the nematode phylogenetic backbone were performed, either through the addition of sequences or morphological/ecological interpretations. Holterman et al. (2006) expanded the number of taxa to 339 and suggested a division of the phylum into 12 clades (Holterman et al., 2006). These 12 clades correspond to the five clades previously proposed by Blaxter et al. (1998), except that there are additional clades due to the increase in taxa sampled. Additionally, the phylogeny revealed that Enoplida (part of clade II as proposed by Blaxter et al.) dominated the most basal clade, differing from Blaxter et al. (1998), where clades I and II were a polytomy at the base of the phylogenetic tree. Meldal et al. (2007) and van Megan et al. (2009) expanded the phylum-wide sampling of nematode species up to 1200 species to include more marine and free-living nematodes (Meldal et al., 2007; van Megen et al., 2009). They aimed to increase the resolution of nematode phylogeny by including more species and tried to resolve the inconsistencies involving the basal clades. The results were generally congruent with previous phylogenies, with the monophyly of the three parasitic clades consisting of human and animal parasites (clades III to V) (Meldal et al., 2007; van Megen et al., 2009). However, the basal relationships between clade I and clade II within Nematoda were still unclear. The relationships varied depending on the phylogenetic analysis method (Bik et al., 2010; Blaxter & Koutsovoulos, 2015; Kampfer et al., 1998; Meldal et al., 2007; van Megen et al., 2009). Despite being unable to resolve which was the most basal clade, the results still support the classification of Blaxter et al. (1998), where the orders within clade I (Dorylaimida, Trichocephalida, Mermithida, and Monochida) and clade II (Enoplida and Triplonchida) were well supported.

In addition to phylum-wide level studies, molecular phylogenies using the 18S rRNA gene were also conducted on specific nematode groups. Examples include studies on marine nematodes, animal-parasitic nematodes in clade III, and plant nematodes (Bert et al., 2008; Holterman et al., 2009; Nadler et al., 2007; Smythe et al., 2019). Molecular advances also brought about the sequencing and availability of transcriptomes and complete mitochondrial genomes for phylogenetic analysis of nematodes (Kang et al., 2009; Smythe et al., 2019; Sultana et al., 2013). The majority of the nematode phylum-wide studies were performed using mitochondrial genomes compared to transcriptomes, and thus, we will discuss the phylogenies obtained using mitochondrial genomes in the following subsection.

10.1.2 Mitochondrial Genome Phylogeny

Mitochondrial genomes can serve as an alternative molecular genetic marker at a different locus to aid in studying nematode molecular systematics, aside from the nuclear 18S rRNA gene (Kang et al., 2009; Sultana et al., 2013). Alternative revelations about the relationships among nematode clades were revealed using complete mitochondrial genomes. Figure 10.1b depicts the general topology using mitochondrial genomes. Inferences based on 25 species generally supported the clades proposed by Blaxter et al. (1998) using the 18S rRNA gene (Blaxter et al., 1998; Kang et al., 2009). However, some differences were observed using complete mitochondrial genomes compared to the 18S rRNA gene phylogenies (Kang et al., 2009; Kern et al., 2020; Sultana et al., 2013). Firstly, clade III, which consists of orders Oxyurida, Ascaridida, and Spirurida, was not recovered as monophyletic. Secondly, the phylogeny inferred supported a close relationship between Ascaridida and Rhabditida (clade V in Blaxter et al. (1998)). Oxyurida was recovered as a sister clade to the Ascaridida + Rhabditida clade. Spirurida, on the other hand, was placed as a sister to the remaining species (except *Trichinella*, which belongs in class Adenophorea). Thirdly, superfamily Strongyloidoidea (clade IV in Blaxter et al.) was not recovered as monophyletic, in contrast to the 18S phylogenies obtained. The species sampled using mitochondrial genomes for this superfamily consist of the insect-parasitic *Steinernema* and animal-parasitic *Strongyloides*.

More studies using an increased number of species were performed, with both complete mitochondrial genomes and partial genes from the mitochondrial DNA (Kern et al., 2020; Park et al., 2011; Sultana et al., 2013). Using 36 and 62 species, respectively, Park et al. (2011) and Liu et al. (2013) substantiated that the mitochondrial phylogenies do not support the monophyly of clade III and clade IV nematodes (Liu et al., 2013; Park et al., 2011). Additionally, depending on the phylogenetic analysis chosen, the phylogenetic relationships of Oxyurida and Spirurida differed. Similarly, using the partial sequences of the mitochondrial 12S and 16S rRNA genes, the non-monophyly of clade III nematodes was supported in a phylum-wide analysis of parasitic nematodes of medical importance to humans and animals (Chan et al. 2020b). However, despite the incongruencies between the mitochondrial and nuclear 18S rRNA gene phylogenies, these phylogenies obtained are an invaluable source of information that we can use for inferences of evolutionary hypotheses.

10.1.3 Whole-Genome Phylogeny

The nematode *C. elegans* was the first metazoan to have its genome sequenced in 1998, and *Brugia malayi* was the first parasitic nematode. Since then, many other genomes of nematodes have been published and sequenced (Howe et al., 2017; Kumar et al., 2012; Mitreva et al., 2005). The International Helminth Genomes Consortium did a study using 81 helminth genomes, combining 36 published genomes and 31 nematodes and 14 platyhelminth species that were newly produced (International Helminth Genomes Consortium, 2019). The phylogeny obtained for nematodes is depicted in Fig. 10.1c, where the main clades proposed by Blaxter et al. (1998) were supported. The main difference was in the phylogenetic relationships between the clades, where clade IV (comprising the animal-parasitic *Strongyloides*) is now sister to clades III and V.

Additionally, clades III and V were revealed to be more closely related, congruent with mitochondrial genome phylogenies but different from the nuclear 18S rRNA gene phylogenies (where clade V shares a closer relationship with clade IV as compared to clade III). Clades III and V were subdivided into three subclades each, based on their phylogenetic relationships, as indicated by the alphabets a to c.

With the increase in the speed and convenience of using next-generation sequencing techniques, the number of available nematode sequences will inevitably increase. Not only will partial genes and complete mitochondrial genomes be useful, but the sequencing of whole genomes will be a linchpin in shedding light on phylogenetic relationships within Nematoda. The following section will focus on the molecular systematics studies of the four main clades in phylum Nematoda that contain parasitic nematodes of medical importance to humans and animals.

10.2 The Four Parasitic Clades of Medical Importance Within Phylum Nematoda

As we introduced the nematodes that are of medical importance and their taxonomic classification in Chap. 2, those species can be classified based on the molecular phylogenetic framework for nematodes in clades I, III, IV, and V (Blaxter et al., 1998). In the following section, we will go through the four clades in detail, with respect to some of the molecular studies conducted, including the key findings and the relationships among species within the clades.

10.2.1 Clade I

Clade I nematodes consist of orders Trichocephalida (also known as Trichinellida in some classifications), Mermithida, Dorylaimida, and Monochida. Of the four orders within clade I, Trichocephalida contains medically important nematodes to humans and animals. Trichuridae, Capillariidae, and Trichinellidae are the three families that will be discussed further.

Trichuridae

Trichuris is the type genus within family Trichuridae, and they are nematodes that inhabit the caecum of mammals (Anderson, 2009). *Trichuris* have a cosmopolitan distribution, and they parasitize a wide range of mammalian hosts, ranging from rodents, ruminants, marsupials, primates, and humans (Robles Mdel et al., 2014). *Trichuris trichiura*, also commonly known as the human whipworm, is of particular importance to humans as they are one of the soil-transmitted helminths affecting billions of people worldwide (WHO, 2020). Currently, up to 60 species of *Trichuris* have been described, and they are usually identified based on the species of host and some morphological features of adults (Callejon et al., 2015). However, due to the lack of reliable morphological characters, species identification of *Trichuris* has heavily relied on their host origin.

Molecular genetic markers have been utilized to characterize *Trichuris* species to aid species identification and resolve phylogenetic relationships. Examples include the molecular identification of *Trichuris* from various non-human primate hosts, investigating whether *Trichuris* obtained from various hosts are indeed different species and assessing the correlation of host specificity and phylogenetic relationships within *Trichuris* (Callejon et al., 2015; Guardone et al., 2013; Hawash et al., 2016; Liu et al., 2012; WHO, 2020).

Using complete mitochondrial genomes, primate-derived *Trichuris*, pig-derived *Trichuris*, ruminant-derived *Trichuris*, and rodent-derived *Trichuris* were clearly distinguished into their respective clades based on host species (Hawash et al.,

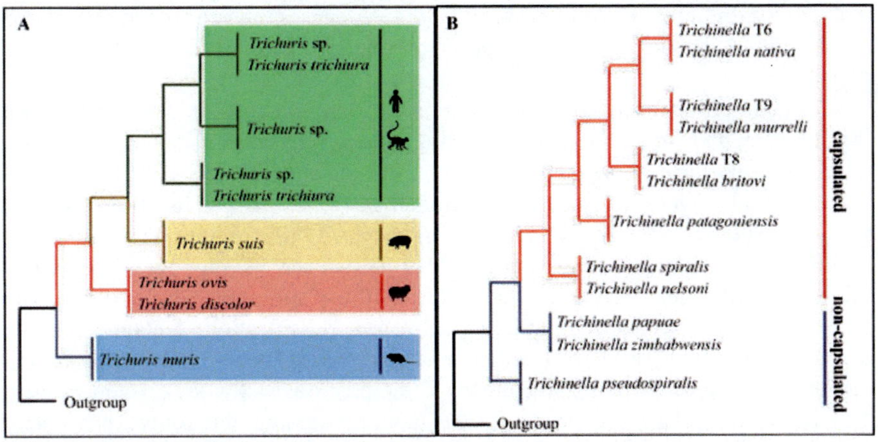

Fig. 10.2 Phylogeny of species within (**a**) *Trichuris* and (**b**) *Trichinella*, adapted from various genetic markers and publications (Callejon et al., 2015; Guardone et al., 2013; Hawash et al., 2015; Mohandas et al., 2014; Pozio & Zarlenga, 2005; Rivero et al., 2021; Zarlenga et al., 2006). In (**a**), the colors indicate *Trichuris* sp. and their respective hosts (green, human and non-human primates; yellow, pig; red, sheep; blue, rodent). In (**b**), the division of *Trichinella* into the capsulated and non-encapsulated clade is indicated in the red and blue lines, respectively

2015). Figure 10.2a depicts the phylogeny that was obtained. Additionally, the phylogeny obtained supports the close relationship between *T. trichiura* and other *Trichuris* sp. obtained from primates. Rivero et al. (2021) further divided the clade consisting of the primate-derived *Trichuris* into three subclades, where clade 1 contains *T. trichiura* from humans, clade 2 contains *Trichuris* sp. from non-human primates, and clade 3 contains *T. trichiura* from humans and non-human primates (Rivero et al., 2021). The study provides evidence of the close relationship between primate and human-derived *Trichuris* and the possibility of cryptic species. Aside from complete mitochondrial genomes, partial mitochondrial and nuclear genes have also been utilized for molecular phylogenetic studies of *Trichuris*. Support for the phylogenetic division based on host species was obtained using the combined genetic markers of the 18S rRNA, *Triose phosphate isomerase*, cytochrome c oxidase subunit I (*COI*), and cytochrome b (*cytB*) genes (Callejon et al., 2015). Additionally, the use of molecular genetic markers such as the nuclear internal transcribed spacer 2 (ITS2) region has also revealed new species of *Trichuris* in small mammals and rodents (Eberhardt et al., 2019; Robles Mdel et al., 2014).

Capillariidae

Their thin, threadlike body morphologically characterizes Capillariidae along with a long esophagus comprising a short, muscular, and long glandular part (Anderson, 2009; Sakaguchi et al., 2020). An estimated 390 species have been recorded from various hosts such as fishes, amphibians, birds, mammals, and humans (Sakaguchi et al., 2020). In humans, *Capillaria philippinensis* causes intestinal capillariasis and is transmitted through the consumption of undercooked freshwater fish (CDC, 2020a).

Molecular phylogenetic analysis supports the close relationship between families Capillariidae and Trichuridae and is congruent with taxonomical classification. Within Capillariidae, Borba et al. (2019) used the nuclear 18S rRNA gene and mitochondrial *COI* gene to infer phylogenetic relationships for *Aonchotheca*, *Pseudocapillaria*, *Baruscapillaria*, *Pearsonema*, *Paratrichosoma*, *Calodium*, *Capillaria*, and *Eucoleus* (Borba et al., 2019). However, the phylogenies obtained did not provide a clear picture of relationships within Capillariidae due to unresolved genera and conflicts from morphological-based classification. Sakaguchi et al. (2020) also used the 18S rRNA gene to examine phylogenetic relationships among Capillariidae for six genera, and contrary to the previous study, the six genera were well resolved (Sakaguchi et al., 2020). Notwithstanding the robust phylogeny obtained, the authors acknowledged that the coverage of phylogenetic studies within Capillariidae is minimal and more studies should be performed using a greater number of species.

Aside from phylogenetic studies, species identification using molecular genetic markers has also been performed. *C. philippinensis* was detected in human fecal samples using a species-specific nested PCR assay with the 18S rRNA gene (El-Dib et al., 2015). Also, *Capillaria* in canine and feline fecal samples were detected using the *COI* gene in a semi-nested PCR assay (Di Cesare et al., 2012).

Trichinellidae

Family Trichinellidae consists of the genus *Trichinella*, which is characterized by having an infective first stage larva and having two generations in the same host species (Anderson, 2009). The life cycle of *Trichinella* is maintained in animal hosts, and their natural reservoirs are wild carnivores and omnivores (Pozio et al., 2009). Human infection of *Trichinella* can occur when the infected tissues are ingested, and risk factors for transmission include a reduction in the interface between humans and animals, resulting in increased human-animal contact (Pozio et al., 2009). Based on classical taxonomy, *Trichinella* is divided into two main clades based on the presence or absence of a host-derived collagen sheath around the infected muscle cell. They are also known as the encapsulated clade and the non-encapsulated clade. To date, six species and three genotypes have been recognized in the encapsulated clade: *Trichinella spiralis*, *Trichinella nativa*, *Trichinella britovi*, *Trichinella murelli*, *Trichinella nelsoni*, *Trichinella patagoniensis*, T6, T8, and T9. *Trichinella pseudospiralis*, *Trichinella papuae*, and *Trichinella zimbabwensis* make up the non-encapsulated clade (Zarlenga et al., 2020).

Without molecular tools, identification of *Trichinella* to the species level is not possible, as they can only be distinguished on whether they are encapsulated or not encapsulated (Mohandas et al., 2014). Using molecular genetic markers such as the nuclear 18S rRNA gene, ITS2 region, complete mitochondrial genomes, *COI*, and

16S rRNA genes, the molecular phylogenies obtained supports a clear distinction of the two clades within *Trichinella*, consistent with morphology (Fig. 10.2b) (Mohandas et al., 2014; Pozio & Zarlenga, 2005; Zarlenga et al., 2006). The non-encapsulated clade was also phylogenetically placed at the base of the tree. Additionally, among the non-encapsulated species, *T. pseudospiralis* (infecting mammals and birds) was separated from *T. zimbabwensis* and *T. papuae* (both infecting mammals and reptiles).

10.2.2 Clade III

Blaxter et al. (1998) collectively placed Oxyurida, Ascaridida, Spirurida, and Rhigonematida into a monophyletic clade III using the 18S rRNA gene. Of the four orders, Oxyurida, Ascaridia, and Spirurida contain nematodes of medical importance to animals and humans. These three orders have a large number of nematodes, with a wide range of life cycle patterns and host types.

As discussed in Sect. 10.1, differences in the phylogenies resulted when mitochondrial genomes and mitochondrial genes were used. Mitochondrial phylogenies never recovered clade III as monophyletic, unlike the 18S rRNA gene phylogenies. Due to the incongruences in the phylogenies obtained from different genetic markers, researchers have undertaken various studies to take a more in-depth look within clade III nematodes (Fig. 10.3).

Oxyurida

Oxyurida, also commonly known as pinworms, are a group of nematodes that parasitize the cecum and colon of mammals and humans (Anderson, 2009). They have a simple direct life cycle, requiring a single definitive host. In humans, *Enterobius vermicularis* causes disease, and transmission occurs through direct contact of eggs to humans (CDC, 2020b).

Within clade III, the phylogenetic position of Oxyurida differed depending on the genetic marker used and the type of analysis. In some 18S rRNA gene phylogenies, Oxyurida shared a closer relationship to superfamily Heterakoidea (in Ascaridida) (Nadler et al., 2007). Additionally, traditional classification based on morphology has also supported the close relationship between Oxyurida and Heterakoidea. In another 18S rRNA gene phylogeny, De Ley and Blaxter (2002) showed that Oxyurida was sister to the clade consisting of Ascaridida + Spirurida (De Ley & Blaxter, 2002). Also, using mitochondrial genomes, Oxyurida was sister to the clade consisting of Rhabditida + Ascaridida, with Spirurida at the base relative to Oxyurida (Liu et al. 2016a; Wang et al., 2016; Zhang et al., 2015).

Within Oxyurida, differences in relationships were also observed. Based on six species in Oxyurida, Wang et al. (2016) showed that the clade of *Syphacia obvelata* + *Aspiculuris tetraptera* shares a close relationship to the clade of

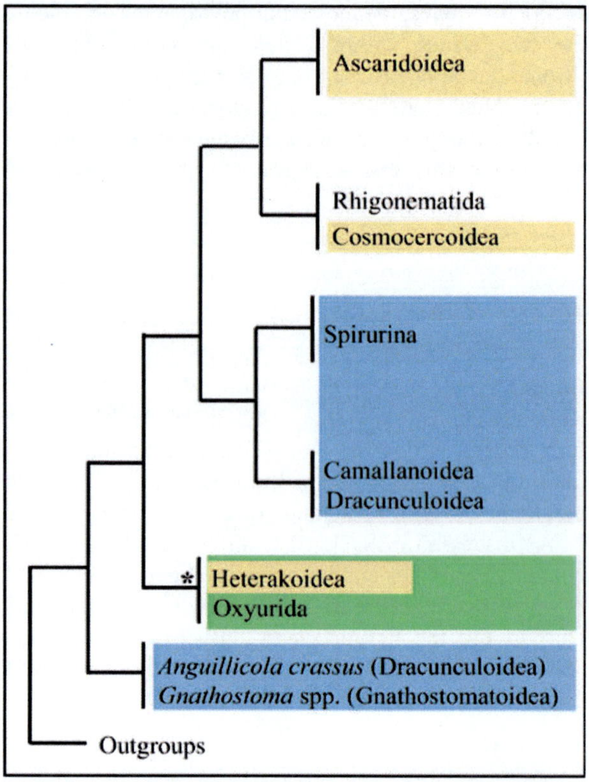

Fig. 10.3 Phylogeny adapted from Nadler et al. (2007) (Nadler et al., 2007) using the 18S rRNA gene showing (1) the sister relationship between Oxyurida and Heterakoidea (highlighted in green), (2) the non-monophyly of Ascaridida (highlighted in yellow), and (3) the non-monophyly of Spirurida (highlighted in blue). An asterisk "*" indicates that the clade position differed according to the method of phylogenetic analysis

Passalurus ambiguous + *Oxyuris equi* and *Enterobius vermicularis* is sister to these two clades (Wang et al., 2016). However, in concatenated phylogenies of the nuclear 18S rRNA, 28S rRNA, and the mitochondrial *COI* gene, the clade consisting of *Oxyuris equi* was sister to *Syphacia* + *Enterobius* (Li et al., 2019). Despite the differences in phylogenies concerning the phylogenetic position of Oxyurida within clade III and Oxyurida, a limiting factor could be the small number of species that have been sequenced to date. There are only six mitochondrial genomes publicly available for Oxyurida, and a limited number of studies have been performed for this order. The availability of more species could perhaps aid in resolving the inconsistencies for Oxyurida.

Ascaridida

With more than 800 described species that parasitize all the major groups of vertebrates, order Ascaridida contains important species of medical importance to humans and animals (Anderson, 2009). Superfamily Ascaridoidea, a superfamily within Ascaridida, contains the most medically important nematode species. These families include Ascarididae, Toxocaridae, and Anisakidae.

From the 18S rRNA gene phylogenies, Ascaridida belonged to a monophyletic group with Oxyurida, Spirurida, and Rhigonematida. Ascaridida shared a sister relationship with Spirurida + Rhigonematida, with Oxyurida sister to Ascaridida + Spirurida + Rhigonematida (Blaxter et al., 1998; De Ley, 2006; De Ley & Blaxter, 2002). Contrarily, mitochondrial genome phylogenies revealed that Ascaridida was sister to Rhabditida, resulting in the non-monophyly of the orders in clade III (Kang et al., 2009; Kern et al., 2020; Liu et al., 2013; Park et al., 2011; Sultana et al., 2013). However, the monophyly of the orders was supported in the phylogenies from both the 18S rRNA gene and mitochondrial genome. With the increased number of taxa, Nadler et al. (2007) sampled 113 taxa in clade III using the 18S rRNA gene to produce a more representative phylogeny. The results revealed that Ascaridida (superfamilies Ascaridoidea and Heterakoidea were sampled) was not monophyletic. Heterakoidea was instead either sister to Oxyurida or had Rhigonematida, and Cosmocercoidea nested between Heterakoidea and Ascaridoidea, dependent on the method of analysis (Nadler et al., 2007).

Within Ascaridoidea, the phylogenetic positions of the families have also been of considerable debate. Using mitochondrial genomes, Liu et al. (2016b) showed that Ascarididae and Anisakidae share a sister relationship, and Toxocaridae was sister to the Ascarididae + Anisakidae clade (Liu et al., 2016b). However, other phylogenies using mitochondrial genomes support the closer relationship of Ascarididae + Toxocaridae, with Anisakidae now sister to the Ascarididae + Toxocaridae clade (Meng et al., 2019). The nuclear 18S rRNA gene phylogenies also showed differences in relationships among Ascaridoidea, with some depicting a closer relationship of Ascarididae + Toxocaridae and some depicting Ascarididae + Anisakidae (Li et al., 2012). Using concatenated genes from five nuclear and three mitochondrial genes from six families within Ascaridoidea, the final phylogeny revealed a closer relationship between Ascarididae and Toxocaridae, with Anisakidae as the basal clade with respect to the clade of Ascarididae + Toxocaridae + Acanthochelidae + Raphidascarididae (Li et al., 2018b). Although the relationships with Ascaridoidea varied among studies, the potential of utilizing concatenated genetic markers from two different loci can be considered to increase phylogenetic resolution.

Spirurida

Order Spirurida consists of nematodes that parasitize terrestrial vertebrates and aquatic vertebrates. They are highly diverse, with life cycles that utilize various host types and transmission routes (Anderson, 2009). Within Spirurida, they are divided into suborders Spirurina and Camallanina (Wijova et al., 2006). Spirurina comprises ten superfamilies, and of note are superfamilies Filarioidea, Gnathostomatoidea, Spiruroidea, and Thelazoidea. Camallanina comprises superfamilies Camallonoidea and Dracunculoidea.

Based on the 18S rRNA gene phylogeny by Blaxter et al. (1998), Spirurida is monophyletic and robustly supported. Spirurida was depicted to share a close relationship with Ascaridida (Blaxter et al., 1998). Focusing solely on clade III nematodes, Nadler et al. (2007) revealed the non-monophyly of Spirurida (Nadler et al., 2007). Three species of *Gnathostoma* (Gnathostomatoidea) and *Anguillicola crassus* (Dracunculoidea) were not placed together with the rest of Spirurida (comprising of Spirurina, Dracunculoidea, and Camallanoidea). Further evidence for the non-monophyly of Spirurida was also revealed using a larger dataset of 18S rRNA gene sequences, where, again, *Gnathostoma* and *Anguillicola* were both placed at the base of the phylogenetic tree, sister to the other clade III nematodes (Laetsch et al., 2012; Wijova et al., 2006). The other sampled species belonging to Spirurida formed a monophyletic clade sister to Ascaridida. Using mitochondrial genomes and the partial 12S and 16S rRNA genes, Spirurida was also not recovered as monophyletic, with *Gnathostoma* placed separately from the rest of Spirurida (Chan et al., 2020b; Liu et al., 2015; Zou et al., 2017). Zou et al. (2017) sampled 22 species belonging to nine families within Spirurida, and the results revealed two main groups within Spirurida that did not conform to the traditional morphological division (Zou et al., 2017). Based on mitochondrial genomes, the first group includes families in suborder Spirurina (Onchocercidae, Setariidae, Gongylonematidae, Thelaziidae, and Physalopteridae), whereas the second group includes families from both Spirurina (Gnathostomatidae) and Camallanina (Dracunculidae, Philometridae, and Camallanidae) suborders.

The complexities surrounding the molecular systematics of clade III nematodes leave room for further research. An approach that holds promise could be in the utility of different loci to resolve the incongruences and may provide a complete picture for phylogenetic relationships among clade III nematodes.

10.2.3 Clade IV

Clade IV, according to the classification from Blaxter et al. (1998), consists of the plant-parasitic Tylenchida, insect-parasitic *Steinernema*, free-living bacterivores, and vertebrate-parasitic *Strongyloides* (Blaxter et al., 1998). Although Blaxter et al. (1998) showed the monophyly of clade IV, other mitochondrial phylogenies could not fully resolve the relationships within them. Using complete mitochondrial genomes, the monophyly of clade IV nematodes was not supported, with a clade comprising of representatives belonging to Tylenchida (Meloidogynidae + Hoplolaimidae + Pratylenchidae) and another clade consisting of *Strongyloides*, *Steinernema*, and Aphelenchoididae (Hu et al., 2003; Kern et al., 2020; Kim et al., 2015; Sun et al., 2014). Additionally, *Strongyloides* and *Steinernema*, both belonging to superfamily Panagrolaimoidea, were mostly not nested together according to mitochondrial genomes (Kikuchi et al., 2016). Sultana et al. (2013) showed that

Bursaphelenchus, belonging to the superfamily Aphelenchoidea, was nested between *Strongyloides* and *Steinernema*. Previous nuclear phylogenies also revealed the unstable relationship between *Steinernema* and *Strongyloides*, showing different sister relationships with respect to Panagrolaimoidea (Holterman et al., 2006; Nadler et al., 2006).

Strongyloididae

In clade IV, *Strongyloides* is within the family Strongyloididae and contains medically important species to humans and animals. More than 50 species parasitize wild animals, including birds; reptiles; amphibians; livestock, domesticated companion animals, including dogs and cats; and humans (Nagayasu et al., 2017). The life cycle of *Strongyloides* alternates between free-living and parasitic generations, with the parthenogenetic female living in the gastrointestinal tract of hosts (Dorris et al., 2002). *Strongyloides stercoralis*, *Strongyloides fuelleborni fuelleborni*, and *Strongyloides fuelleborni kelleyi* are the three species infecting humans.

Without molecular tools, species discrimination of the rhabditiform and filariform larvae of *Strongyloides* based on morphology can be challenging. The nuclear 18S rRNA gene was one of the first genetic markers to be used for species discrimination of *Strongyloides*, where Hasegawa et al. (2009) revealed the utility of the hypervariable regions in the 18S rRNA genes (Hasegawa et al., 2009). Aside from *Strongyloides*, *Parastrongyloides* and *Rhabditophanes* are closely related. Based on life cycle traits, *Strongyloides* and *Parastrongyloides* are more closely related than *Rhabditophanes*, due to the former having both free-living and parasitic generations, whereas *Rhabditophanes* are free-living (Dorris et al., 2002). Molecular phylogenetics using whole genomes, 18S rRNA gene, complete mitochondrial genomes, and the *COI* gene also revealed congruence in relationships between the three genera (Fig. 10.4) (Hasegawa et al., 2009; Hunt et al., 2016; Kikuchi et al., 2016; Ko et al., 2019). Within *Strongyloides*, two distinct clades result, with *S. stercoralis* and *Strongyloides ratti* in one clade while *Strongyloides venezuelensis*, *Strongyloides papillosus*, and *Strongyloides vituli* in another.

Additionally, studies focusing solely on *S. stercoralis* using the *COI* gene revealed a division into two clades, with the first clade comprising *S. stercoralis* from humans and dogs and the second clade consisting of solely *S. stercoralis* from dogs (Nagayasu et al., 2017). The study also revealed that the closest species to *S. stercoralis* was *Strongyloides procyonis*, a species found in raccoons. The findings possibly indicate that *S. stercoralis* originally evolved as a parasite of canines and then later spread to humans.

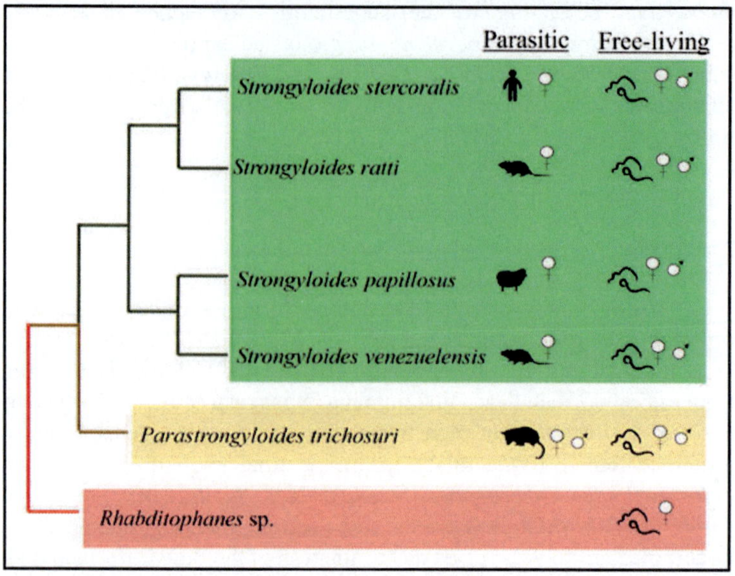

Fig. 10.4 Phylogeny of *Strongyloides* and related nematodes, showing parasitic and free-living characteristics (Hunt et al., 2018; Hunt et al., 2016). The male and female symbols indicate the sexes with parasitic or free-living stages, while the hosts are shown in black icons. For example, for *S. stercoralis*, both males and females are found in free-living stages, while only females are found parasitizing humans. The colors indicate the different genera

10.2.4 Clade V

Based on Blaxter et al. (1998) classification, clade V brings together the vertebrate-parasitic Strongylida, insect-parasitic Diplogasterida, and Heterorhabditis, along with the model organism *C. elegans* and suborder Rhabditina (Blaxter et al., 1998). Among them, order Strongylida contains medically important nematodes to humans and animals. Within Strongylida, based on morphological characters such as the mouth and male caudal bursa and their habitat in definitive hosts, they can be further divided into four suborders (Chilton et al., 2006; Gouy de Bellocq et al., 2001). The four suborders are Trichostrongylina, Metastrongylina, Strongylina, and Ancylostomatina. Hypothesis based on morphological classification places Metastrongylina as the most basal suborder relative to the other three orders, where nematodes belonging to Metastrongylina reside in the lungs of definitive hosts and the other three orders in the gastrointestinal tract (Chilton et al., 2006). Additionally, the reduced bursa rays in males of Metastrongylids and reduced ovijector further supports this hypothesis.

Molecular phylogenies using the nuclear 18S and 28S rRNA genes support the basal positioning of Metastrongylina, with Strongylina and Trichostrongylina as sister clades, congruent with morphology (Chilton et al., 2006). On the other hand, the whole-genome phylogeny shows a closer relationship between Trichostrongylina and Metastrongylina, as compared to Strongylina and Ancylostomatina (International Helminth Genomes Consortium, 2019). Additionally, mitochondrial genome phylogenies also revealed inconsistencies of relationships within Strongylida, where Liu et al. (2013) showed that Metastrongylina and Trichostrongylina are sister clades, in contrast to Jex et al. (2010) where Trichostrongylina showed a closer relationship to Strongylina and Ancylostomatina (Jex et al., 2010; Liu et al., 2013). Additionally, the monophyly of the suborders was not recovered in some studies and is dependent on the species sampled. We will take a further look at each of the four suborders in the paragraphs below.

Trichostrongylina

Nematodes in Trichostrongylina are major parasites of veterinary importance as they parasitize the gastrointestinal tracts of ruminants and livestock (Hoste et al., 1998; Paguem et al., 2020). In particular, species belonging to *Haemonchus*, *Cooperia*, *Teladorsagia*, and *Trichostrongylus* are notable genera. Additionally, three species of Trichostrongylus have been reported to cause infection in humans (*Trichostrongylus orientalis*, *Trichostrongylus colubriformis*, and *Trichostrongylus axei*) (Paguem et al., 2020).

Molecular studies on Trichostrongylina have mainly focused on species identification and detection for disease diagnosis, instead of molecular phylogenetics, probably due to the need for accurate species detection and disease diagnosis for the control and intervention of livestock parasites. The ITS regions are popular genetic markers that have been used for species identification of trichostrongylids (Hoste et al., 1998; Paguem et al., 2020; Sharifdini et al. 2017a, b). Within Trichostrongylina, families include Dictyocaulidae, Mollineidae, Haemonchidae, Trichostrongylidae, and Cooperiidae. Based on molecular phylogenies from whole genomes, mitochondrial genomes, and nuclear genes, Dictyocaulidae showed a closer relationship with Metastrongylina than with the rest of the families within Trichostrongylina (Chilton et al., 2006; International Helminth Genomes Consortium, 2019; Sun et al., 2018). Dictyocaulidae, more commonly known as large lungworms, are classified within Trichostrongylina. In contrast, other lungworms belonging to families Angiostrongylidae and Metastrongylidae are in suborder Metastrongylina. Thus, it might be no surprise that Dictyocaulidae and metastrongylids share a closer phylogenetic relationship. Within Trichostrongylina, the non-monophyly of Trichostrongylidae has also been observed using mitochondrial genomes and nuclear genes, with species grouping with both Cooperiidae and Haemonchidae (Ahmad et al., 2019; Chilton et al., 2006; Sun et al., 2018). However, it is important to note that the discrepancies in phylogenetic relationships could be affected by the limited number of sequences available for this group of nematodes.

Metastrongylina

Metastrongylids can parasitize various mammalian hosts, and humans are no exception. Based on habitat, nematodes belonging to suborder Metastrongylina parasitize the lungs of mammal definitive hosts (Carreno & Nadler, 2003). They can also be distinguished morphologically from other strongylids with reduced male copulatory bursa. *Angiostrongylus cantonensis*, the rat lungworm, causes eosinophilic meningitis in humans, where humans are accidental dead-end hosts (Barratt et al., 2016).

Carreno & Nadler (2003) performed phylogenetic analysis for seven families within Metastrongylina and could not resolve all using the nuclear rRNA genes (Carreno & Nadler, 2003). Similarly, Eamsobhana et al. (2015) showed that five out of the seven families were not recovered as monophyletic, only obtaining the monophyly of Metastrongylidae and Crenosomatidae (Eamsobhana et al., 2015). For mitochondrial genomes, limited species have been sequenced, and phylogenetic analysis usually only includes species belonging to Angiostrongylidae and Metastrongylidae, and thus no comparison can be inferred (Li et al. 2018a; Yong et al., 2016). Within Angiostrongylidae, mitochondrial and nuclear phylogenies support the sister relationship between *A. cantonensis* and *Angiostrongylus malaysiensis*, with *Angiostrongylus vasorum* and *Angiostrongylus costaricensis* sister to the former (Chan et al., 2020a; Dusitsittipon et al., 2018; Eamsobhana et al., 2015; Yong et al., 2016). More recently, the mitochondrial genome of *Angiostrongylus mackerrase*, which is part of the *A. cantonensis* lineage (together with *A. cantonensis* and *A. malaysiensis*), was sequenced, showing the close relationship between these three morphologically similar species (Valentyne et al., 2020). Similarly, using the mitochondrial rRNA genes, the three species within *A. cantonensis* lineage formed a distinct clade, supporting the close relationship between them (Chan et al., 2020a).

Strongylidae

Family Strongylidae, which is in suborder Strongylina, are a diverse group of nematodes parasitizing the gastrointestinal tract of livestock, such as horses and cattle (Hung et al., 2000). Morphological classification divides Strongylidae into two subfamilies—Strongylinae and Cyathostominae, based on the size and shape of buccal capsules. Species with subglobular or funnel-shaped buccal capsules are classified under Strongylinae, whereas species with cylindrical or ring-shaped buccal capsules are under Cyathostominae (Hung et al., 2000). However, molecular phylogenies do not support the division based on morphology. Using the internal transcribed spacer 1 (ITS1) and ITS2 regions for 30 species, species in the two subfamilies were never recovered as monophyletic regardless of phylogenetic tree building methods (Hung et al., 2000). *Craterostomum acuticaudatum*, *Triodontophorus serratus*, and *Oesophagostomum robustus*, in Strongylina based on their subglobular buccal capsules, were phylogenetically placed with

Cyathostominae. Mitochondrial genome phylogenies also corroborate with the results obtained using the nuclear spacer regions, where monophyly of both subfamilies was not obtained (Gao et al. 2017a, b; Hu et al., 2020). *Triodontophorus* was placed within Cyathostominae instead of having a closer relationship to *Strongylus*.

Ancylostomatidae

Nematodes belonging to Ancylostomatidae, more commonly recognized as hookworms, parasitize mammals including dogs, cats, pigs, ruminants, and humans (Blaxter, 2000; Wang et al., 2012). Residing in the gastrointestinal tract of hosts, Ancylostomatidae has a direct life cycle and can get transmitted either through oral ingestion or skin penetration. Infection by *Necator americanus*, *Ancylostoma duodenale*, and *Ancylostoma ceylanicum* causes disease in humans, infecting over one billion people yearly (Blaxter, 2000).

Ancylostomatidae shares a sister clade relationship to Strongylidae, with Metastrongylina and Trichostrongylina, placed further away based on phylogenetic relationships (Chilton et al., 2006). However, Ancylostomatidae was not recovered as monophyletic from nuclear rRNA genes and mitochondrial genomes. Some mitochondrial genome phylogenies revealed that *Ancylostoma* had a closer relationship to members of Strongylidae, while *Bunostomum* and *Necator* formed another clade (Jex et al., 2010). According to the genetic marker and analysis method, the relationship among the three genera within Ancylostomatidae differed. Using the ITS regions and some mitochondrial genomes, phylogenies revealed that *Ancylostoma* is sister to *Bunostomum*, with *Necator* sister to *Ancylostoma* + *Bunostomum* (Blaxter, 2000; Jex et al., 2010; Wang et al., 2012). However, some mitochondrial genome phylogenies showed a closer relationship between *Ancylostoma* and *Necator* than *Bunostomum* (Gao et al., 2014; Shi et al., 2018).

Aside from molecular phylogenetic studies, many molecular-based studies on clade V nematodes have focused on species identification and disease detection. The focus is partly due to the similar morphological features of the eggs of clade V nematodes in fecal samples, which hinder accurate species identification, and within some groups, species identification is not possible. Proper species identification is without a doubt intricately tied to successful disease diagnosis and administering appropriate treatments and control measures, especially among livestock.

10.3 Discordances and Improvements for Nematode Molecular Systematics

Decoding the complexities behind the relationships within phylum Nematoda is no easy feat. As seen in Sections 10.1 and 10.2, discordances between nuclear and mitochondrial phylogenies are aplenty between the orders and at lower taxonomic levels within families and subfamilies. Here, we summarize the main disagreements between the nuclear and mitochondrial phylogenies and provide an alternative proposition for increased phylogenetic resolution for parasitic nematodes.

10.3.1 Discordances Between Nuclear and Mitochondrial Phylogenies

Firstly, of the four clades containing parasitic nematodes of medical importance to humans and animals, clade III, containing orders Oxyurida, Ascaridida, and Spirurida, have been particularly challenging. Clade III was recovered as monophyletic in nuclear phylogenies, whereas mitochondrial phylogenies never recovered the monophyly of clade III. In mitochondrial phylogenies, Ascaridida shared a close relationship with clade V nematodes, whereas the Oxyurida + Spirurida clade is placed sister to clade IV nematodes (Kang et al., 2009; Kern et al., 2020; Liu et al., 2013; Park et al., 2011; Sultana et al., 2013). The phylogenetic position of Oxyurida and Spirurida also differed, depending on the taxa represented and the type of phylogenetic analysis. Additionally, within each of the three orders, sampling of more taxa indicated the non-monophyly of Ascaridida and Spirurida using both nuclear and mitochondrial genetic markers. Of note, using the nuclear 18S rRNA gene, the inclusion of superfamily Heterakoidea revealed the non-monophyly of Ascaridida, where the position of Heterakoidea differed depending on the analysis method (Nadler et al., 2007). Another example is with Gnathostomatoidea, a superfamily in Spirurida, where Gnathostomatoidea was consistently not nested with the rest of Spirurida (Chan et al. 2020b; Laetsch et al., 2012; Liu et al., 2015; Nadler et al., 2007; Wijova et al., 2006; Zou et al., 2017).

Secondly, clade IV, consisting of the plant-parasitic Tylenchida, insect-parasitic *Steinernema*, free-living bacterivores, and vertebrate-parasitic *Strongyloides* according to Blaxter et al. (1998), was recovered as monophyletic (Blaxter et al., 1998). Other nuclear phylogenies also recovered clade IV as monophyletic (De Ley & Blaxter, 2002). Contrastingly, mitochondrial phylogenies were unable to resolve the relationships among them. Two clades resulted, with one sharing a closer relationship to the Ascaridida + clade V nematodes and another placed at the basal of the class Secernentea (sister to the rest of Secernentea) with Spirurida and Oxyurida nested in between the two clades. The first clade comprises nematodes belonging to *Strongyloides*, *Steinernema*, and Aphelenchoididae, whereas the second clade comprises Tylenchida representatives of Meloidgynidae + Hoplolaimidae

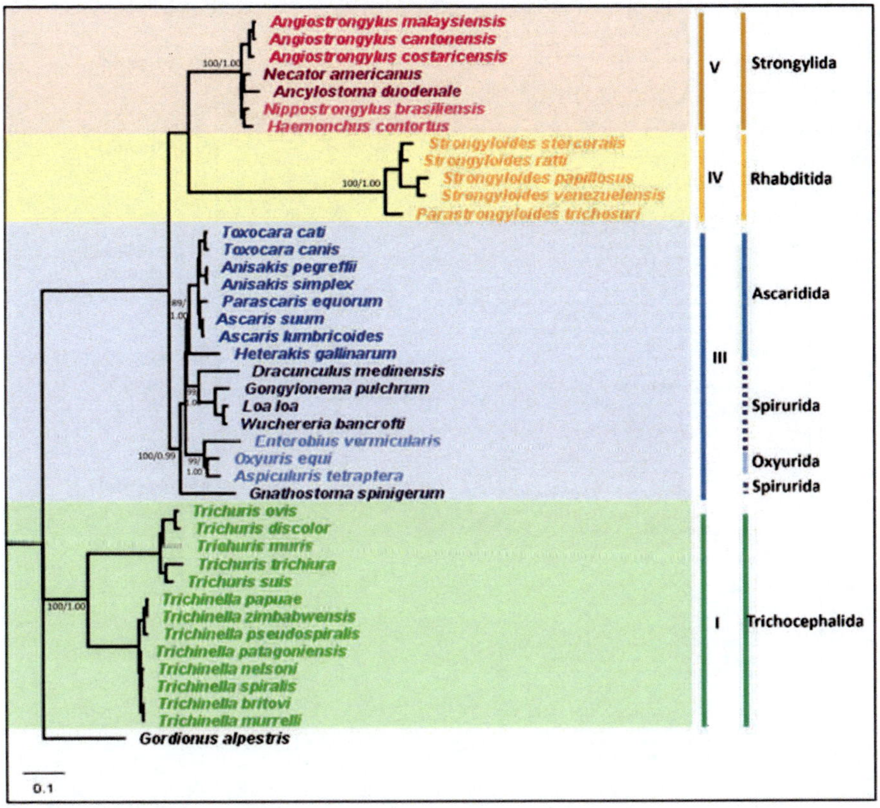

Fig. 10.5 Molecular phylogeny of nematodes inferred with Bayesian inference and maximum likelihood using the combined 12S and 18S rRNA genes

+ Pratylenchidae (Hu et al., 2003; Kern et al., 2020; Kim et al., 2015; Sun et al., 2014).

10.3.2 Utilizing the Concatenated Sequences of the Nuclear 18S and Mitochondrial 12S rRNA Genes

Utilizing genetic markers from two different loci can potentially be helpful to resolve discordances between nuclear and mitochondrial phylogenies, serving as alternative genetic markers. Additionally, the phylogenies obtained through the concatenation of two loci provide more phylogenetically informative sites and can generally improve phylogenetic resolution. Here, in an attempt to improve the current phylogeny for parasitic nematodes, we concatenated the nuclear 18S rRNA gene with

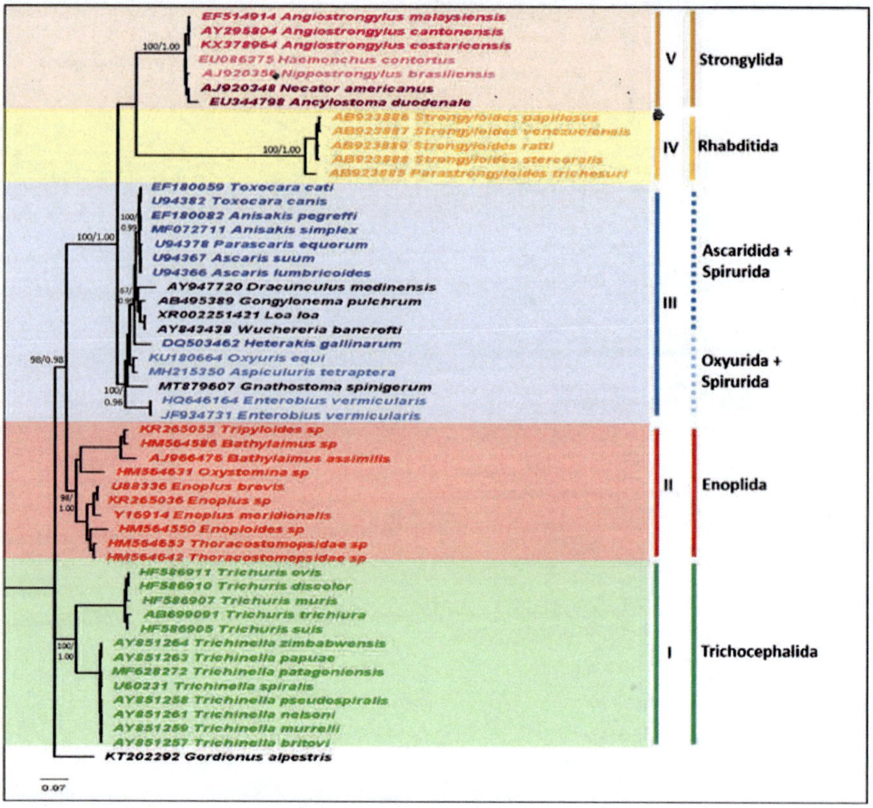

Fig. 10.6 Molecular phylogeny of nematodes inferred with Bayesian inference and maximum likelihood using the 18S rRNA gene

the mitochondrial 12S rRNA gene for parasitic nematodes belonging to clades I, III, IV, and V, as an alternative proposition for nematode systematics.

As indicated in Fig. 10.5, species belonging to clade III were recovered as monophyletic when the two genetic markers were concatenated, agreeing with the nuclear 18S rRNA gene phylogeny (Fig. 10.6). Contrarily, clade III was not recovered as monophyletic in the mitochondrial 12S rRNA gene phylogeny (Fig. 10.7). Thus, the concatenated gene phylogeny shows an improved phylogeny compared to the mitochondrial phylogeny, where previous studies had consistently demonstrated the non-monophyly of clade III nematodes when mitochondrial genes were used. Another improvement using the concatenated genes was observed within clade III, where orders Ascaridida and Oxyurida were recovered as monophyletic. The nuclear 18S rRNA gene, in contrast, was not able to recover any of the three orders as monophyletic. Additionally, increased taxonomic resolution at the species level can also be achieved using the concatenated genes, as the mitochondrial 12S rRNA gene can provide sufficient genetic variation for species discrimination, in contrast to the

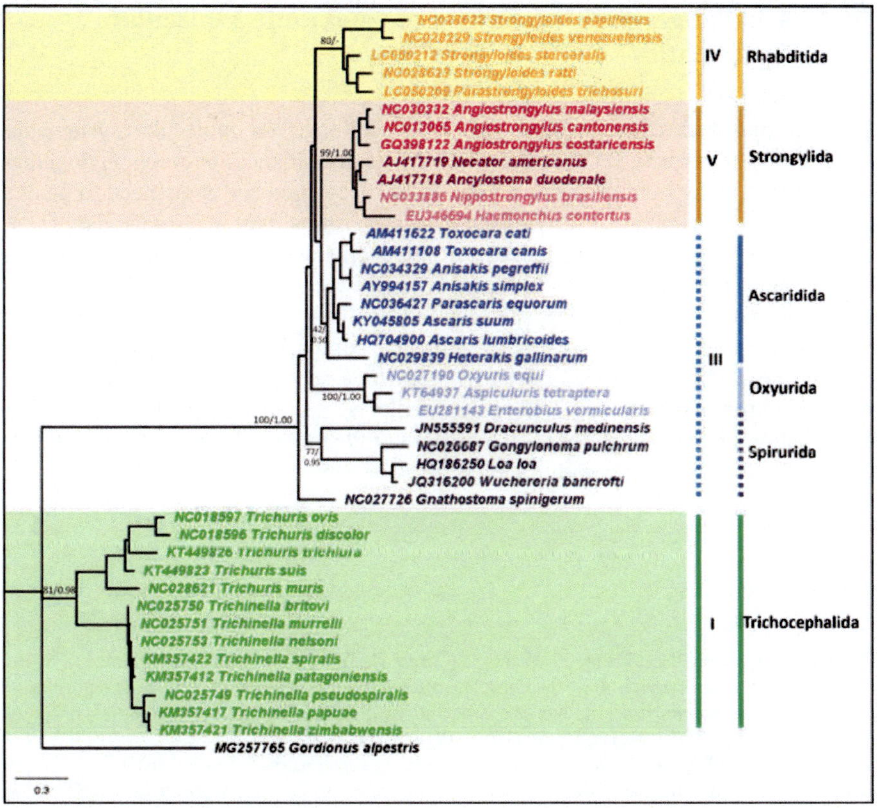

Fig. 10.7 Molecular phylogeny of nematodes inferred with Bayesian inference and maximum likelihood using the 12S rRNA gene

limited ability of the nuclear 18S rRNA gene to discriminate between some closely related species.

Through concatenating the nuclear 18S and mitochondrial 12S rRNA genes, the improved nematode phylogeny result was obtained, possibly due to the property of each molecular genetic marker. The nuclear 18S rRNA gene can provide good phylogenetic resolution at a higher taxonomic level (as indicated through the monophyly of the clades). In contrast, the mitochondrial 12S rRNA gene can provide good phylogenetic resolution at slightly lower taxonomic levels (as revealed through sufficient variation for species discrimination). The method of utilizing combined genetic markers can therefore be helpful for nematode molecular systematics and provide alternative phylogenies.

10.4 Challenges and the Future of Nematode Molecular Systematics

The vast numbers of nematodes present a challenge for molecular systematics studies, with many taxa yet to be discovered. The discordances between phylogenies indicate that many more taxa are needed to be discovered and sequenced to paint a comprehensive picture of phylogenetic relationships within Nematoda. These undiscovered or not yet sequenced taxa are crucial to fill in the gaps in the phylogenetic trees for robust phylogenetic inferences. Additionally, the choice of genetic markers, representative species, and analysis methods differed across the various studies, rendering it challenging to make comparisons for the whole phylum.

With greater technological advances, nematode species discovery and costs for high throughput sequencing will undoubtedly be accelerated in the near future. Sequencing increased numbers of complete mitochondrial genomes and whole genomes will significantly aid in the progress of nematode systematics.

References

Ahmad, A. A., Yang, X., Zhang, T., Wang, C., Zhou, C., Yan, X., Hassan, M., Ikram, M., & Hu, M. (2019). Characterization of the complete mitochondrial genome of *Ostertagia trifurcata* of small ruminants and its phylogenetic associations for the Trichostrongyloidea superfamily. *Genes (Basel), 10*(2). https://doi.org/10.3390/genes10020107

Anderson, R. (2009). *Keys to the nematode parasites of vertebrates*. CABI Publishing.

Barratt, J., Chan, D., Sandaradura, I., Malik, R., Spielman, D., Lee, R., Marriott, D., Harkness, J., Ellis, J., & Stark, D. (2016). *Angiostrongylus cantonensis*: A review of its distribution, molecular biology and clinical significance as a human pathogen. *Parasitology, 143*(9), 1087–1118. https://doi.org/10.1017/s0031182016000652

Bert, W., Leliaert, F., Vierstraete, A. R., Vanfleteren, J. R., & Borgonie, G. (2008). Molecular phylogeny of the Tylenchina and evolution of the female gonoduct (Nematoda: Rhabditida). *Molecular Phylogenetics and Evolution, 48*(2), 728–744. https://doi.org/10.1016/j.ympev.2008.04.011

Bik, H. M., Lambshead, P. J., Thomas, W. K., & Lunt, D. H. (2010). Moving towards a complete molecular framework of the Nematoda: A focus on the Enoplida and early-branching clades. *BMC Evolutionary Biology, 10*, 353. https://doi.org/10.1186/1471-2148-10-353

Blaxter, M. (2000). Genes and genomes of *Necator americanus* and related hookworms. *International Journal for Parasitology, 30*(4), 347–355. https://doi.org/10.1016/s0020-7519(99)00198-8

Blaxter, M., & Koutsovoulos, G. (2015). The evolution of parasitism in Nematoda. *Parasitology, 142*(Suppl 1), S26–S39. https://doi.org/10.1017/S0031182014000791

Blaxter, M. L., De Ley, P., Garey, J. R., Liu, L. X., Scheldeman, P., Vierstraete, A., Vanfleteren, J. R., Mackey, L. Y., Dorris, M., Frisse, L. M., Vida, J. T., & Thomas, W. K. (1998). A molecular evolutionary framework for the phylum Nematoda. *Nature, 392*(6671), 71–75. https://doi.org/10.1038/32160

Borba, V. H., Machado-Silva, J. R., Le Bailly, M., & Iniguez, A. M. (2019). Worldwide paleodistribution of capillariid parasites: Paleoparasitology, current status of phylogeny and taxonomic perspectives. *PLoS One, 14*(4), e0216150. https://doi.org/10.1371/journal.pone.0216150

Callejon, R., Cutillas, C., & Nadler, S. A. (2015). Nuclear and mitochondrial genes for inferring *Trichuris* phylogeny. *Parasitology Research, 114*(12), 4591–4599. https://doi.org/10.1007/s00436-015-4705-7

Carreno, R. A., & Nadler, S. A. (2003). Phylogenetic analysis of the Metastrongyloidea (Nematoda: Strongylida) inferred from ribosomal RNA gene sequences. *The Journal of Parasitology, 89*(5), 965–973. https://doi.org/10.1645/ge-76r

CDC. (2020a). *Biology - Life cycle of Capillaria philippinensis*. https://www.cdc.gov/parasites/capillaria/biology_c_philippinensis.html

CDC. (2020b). *Parasites - Enterobiasis (also known as Pinworm infection)*. https://www.cdc.gov/parasites/pinworm/

Chan, A. H. E., Chaisiri, K., Dusitsittipon, S., Jakkul, W., Charoennitiwat, V., Komalamisra, C., & Thaenkham, U. (2020a). Mitochondrial ribosomal genes as novel genetic markers for discrimination of closely related species in the *Angiostrongylus cantonensis* lineage. *Acta Tropica, 211*, 105645. https://doi.org/10.1016/j.actatropica.2020.105645

Chan, A. H. E., Chaisiri, K., Morand, S., Saralamba, N., & Thaenkham, U. (2020b). Evaluation and utility of mitochondrial ribosomal genes for molecular systematics of parasitic nematodes. *Parasites & Vectors, 13*(1), 364. https://doi.org/10.1186/s13071-020-04242-8

Chilton, N. B., Huby-Chilton, F., Gasser, R. B., & Beveridge, I. (2006). The evolutionary origins of nematodes within the order Strongylida are related to predilection sites within hosts. *Molecular Phylogenetics and Evolution, 40*(1), 118–128. https://doi.org/10.1016/j.ympev.2006.01.003

Chitwood, B. G. (1937). A revised classification of the Nematoda. In *Papers on helminthology, 30 year jubileum* (pp. 67–79). Lenin Academy of Agricultural Sciences.

De Ley, P. (2006). A quick tour of nematode diversity and the backbone of nematode phylogeny. *WormBook*, 1–8. https://www.ncbi.nlm.nih.gov/books/NBK19684/

De Ley, P., & Blaxter, M. (2002). Systematic position and phylogeny. In *The biology of nematodes* (pp. 1–30). Taylor & Francis.

Di Cesare, A., Castagna, G., Otranto, D., Meloni, S., Milillo, P., Latrofa, M. S., Paoletti, B., Bartolini, R., & Traversa, D. (2012). Molecular detection of *Capillaria aerophila*, an agent of canine and feline pulmonary capillariosis. *Journal of Clinical Microbiology, 50*(6), 1958–1963. https://doi.org/10.1128/JCM.00103-12

Dorris, M., De Ley, P., & Blaxter, M. L. (1999). Molecular analysis of nematode diversity and the evolution of parasitism. *Parasitology Today, 15*(5), 188–193. https://doi.org/10.1016/s0169-4758(99)01439-8

Dorris, M., Viney, M. E., & Blaxter, M. L. (2002). Molecular phylogenetic analysis of the genus *Strongyloides* and related nematodes. *International Journal for Parasitology, 32*(12), 1507–1517. https://doi.org/10.1016/s0020-7519(02)00156-x

Dusitsittipon, S., Criscione, C. D., Morand, S., Komalamisra, C., & Thaenkham, U. (2018). Hurdles in the evolutionary epidemiology of *Angiostrongylus cantonensis*: Pseudogenes, incongruence between taxonomy and DNA sequence variants, and cryptic lineages. *Evolutionary Applications, 11*(8), 1257–1269. https://doi.org/10.1111/eva.12621

Eamsobhana, P., Lim, P. E., & Yong, H. S. (2015). Phylogenetics and systematics of *Angiostrongylus* lungworms and related taxa (Nematoda: Metastrongyloidea) inferred from the nuclear small subunit (SSU) ribosomal DNA sequences. *Journal of Helminthology, 89*(3), 317–325. https://doi.org/10.1017/s0022149x14000108

Eberhardt, A. T., Robles, M. D. R., Monje, L. D., Beldomenico, P. M., & Callejón, R. (2019). A new *Trichuris* species (Nematoda: Trichuridae) from capybaras: Morphological-molecular characterization and phylogenetic relationships. *Acta Tropica, 190*, 244–252. https://doi.org/10.1016/j.actatropica.2018.11.029

El-Dib, N. A., El-Badry, A. A., Ta-Tang, T. H., & Rubio, J. M. (2015). Molecular detection of *Capillaria philippinensis*: An emerging zoonosis in Egypt. *Experimental Parasitology, 154*, 127–133. https://doi.org/10.1016/j.exppara.2015.04.011

Gao, J. F., Liu, G. H., Duan, H., Gao, Y., Zhang, Y., Chang, Q. C., Fang, M., & Wang, C. R. (2017a). Complete mitochondrial genomes of *Triodontophorus serratus* and *Triodontophorus nipponicus*, and their comparison with *Triodontophorus brevicauda*. *Experimental Parasitology, 181*, 88–93. https://doi.org/10.1016/j.exppara.2017.08.002

Gao, J. F., Zhao, Q., Liu, G. H., Zhang, Y., Zhang, Y., Wang, W. T., Chang, Q. C., Wang, C. R., & Zhu, X. Q. (2014). Comparative analyses of the complete mitochondrial genomes of the two ruminant hookworms *Bunostomum trigonocephalum* and *Bunostomum phlebotomum*. *Gene, 541*(2), 92–100. https://doi.org/10.1016/j.gene.2014.03.017

Gao, Y., Qiu, J. H., Zhang, B. B., Su, X., Fu, X., Yue, D. M., & Wang, C. R. (2017b). Complete mitochondrial genome of parasitic nematode *Cylicocyclus nassatus* and comparative analyses with *Cylicocyclus insigne*. *Experimental Parasitology, 172*, 18–22. https://doi.org/10.1016/j.exppara.2016.11.005

Gouy de Bellocq, J., Ferte, H., Depaquit, J., Justine, J. L., Tillier, A., & Durette-Desset, M. C. (2001). Phylogeny of the Trichostrongylina (Nematoda) inferred from 28S rDNA sequences. *Molecular Phylogenetics and Evolution, 19*(3), 430–442. https://doi.org/10.1006/mpev.2001.0925

Guardone, L., Deplazes, P., Macchioni, F., Magi, M., & Mathis, A. (2013). Ribosomal and mitochondrial DNA analysis of Trichuridae nematodes of carnivores and small mammals. *Veterinary Parasitology, 197*(1–2), 364–369. https://doi.org/10.1016/j.vetpar.2013.06.022

Hasegawa, H., Hayashida, S., Ikeda, Y., & Sato, H. (2009). Hyper-variable regions in 18S rDNA of *Strongyloides* spp. as markers for species-specific diagnosis. *Parasitology Research, 104*(4), 869–874. https://doi.org/10.1007/s00436-008-1269-9

Hawash, M. B., Andersen, L. O., Gasser, R. B., Stensvold, C. R., & Nejsum, P. (2015). Mitochondrial genome analyses suggest multiple *Trichuris* species in humans, baboons, and pigs from different geographical regions. *PLoS Neglected Tropical Diseases, 9*(9), e0004059. https://doi.org/10.1371/journal.pntd.0004059

Hawash, M. B., Betson, M., Al-Jubury, A., Ketzis, J., LeeWillingham, A., Bertelsen, M. F., Cooper, P. J., Littlewood, D. T., Zhu, X. Q., & Nejsum, P. (2016). Whipworms in humans and pigs: Origins and demography. *Parasites & Vectors, 9*, 37. https://doi.org/10.1186/s13071-016-1325-8

Holterman, M., Karssen, G., van den Elsen, S., van Megen, H., Bakker, J., & Helder, J. (2009). Small subunit rDNA-based phylogeny of the Tylenchida sheds light on relationships among some high-impact plant-parasitic nematodes and the evolution of plant feeding. *Phytopathology, 99*(3), 227–235. https://doi.org/10.1094/phyto-99-3-0227

Holterman, M., van der Wurff, A., van den Elsen, S., van Megen, H., Bongers, T., Holovachov, O., Bakker, J., & Helder, J. (2006). Phylum-wide analysis of SSU rDNA reveals deep phylogenetic relationships among nematodes and accelerated evolution toward crown clades. *Molecular Biology and Evolution, 23*(9), 1792–1800. https://doi.org/10.1093/molbev/msl044

Hoste, H., Chilton, N. B., Beveridge, I., & Gasser, R. B. (1998). A comparison of the first internal transcribed spacer of ribosomal DNA in seven species of *Trichostrongylus* (Nematoda: Trichostrongylidae). *International Journal for Parasitology, 28*(8), 1251–1260. https://doi.org/10.1016/s0020-7519(98)00093-9

Howe, K. L., Bolt, B. J., Shafie, M., Kersey, P., & Berriman, M. (2017). WormBase ParaSite - A comprehensive resource for helminth genomics. *Molecular and Biochemical Parasitology, 215*, 2–10. https://doi.org/10.1016/j.molbiopara.2016.11.005

Hu, L., Zhang, M., Sun, Y., & Bu, Y. (2020). Characterization and phylogenetic analysis of the first complete mitochondrial genome of *Cylicocyclus radiatus*. *Veterinary Parasitology, 281*, 109097. https://doi.org/10.1016/j.vetpar.2020.109097

Hu, M., Chilton, N. B., & Gasser, R. B. (2003). The mitochondrial genome of *Strongyloides stercoralis* (Nematoda) – Idiosyncratic gene order and evolutionary implications. *International Journal for Parasitology, 33*(12), 1393–1408. https://doi.org/10.1016/s0020-7519(03)00130-9

Hung, G. C., Chilton, N. B., Beveridge, I., & Gasser, R. B. (2000). A molecular systematic framework for equine strongyles based on ribosomal DNA sequence data. *International Journal for Parasitology, 30*(1), 95–103. https://doi.org/10.1016/s0020-7519(99)00166-6

Hunt, V. L., Hino, A., Yoshida, A., & Kikuchi, T. (2018). Comparative transcriptomics gives insights into the evolution of parasitism in *Strongyloides* nematodes at the genus, subclade and species level. *Scientific Reports, 8*(1), 5192. https://doi.org/10.1038/s41598-018-23514-z

Hunt, V. L., Tsai, I. J., Coghlan, A., Reid, A. J., Holroyd, N., Foth, B. J., Tracey, A., Cotton, J. A., Stanley, E. J., Beasley, H., Bennett, H. M., Brooks, K., Harsha, B., Kajitani, R., Kulkarni, A., Harbecke, D., Nagayasu, E., Nichol, S., Ogura, Y., ... Berriman, M. (2016). The genomic basis of parasitism in the *Strongyloides* clade of nematodes. *Nature Genetics, 48*(3), 299–307. https://doi.org/10.1038/ng.3495

International Helminth Genomes Consortium. (2019). Comparative genomics of the major parasitic worms. *Nature Genetics, 51*(1), 163–174. https://doi.org/10.1038/s41588-018-0262-1

Jex, A. R., Hall, R. S., Littlewood, D. T., & Gasser, R. B. (2010). An integrated pipeline for next-generation sequencing and annotation of mitochondrial genomes. *Nucleic Acids Research, 38*(2), 522–533. https://doi.org/10.1093/nar/gkp883

Kampfer, S., Sturmbauer, C., & Ott, J. (1998). Phylogenetic analysis of rDNA sequences from adenophorean nematodes and implications for the Adenophorea-Secernentea controversy. *Invertebrate Biology, 117*(1), 29–36. https://doi.org/10.2307/3226849

Kang, S., Sultana, T., Eom, K. S., Park, Y. C., Soonthornpong, N., Nadler, S. A., & Park, J. K. (2009). The mitochondrial genome sequence of *Enterobius vermicularis* (Nematoda: Oxyurida)--an idiosyncratic gene order and phylogenetic information for chromadorean nematodes. *Gene, 429*(1–2), 87–97. https://doi.org/10.1016/j.gene.2008.09.011

Kern, E. M. A., Kim, T., & Park, J.-K. (2020). The mitochondrial genome in nematode phylogenetics. *Frontiers in Ecology and Evolution, 8*. https://doi.org/10.3389/fevo.2020.00250

Kikuchi, T., Afrin, T., & Yoshida, M. (2016). Complete mitochondrial genomes of four entomopathogenic nematode species of the genus *Steinernema*. *Parasites & Vectors, 9*(1), 430. https://doi.org/10.1186/s13071-016-1730-z

Kim, J., Lee, S.-H., Gazi, M., Kim, T., Jung, D., Chun, J.-Y., Kim, S., Seo, T.-K., Park, C., Baldwin, J. G., Nadler, S. A., & Park, J.-K. (2015). Mitochondrial genomes advance phylogenetic hypotheses for Tylenchina (Nematoda: Chromadorea). *Zoologica Scripta, 44*(4), 446–462. https://doi.org/10.1111/zsc.12112

Ko, P. P., Sakaguchi, K., Yoshida, A., Maruyama, H., Nonaka, N., & Nagayasu, E. (2019). First molecular identification of *Strongyloides vituli* in cattle in Japan and insights into the evolutionary history of *Strongyloides* parasites of ruminants. *Parasitology International, 72*, 101937. https://doi.org/10.1016/j.parint.2019.101937

Kumar, S., Koutsovoulos, G., Kaur, G., & Blaxter, M. (2012). Toward 959 nematode genomes. *Worm, 1*(1), 42–50. https://doi.org/10.4161/worm.19046

Laetsch, D. R., Heitlinger, E. G., Taraschewski, H., Nadler, S. A., & Blaxter, M. L. (2012). The phylogenetics of Anguillicolidae (Nematoda: Anguillicoloidea), swimbladder parasites of eels. *BMC Evolutionary Biology, 12*, 60. https://doi.org/10.1186/1471-2148-12-60

Li, K., Shahzad, M., Zhang, H., Mehmood, K., Jiang, X., Luo, H., Zhang, L., Dong, X., & Li, J. (2018b). Characterization of the complete mitochondrial genome of *Metastrongylus salmi* (M. salmi) derived from Tibetan pigs in Tibet, China. *Acta Parasitologica, 63*(2), 280–286. https://doi.org/10.1515/ap-2018-0032

Li, L., Lu, L., Nadler, S. A., Gibson, D. I., Zhang, L. P., Chen, H. X., Zhao, W. T., & Guo, Y. N. (2018a). Molecular phylogeny and dating reveal a terrestrial origin in the early carboniferous for ascaridoid nematodes. *Systematic Biology, 67*(5), 888–900. https://doi.org/10.1093/sysbio/syy018

Li, Y., Chen, H. X., Yang, X. L., & Li, L. (2019). Morphological and genetic characterization of *Syphabulea tjanschani* (Nematoda: Oxyuridae), with phylogenetic position of Syphabulea in Oxyuridae. *Infection, Genetics and Evolution, 67*, 159–166. https://doi.org/10.1016/j.meegid.2018.11.016

Li, Y., Niu, L., Wang, Q., Zhang, Z., Chen, Z., Gu, X., Xie, Y., Yan, N., Wang, S., Peng, X., & Yang, G. (2012). Molecular characterization and phylogenetic analysis of ascarid nematodes from twenty-one species of captive wild mammals based on mitochondrial and nuclear sequences. *Parasitology, 139*(10), 1329–1338. https://doi.org/10.1017/S003118201200056X

Liu, G. H., Gasser, R. B., Su, A., Nejsum, P., Peng, L., Lin, R. Q., Li, M. W., Xu, M. J., & Zhu, X. Q. (2012). Clear genetic distinctiveness between human- and pig-derived *Trichuris* based on analyses of mitochondrial datasets. *PLoS Neglected Tropical Diseases, 6*(2), e1539. https://doi.org/10.1371/journal.pntd.0001539

Liu, G. H., Li, S., Zou, F. C., Wang, C. R., & Zhu, X. Q. (2016a). The complete mitochondrial genome of rabbit pinworm *Passalurus ambiguus*: Genome characterization and phylogenetic analysis. *Parasitology Research, 115*(1), 423–429. https://doi.org/10.1007/s00436-015-4778-3

Liu, G. H., Nadler, S. A., Liu, S. S., Podolska, M., D'Amelio, S., Shao, R., Gasser, R. B., & Zhu, X. Q. (2016b). Mitochondrial phylogenomics yields strongly supported hypotheses for ascaridomorph nematodes. *Scientific Reports, 6*, 39248. https://doi.org/10.1038/srep39248

Liu, G. H., Shao, R., Cai, X. Q., Li, W. W., & Zhu, X. Q. (2015). *Gnathostoma spinigerum* mitochondrial genome sequence: A novel gene arrangement and its phylogenetic position within the class Chromadorea. *Scientific Reports, 5*, 12691. https://doi.org/10.1038/srep12691

Liu, G. H., Shao, R., Li, J. Y., Zhou, D. H., Li, H., & Zhu, X. Q. (2013). The complete mitochondrial genomes of three parasitic nematodes of birds: A unique gene order and insights into nematode phylogeny. *BMC Genomics, 14*(1), 414. https://doi.org/10.1186/1471-2164-14-414

Lorenzen, S. (1994). *The phylogenetic systematics of free-living nematodes*. Ray Society.

Maggenti, A. R. (1983). Nematode higher classification as influenced by species and family concepts. In *Concepts in nematode systematics* (pp. 25–40). Academic press.

Meldal, B. H., Debenham, N. J., De Ley, P., De Ley, I. T., Vanfleteren, J. R., Vierstraete, A. R., Bert, W., Borgonie, G., Moens, T., Tyler, P. A., Austen, M. C., Blaxter, M. L., Rogers, A. D., & Lambshead, P. J. (2007). An improved molecular phylogeny of the Nematoda with special emphasis on marine taxa. *Molecular Phylogenetics and Evolution, 42*(3), 622–636. https://doi.org/10.1016/j.ympev.2006.08.025

Meng, X., Xie, Y., Gu, X., Zheng, Y., Liu, Y., Li, Y., Wang, L., Zhou, X., Zuo, Z., & Yang, G. (2019). Sequencing and analysis of the complete mitochondrial genome of dog roundworm *Toxocara canis* (Nematoda: Toxocaridae) from USA. *Mitochondrial DNA. Part B, Resources, 4*(2), 2999–3001. https://doi.org/10.1080/23802359.2019.1666042

Mitreva, M., Blaxter, M. L., Bird, D. M., & McCarter, J. P. (2005). Comparative genomics of nematodes. *Trends in Genetics, 21*(10), 573–581. https://doi.org/10.1016/j.tig.2005.08.003

Mohandas, N., Pozio, E., La Rosa, G., Korhonen, P. K., Young, N. D., Koehler, A. V., Hall, R. S., Sternberg, P. W., Boag, P. R., Jex, A. R., Chang, B. C., & Gasser, R. B. (2014). Mitochondrial genomes of *Trichinella* species and genotypes - A basis for diagnosis, and systematic and epidemiological explorations. *International Journal for Parasitology, 44*(14), 1073–1080. https://doi.org/10.1016/j.ijpara.2014.08.010

Nadler, S. A., Carreno, R. A., Mejia-Madrid, H., Ullberg, J., Pagan, C., Houston, R., & Hugot, J. P. (2007). Molecular phylogeny of clade III nematodes reveals multiple origins of tissue parasitism. *Parasitology, 134*(Pt 10), 1421–1442. https://doi.org/10.1017/S0031182007002880

Nadler, S. A., De Ley, P., Mundo-Ocampo, M., Smythe, A. B., Patricia Stock, S., Bumbarger, D., Adams, B. J., De Ley, I. T., Holovachov, O., & Baldwin, J. G. (2006). Phylogeny of Cephalobina (Nematoda): Molecular evidence for recurrent evolution of probolae and incongruence with traditional classifications. *Molecular Phylogenetics and Evolution, 40*(3), 696–711. https://doi.org/10.1016/j.ympev.2006.04.005

Nagayasu, E., Aung, M., Hortiwakul, T., Hino, A., Tanaka, T., Higashiarakawa, M., Olia, A., Taniguchi, T., Win, S. M. T., Ohashi, I., Odongo-Aginya, E. I., Aye, K. M., Mon, M., Win, K. K., Ota, K., Torisu, Y., Panthuwong, S., Kimura, E., Palacpac, N. M. Q., . . . Maruyama, H. (2017). A possible origin population of pathogenic intestinal nematodes, *Strongyloides stercoralis*, unveiled by molecular phylogeny. *Scientific Reports, 7*(1), 4844. https://doi.org/10.1038/s41598-017-05049-x

Paguem, A., Abanda, B., Ngwasiri, N. N., Eisenbarth, A., Renz, A., Streit, A., & Achukwi, M. D. (2020). Host specificity and phylogeny of Trichostrongylidae of domestic ruminants in the Guinea savannah of the Adamawa plateau in Cameroon. *Veterinary Parasitology, Regional Studies and Reports, 21*, 100412. https://doi.org/10.1016/j.vprsr.2020.100412

Park, J. K., Sultana, T., Lee, S. H., Kang, S., Kim, H. K., Min, G. S., Eom, K. S., & Nadler, S. A. (2011). Monophyly of clade III nematodes is not supported by phylogenetic analysis of complete mitochondrial genome sequences. *BMC Genomics, 12*, 392. https://doi.org/10.1186/1471-2164-12-392

Perkins, S. L., Martinsen, E. S., & Falk, B. G. (2011). Do molecules matter more than morphology? Promises and pitfalls in parasites. *Parasitology, 138*(13), 1664–1674. https://doi.org/10.1017/s0031182011000679

Pozio, E., Hoberg, E., La Rosa, G., & Zarlenga, D. S. (2009). Molecular taxonomy, phylogeny and biogeography of nematodes belonging to the *Trichinella* genus. *Infection, Genetics and Evolution, 9*(4), 606–616. https://doi.org/10.1016/j.meegid.2009.03.003

Pozio, E., & Zarlenga, D. S. (2005). Recent advances on the taxonomy, systematics and epidemiology of *Trichinella*. *International Journal for Parasitology, 35*(11–12), 1191–1204. https://doi.org/10.1016/j.ijpara.2005.07.012

Ricciardi, A., & Ndao, M. (2015). Diagnosis of parasitic infections: What's going on? *Journal of Biomolecular Screening, 20*(1), 6–21. https://doi.org/10.1177/1087057114548065

Rivero, J., Callejon, R., & Cutillas, C. (2021). Complete mitochondrial genome of *Trichuris trichiura* from *Macaca sylvanus* and *Papio papio*. *Life (Basel), 11*(2). https://doi.org/10.3390/life11020126

Robles Mdel, R., Cutillas, C., Panei, C. J., & Callejón, R. (2014). Morphological and molecular characterization of a new *Trichuris* species (Nematoda- Trichuridae), and phylogenetic relationships of *Trichuris* species of Cricetid rodents from Argentina. *PLoS One, 9*(11), e112069. https://doi.org/10.1371/journal.pone.0112069

Sakaguchi, S., Yunus, M., Sugi, S., & Sato, H. (2020). Integrated taxonomic approaches to seven species of capillariid nematodes (Nematoda: Trichocephalida: Trichinelloidea) in poultry from Japan and Indonesia, with special reference to their 18S rDNA phylogenetic relationships. *Parasitology Research, 119*(3), 957–972. https://doi.org/10.1007/s00436-019-06544-y

Sharifdini, M., Derakhshani, S., Alizadeh, S. A., Ghanbarzadeh, L., Mirjalali, H., Mobedi, I., & Saraei, M. (2017a). Molecular identification and phylogenetic analysis of human *Trichostrongylus* species from an endemic area of Iran. *Acta Tropica, 176*, 293–299. https://doi.org/10.1016/j.actatropica.2017.07.001

Sharifdini, M., Heidari, Z., Hesari, Z., Vatandoost, S., & Kia, E. B. (2017b). Molecular phylogenetics of *Trichostrongylus* species (Nematoda: Trichostrongylidae) from humans of Mazandaran Province, Iran. *Korean Journal of Parasitology, 55*(3), 279–285. https://doi.org/10.3347/kjp.2017.55.3.279

Shi, X., Wang, M., Abdullahi, A. Y., Fu, Y., Yang, F., Yu, X., Pan, W., Yan, X., Hang, J., Zhang, P., & Li, G. (2018). Comparative analysis of *Ancylostoma ceylanicum* mitochondrial genome with other *Ancylostoma* species. *Infection, Genetics and Evolution, 62*, 40–45. https://doi.org/10.1016/j.meegid.2018.04.012

Smythe, A. B., Holovachov, O., & Kocot, K. M. (2019). Improved phylogenomic sampling of free-living nematodes enhances resolution of higher-level nematode phylogeny. *BMC Evolutionary Biology, 19*(1), 121. https://doi.org/10.1186/s12862-019-1444-x

Sultana, T., Kim, J., Lee, S.-H., Han, H., Kim, S., Min, G.-S., Nadler, S. A., & Park, J.-K. (2013). Comparative analysis of complete mitochondrial genome sequences confirms independent origins of plant-parasitic nematodes. *BMC Evolutionary Biology, 13*(1), 12. https://doi.org/10.1186/1471-2148-13-12

Sun, L., Zhuo, K., Lin, B., Wang, H., & Liao, J. (2014). The complete mitochondrial genome of *Meloidogyne graminicola* (Tylenchina): A unique gene arrangement and its phylogenetic implications. *PLoS One, 9*(6), e98558. https://doi.org/10.1371/journal.pone.0098558

Sun, M. M., Han, L., Zhang, F. K., Zhou, D. H., Wang, S. Q., Ma, J., Zhu, X. Q., & Liu, G. H. (2018). Characterization of the complete mitochondrial genome of *Marshallagia marshalli* and phylogenetic implications for the superfamily Trichostrongyloidea. *Parasitology Research, 117*(1), 307–313. https://doi.org/10.1007/s00436-017-5669-6

Valentyne, H., Spratt, D. M., Aghazadeh, M., Jones, M. K., & Šlapeta, J. (2020). The mitochondrial genome of *Angiostrongylus mackerrasae* is distinct from *A. cantonensis* and *A. malaysiensis*. *Parasitology, 147*(6), 681–688. https://doi.org/10.1017/s0031182020000232

van Megen, H., van den Elsen, S., Holterman, M., Karssen, G., Mooyman, P., Bongers, T., Holovachov, O., Bakker, J., & Helder, J. (2009). A phylogenetic tree of nematodes based on about 1200 full-length small subunit ribosomal DNA sequences. *Nematology, 11*(6), 927–950. https://doi.org/10.1163/156854109x456862

Wang, C. R., Gao, J. F., Zhu, X. Q., & Zhao, Q. (2012). Characterization of *Bunostomum trigonocephalum* and *Bunostomum phlebotomum* from sheep and cattle by internal transcribed spacers of nuclear ribosomal DNA. *Research in Veterinary Science, 92*(1), 99–102. https://doi.org/10.1016/j.rvsc.2010.10.024

Wang, C. R., Lou, Y., Gao, J. F., Qiu, J. H., Zhang, Y., Gao, Y., & Chang, Q. C. (2016). Comparative analyses of the complete mitochondrial genomes of the two murine pinworms *Aspiculuris tetraptera* and *Syphacia obvelata*. *Gene, 585*(1), 71–75. https://doi.org/10.1016/j.gene.2016.03.037

WHO. (2020). *Soil-transmitted helminth infections*. https://www.who.int/news-room/fact-sheets/detail/soil-transmitted-helminth-infections

Wijova, M., Moravec, F., Horak, A., & Lukes, J. (2006). Evolutionary relationships of Spirurina (Nematoda: Chromadorea: Rhabditida) with special emphasis on dracunculoid nematodes inferred from SSU rRNA gene sequences. *International Journal for Parasitology, 36*(9), 1067–1075. https://doi.org/10.1016/j.ijpara.2006.04.005

Yong, H. S., Song, S. L., Eamsobhana, P., & Lim, P. E. (2016). Complete mitochondrial genome of *Angiostrongylus malaysiensis* lungworm and molecular phylogeny of Metastrongyloid nematodes. *Acta Tropica, 161*, 33–40. https://doi.org/10.1016/j.actatropica.2016.05.002

Zarlenga, D., Thompson, P., & Pozio, E. (2020). *Trichinella* species and genotypes. *Research in Veterinary Science, 133*, 289–296. https://doi.org/10.1016/j.rvsc.2020.08.012

Zarlenga, D. S., Rosenthal, B. M., La Rosa, G., Pozio, E., & Hoberg, E. P. (2006). Post-Miocene expansion, colonization, and host switching drove speciation among extant nematodes of the archaic genus *Trichinella*. *Proceedings of the National Academy of Sciences of the United States of America, 103*(19), 7354–7359. https://doi.org/10.1073/pnas.0602466103

Zhang, Y., Xu, W. W., Guo, D. H., Liu, Z. X., Duan, H., Su, X., Fu, X., Yue, D. M., Gao, Y., & Wang, C. R. (2015). The complete mitochondrial genome of *Oxyuris equi*: Comparison with other closely related species and phylogenetic implications. *Experimental Parasitology, 159*, 215–221. https://doi.org/10.1016/j.exppara.2015.09.013

Zou, H., Jakovlic, I., Chen, R., Zhang, D., Zhang, J., Li, W. X., & Wang, G. T. (2017). The complete mitochondrial genome of parasitic nematode *Camallanus cotti*: Extreme discontinuity in the rate of mitogenomic architecture evolution within the Chromadorea class. *BMC Genomics, 18*(1), 840. https://doi.org/10.1186/s12864-017-4237-x

Chapter 11
Molecular Systematics of Parasitic Trematodes

Abstract In Chap. 10, we saw the progress and discoveries made in nematode molecular systematics. This chapter focused on parasitic trematodes (subclass Digenea), another major helminth group that is medically important to humans and animals. We discussed their phylogenetic relationships and significance based on research that has been ongoing throughout the years. Subclass Digenea, characterized by the presence of 1 or 2 muscular suckers, is a large and speciose group comprising over 18,000 species in 1748 genera. Examples of prominent parasitic trematodes include the liver flukes *Opisthorchis viverrini*, the blood flukes *Schistosoma* spp., the minute intestinal flukes belonging to family Heterophyidae, and the giant liver flukes of *Fasciola* spp. Globally, the World Health Organization estimates that infections due to food-borne trematodes result in two million life-years lost due to disability and death per year, with more than 40 million people affected. The disease burden caused by parasitic trematodes reinforces the importance of accurate disease diagnosis and control. Species identification and classification, phylogenetic relationships, and molecular systematics are the core to understanding the parasite's life cycle, distribution, and role in disease transmission.

11.1 The Molecular Phylogenetic Backbone of Digenea and Their Classification

Classification of Digenea relies on traditional taxonomy methods through morphological characters. Early studies utilized variations in morphological characters like the sucker arrangements and form, digestive tract, and reproductive systems of adults. Cercaria morphology, types of hosts, and life cycle patterns were subsequently incorporated into the morphological matrix (Brooks et al., 1985; Kostadinova & Pérez-del-Olmo, 2014; Pearson, 1992). The publication of the three volumes of the *Keys to the Trematoda* was a prominent milestone for the classification and identification of Digenea, with the division of Digenea into 3 orders comprising of 25 superfamilies and 1748 valid genera (Bray et al., 2008; Gibson et al., 2002; Jones et al., 2005). The three orders are Strigeida, Echinostomida, and Plagiorchiida. Strigeida is characterized by having furcocercous cercariae that

U. Thaenkham et al., *Molecular Systematics of Parasitic Helminths*,
https://doi.org/10.1007/978-981-19-1786-8_11

penetrate hosts, in contrast to Plagiorchiida; Echinostomatida is characterized by cercariae which tend to encyst in the open rather than penetrating hosts (Bray et al., 2008; Gibson & Bray, 1994; Gibson et al., 2002; Jones et al., 2005). However, the classification of Digenea remains a conundrum and is unstable, partly due to the highly ambiguous and homoplastic morphological characters of digeneans, their small size, and complex life cycles (Nadler & de Leon, 2011; Perkins et al., 2011). Consequently, several times, taxonomic revisions have been made through the elevation and appraisal of the taxonomic status of species within Digenea.

The utility of molecular-based techniques has shed light on phylogenetic relationships within Digenea, allowing verifications and reassessments at both high and low taxonomic levels. One of the earliest molecular studies was performed by Cribb et al. (2003), where the nuclear 18S ribosomal RNA (rRNA) gene and morphological data were used to analyze phylogenetic relationships among 75 species representing 55 families (Cribb et al., 2003). They showed that Diplostomoidea and Schistosomatoidea represented the earliest diverged digeneans. Olson et al. (2003) used the nuclear 18S and 28S rRNA genes to provide a comprehensive phylogeny of Digenea with 170 taxa from 77 families being represented (Olson et al., 2003). The results from Olson et al. (2003) revealed a classification of Digenea into two main clades, the first clade (Diplostomida) comprising of Brachylaimoidea, Diplostomoidea, and Schistosomatoidea and the second clade (Plagiorchiida) comprising of Pronocephalata, Echinostoma, Xiphidiata, and Opisthorchioidea. Hence, the molecular phylogenies obtained did not agree with the traditional division of Digenea into the three main orders.

Advances in sequencing technologies enabled the sequencing of complete mitochondrial genomes, which also serves as an additional genetic marker in elucidating phylogenetic relationships (Le et al., 2016; Ma et al., 2016; Xu et al., 2021). Similar to nematodes, discrepancies between nuclear rRNA gene phylogenies and mitochondrial genome phylogenies are also present for trematodes (Brabec et al., 2015; Cribb et al., 2003; Littlewood et al., 2015; Locke et al., 2018; Olson et al., 2003). An increase in the number of genetic markers used and more trematode species being studied and sequenced also led to contrasting phylogenies obtained from the various studies. In addition to molecular phylogenetic studies being conducted at higher taxonomic levels (among Digenea), many studies have also been undertaken at lower taxonomic levels. Studies on the phylogenetic relationships of species with families Opisthorchiidae, Heterophyidae, and Schistosomatidae are among other examples (Lockyer et al., 2003; Thaenkham et al., 2011, 2012).

The major molecular systematics studies conducted for Digenea using the nuclear rRNA genes and mitochondrial genomes will be discussed in the following subsections emphasizing medically important trematodes and discordances observed between nuclear and mitochondrial phylogenies.

11.1.1 The Nuclear 18S and 28S rRNA Gene Phylogeny

The first molecular comprehensive phylogeny was performed by Cribb et al. (2003), encompassing a total of 75 taxa from 55 families (Cribb et al., 2003). Using both the 18S rRNA gene and a morphological matrix incorporating all stages of the digenean life cycle, the phylogeny revealed that Diplostomoidea and Schistosomatoidea are the earliest divergent digeneans. Additionally, the phylogeny obtained using both molecular and morphological matrices produced greater resolution than morphology alone.

Including 170 taxa representing 77 families using both the 18S and 28S rRNA genes, Olson et al. (2003) increased the number of digenean species for phylogenetic analysis (Olson et al., 2003). Figure 11.1a depicts the phylogeny obtained by Olson et al. (2003) using the nuclear rRNA genes, with the major groupings. The molecular phylogeny obtained revealed a division of Digenea into two main suborders instead of the traditional three orders (Strigeida, Echinostomida, and Plagiorchiida). The two main suborders are Diplostomida and Plagiorchiida. Diplostomida, representing the lineage where digeneans diversified, comprises three main lineages—Brachylaimoidea, Diplostomoidea, and another clade with the blood-dwelling trem atodes of Sanguinicolidae + Schistosomatidae + Spirochidae + Clinostomidae. Plagiorchiida, the larger of the two groups, represents all of the remaining taxa within Digenea, with 13 orders—Bivesiculata, Transversotremata, Hemiurata,

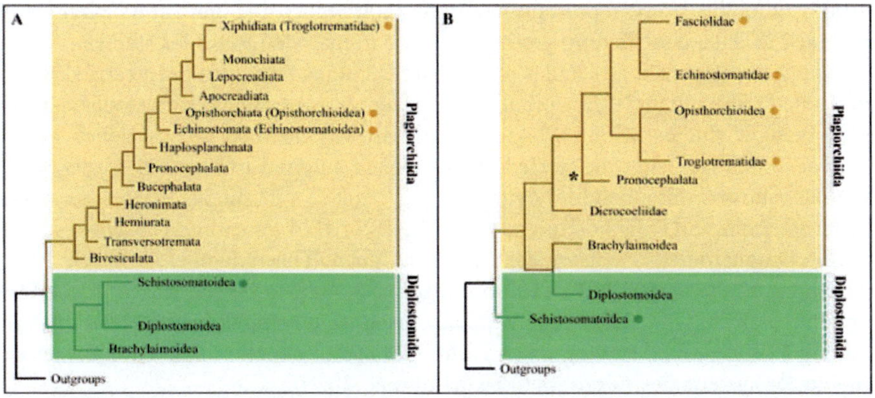

Fig. 11.1 Phylogeny of Digenea with the (**a**) nuclear 18S and 28S rRNA genes and (**b**) complete mitochondrial genome. The nuclear rRNA gene phylogeny was adapted from Olson et al. (2003) and Littlewood et al. (2015) (Littlewood et al., 2015; Olson et al., 2003). Superfamilies and families in parentheses are the representative families in the orders that have been used in mitochondrial genome phylogenies. The mitochondrial phylogeny was adapted based on a compilation of studies utilizing complete mitochondrial genomes (Brabec et al., 2015; Chang et al., 2016; Fu et al., 2019; Landeryou et al., 2020; Le et al., 2016; Li et al. 2019a; Liu et al., 2016; Locke et al., 2018; Ma et al., 2016; Rajapakse et al., 2020; Suleman et al., 2020, 2021; Xu et al., 2021; Yang et al., 2016). Medically important trematodes are indicated in colored circles. An asterisk "*" indicates that the position of the group differed among studies

Heronimata, Bucephalata, Pronocephalata, Haplosplanchnata, Echinostomata, Opisthorchiata, Apocreadiata, Lepocreadiata, Monorchiata, and Xiphidiata. Echinostomida (the third order based on traditional classification) was found to represent a polyphyletic assemblage with members scattered throughout Plagiorchiida. Thus, Olson et al. (2003) suggested a revision of the classification of Digenea to contain two main groups instead of the traditional three (Olson et al., 2003). The phylogeny obtained by Olson et al. (2003) has thus served as the molecular phylogenetic backbone for Digenea, the essence for molecular systematics studies of trematodes (Olson & Tkach, 2005).

The diversity and diversification patterns of Digenea were further studied by Littlewood et al. (2015), where definitive host types and habitats were incorporated into the nuclear rRNA gene phylogeny (Littlewood et al., 2015). Even more recently, incorporating 1077 taxa representing 106 families, de Leon and Hernandez-Mena (2019) updated the database for digeneans with the 18S and 28S rRNA genes (de Leon & Hernandez-Mena, 2019). The obtained phylogeny supports the division of Digenea into Diplostomida and Plagiorchiida. Relationships within Diplostomida were consistent between both genetic markers and analysis methods, with the three major divisions into Diplostomoidea, Brachylaimoidea, and the blood-dwelling trematodes. Relationships within Plagiorchiida were less consistent, with variations in phylogenies obtained from the nuclear rRNA genes and analysis methods. Relationships obtained using the nuclear rRNA genes were not fully resolved at lower taxonomic levels between and within families, with the intermingling of taxa. However, overall, the major groupings obtained were congruent with Olson et al. (2003), and the study represents the most comprehensive trematode phylogeny obtained to date, contributing greatly to future trematode molecular studies.

Many smaller-scale molecular phylogenetic studies have also been carried out to understand and untangle the complicated molecular web within trematodes. Examples include studies within Plagiorchiata utilizing the 28S rRNA gene; within Schistosomatidae using the nuclear rRNA genes, internal transcribed spacer (ITS) regions, and the mitochondrial cytochrome *c* oxidase subunit I (*COI*) gene; within Heterophyidae and Opisthorchiidae using the 18S rRNA genes and ITS2 region; and within Omphalometridae using the 28S rRNA gene (Thaenkham et al., 2011, 2012; Tkach et al., 2000, 2001). In addition to primary nucleotide sequences, secondary structures of the 18S rRNA gene were also used to infer phylogenetic relationships among Plagiorchiida, serving as an additional useful genetic marker for trematode molecular systematics (Voronova & Chelomina, 2020).

11.1.2 Mitochondrial Genome Phylogeny

Mitochondrial genomes are also valuable sources of information, aiding in the reconstruction of evolutionary relationships within Digenea. Complete mitochondrial genomes of parasitic trematodes of medical importance to humans have mainly been sequenced, with at least 40 species in the NCBI database to date. Figure 11.1b depicts the phylogeny obtained using mitochondrial genomes. Generally, the tree topology is concordant with that of the phylogeny obtained using the nuclear rRNA genes, with Diplostomida at the base of Digenea. Le et al. (2016) sequenced the complete mitochondrial genome of Echinochasmidae and performed phylogenetic analysis using 40 species representing 14 families (Le et al., 2016). Families Schistosomatidae, Diplostomidae, and Clinostomidae represent the earlier families within Digenea. Families Opisthorchiidae and Heterophyidae were sisters to another clade comprising Troglotrematidae, Echinostomatidae, Echinochasmidae, Fasciolidae, and Pronocephalata. Dicrocoeliidae was then nested basal to this clade. Other mitochondrial genome phylogenies also gave a similar picture, substantiating the power of using mitochondrial genomes for trematode molecular systematics (Brabec et al., 2015; Landeryou et al., 2020; Le et al., 2016; Li et al. 2019b; Liu et al., 2016; Ma et al., 2016; Rajapakse et al., 2020; Suleman et al., 2021; Xu et al., 2021; Yang et al., 2016).

However, like nematodes, differences exist between nuclear and mitochondrial genome phylogenies. The first difference is order Diplostomida, where nuclear rRNA gene phylogenies support a monophyletic lineage of Diplostomoidea, Brachylaimoidea, and Schistosomatoidea (consisting of Sanguinicolidae, Spirochidae, Clinostomidae, Schistosomatidae), while mitochondrial phylogenies could not recover Diplostomida as monophyletic. With a phylogeny representing 53 species, Fu et al. (2019) showed that superfamilies Brachylaimoidea and Diplostomoidea share a closer relationship to members of Plagiorchiida instead of Schistosomatoidea (Fu et al., 2019). Other mitochondrial phylogenies have also never recovered Diplostomida as monophyletic (Brabec et al., 2015; Littlewood et al., 2015). The second difference is the phylogenetic position of Dicrocoeliidae within Plagiorchiida. In mitochondrial genome phylogenies, Dicrocoeliidae is recovered as basal within Plagiorchiida, sister to the remainder comprising Pronocephalata, Opisthorchiidae, Heterophyidae, Troglotrematidae, Echinostomatidae, Echinochasmidae, and Fasciolidae (Chang et al., 2016; Suleman et al., 2020, 2021). In contrast, in the nuclear rRNA gene phylogeny by Olson et al. (2003), Dicrocoeliidae shows a closer relationship to Troglotrematidae (belonging to superfamily Gorgoderoidea), and Pronocephalata was nested as basal within Plagiorchiida.

Aside from the discrepancies between mitochondrial and nuclear phylogenies, differences in phylogenetic relationships between studies using mitochondrial genomes have also been observed, in particular, the phylogenetic position of Pronocephalata; in some, Pronocephalata shared a sister relationship to the clade of Fasciolidae + Echinostomatidae, with Opisthorchiidae + Heterophyidae sister to the Pronocephalata + Fasciolidae + Echinostomatidae clade (Le et al., 2016), while in others, Pronocephalata has recovered sister to Fasciolidae + Echinostomatidae + Opisthorchiidae + Heterophyidae (Suleman et al., 2021; Xu et al., 2021). However, despite the differences obtained in the various studies, mitochondrial genomes remain a precious data source for phylogenetic inferences of evolutionary relationships among trematodes.

The addition of more sequences from various genetic markers to fill the gaps in the phylogeny of Digenea will no doubt be crucial in our understanding of trematode molecular systematics. It is definite that the puzzle pieces making up the picture for trematode phylogenies will gradually be put together in the coming years, with advanced sequencing technologies and wider-scale biodiversity surveys.

11.2 Molecular Systematics of Medically Important Orders Within Digenea

Olson et al. (2003) divided Digenea into Diplostomida and Plagiorchiida based on the relationships obtained with the nuclear rRNA genes (Olson et al., 2003). In Chap. 2, trematode groups medically important to humans globally were discussed. This section will discuss some of the medically important groups in detail in the context of molecular phylogenetics and their relationships.

11.2.1 Diplostomida

Suborder Diplostomida represents the lesser diverse suborder of the two within Digenea, consisting of an estimated 1477 species in 210 genera and 3 superfamilies (Hernandez-Mena et al., 2017; Littlewood et al., 2015). The three superfamilies within Diplostomida are Schistosomatoidea, Brachylaimoidea, and Diplostomoidea. The total number of species in Diplostomida contains just 12.7% of all digenean species.

Phylogenetic relationships within Diplostomida have been of considerable debate, with differences between nuclear and mitochondrial phylogenies. As mentioned in Sect. 11.1, the nuclear phylogenies recovered the monophyly of Diplostomida, where representatives belonging to Schistosomatoidea, Brachylaimoidea, and Diplostomoidea were grouped together. Schistosomatoidea shared a closer relationship to Diplostomoidea, with Brachylaimoidea sister to the two (Brabec et al., 2015; Locke et al., 2018). Contrarily, mitochondrial genome phylogenies could not recover the monophyly of Diplostomida. Representatives of Diplostomoidea and Clinostomidae were more closely related to suborder Plagiorchiida. Schistosomatoidea was, in turn, the most basal group in Digenea (Brabec et al., 2015; Fu et al., 2019; Locke et al., 2018). The differences in phylogenies obtained from the two loci certainly call for a sampling of more representative species and a closer look at phylogenetic relationships of the members within Diplostomida.

Schistosomatidae

Family Schistosomatidae, belonging to superfamily Schistosomatoidea, contains the medically important schistosomes. Globally, at least 200 million people are afflicted with the disease caused by schistosomes (WHO, 2021b). Transmission has also been reported from 78 countries, where a majority of the countries are in Africa. Six species have been reported to parasitize humans as definitive hosts: *Schistosoma japonicum*, *Schistosoma mansoni*, *Schistosoma haematobium*, *Schistosoma mekongi*, *Schistosoma intercalatum*, and *Schistosoma guineensis* (WHO, 2021b; Zarowiecki et al., 2007). Aside from humans, other species parasitize other hosts such as birds, mammals, and reptiles. Additionally, other schistosome species which do not utilize humans as definitive hosts can also affect humans, causing cercarial dermatitis (Brant & Loker, 2009).

Based on egg morphology, intermediate host specificity, and biogeography, *Schistosoma* is divided into four groups—*S. mansoni* group, *S. haematobium* group, *S. japonicum* group, and *S. indicum* group (Lockyer et al., 2003). The species in *S. mansoni* group have eggs with a lateral spine, utilize snails belonging to *Biomphalaria*, and are widespread in Africa and present in South America. Species in *S. haematobium* group have eggs with a terminal spine, utilize snails belonging to *Bulinus*, and are mainly in Africa. For species in *S. japonicum* group, their eggs have a rounded minute spine and are widespread throughout East Asia and Southeast Asia. For species in *S. indicum* group, their egg morphology differs and mainly contain species from India, with none of them infecting humans as the definitive host.

Various genetic markers have been used to study the molecular phylogenetics of *Schistosoma*, and they include the nuclear rRNA genes, ITS regions, and the mitochondrial genes (Brant & Loker, 2009; Littlewood et al., 2006; Lockyer et al., 2003; Snyder, 2004; Zhao et al., 2012). Using the nuclear 18S and 28S rRNA genes, ITS regions, and the *COI* gene, Brant and Loker (2009) performed a phylogenetic analysis of schistosomes in North American birds and snails. They showed that *Schistosoma* and *Trichobilharzia* are sister clades (Brant & Loker, 2009). Lockyer et al. (2003) used a combination of the nuclear rRNA genes and the *COI* gene from 30 species within Schistosomatidae for phylogenetic analysis (Lockyer et al., 2003). The genetic markers revealed that the bird schistosomes (*Ornithobilharzia* + *Austrobilharzia*) is sister to *Schistosoma* + *Orientobilharzia*. This clade is in turn sister to the BTDG clade, containing *Bilharziella*, *Trichobilharzia*, *Dendritobilharzia*, and *Gigantobilharzia*. Figure 11.2 depicts the molecular relationship between the genera within Schistosomatidae. *Schistosoma* was revealed to be not monophyletic, with genus *Orientobilharzia* nested in between the *Schistosoma* species from East Asia and other Schistosoma species. The non-monophyly of *Schistosoma* was also revealed in a study performed by Morgan et al. (2003), where *Orientobilharzia* was nested in between (Morgan et al., 2003). Within *Schistosoma*, mitochondrial and nuclear phylogenies revealed two distinct clades, one consisting of the species from East Asia and Southeast Asia and the other

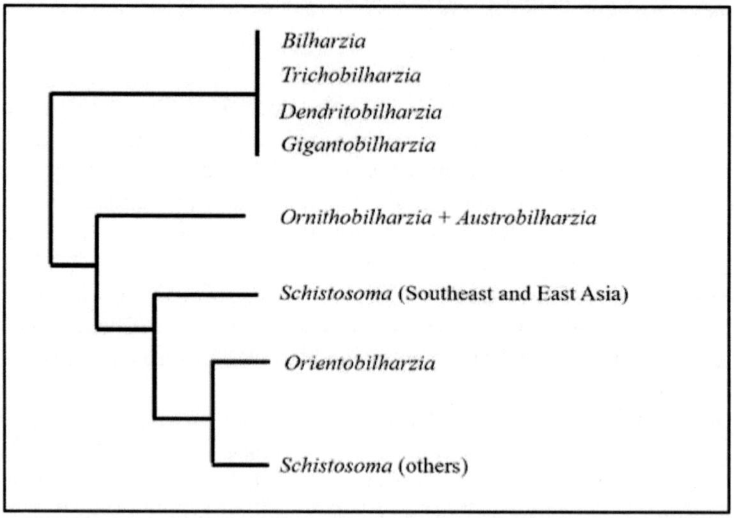

Fig. 11.2 Phylogeny within Schistosomatidae, based on the nuclear rRNA and *COI* genes. (Phylogeny adapted from Lockyer et al. (2003)) (Lockyer et al., 2003)

consisting of the African and Indian species (Littlewood et al., 2006; Morgan et al., 2003; Zhao et al., 2012): *S. haematobium* + *Schistosoma spindale* sister to *S. mansoni* in one clade and the second clade containing *S. mekongi* + *Schistosoma malayensis* sister to *S. japonicum*. Additionally, phylogenetic analysis revealed that the species from East Asia were the earliest diverged *Schistosoma* species, compared to the African and Indian species (Littlewood et al., 2006; Lockyer et al., 2003; Morgan et al., 2003).

Information obtained from molecular data, morphological characters, host use, and geographical distribution of schistosomes helps elucidate relationships within Schistosomatidae.

11.2.2 Plagiorchiida

Suborder Plagiorchiida represents the remainder of the species belonging to Digenea and is the larger of the two subclasses with 13 orders (Hernandez-Mena et al., 2017; Littlewood et al., 2015). The relationships within Plagiorchiida were less consistent than Diplostomida across the various molecular systematics studies, clearly indicating the challenges for trematode systematics. Within Plagiorchiida, the three major groups containing medically important species to humans are Echinostomatoidea, Troglotrematidae, and Opisthorchioidea. Each of these three groups will be further discussed below, in the context of molecular phylogenetic studies performed, emphasizing medically important species.

Echinostomatoidea

Superfamily Echinostomatoidea is a large group of digeneans, consisting of nine families with more than 100 genera (Tkach et al., 2016). The nine families are Echinostomatidae, Echinochasmidae, Fasciolidae, Philophthalmidae, Psilostomidae, Himasthlidae, Cathaemasiidae, Rhopaliidae, and Cyclocoelidae. Of the nine, families Echinostomatidae and Fasciolidae contain trematodes that are medically important to humans.

Echinostomatidae

Family Echinostomatidae is intestinal trematodes that parasitize birds and mammals, including humans. Infection in humans occurs when they ingest the infective stage metacercariae present in intermediate hosts, resulting in echinostomiasis (Saijuntha et al., 2011). The eight genera attributed to causing disease in humans are *Echinostoma*, *Echinochasmus*, *Acanthoparyphium*, *Artyfechinostomum*, *Episthmium*, *Himasthla*, *Hypoderaeum*, and *Isthmiophora* (Nolkong et al., 2014). Echinostomes are morphologically characterized by having a distinctive cephalic crown of collar spines, ventral sucker larger than oral sucker, two testes tandemly or symmetrically arranged, a pre-testicular ovary, and a cirrus sac (Laidemitt et al., 2019). Although echinostomes can be distinguished by the number of collar spines in immature and mature stages, they are still prone to misidentification due to the small size and possible loss and retraction of spines.

Molecular phylogenetics has chiefly been performed on *Echinostoma* and related species in defining species limits of *Echinostoma* species complexes and aiding in molecular identification of the cosmopolitan parasites from various hosts. Using both the mitochondrial nicotinamide adenine dinucleotide 1 (*ND1*) and *COI* genes for echinostomes found in North America, two distinct clades resulted, with *Echinostoma* in one clade and a second clade consisting of *Echinoparyphium* and *Hypoderaeum* (Detwiler et al., 2010). Similar clade groupings also resulted in a study that characterized echinostomes from East Africa using *ND1* and the nuclear 28S rRNA gene (Laidemitt et al., 2019). Within *Echinostoma*, Georgieva et al. (2014) performed molecular prospecting for the European species of *Echinostoma* using *ND1* and 28S rRNA genes for five of the six European species of *Echinostoma* (Georgieva et al., 2014). Two clades resulted, with the first consisting of *Echinostoma revolutum* from the USA and Europe, *Echinostoma paraulum*, and *Echinostoma miyagawai* and the second consisting of *Echinostoma caproni*, *Echinostoma trivolis*, *Echinostoma bolschewense*, and other *Echinostoma* species. Using the mitochondrial *COI* gene, Chai et al. (2020) showed that *E. revolutum* was sister to a clade consisting of *Echinostoma robustum* and *E. miyagawai* (Chai et al., 2020). *E. revolutum* have also been shown to be phylogenetically divided based on their lineages—the Eurasian (consisting of Southeast Asian) and American lineages (Mohanta et al., 2019; Nagataki et al., 2015). Also, molecular phylogenetics provided support for the elevation of *Echinochasmus* to distinct family status—Echinochasmidae. Mitochondrial genomes and nuclear genetic markers revealed that family Echinochasmidae is phylogenetically distinct from Echinostomatidae, where Echinochasmidae was basal to the clade consisting of Echinostomatidae + Fasciolidae (Le et al., 2020; Li et al. 2019b; Tkach et al., 2016).

Fasciolidae

Family Fasciolidae, notable for their large size and considerable public health and veterinary significance, is a relatively small family comprising only nine recognized species in six genera (Lotfy et al., 2008). Of note, *Fasciola hepatica*, *Fasciola gigantica*, and *Fasciolopsis buski* result in fasciolosis in humans. The two closely related species, *F. hepatica* and *F. gigantica*, are generally distributed based on their geographical location and cattle hosts (Amer et al., 2011; Peng et al., 2009). *F. hepatica* is usually found in the Americas, Europe, Western and North Asia, coinciding with the distribution of *Bos taurus* and *Bos indicus*, while *F. gigantica* is the known species in South Asia and Africa, coinciding with the presence of *Bos indicus*. Additionally, morphological intermediate forms between *F. hepatica* and *F. gigantica* have been discovered, complicating morphological-based identification (Peng et al., 2009).

Molecular-based approaches generally focus on species identification of *Fasciola* from humans and other definitive hosts and distinguishing the intermediate forms of *Fasciola*. Using the nuclear ITS regions and the mitochondrial *ND1* and *COI* genes, *Fasciola* from sheep and cattle hosts have been successfully identified in Egypt, Iran, India, China, and Thailand (Amer et al., 2011; Peng et al., 2009; Prasad et al., 2008; Shafiei et al., 2014; Wannasan et al., 2014). Aside from species identification using genetic markers, phylogenetic studies within Fasciolidae have also been conducted. Using the 28S rRNA gene, Lotfy et al. (2008) showed that *Fascioloides magna* was nested between *Fasciola jacksoni* and *F. gigantica* and *F. hepatica*, rendering the non-monophyly of genus *Fasciola* (Lotfy et al., 2008). *Parafasciolopsis fasciolaemorpha* and *Fasciolopsis buski* were in turn sister to the *Fasciola* + *Fascioloides* clade, and *Protofasciola robusta* was placed basal to Fasciolidae. The close phylogenetic relationship of *Fasciola jacksoni* and *Fascioloides magna* were also observed using the nuclear rRNA genes, ITS regions, complete mitochondrial genomes, and mitochondrial genes, suggesting a re-evaluation of the taxonomic status of *Fasciola jacksoni* and *Fascioloides magna* (Amer et al., 2011; Heneberg, 2013; Le et al., 2020; Li et al. 2019b; Lotfy et al., 2008).

Troglotrematidae

Genus *Paragonimus*, within the family Troglotrematidae, are also commonly known as lung flukes. Consisting of over 50 species, they constitute an ancient and

Table 11.1 Geographical distribution of *Paragonimus* species

Species/species complex	Geographical distribution	Species within a species complex
P. westermani species complex	Southeast Asia, South Asia	*P. westermani*, *P. siamensis*
P. heterotremus species complex	Southeast Asia, East Asia	*P. heterotremus*, *P. pseudoheterotremus*
P. skrjabini species complex	Southeast Asia, East Asia	*P. skrjabini*, *P. miyazakii*, *P. proliferus*
P. ohirai species complex	East Asia	*P. ohirai*, *P. bangkokensis*, *P. harinasutai*
P. africanus	West Africa	–
P. uterobilateralis	West Africa	–
P. kellicotti	North America	–
P. mexicanus	Central and South America	–

species-rich group of parasites in Asia, Africa, and the Americas. Several species and species complexes are known to afflict humans with paragonimiasis (WHO, 2020). Across the countries, approximately 23 million people are infected, and 292 million people are estimated to be at risk (Keiser & Utzinger, 2009). There are four major species complexes in Asia—*Paragonimus heterotremus*, *Paragonimus skrjabini*, *Paragonimus ohirai*, and *Paragonimus westermani* (Procop, 2009) *Paragonimus kellicotti* is endemic to North America, while *Paragonimus mexicanus* is found in Central and South America. Table 11.1 shows the geographical distribution of *Paragonimus* and their species within the species complexes. The diversity of the populations and species complexes, together with the broad geographical range of the parasites, complicates the taxonomy and identification of *Paragonimus*. Morphological features in adults of *Paragonimus*, such as lobation patterns of the ovary and testes and the appearance of the cuticular spines, are relied upon for identification (Procop, 2009). However, evidence demonstrates that morphological differences do not necessarily correlate with phylogenetic distances and the closeness of relationships genetically (Doanh et al., 2013). For example, although *Paragonimus bangkokensis* and *Paragonimus harinasutai* are genetically related, they are morphologically distinct at adult and metacercariae stages.

Molecular studies have greatly aided in identifying species within *Paragonimus* species complexes, especially in areas where morphological-based identification does not provide any answers. Additionally, molecular studies have provided evidence revealing the complexity of species in *Paragonimus*, proposing new species limits and re-appraising taxonomic statuses, especially for species complexes (Doanh et al., 2009, 2013, 2015; Sanpool et al., 2015). Studying the species complexes in Asia using the ITS2 region, *P. heterotremus* species complex (*P. heterotremus* + *P. pseudoheterotremus*) was sister to *Paragonimus macrochis*. This clade was in turn sister to *P. skrjabini* species complex (*P. skrjabini* + *P. miyazakii* + *P. proliferus*). *P. ohirai* species complex, consisting of *P. ohirai* + *P. harinasutai* + *P. bangkokensis*, was sister to the *P. heterotremus* and *P. skrjabini* species complexes. The *P. westermani* species complex (*P. westermani* + *Paragonimus siamensis*) was sister to all species sampled (Sanpool et al., 2015). The basal relationship of *P. westermani* to the rest of *Paragonimus* was also supported using complete genomes, where they were sister to *Paragonimus kellicotti*, *P. ohirai*, *P. heterotremus*, and *P. miyazakii* (Rosa et al., 2020).

Within *Paragonimus* species complexes, the *P. westermani* and *P. heterotremus* species complexes are the most widely studied. For *P. westermani*, molecular evidence using the ITS2 region, the mitochondrial 16S rRNA, and *COI* genes showed distinct clustering based on geographical region (East Asia, Southeast Asia, and South Asia) (Doanh et al., 2009, 2016). The first group consisted of isolates from India and Sri Lanka, the second group consisted of isolates from Malaysia and the Philippines, and the third group consisted of isolates from Vietnam, Thailand, and East Asian countries. Rekha Devi et al. (2013) and Blair et al. (2001) proposed that the radiation of *P. westermani* arose from one locality through the utilization of snail hosts and then subsequently radiated to other localities through the utilization of more snail species (Blair et al., 2001; Rekha Devi et al., 2013). However, the actual evolutionary scenario could be more complicated than the proposed one due to species complex in different localities.

For *P. heterotremus*, phylogenetic evidence points to distinct lineages based on geographical regions. Isolates from Vietnam, China, and Thailand clustered in one clade, while isolates from India were in another, using the ITS2 region and 28S rRNA gene (Voronova et al., 2020). Molecular data was also used to clarify the taxonomic status of *P. pseudoheterotremus* and their phylogenetic relationship with *P. heterotremus*. Results using the *COI* gene revealed genetic differences between *P. heterotremus* and *P. pseudoheterotremus*, whereas no difference was present using the ITS2 region (Thaenkham & Waikagul, 2008).

Opisthorchioidea

Opisthorchiidae, Heterophyidae, and Cryptogonimidae are the three families in Opisthorchioidea (Keiser & Utzinger, 2009). They share similarities in morphological characters and life cycles, with Opisthorchiidae and Heterophyidae utilizing similar second intermediate hosts and definitive hosts (Chai et al., 2005; Waikagul & Thaenkham, 2014). Opisthorchiidae and Heterophyidae contain trematodes of medical importance to humans, with notable species belonging to *Opisthorchis viverrini* and *Clonorchis sinensis* (in Opisthorchiidae) and *Haplorchis* (in Heterophyidae). Taxonomy and classification of species within Opisthorchioidea have been reassessed many times due to uncertainties in relationships at family and genera levels.

Molecular systematics studies for Opisthorchioidea have mainly been performed using the nuclear rRNA genes to aid in their classification. The paraphyly of Heterophyidae has been shown by several studies, using the nuclear rRNA genes individually, concatenated nuclear rRNA genes, and the ITS2 region (Le et al., 2017;

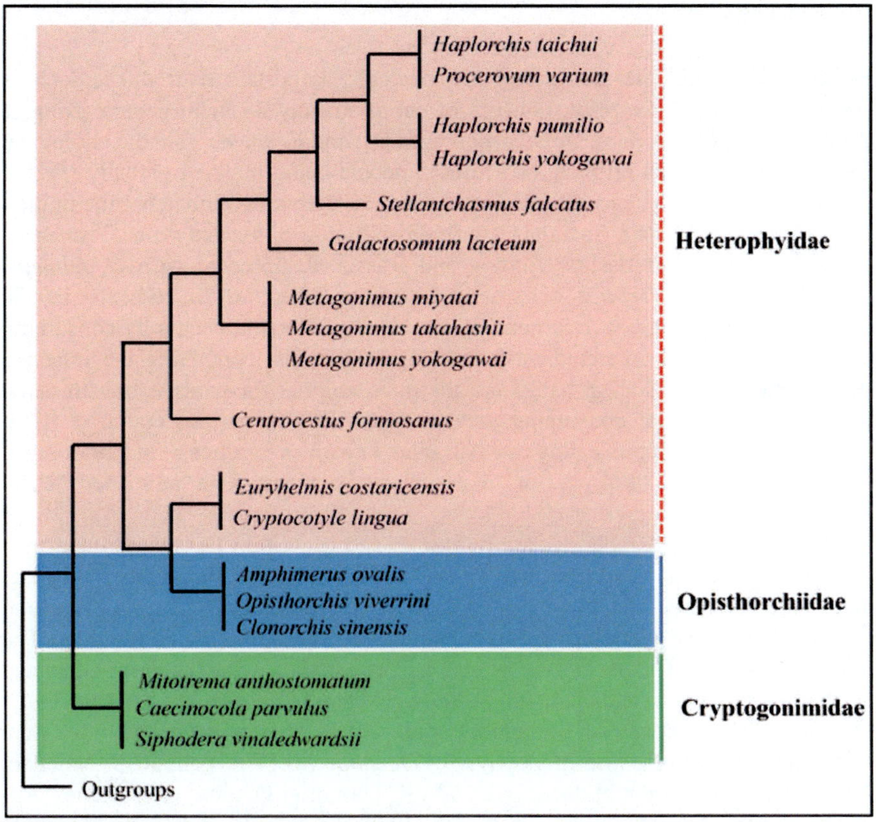

Fig. 11.3 Phylogeny within Opisthorchioidea, based on the concatenated nuclear rRNA genes. (Phylogeny adapted from Thaenkham et al. (2011))

Martinez-Aquino et al., 2017; Thaenkham et al., 2011, 2012). Figure 11.3 illustrates the phylogenetic relationships within Opisthorchioidea. Thaenkham et al. (2011, 2012) revealed that *Euryhelmis costaricensis* and *Cryptocotyle lingua* were nested together with Opisthorchiidae, showing a closer relationship to members of Opisthorchiidae rather than the rest of Heterophyidae (Thaenkham et al., 2011, 2012). The clade with the rest of Heterophyidae consists of *Metagonimus*, *Haplorchis*, *Procerovum*, and *Stellantchasmus*, as examples of medically important species within Heterophyidae. Additionally, Hernandez-Orts et al. (2019) showed that *Centrocestus formosanus* (belonging to Heterophyidae) was revealed to be basal to the Opisthorchiidae + Heterophyidae clade. Among the three families, Cryptogonimidae was basal to Opisthorchiidae + Heterophyidae. Clearly, Opisthorchioidea presents broad systematic and taxonomic challenges.

Opisthorchiidae

Family Opisthorchiidae, containing the medically important liver flukes such as *Opisthorchis* and *Clonorchis*, consists of approximately 33 valid genera (King & Scholz, 2001). *O. viverrini*, *Opisthorchis felineus*, and *C. sinensis* are the etiological agents, resulting in an estimated 45 million people being infected globally (WHO, 2021a). *O. viverrini* is endemic in Southeast Asia, with approximately nine million people infected. In Asia (mainly China), 35 million are infected with *C. sinensis*, while one million in Eastern Europe and Russia are infected with *O. felineus*. Additionally, *O. viverrini* is classified as a group 1 carcinogen, resulting in bile duct cancer or cholangiocarcinoma (CCA). Human infection occurs by consuming undercooked second intermediate hosts (cyprinid fishes) containing the infective stage metacercariae. Eating habits are the main culprits for parasite transmission, with the popularity of consuming undercooked fishes in various countries (Chai et al., 2005), for example, eating raw fish salad known as "yusheng" in China, sushi in Korea and Japan, "koi pla" or "pla som" in Southeast Asia, and dried fish in Russia.

Despite the considerable disease burden of species within Opisthorchiidae, systematic relationships have remained an issue of controversy. Variations in phylogenies have been observed, emphasizing the complexity of relationships in this group of trematodes and defining species boundaries for species delimitation. The non-monophyly of *Opisthorchis* was seen in some studies, with *O. felineus* depicted as sharing a closer relationship to *C. sinensis* than *O. felineus* to *O. viverrini* + *O. lobatus* (Kiyan et al., 2018; Sanpool et al., 2018). Mitochondrial genomes were also congruent with the non-monophyly of *Opisthorchis* (Ma et al., 2019; Na et al., 2016). Within Opisthorchiidae, ITS2 and mitochondrial genome phylogenies usually show that two main clades are present, with *Opisthorchis* + *Clonorchis* in one and *Metorchis* in another. *Amphimerus* is sister to *Opisthorchis* + *Clonorchis* + *Metorchis*. Studies on population genetics of various geographical isolates of *O. viverrini* have also revealed that *O. viverrini* might be a species complex, expanding their species boundaries. Saijuntha et al. (2007) showed that *O. viverrini* in Lao PDR and Thailand consists of at least two genetically distinct yet morphologically similar species, with five distinct groups detected (Saijuntha et al., 2007).

Heterophyidae

Family Heterophyidae contains trematodes that inhabit the intestinal tract of mammals and birds. They share similar second intermediate hosts and definitive hosts with Opisthorchiidae and are also similar in morphology to other species within superfamily Opisthorchioidea (Chai et al., 2005; Keiser & Utzinger, 2009). Containing five subfamilies and over 50 genera, a substantial number of species can be found parasitizing humans (Chai et al., 2005; Keiser & Utzinger, 2009). The notable species, mainly located in East Asia and Southeast Asia, include *Haplorchis taichui*, *Haplorchis pumilio*, *Haplorchis yokogawai*, *Procerovum varium*, *Stellantchasmus falcatus*, and *Metagonimus yokogawai*. Like Opisthorchiidae, human infection is acquired through the consumption of undercooked freshwater fishes containing the infective stage metacercariae (Chai et al., 2005). Generally, Heterophyidae can be distinguished from Opisthorchiidae and Cryptogonimidae through the presence of a genital or ventrogenital sac in adult worms. Heterophyidae can be further classified based on the shape of the anterior part of the adult body, the presence or absence of spines around the oral aperture, and the number and position of testes.

However, like Opisthorchiidae, the molecular systematics of Heterophyidae has been uncertain, especially among genera. The paraphyly of Heterophyidae has been confirmed from various studies using the nuclear rRNA genes and the ITS regions (Le et al., 2017; Martinez-Aquino et al., 2017; Thaenkham et al., 2011, 2012). Thaenkham et al. (2012) revealed that *Euryhelmis costaricensis* and *Cryptocotyle lingua*, both belonging to Heterophyidae, were found nested within Opisthorchiidae, while other species in Heterophyidae were in another clade (Thaenkham et al., 2012). Within the clade containing the other species in Heterophyidae, *Haplorchis* + *Procerovum* + *Haplorchoides* were closely related, while *Metagonimus* is the sister clade, congruent with the subfamily division into Haplorchinae. In another study, using the 28S rRNA gene for 47 species within Opisthorchioidea, in addition to the paraphyly of Heterophyidae, *Centrocestus formosanus* was found to be basal to the Opisthorchiidae + Heterophyidae clade (Hernandez-Orts et al., 2019). Another inconsistency in a phylogenetic relationship is the close relationship between *Procerovum* and *Haplorchis*. *Haplorchis* was rendered non-monophyletic due to *Procerovum* sp. nested within. Using the concatenated and individual nuclear rRNA genes, *Procerovum* was sister to *H. taichui*, with *H. pumilio* and *H. yokogawai* sister to *Procerovum* + *H. taichui*. Contrarily, with the ITS2 region, *Procerovum* was sister to *H. pumilio* instead (Tatonova & Besprozvannykh, 2019; Thaenkham et al., 2010). The mitochondrial 16S rRNA gene also showed a different phylogenetic relationship, where *Procerovum* + *H. yokogawai* now share a closer relationship than *H. taichui* + *H. pumilio*. The close relationship between *Procerovum* and *Haplorchis* is supported with similar morphological characters, where species in both genera possess a single testis (Chan et al., 2022).

Albeit the inconsistencies in the molecular systematics of Heterophyidae and Opisthorchiidae, genetic markers and molecular phylogenetics have undoubtedly aided in the taxonomy and identification of medically important species within Opisthorchioidea. Molecular methods have been helpful, especially where the similar morphological characters of heterophyid and opisthorchiid eggs and immatures hindered accurate identification of medically important species, limiting disease diagnosis and the geographical distribution of the parasites.

11.3 Challenges of Trematode Molecular Systematics

Although the medically important species of trematodes are well known and have been comprehensively studied in terms of tools for molecular identification and disease diagnosis and focusing on specific groups of trematodes, the molecular systematics and evolutionary history of trematodes are only understood in sketchy outline. Much work is still needed to fully complete the picture for trematode molecular systematics.

The molecular phylogenetic framework of Digenea by Olson et al. (2003) based on the 18S and 28S rRNA genes, and even more recently by de Leon and Hernandez-Mena (2019) with 1077 taxa, was a boost in aiding to untangle the evolutionary relationships within (de Leon & Hernandez-Mena, 2019; Olson et al., 2003). Since the last century, the number of trematode genera has exponentially increased to close to 1800 genera, with much effort being performed to taxonomically describe new species. However, the end for trematode molecular systematics and taxonomy is not yet in sight. First, many trematode taxa remain undiscovered, especially in hosts that might not be easily caught and sampled. Along with morphologically similar species and cryptic species that are common for trematodes, the current species diversity might be greatly underinflated. The changing global climate, increasing environmental pollution, and human anthropogenic activity could be major forces affecting trematode species diversity and driving evolutionary processes in the near future (more details in Chap. 3). Second, obtaining intact adult specimens poses another barrier, as more often than not, fecal or intermediate host species are sampled, where usually only immatures are found. Third, many discovered species have yet to have their DNA sequences available, despite the tremendous effort in molecular characterization of new species. Fourth, using different genetic markers and taxa can result in different phylogenetic relationships and might confuse researchers in choosing a suitable genetic marker for the appropriate application. Last, the paucity of universal primers that can provide taxonomic resolution at both higher and lower taxa complicates undertaking large-scale molecular systematics studies for Digenea. By using common genetic markers for the appropriate application with similar trematode species, a more coordinated effort will be a significant step in synchronizing the global molecular data for trematodes.

11.4 What the Future Holds for Trematode Molecular Systematics

Although a key aspect of molecular systematics is understanding evolutionary relationships, the molecular processes occurring at the population and species level are the basis of evolution and diversity at a larger-scale level. DNA barcoding and integrated taxonomy, using both molecular and morphological data, will aid in proper cataloguing of trematode species, with both morphological, life history traits, and molecular data. A suitable genetic marker with universal primers that are readily available for broad taxa range in Digenea will undoubtedly amplify the number of available sequences, setting the stage for large-scale phylogenetic studies and bringing together a uniform system for trematode molecular-based studies. To date, DNA metabarcoding studies specifically targeting Digenea have not been performed. Like DNA barcoding, a key ingredient for successful DNA metabarcoding of Digenea success lies in selecting appropriate genetic markers and primers that can encompass a wide range of species in Digenea. DNA metabarcoding studies in the future will aid in uncovering more species of trematodes, which can help fill the gaps for the molecular systematics of trematodes. Additionally, advancements in DNA sequencing technologies enabled a genomics era for parasites, allowing for the availability of draft genomes and mitochondrial genomes in public databases. The improvements in bioinformatics tools for analytical inferences and the emphasis on data science in the recent decade will also be an influential ingredient in the recipe for untangling trematode phylogenics.

The molecular systematics of trematodes remains a work in progress, and it will be interesting to see what will come to light with the availability of more advanced sequencing technologies coupled with tools for data analytics in the near future.

References

Amer, S., Dar, Y., Ichikawa, M., Fukuda, Y., Tada, C., Itagaki, T., & Nakai, Y. (2011). Identification of *Fasciola* species isolated from Egypt based on sequence analysis of genomic (ITS1 and ITS2) and mitochondrial (*NDI* and *COI*) gene markers. *Parasitology International, 60*(1), 5–12. https://doi.org/10.1016/j.parint.2010.09.003

Blair, D., Davis, G. M., & Wu, B. (2001). Evolutionary relationships between trematodes and snails emphasizing schistosomes and paragonimids. *Parasitology, 123*(Suppl), S229–S243. https://doi.org/10.1017/s003118200100837x

Brabec, J., Kostadinova, A., Scholz, T., & Littlewood, D. T. (2015). Complete mitochondrial genomes and nuclear ribosomal RNA operons of two species of *Diplostomum* (Platyhelminthes: Trematoda): A molecular resource for taxonomy and molecular epidemiology of important fish pathogens. *Parasites & Vectors, 8*, 336. https://doi.org/10.1186/s13071-015-0949-4

Brant, S. V., & Loker, E. S. (2009). Molecular systematics of the avian schistosome genus *Trichobilharzia* (Trematoda: Schistosomatidae) in North America. *The Journal of Parasitology, 95*(4), 941–963. https://doi.org/10.1645/ge-1870.1

Bray, R. A., Gibson, D. I., & Jones, A. (2008). *Keys to the Trematoda* (Vol. 3). CABI Publishing.

Brooks, D. R., O'Grady, R. T., & Glen, D. R. (1985). Phylogenetic analysis of the Digenea (Platyhelminthes: Cercomeria) with comments on their adaptive radiation. *Canadian Journal of Zoology, 63*(2), 411–443. https://doi.org/10.1139/z85-062

Chai, J. Y., Cho, J., Chang, T., Jung, B. K., & Sohn, W. M. (2020). Taxonomy of *Echinostoma revolutum* and 37-collar-Spined *Echinostoma* spp.: A historical review. *The Korean Journal of Parasitology, 58*(4), 343–371. https://doi.org/10.3347/kjp.2020.58.4.343

Chai, J. Y., Darwin Murrell, K., & Lymbery, A. J. (2005). Fish-borne parasitic zoonoses: Status and issues. *International Journal for Parasitology, 35*(11–12), 1233–1254. https://doi.org/10.1016/j.ijpara.2005.07.013

Chan, A. H. E., Saralamba, N., Saralamba, S., Ruangsittichai, J., & Thaenkham, U. (2022). The potential use of mitochondrial ribosomal genes (12S and 16S) in DNA barcoding and phylogenetic analysis of trematodes. *BMC Genomics, 23*, 104.

Chang, Q. C., Liu, G. H., Gao, J. F., Zheng, X., Zhang, Y., Duan, H., Yue, D. M., Fu, X., Su, X., Gao, Y., & Wang, C. R. (2016). Sequencing and characterization of the complete mitochondrial genome from the pancreatic fluke *Eurytrema pancreaticum* (Trematoda: Dicrocoeliidae). *Gene, 576*(1 Pt 1), 160–165. https://doi.org/10.1016/j.gene.2015.09.081

Cribb, T., Bray, R., Olson, P., Timothy, D., & Littlewood, J. (2003). Life cycle evolution in the Digenea: A new perspective from phylogeny. *Advances in Parasitology, 54*, 197–254. https://doi.org/10.1016/s0065-308x(03)54004-0

de Leon, G. P. P., & Hernandez-Mena, D. I. (2019). Testing the higher-level phylogenetic classification of Digenea (Platyhelminthes, Trematoda) based on nuclear rDNA sequences before entering the age of the 'next-generation' tree of life. *Journal of Helminthology, 93*(3), 260–276. https://doi.org/10.1017/S0022149X19000191

Detwiler, J. T., Bos, D. H., & Minchella, D. J. (2010). Revealing the secret lives of cryptic species: Examining the phylogenetic relationships of echinostome parasites in North America. *Molecular Phylogenetics and Evolution, 55*(2), 611–620. https://doi.org/10.1016/j.ympev.2010.01.004

Doanh, N. P., Tu, A. L., Bui, T. D., Loan, T. H., Nonaka, N., Horii, Y., Blair, D., & Nawa, Y. (2016). Molecular and morphological variation of *Paragonimus westermani* in Vietnam with records of new second intermediate crab hosts and a new locality in a northern province. *Parasitology, 143*(12), 1639–1646. https://doi.org/10.1017/S0031182016001219

Doanh, P. N., Hien, H. V., Nonaka, N., Horii, Y., & Nawa, Y. (2013). Discovery of *Paragonimus skrjabini* in Vietnam and its phylogenetic status in the *Paragonimus skrjabini* complex. *Journal of Helminthology, 87*(4), 450–456. https://doi.org/10.1017/S0022149X1200065X

Doanh, P. N., Shinohara, A., Horii, Y., Habe, S., & Nawa, Y. (2009). Discovery of *Paragonimus westermani* in Vietnam and its molecular phylogenetic status in *P. westermani* complex. *Parasitology Research, 104*(5), 1149–1155. https://doi.org/10.1007/s00436-008-1302-z

Doanh, P. N., Thaenkham, U., An, P. T., Hien, H. V., Horii, Y., & Nawa, Y. (2015). Metacercarial polymorphism and genetic variation of *Paragonimus heterotremus* (Digenea: Paragonimidae), and a re-appraisal of the taxonomic status of *Paragonimus pseudoheterotremus*. *Journal of Helminthology, 89*(2), 182–188. https://doi.org/10.1017/S0022149X13000734

Fu, Y. T., Jin, Y. C., & Liu, G. H. (2019). The complete mitochondrial genome of the Caecal fluke of poultry, *Postharmostomum commutatum*, as the first representative from the superfamily Brachylaimoidea. *Frontiers in Genetics, 10*, 1037. https://doi.org/10.3389/fgene.2019.01037

Georgieva, S., Faltýnková, A., Brown, R., Blasco-Costa, I., Soldánová, M., Sitko, J., Scholz, T., & Kostadinova, A. (2014). Echinostoma 'revolutum' (Digenea: Echinostomatidae) species complex revisited: Species delimitation based on novel molecular and morphological data gathered in Europe. *Parasites & Vectors, 7*, 520. https://doi.org/10.1186/s13071-014-0520-8

Gibson, D. I., & Bray, R. A. (1994). The evolutionary expansion and host-parasite relationships of the Digenea. *International Journal for Parasitology, 24*(8), 1213–1226. https://doi.org/10.1016/0020-7519(94)90192-9

Gibson, D. I., Jones, A., & Bray, R. A. (2002). *Keys to the Trematoda* (Vol. 1). CABI Publishing.

Heneberg, P. (2013). Phylogenetic data suggest the reclassification of *Fasciola jacksoni* (Digenea: Fasciolidae) as *Fascioloides jacksoni* comb. nov. *Parasitology Research, 112*(4), 1679–1689. https://doi.org/10.1007/s00436-013-3326-2

Hernandez-Mena, D. I., Garcia-Varela, M., & Perez-Ponce de Leon, G. (2017). Filling the gaps in the classification of the Digenea Carus, 1863: Systematic position of the Proterodiplostomidae Dubois, 1936 within the superfamily Diplostomoidea Poirier, 1886, inferred from nuclear and mitochondrial DNA sequences. *Systematic Parasitology, 94*(8), 833–848. https://doi.org/10.1007/s11230-017-9745-1

Hernandez-Orts, J. S., Georgieva, S., Landete, D. N., & Scholz, T. (2019). Heterophyid trematodes (Digenea) from penguins: A new species of *Ascocotyle* Looss, 1899, first description of metacercaria of *Ascocotyle* (A.) *patagoniensis* Hernandez-Orts et al. (2012), and first molecular data. *Int J Parasitol Parasites Wildl, 8*, 94–105. https://doi.org/10.1016/j.ijppaw.2018.12.008

Jones, A., Bray, R. A., & Gibson, D. I. (2005). *Keys to the Trematoda* (Vol. 2). CABI Publishing.

Keiser, J., & Utzinger, J. (2009). Food-borne trematodiases. *Clinical Microbiology Reviews, 22*(3), 466–483. https://doi.org/10.1128/CMR.00012-09

King, S., & Scholz, T. (2001). Trematodes of the family Opisthorchiidae: A minireview. *The Korean Journal of Parasitology, 39*(3), 209–221. https://doi.org/10.3347/kjp.2001.39.3.209

Kiyan, V. S., Bulashev, A. K., & Katokhin, A. V. (2018). *Opisthorchis felineus* and *Metorchis bilis* Metacercariae in cyprinid fish Leuciscus idus in Nura-Sarysu River, Kazakhstan. *Korean Journal of Parasitology, 56*(3), 267–274. https://doi.org/10.3347/kjp.2018.56.3.267

Kostadinova, A., & Pérez-del-Olmo, A. (2014). The systematics of the Trematoda. In R. Toledo & B. Fried (Eds.), *Digenetic trematodes* (pp. 21–44). Springer. https://doi.org/10.1007/978-1-4939-0915-5_2

Laidemitt, M. R., Brant, S. V., Mutuku, M. W., Mkoji, G. M., & Loker, E. S. (2019). The diverse echinostomes from East Africa: With a focus on species that use *Biomphalaria* and *Bulinus* as intermediate hosts. *Acta Tropica, 193*, 38–49. https://doi.org/10.1016/j.actatropica.2019.01.025

Landeryou, T., Kett, S. M., Ropiquet, A., Wildeboer, D., & Lawton, S. P. (2020). Characterization of the complete mitochondrial genome of *Diplostomum baeri*. *Parasitology International, 79*, 102166. https://doi.org/10.1016/j.parint.2020.102166

Le, T. H., Nguyen, K. T., Nguyen, N. T., Doan, H. T., Dung, D. T., & Blair, D. (2017). The ribosomal transcription units of *Haplorchis pumilio* and *H. taichui* and the use of 28S rDNA sequences for phylogenetic identification of common heterophyids in Vietnam. *Parasites & Vectors, 10*(1), 17. https://doi.org/10.1186/s13071-017-1968-0

Le, T. H., Nguyen, N. T. B., Nguyen, K. T., Doan, H. T. T., Dung, D. T., & Blair, D. (2016). A complete mitochondrial genome from *Echinochasmus japonicus* supports the elevation of Echinochasminae Odhner, 1910 to family rank (Trematoda: Platyhelminthes). *Infection, Genetics and Evolution, 45*, 369–377. https://doi.org/10.1016/j.meegid.2016.09.024

Le, T. H., Pham, K. L. T., Doan, H. T. T., Le, T. K. X., Nguyen, K. T., & Lawton, S. P. (2020). Description and phylogenetic analyses of ribosomal transcription units from species of Fasciolidae (Platyhelminthes: Digenea). *Journal of Helminthology, 94*, e136. https://doi.org/10.1017/S0022149X20000164

Li, Y., Ma, X. X., Lv, Q. B., Hu, Y., Qiu, H. Y., Chang, Q. C., & Wang, C. R. (2019a). Characterization of the complete mitochondrial genome sequence of *Tracheophilus cymbius* (Digenea), the first representative from the family Cyclocoelidae. *Journal of Helminthology, 94*, e101. https://doi.org/10.1017/S0022149X19000932

Li, Y., Qiu, Y. Y., Zeng, M. H., Diao, P. W., Chang, Q. C., Gao, Y., Zhang, Y., & Wang, C. R. (2019b). The complete mitochondrial genome of *Echinostoma miyagawai*: Comparisons with closely related species and phylogenetic implications. *Infection, Genetics and Evolution, 75*, 103961. https://doi.org/10.1016/j.meegid.2019.103961

Littlewood, D. T., Lockyer, A. E., Webster, B. L., Johnston, D. A., & Le, T. H. (2006). The complete mitochondrial genomes of *Schistosoma haematobium* and *Schistosoma spindale* and the evolutionary history of mitochondrial genome changes among parasitic flatworms. *Molecular Phylogenetics and Evolution, 39*(2), 452–467. https://doi.org/10.1016/j.ympev.2005.12.012

Littlewood, D. T. J., Bray, R. A., & Waeschenbach, A. (2015). Phylogenetic patterns of diversity in cestodes and trematodes. In B. R. Krasnov, D. T. J. Littlewood, & S. Morand (Eds.), *Parasite diversity and diversification: Evolutionary ecology meets phylogenetics* (pp. 304–319). Cambridge University Press. https://doi.org/10.1017/CBO9781139794749.020

Liu, Z. X., Zhang, Y., Liu, Y. T., Chang, Q. C., Su, X., Fu, X., Yue, D. M., Gao, Y., & Wang, C. R. (2016). Complete mitochondrial genome of *Echinostoma hortense* (Digenea: Echinostomatidae). *The Korean Journal of Parasitology, 54*(2), 173–179. https://doi.org/10.3347/kjp.2016.54.2.173

Locke, S. A., Van Dam, A., Caffara, M., Pinto, H. A., Lopez-Hernandez, D., & Blanar, C. A. (2018). Validity of the Diplostomoidea and Diplostomida (Digenea, Platyhelminthes) upheld in phylogenomic analysis. *International Journal for Parasitology, 48*(13), 1043–1059. https://doi.org/10.1016/j.ijpara.2018.07.001

Lockyer, A. E., Olson, P. D., Ostergaard, P., Rollinson, D., Johnston, D. A., Attwood, S. W., Southgate, V. R., Horak, P., Snyder, S. D., Le, T. H., Agatsuma, T., McManus, D. P., Carmichael, A. C., Naem, S., & Littlewood, D. T. (2003). The phylogeny of the Schistosomatidae based on three genes with emphasis on the interrelationships of *Schistosoma* Weinland, 1858. *Parasitology, 126*(Pt 3), 203–224. https://doi.org/10.1017/s0031182002002792

Lotfy, W. M., Brant, S. V., DeJong, R. J., Le, T. H., Demiaszkiewicz, A., Rajapakse, R. P., Perera, V. B., Laursen, J. R., & Loker, E. S. (2008). Evolutionary origins, diversification, and biogeography of liver flukes (Digenea, Fasciolidae). *The American Journal of Tropical Medicine and Hygiene, 79*(2), 248–255.

Ma, J., He, J. J., Liu, G. H., Leontovyc, R., Kasny, M., & Zhu, X. Q. (2016). Complete mitochondrial genome of the giant liver fluke *Fascioloides magna* (Digenea: Fasciolidae) and its comparison with selected trematodes. *Parasites & Vectors, 9*(1), 429. https://doi.org/10.1186/s13071-016-1699-7

Ma, J., He, J. J., Zhou, C. Y., Sun, M. M., Cevallos, W., Sugiyama, H., Zhu, X. Q., & Calvopina, M. (2019). Characterization of the mitochondrial genome sequences of the liver fluke *Amphimerus* sp. (Trematoda: Opisthorchiidae) from Ecuador and phylogenetic implications. *Acta Tropica, 195*, 90–96. https://doi.org/10.1016/j.actatropica.2019.04.025

Mahmud, R., Lim, Y. A. L., & Amir, A. (2017). *Medical parasitology* (1st ed.). Springer.

Martinez-Aquino, A., Vidal-Martinez, V. M., & Aguirre-Macedo, M. L. (2017). A molecular phylogenetic appraisal of the acanthostomines *Acanthostomum* and *Timoniella* and their position within Cryptogonimidae (Trematoda: Opisthorchioidea). *PeerJ, 5*, e4158. https://doi.org/10.7717/peerj.4158

Mohanta, U. K., Watanabe, T., Anisuzzaman, O., & Y., & Itagaki, T. (2019). Characterization of *Echinostoma revolutum* and *Echinostoma robustum* from ducks in Bangladesh based on morphology, nuclear ribosomal ITS2 and mitochondrial *nad1* sequences. *Parasitology International, 69*, 1–7. https://doi.org/10.1016/j.parint.2018.11.002

Morgan, J. A. T., DeJong, R. J., Kazibwe, F., Mkoji, G. M., & Loker, E. S. (2003). A newly-identified lineage of *Schistosoma*. *International Journal for Parasitology, 33*(9), 977–985. https://doi.org/10.1016/s0020-7519(03)00132-2

Na, L., Gao, J. F., Liu, G. H., Fu, X., Su, X., Yue, D. M., Gao, Y., Zhang, Y., & Wang, C. R. (2016). The complete mitochondrial genome of *Metorchis orientalis* (Trematoda: Opisthorchiidae): Comparison with other closely related species and phylogenetic implications. *Infection, Genetics and Evolution, 39*, 45–50. https://doi.org/10.1016/j.meegid.2016.01.010

Nadler, S. A., & de Leon, G. P. P. (2011). Integrating molecular and morphological approaches for characterizing parasite cryptic species: Implications for parasitology. *Parasitology, 138*(13), 1688–1709. https://doi.org/10.1017/S003118201000168X

Nagataki, M., Tantrawatpan, C., Agatsuma, T., Sugiura, T., Duenngai, K., Sithithaworn, P., Andrews, R. H., Petney, T. N., & Saijuntha, W. (2015). Mitochondrial DNA sequences of 37 collar-spined echinostomes (Digenea: Echinostomatidae) in Thailand and Lao PDR reveals presence of two species: *Echinostoma revolutum* and *E. miyagawai*. *Infection, Genetics and Evolution, 35*, 56–62. https://doi.org/10.1016/j.meegid.2015.07.022

Noikong, W., Wongsawad, C., Chai, J. Y., Saenphet, S., & Trudgett, A. (2014). Molecular analysis of echinostome metacercariae from their second intermediate host found in a localised geographic region reveals genetic heterogeneity and possible cryptic speciation. *PLoS Neglected Tropical Diseases, 8*(4), e2778. https://doi.org/10.1371/journal.pntd.0002778

Olson, P. D., Cribb, T. H., Tkach, V. V., Bray, R. A., & Littlewood, D. T. J. (2003). Phylogeny and classification of the Digenea (Platyhelminthes: Trematoda). *International Journal for Parasitology, 33*(7), 733–755. https://doi.org/10.1016/s0020-7519(03)00049-3

Olson, P. D., & Tkach, V. V. (2005). Advances and trends in the molecular systematics of the parasitic Platyhelminthes. *Advances in Parasitology, 60*, 165–243. https://doi.org/10.1016/s0065-308x(05)60003-6

Pearson, J. C. (1992). On the position of the digenean family Heronimidae: An inquiry into a cladistic classification of the Digenea. *Systematic Parasitology, 21*(2), 81–166. https://doi.org/10.1007/BF00010255

Peng, M., Ichinomiya, M., Ohtori, M., Ichikawa, M., Shibahara, T., & Itagaki, T. (2009). Molecular characterization of *Fasciola hepatica*, *Fasciola gigantica*, and aspermic *Fasciola* sp. in China based on nuclear and mitochondrial DNA. *Parasitology Research, 105*(3), 809–815. https://doi.org/10.1007/s00436-009-1459-0

Peoples, R. C., & Fried, B. (2014). Form and function in the Digenea. In R. Toledo & B. Fried (Eds.), *Digenetic trematodes* (pp. 3–20). Springer. https://doi.org/10.1007/978-1-4939-0915-5_1

Perkins, S. L., Martinsen, E. S., & Falk, B. G. (2011). Do molecules matter more than morphology? Promises and pitfalls in parasites. *Parasitology, 138*(13), 1664–1674. https://doi.org/10.1017/s0031182011000679

Prasad, P. K., Tandon, V., Biswal, D. K., Goswami, L. M., & Chatterjee, A. (2008). Molecular identification of the Indian liver fluke, *Fasciola* (Trematoda: Fasciolidae) based on the ribosomal internal transcribed spacer regions. *Parasitology Research, 103*(6), 1247–1255. https://doi.org/10.1007/s00436-008-1121-2

Procop, G. W. (2009). North American paragonimiasis (caused by *Paragonimus kellicotti*) in the context of global paragonimiasis. *Clinical Microbiology Reviews, 22*(3), 415–446. https://doi.org/10.1128/CMR.00005-08

Rajapakse, R. P. V. J., Pham, K. L. T., Karunathilake, K. J. K., Lawton, S. P., & Le, T. H. (2020). Characterization and phylogenetic properties of the complete mitochondrial genome of *Fascioloides jacksoni* (syn. *Fasciola jacksoni*) support the suggested intergeneric change from *Fasciola* to *Fascioloides* (Platyhelminthes: Trematoda: Plagiorchiida). *Infection, Genetics and Evolution, 82*, 104281. https://doi.org/10.1016/j.meegid.2020.104281

Rekha Devi, K., Narain, K., Mahanta, J., Nirmolia, T., Blair, D., Saikia, S. P., & Agatsuma, T. (2013). Presence of three distinct genotypes within the *Paragonimus westermani* complex in northeastern India. *Parasitology, 140*(1), 76–86. https://doi.org/10.1017/s0031182012001229

Rosa, B. A., Choi, Y. J., McNulty, S. N., Jung, H., Martin, J., Agatsuma, T., Sugiyama, H., Le, T. H., Doanh, P. N., Maleewong, W., Blair, D., Brindley, P. J., Fischer, P. U., & Mitreva, M. (2020). Comparative genomics and transcriptomics of 4 *Paragonimus* species provide insights into lung fluke parasitism and pathogenesis. *Gigascience, 9*(7), giaa073. https://doi.org/10.1093/gigascience/giaa073

Saijuntha, W., Sithithaworn, P., Wongkham, S., Laha, T., Pipitgool, V., Tesana, S., Chilton, N. B., Petney, T. N., & Andrews, R. H. (2007). Evidence of a species complex within the food-borne trematode *Opisthorchis viverrini* and possible co-evolution with their first intermediate hosts. *International Journal for Parasitology, 37*(6), 695–703. https://doi.org/10.1016/j.ijpara.2006.12.008

Saijuntha, W., Tantrawatpan, C., Sithithaworn, P., Andrews, R. H., & Petney, T. N. (2011). Genetic characterization of *Echinostoma revolutum* and *Echinoparyphium recurvatum* (Trematoda: Echinostomatidae) in Thailand and phylogenetic relationships with other isolates inferred by ITS1 sequence. *Parasitology Research, 108*(3), 751–755. https://doi.org/10.1007/s00436-010-2180-8

Sanpool, O., Aung, W. P. P., Rodpai, R., Maleewong, W., & Intapan, P. M. (2018). Human liver fluke *Opisthorchis viverrini* (Trematoda, Opisthorchiidae) in Central Myanmar: New records of adults and metacercariae identified by morphology and molecular analysis. *Acta Tropica, 185*, 149–155. https://doi.org/10.1016/j.actatropica.2018.05.009

Sanpool, O., Intapan, P. M., Thanchomnang, T., Janwan, P., Laymanivong, S., Sugiyama, H., & Maleewong, W. (2015). Morphological and molecular identification of a lung fluke,

Paragonimus macrorchis (Trematoda, Paragonimidae), found in central Lao PDR and its molecular phylogenetic status in the genus *Paragonimus*. *Parasitology International, 64*(6), 513–518. https://doi.org/10.1016/j.parint.2015.06.013

Shafiei, R., Sarkari, B., Sadjjadi, S. M., Mowlavi, G. R., & Moshfe, A. (2014). Molecular and morphological characterization of *Fasciola* spp. isolated from different host species in a newly emerging focus of human fascioliasis in Iran. *Vet Med Int, 2014*, 405740. https://doi.org/10.1155/2014/405740

Snyder, S. D. (2004). Phylogeny and paraphyly among tetrapod blood flukes (Digenea: Schistosomatidae and Spirorchiidae). *International Journal for Parasitology, 34*(12), 1385–1392. https://doi.org/10.1016/j.ijpara.2004.08.006

Suleman, Khan, M. S., Tkach, V. V., Muhammad, N., Zhang, D., Zhu, X. Q., & Ma, J. (2020). Molecular phylogenetics and mitogenomics of three avian dicrocoeliids (Digenea: Dicrocoeliidae) and comparison with mammalian dicrocoeliids. *Parasites & Vectors, 13*(1), 74. https://doi.org/10.1186/s13071-020-3940-7

Suleman, Muhammad, N., Khan, M. S., Tkach, V. V., Ullah, H., Ehsan, M., Ma, J., & Zhu, X. Q. (2021). Mitochondrial genomes of two eucotylids as the first representatives from the superfamily Microphalloidea (Trematoda) and phylogenetic implications. *Parasites & Vectors, 14*(1), 48. https://doi.org/10.1186/s13071-020-04547-8

Tatonova, Y. V., & Besprozvannykh, V. V. (2019). Description of a new species, *Cryptocotyle lata* sp. nov., and discussion of the phylogenetic relationships in Opisthorchioidea. *Parasitology International, 72*, 101939. https://doi.org/10.1016/j.parint.2019.101939

Thaenkham, U., Blair, D., Nawa, Y., & Waikagul, J. (2012). Families Opisthorchiidae and Heterophyidae: Are they distinct? *Parasitology International, 61*(1), 90–93. https://doi.org/10.1016/j.parint.2011.06.004

Thaenkham, U., Dekumyoy, P., Komalamisra, C., Sato, M., Dung Do, T., & Waikagul, J. (2010). Systematics of the subfamily Haplorchiinae (Trematoda: Heterphyidae), based on nuclear ribosomal DNA genes and ITS2 region. *Parasitology International, 59*(3), 460–465. https://doi.org/10.1016/j.parint.2010.06.009

Thaenkham, U., Nawa, Y., Blair, D., & Pakdee, W. (2011). Confirmation of the paraphyletic relationship between families Opisthorchiidae and Heterophyidae using small and large subunit ribosomal DNA sequences. *Parasitology International, 60*(4), 521–523. https://doi.org/10.1016/j.parint.2011.07.015

Thaenkham, U., & Waikagul, J. (2008). Molecular phylogenetic relationship of *Paragonimus pseudoheterotremus*. *The Southeast Asian Journal of Tropical Medicine and Public Health, 39*(2), 217–221.

Tkach, V., Grabda-Kazubska, B., & Swiderski, Z. (2001). Systematic position and phylogenetic relationships of the family Omphalometridae (Digenea, Plagiorchiida) inferred from partial lsrDNA sequences. *International Journal for Parasitology, 31*(1), 81–85. https://doi.org/10.1016/S0020-7519(00)00154-5

Tkach, V., Pawlowski, J., & Mariaux, J. (2000). Phylogenetic analysis of the suborder plagiorchiata (Platyhelminthes, Digenea) based on partial lsrDNA sequences. *International Journal for Parasitology, 30*(1), 83–93. https://doi.org/10.1016/s0020-7519(99)00163-0

Tkach, V. V., Kudlai, O., & Kostadinova, A. (2016). Molecular phylogeny and systematics of the Echinostomatoidea Looss, 1899 (Platyhelminthes: Digenea). *International Journal for Parasitology, 46*(3), 171–185. https://doi.org/10.1016/j.ijpara.2015.11.001

Voronova, A. N., Besprozvannykh, V. V., Ngo, H. D., Plekhova, N. G., Hung, N. M., & Tatonova, Y. V. (2020). *Paragonimus heterotremus* Chen et Hsia, 1964 (Digenea: Paragonimidae): Species identification based on the biological and genetic criteria, and pathology of infection. *Parasitology Research, 119*(12), 4073–4088. https://doi.org/10.1007/s00436-020-06929-4

Voronova, A. N., & Chelomina, G. N. (2020). The SSU rRNA secondary structures of the Plagiorchiida species (Digenea), its applications in systematics and evolutionary inferences. *Infection, Genetics and Evolution, 78*, 104042. https://doi.org/10.1016/j.meegid.2019.104042

Waikagul, J., & Thaenkham, U. (2014). *Approaches to the research on the systematics of fish-borne trematodes*. Academic Press.

Wannasan, A., Khositharattanakool, P., Chaiwong, P., Piangjai, S., Uparanukraw, P., & Morakote, N. (2014). Identification of *Fasciola* species based on mitochondrial and nuclear DNA reveals the co-existence of intermediate *Fasciola* and *Fasciola gigantica* in Thailand. *Experimental Parasitology, 146*, 64–70. https://doi.org/10.1016/j.exppara.2014.09.006

WHO. (2020). *Neglected tropical diseases: Paragonimiasis*. https://www.who.int/news-room/q-a-detail/neglected-tropical-diseases-paragonimiasis

WHO. (2021a). *Foodborne trematode infections*. http://www.who.int/news-room/fact-sheets/detail/foodborne-trematode-infections

WHO. (2021b). *Schistosomiasis*. https://www.who.int/news-room/fact-sheets/detail/schistosomiasis

Xu, G., Zhu, P., Zhu, W., Ma, B., Li, X., & Li, W. (2021). Characterization of the complete mitochondrial genome of *Notocotylus* sp. (Trematoda, Notocotylidae) and its phylogenetic implications. *Parasitology Research, 120*(4), 1291–1301. https://doi.org/10.1007/s00436-021-07075-1

Yang, X., Wang, L., Feng, H., Qi, M., Zhang, Z., Gao, C., Wang, C., Hu, M., Fang, R., & Li, C. (2016). Characterization of the complete mitochondrial genome sequence of *Homalogaster paloniae* (Gastrodiscidae, Trematoda) and comparative analyses with selected digeneans. *Parasitology Research, 115*(10), 3941–3949. https://doi.org/10.1007/s00436-016-5160-9

Zarowiecki, M. Z., Huyse, T., & Littlewood, D. T. (2007). Making the most of mitochondrial genomes--markers for phylogeny, molecular ecology and barcodes in *Schistosoma* (Platyhelminthes: Digenea). *International Journal for Parasitology, 37*(12), 1401–1418. https://doi.org/10.1016/j.ijpara.2007.04.014

Zhao, G. H., Li, J., Mo, X. H., Li, X. Y., Lin, R. Q., Zou, F. C., Weng, Y. B., Song, H. Q., & Zhu, X. Q. (2012). The second transcribed spacer rDNA sequence: An effective genetic marker for inter-species phylogenetic analysis of trematodes in the order Strigeata. *Parasitology Research, 111*(4), 1467–1472. https://doi.org/10.1007/s00436-012-2981-z

Chapter 12
Molecular Systematics of Parasitic Cestodes

Abstract In Chaps. 10 and 11, the molecular systematics of parasitic nematodes and trematodes were discussed, and we saw how molecular phylogenetic studies had contributed significantly to shed light on the relationships among groups of helminths. This chapter elaborated on the molecular systematics of parasitic cestodes, also more commonly known as tapeworms. Phylogenetic relationships among medically important cestodes were discussed through evidence from molecular-based studies. Based on an estimate, there are approximately 5100 species within 680 genera in class Cestoda, with a substantial proportion causing diseases in humans and animals. Examples include *Taenia solium* and *Taenia saginata*, better recognized as the pork and beef tapeworm. Estimates according to the World Health Organization (WHO) and the Food and Agriculture Organization (FAO) indicate that 50 million people are infected with *T. solium*. Additionally, major stakeholders such as the food and livestock industry have to tackle tapeworm infections in animals, leading to economic losses and decreased production capability. Thus, molecular systematics and phylogenetic studies are the cornerstones for disease control by uncovering the parasites' distribution, identity, biology, relationships, host associations, and evolutionary forces driving genetic variation and speciation.

12.1 The Molecular Phylogenetic Backbone of Eucestoda and Their Classification

The classification of Eucestoda was based on traditional taxonomy methods using morphological and ecological traits of both adults and larval stages. Khalil, Jones, and Bray (1994), in the Keys to Cestoda, recognized 12 orders in the subclass Eucestoda (Khalil et al., 1994). Using the general features like the differentiation of reproductive organs, external segmentation, and the type of scolex, Eucestoda was mainly classified into the monofossate, difossate/bothriate, and tetrafossate/actabulate clades (Hoberg, Gardner, & Campbell, 1999; Hoberg, Mariaux, Justine, Brooks, & Weekes, 1997; Olson, Littlewood, Bray, & Mariaux, 2001). Based on these characters, it was hypothesized that the earliest diverging Eucestoda consists of the monofossate lineage, with subclass Cestodaria (containing orders Amphilinidea

U. Thaenkham et al., *Molecular Systematics of Parasitic Helminths*,
https://doi.org/10.1007/978-981-19-1786-8_12

and Gyrocotylidea which were unsegmented and do not possess a scolex) being a common ancestor. The monofossate lineage, consisting of monozoic and polyzoic forms, is characterized by no external segmentation (Hoberg et al., 1997; Hoberg, Gardner, & Campbell, 1999).

Caryophyllidea is typified as an order belonging to the monozoic form, with external segmentation and no differentiation of the reproductive organs, whereas Spathebothriidea belongs to the polyzoic form with no external segmentation with differentiation of the reproductive organs (Hoberg et al., 1997; Hoberg, Gardner, & Campbell, 1999). The next group of diverging eucestodes was the difossate orders, where members possess external segmentation (proglottids) with the development of multiple reproductive organs and a paired bothria at the scolex (Olson et al., 2001). Pseudophyllidea, which contains medically important species to humans and animals, is an example of the difossate order. The tetrafossate lineage comprises orders characterized by the scolex being differentiated into a tetrapartite bearing four bothria or suckers (Olson et al., 2001). Examples include Cyclophyllidea, where this order contains more than 50% of the total species in Eucestoda (Morand, Krasnov, & Poulin, 2006). Although the classification of eucestodes was based mainly on the three lineages, the relationships between the orders went largely unresolved. Variations in the morphological interpretations by various studies, along with the paucity of morphological characters, suggest that further investigations were needed to fully resolve the relationships within Eucestoda (Brooks & McLennan, 1993; Hoberg et al., 1997; Hoberg, Gardner, & Campbell, 1999; Olson et al., 2001).

Studies that utilize molecular methods greatly facilitated the unraveling of the phylogeny of Eucestoda. Molecular data gave rise to the erection of new orders and also resulted in restructuring within some orders. Early molecular-based studies utilized the nuclear 18S ribosomal RNA (rRNA) gene, either on representative species encompassing the different orders in Eucestoda or with a focus on relationships within the specific orders (Caira, Jensen, Waeschenbach, Olson, & Littlewood, 2014; Mariaux, 1998; Olson et al., 2001; Olson & Caira, 1999; Olson, Ruhnke, Sanney, & Hudson, 1999; Palm, Waeschenbach, Olson, & Littlewood, 2009). One of the earliest studies was performed by Mariaux (1998) using 45 taxa representing 10 orders with the 18S rRNA gene (Mariaux, 1998). The results were almost completely congruent with morphological-based hypotheses, showing the split into the three lineages and Caryophyllidea as the earliest diverging order.

Olson and Caira (1999), utilizing both the 18S rRNA gene and *elongation factor 1-alpha (Ef-1α)* gene, revealed that Spathebothriidea, instead of Caryophyllidea, was the earliest diverging order (Olson & Caira, 1999). Perhaps the most prominent studies contributing to the molecular phylogenetic backbone of Eucestoda were by Waeschenbach, Webster, Bray, and Littlewood (2007); Waeschenbach, Webster, and Littlewood (2012). Utilizing both the nuclear 18S and 28S rRNA genes for 13 orders in Eucestoda, it was revealed that both genetic markers provided added resolution to the phylogeny, in accordance with previous findings for trematodes (Waeschenbach et al., 2007). The partial mitochondrial DNA fragment was subsequently added to the nuclear rRNA genes to increase resolution through improved node support and stability, providing the most robust phylogenetic framework for Eucestoda to date (Waeschenbach et al., 2012).

The ability of mitochondrial DNA for molecular studies is unequivocally helpful for providing additional informative characters and is a rich resource for phylogeny reconstruction and interpretation of evolutionary relationships. Studies sequencing the complete mitochondrial genomes of cestode species have risen to be a forerunner in genetic research and will significantly benefit future molecular phylogenetic studies (Feng, Feng, Fang, & Su, 2017; Guo, 2016a; Li et al., 2017; Zhang et al., 2017).

More elaboration on the molecular systematics of cestodes will be presented in the following subsections to describe prominent studies that have been performed to elucidate the major orders of cestodes through the use of the nuclear and mitochondrial genetic markers.

12.1.1 The Nuclear 18S and 28S rRNA Gene Phylogeny

As seen in the previous two chapters, the nuclear rRNA genes, out of the various genetic markers that have been utilized, have played a central role in unraveling the higher taxonomic level relationships and enriched our understanding of relationships within nematodes and trematodes. It is no exception among cestodes.

The first study that took advantage of the properties of the nuclear rRNA genes for molecular systematics of cestodes was performed by Mariaux (1998). Analysis of the partial sequences of the nuclear 18S rRNA gene for 45 species belonging to 10 orders revealed almost complete congruence with morphological hypotheses. Representatives in the monofossate orders of Caryophyllidea and Spathebothridea were the early-diverging lineage, with Caryophyllidea basal to Spathebothridea, corroborating with the hypothesis of the monozoic to polyzoic form. The difossate orders of Pseudophyllidea and Trypanorhyncha were then sister to the clade of tetrafossate orders consisting of Proteocephalidea, Tetrabothriidea, and Cyclophyllidea. However, Diphyllidea, considered a difossate, was found to share a closer relationship with Proteocephalidea (tetrafossate lineage) using molecular data. Additionally, morphological data consider the orders Spathebothridea and Pseudophyllidea to be closely related and often lumped together, but the 18S rRNA data revealed that they should be considered as separate orders (Brooks & McLennan, 1993).

Following the work of Mariaux (1998), Olson and Caira (1999) then utilized the complete 18S rRNA gene, together with the partial *Ef-1α* gene, for phylogenetic analysis within cestodes (Mariaux, 1998; Olson & Caira, 1999). Contrary to the phylogeny obtained by Mariaux (1998), Spathebothridea was found to be the most basal instead of Caryophyllidea. Two clades resulted with the remaining orders, with a clade consisting of the difossate orders of Trypanorhyncha, Pseudophyllidea, and Diphyllidea and another consisting of the tetrafossate orders. The non-monophyly of Pseudophyllidea was also supported, including *Diphyllobothrium stemmacephalum* and *Schistocephalus solidus*. The finding was to be further built upon and supported by subsequent studies, where Pseudophyllidea was eventually to be divided into two new orders—Diphyllobothriidea and Bothriocephalidea (Abdel-Gaber, Alajmi, Quraishy, Morsy, & Rasheid, 2018; Brabec, Kuchta, & Scholz, 2006; Brabec, Waeschenbach, Scholz, Littlewood, & Kuchta, 2015; Kodedová et al., 2000; Waeschenbach & Littlewood, 2017).

Using both the complete nuclear 18S and partial 28S rRNA genes, Olson et al. (2001) further contributed to the molecular systematics of cestodes (Olson et al., 2001). Again, Pseudophyllidea was non-monophyletic, with Diphyllobothriidae sister to the rest of Eucestoda (except Haplobothriidea) and the remainder of Pseudophyllidea in a clade with the other difossate orders of Trypanorhyncha and Diphyllidea. Differences in the most basal taxon within Eucestoda resulted, depending on the type of analysis and genetic marker used. In the minimum evolution (ME) analysis using the 18S rRNA gene, Haplobothriidea was the most basal, while minimum parsimony (MP) analysis resulted in a large polytomy of the monofossate and difossate lineages. Using the 28S rRNA gene, both ME and MP analysis supported Caryophyllidea to be the most basal. With the concatenated 18S and 28S rRNA genes, both ME and MP analyses supported a basal positioning of Haplobothriidea. Additional findings include support for the positioning of Diphyllidea within the difossate lineage and the non-monophyly of Tetraphyllidea and Cyclophyllidea.

Building on the success of using both the nuclear 18S and 28S rRNA genes to unravel ordinal level relationships within Eucestoda, Waeschenbach et al. (2007) improved on previous studies by increasing the number of taxa studied and used complete 28S rRNA gene sequences with complete 18S rRNA gene sequences (Waeschenbach et al., 2007). Spathebothriidea was found to be the earliest diverging order among eucestodes. Corroborating with previous studies, the non-monophyly of Pseudophyllidea was observed, with the division into Diphyllobothriidea and Bothriocephalidea (Mariaux, 1998; Olson et al., 2001; Olson & Caira, 1999). Diphyllobothriidea, Haplobothriidea, and Caryophyllidea formed a clade sister to the remainder of Eucestoda (except Spathebothriidea). Support for Diphyllidea within the difossate lineage was also seen, with Diphyllidea and Trypanorhyncha having a close relationship. Additionally, the non-monophyly of Tetraphyllidea and Cyclophyllidea was also supported, congruent with Olson et al. (2001). The higher tetrafossate species of Cyclophyllidea + Mesocestoides, Tetrabothriidea, and Nippotaeniidea formed a monophyletic clade, supporting them as the most derived order among Eucestoda.

Using the nuclear rRNA genes, it is clear that the classification based on morphological characters might not be sufficient to fully elucidate the relationships within Eucestoda. The ordinal relationships within Eucestoda were improved upon through the concatenated nuclear rRNA genes, providing support for previous studies and erecting and elevating various orders with more phylogenetically informative sites.

12.1.2 Adding Resolution with Partial Mitochondrial DNA

Although the strategy of using the concatenated nuclear rRNA genes successfully provided a phylogenetic backbone for Eucestoda, variations exist between phylogenies with different methods of analysis and with morphological hypotheses. This was especially prominent in the disagreements of the earliest diverging order and therefore calls for an alternative genetic marker to provide additional support. The partial mitochondrial DNA, which spans from the partial nicotinamide adenine dinucleotide 1 (*ND1*) gene to the partial 12S rRNA gene (representing approximately 30% of the complete mitochondrial genome), was added to the nuclear 18S and 28S rRNA gene sequences, allowing for a total of more than 10,000 nucleotides. With the addition of the partial mitochondrial DNA, Waeschenbach et al. (2012) set out to add resolution to the molecular phylogenetic backbone of cestodes for 35 species within 15 orders (Diphyllobothridae + Bothriocephalidea counted as one order) of Eucestoda (Waeschenbach et al., 2012). Figure 12.1 presents the phylogeny that was obtained.

Firstly, for the earliest diverging order, Caryophyllidea was supported as the most basal of Eucestoda, corroborating morphological hypotheses and the 18S rRNA

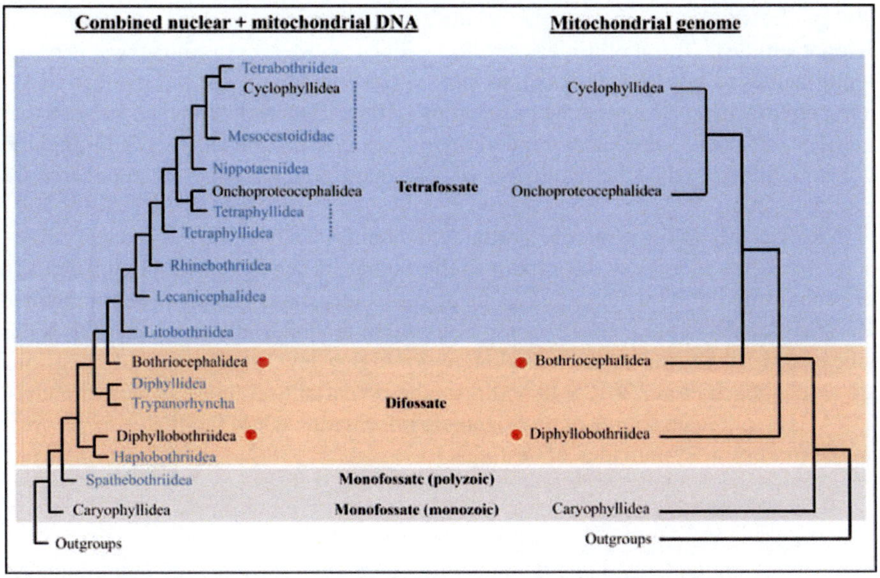

Fig. 12.1 Phylogeny of Eucestoda at the ordinal level with the combined nuclear rRNA genes + partial mtDNA and the phylogeny using complete mitochondrial genomes. The phylogenetic tree was adapted from Waeschenbach et al. (2012) and Li et al. (2019). The division into the monofossate, difossate, and tetrafossate lineages based on morphology is indicated in the highlighted areas in gray, red, and blue, respectively. The division of Pseudophyllidea to Diphyllobothriidea and Bothriocephalidea is indicated in "red circles," and the non-monophyly of Tetraphyllida and Cyclophyllidea is shown in "blue dashed lines"

gene phylogeny obtained by Mariaux (1998). The division of Pseudophyllidea (Diphyllobothridea and Bothriocephalidea) was also seen, with Diphyllobothriidea forming a clade with Haplobothriidea, while Bothriocephalidea shared a closer relationship with Litobothriidea. The evidence of the non-monophyly of Pseudophyllidea, further supported with the addition of mitochondrial DNA, contributed to the amounting evidence for the erection of orders Diphyllobothriidea and Bothriocephalidea (Abdel-Gaber et al., 2018; Brabec et al., 2006, 2015; Kodedová et al., 2000; Olson et al., 2001; Waeschenbach & Littlewood, 2017). Secondly, the phylogenetic position of Diphyllidea being closely related to Trypanorhyncha was also substantiated. Although the initial analysis by Mariaux (1998) using the nuclear 18S rRNA gene and mitochondrial DNA did not support a sister group relationship between Diphyllidea and Trypanorhyncha, the combined nuclear and mitochondrial phylogeny supports the sister relationship (Mariaux, 1998). Thirdly, Tetraphyllidea was rendered non-monophyletic, supporting previous studies. Tetraphyllidea includes the most morphologically diverse tapeworms, but they are also among the most host-specific (Olson et al., 1999). The phylogeny provided support for the elevation of Rhinebothriidea, the erection of Phyllobothriidea and Cathetocephalidea, and the merging of families Onchobothriidae + Proteocephalidae to Onchoproteocephalidea (de Chambrier, Waeschenbach, Fisseha, Scholz, & Mariaux, 2015; Healy, Caira, Jensen, Webster, & Littlewood, 2009; Olson et al., 2001). Lastly, support for the clade of higher tetrafossates consisting of Nippotaeniidea, Tetrabothriidea, Cyclophyllidea + Mesocestoididae was consistently achieved with better nodal support when compared with the nuclear rRNA gene phylogenies. Although the positioning of the orders differed when subjected to various analysis methods, monophyly was still obtained. Additionally, Cyclophyllidea, with the inclusion of Mesocestoididae, was not recovered as monophyletic.

Clearly, the addition of the partial mitochondrial DNA to the nuclear rRNA sequences has increased the phylogenetic signal of the obtained phylogeny and also confirmed previous hypotheses using molecular data. The phylogeny obtained by Waeschenbach et al. (2012) using both the mitochondrial and nuclear DNA has since served as the most robust phylogeny for higher-level systematics of Eucestoda (Waeschenbach et al., 2012). In addition, the potential usefulness of mitochondrial DNA as an alternate source of phylogenetic information could facilitate our progress in the molecular systematics of cestodes.

12.1.3 Mitochondrial Genome Phylogeny

Of the 15 described orders in Eucestoda, complete mitochondrial genomes have been sequenced for 11 of the orders, making up to more than 80 species available in the NCBI database. Mitochondrial genomes are useful phylogenetic information to serve as alternative phylogenies for the molecular systematics of cestodes.

Li et al. (2019), using 33 species within Eucestoda, revealed the most comprehensive mitochondrial genome phylogeny to date (Li et al., 2019). Figure 12.1 depicts the ordinal level relationships congruent with studies that have included the mitochondrial genomes of Caryophyllidea; this order was revealed to be the most basal relative to the remainder of the species (Li et al., 2017, 2018). Diphyllobothriidea was shown to be next, followed by Bothriocephalidea, Onchoproteocephalidea, and lastly, Cyclophyllidea. Phylogenetic relationships among the orders were similar among the mitochondrial genome phylogenies, corroborating also with the phylogenies obtained from the nuclear and combined nuclear + mitochondrial DNA phylogenies (for the orders that were included) (Feng et al., 2017; Guo, 2016a; Li et al., 2017, 2018; Waeschenbach et al., 2012; Zhang, Duan, Shi, et al., 2017). Additionally, mitochondrial genome phylogenies also supported the erection of new orders and restructuring within Eucestoda, as proposed with the molecular data obtained from the nuclear rRNA genes.

The smaller number of sequences available for complete mitochondrial genomes compared to the nuclear rRNA genes is a reminder of the uphill task needed to fully elucidate relationships in Eucestoda. However, with certainty, the number of mitochondrial genomes being sequenced for this group of helminths will increase in the coming years and potentially allow new insights and comparisons to be within reach in the near future.

12.2 Molecular Systematics of Medically Important Orders Within Eucestoda

In Chap. 2, medically important species of cestodes to humans globally were discussed, including their impact on human health. This section will narrow our focus to the molecular systematics of two medically important orders—Cyclophyllidea and Diphyllobothriidea—to dive into the details in the context of molecular phylogenetic relationships and implications for human health.

12.2.1 Cyclophyllidea

Of the orders in Eucestoda, Cyclophyllidea is the most species-rich, containing an estimated 3100 described species distributed among 15 families (Morand et al., 2006). Cyclophyllidea includes prominent parasites of humans, livestock, and wildlife, parasitizing mainly tetrapods. Family Taeniidae, containing genera *Taenia* and *Echinococcus* WHO (2012, 2018), are notable parasites in Cyclophyllidea with a global distribution that are medically important to humans and animals (Mahmud, Lim, & Amir, 2017; Morand et al., 2006).

Molecular studies on the phylogenetic relationships within Cyclophyllidea have been undertaken using genetic markers like the mitochondrial 12S rRNA gene, cytochrome *c* oxidase subunit I (*COI*) gene, and the nuclear 18S rRNA gene (Foronda, Casanova, Valladares, Martinez, & Feliu, 2004; Littlewood, Waeschenbach, & Nikolov, 2008; Sharma, Lyngdoh, Roy, & Tandon, 2016; von Nickisch-Rosenegk, Lucius, & Loos-Frank, 1999). Using the 18S rRNA gene, 15 species from five families (Davaineidae, Dilepididae, Anoplocephalidae, Paruterinidae, Taeniidae) were analyzed (Foronda et al., 2004). Taeniidae was revealed to be sister to the remainder of the four families, with Davaineidae, Dilepididae, Anoplocephalidae, and Paruterinidae sharing a closer relationship with respect to Taeniidae. The mitochondrial 12S rRNA gene phylogeny by von Nickisch-Rosenegk et al. (1999) revealed similar family-level relationships, with Taeniidae in a different clade to Anoplocephalidae and Dilepididae (von Nickisch-Rosenegk et al., 1999).

Additionally, with eight families being represented, the 12S rRNA gene phylogeny also revealed the non-monophyly of Hymenolepididae, with Dilepididae nested in between. Anoplocephalidae formed a sister group relationship with Hymenolepididae + Dilepididae, and this clade was, in turn, sister to a clade containing Dipylidiidae + Taeniidae + *Mesocestoides*. Catenotaeniidae was basal to the families represented in the phylogeny. Slight differences in family-level relationships were observed using the mitochondrial *COI* gene, but the overall topology was similar (Sharma et al., 2016). A first clade containing Anoplocephalidae, Davaineidae, and Hymenolepididae was observed, with *Mesocestoides* + members of Taeniidae (*Echinococcus* and *Versteria*) as a sister. The second clade contained the remainder of Taeniidae (*Taenia*), Dipylidiidae, and Paruterinidae. Dilepididae was found to be basal to the families represented. The *COI* phylogeny could not resolve Taeniidae as monophyletic, with *Echinococcus* and *Versteria* sharing a closer relationship to the clade containing Anoplocephalidae.

A combined approach was undertaken by Littlewood et al. (2008), where nuclear and mitochondrial genetic markers were used to estimate a reference phylogeny for cyclophyllideans in families Davaineidae, Hymenolepididae, and Dilepididae (Littlewood et al., 2008). Nuclear genetic markers were represented by the 18S and 28S rRNA genes, while the mitochondrial genetic markers were represented by the *COI*, *ND1*, and 16S rRNA genes. Similar to previous studies, the phylogenetic position of Dilepididae differed, where the 18S and 28S rRNA genes support a Dilepididae + Hymenolepididae relationship. In contrast, the mitochondrial genes support either a Dilepididae + Davaineidae clade or a closer relationship of Dilepididae to the outgroup (represented by Tetrabothriidea). The combined phylogeny of the nuclear and mitochondrial genes was similar to the nuclear phylogeny, where Dilepididae share a sister relationship with Hymenolepididae. The varying positions of Dilepididae from the genetic markers also serve as evidence of the complexities in uncovering phylogenetic relationships within cyclophyllidean cestodes.

Complete mitochondrial genomes have also shed light on the phylogenetic relationships within Cyclophyllidea. Using five representative families, Guo et al.

(2019) revealed a basal position of Dipylidiidae and a further division of the remaining species into two clades (Guo et al., 2019). The first contains members of Hymenolepididae and Anoplocephalidae, while the second clade contains Taeniidae and Paruterinidae. The close relationship between Hymenolepididae and Anoplocephalidae was also observed in other mitochondrial genome phylogenies and is also consistent with phylogenies obtained using the nuclear and mitochondrial genetic markers as mentioned in the above paragraphs (Guo, 2015, 2016b; Guo et al., 2019; Li et al., 2018, 2019). Furthermore, mitochondrial genome phylogenies support the close relationship between Taeniidae and Paruterinidae, which agrees with morphological data by Hoberg, Jones, and Bray (1999).

Currently, only approximately half of the families are being represented by molecular phylogenetic studies within Cyclophyllidea. There is a need for more extensive and representative sampling to fully elucidate relationships and test hypotheses within this species-rich order. Due to the cosmopolitan distribution of medically important species within Cyclophyllidea, emphasis has been placed on molecular studies for disease diagnosis, species identification, and genetic variation among populations. However, molecular systematics and phylogenetics go hand in hand with species identification and accurate disease diagnosis. The following paragraphs will touch on the molecular phylogenetic relationships within Taeniidae, focusing on the medically important *Taenia* and *Echinococcus* genera.

Taeniidae

Family Taeniidae, belonging to the higher tetrafossates, are the most widely studied due to their global impact on human and animal health. Molecular and phylogenetic studies have been vigorously conducted for species belonging to *Taenia* and *Echinococcus* due to their medical and veterinary importance. Four genera have been proposed: *Taenia*, *Echinococcus*, *Hydatigera*, and *Versteria* (Guo, 2015; Nakao et al., 2013). Phylogenetic relationships between the four genera using mitochondrial genomes support a sister group relationship between *Taenia* + *Hydatigera* and *Echinococcus* + *Versteria* (Cheng, Liu, Song, Lin, & Zhu, 2016; Guo, 2016b). Among the four genera, species belonging to *Taenia* and *Echinococcus* parasitize humans, causing cysticercosis and echinococcosis, respectively (Bobes et al., 2014; Ito et al., 2019).

Taenia

Comprising at least 45 valid species, *Taenia* is a large genus utilizing various intermediate and definitive hosts. Intermediate hosts are usually herbivores and small animals such as rodents, sheep, or pigs, while the definitive hosts are the larger carnivores comprising dogs, foxes, and wolves (Ito, Yanagida, & Nakao, 2016). Table 12.1 presents some of the species with their intermediate and definitive hosts. Of the species in *Taenia*, *T. solium*, *T. saginata*, and *Taenia asiatica* are the

Table 12.1 The definitive and intermediate hosts of *Taenia* and *Echinococcus*

Genus	Species	Intermediate hosts	Definitive hosts
Taenia	*T. solium*	Pig, dog, human	Human
	T. saginata	Cattle	Human
	T. asiatica	Pig, cattle	Human
	T. crassiceps	Rodent	Dog, fox
	T. ovis	Sheep	Dog
	T. hydatigena	Pig, sheep, goat	Dog
	T. pisiformis	Rabbit	Dog, fox
	T. taeniaeformis	Rodent	Cat
	T. multiceps	Sheep, rabbit	Dog, cat, wolf
	T. serialis	Rabbit, rodent	Dog, fox, wolf
Echinococcus	*E. granulosus* s.s. (G1, G2, G3)*	Sheep, goat, buffalo	Dog
	E. equinus (G4)*		
	E. ortleppi (G5)*	Cattle	Dog
	E. canadensis (G6/G7)*	Camel, pig, cattle, goat, sheep, reindeer	Dog, wolf
	E. canadensis (G8)*	Moose, elk, musk ox	Wolf
	E. canadensis (G10)*	Moose, elk, reindeer	Wolf, dog
	*E. felidis**	Warthog	Lion, hyena
	E. shiquicus	Pika (lagomorph)	Tibetan fox
	E. multilocularis	Rodent	Artic fox, dog, cat
	E. vogeli	Paca (rodent)	Bush dog
	E. oligarthra	Agouti (rodent)	Wild felines

An asterisk "*" indicates species in *E. granulosus* s.l. species complex

three species that utilize humans as definitive hosts (Bobes et al., 2014; Ito et al., 2019). Humans are also accidental hosts for *T. solium* due to the ingestion of embryonated eggs, resulting in neurocysticercosis. *T. solium* has been identified as a leading cause of death from foodborne diseases, and an estimated 8 million people are affected (WHO, 2021b).

Although the traditional approach of using morphological characters is able to distinguish the adults of *T. solium* and *T. saginata* or *T. asiatica*, it is not easy to morphologically differentiate between *T. saginata* and *T. asiatica* (Ito et al., 2019). Moreover, species identification based on egg morphology is impossible (Jimenez, Rodriguez, Moyano, Castillo, & García, 2010). Molecular identification, through the use of genetic markers like the mitochondrial *COI*, *ND1*, and nuclear protein-coding genes, has greatly assisted in species identification within *Taenia* (Knapp et al., 2011; Solano, Navarro, Leon-Reyes, Benitez-Ortiz, & Rodriguez-Hidalgo, 2016; Yan et al., 2013; Zhang et al., 2018). The sister relationship between *T. saginata* and *T. asiatica*, despite having different intermediate hosts, and the species status of *T. asiatica* have been recognized due to molecular-based investigations. Using both mitochondrial and nuclear genetic markers, *T. asiatica* has been consistently closer to *T. saginata*, relative to *T. solium* (although both *T. asiatica* and *T. solium* utilize pigs as hosts). Additionally, nuclear genetic markers revealed that *T. asiatica* sampled in Korea, Lao PDR, Indonesia, China, and Thailand were hybrid derived descendants of *T. saginata* and *T. asiatica*, while pure *T. asiatica* has only been found in the Philippines and Taiwan (Sato et al., 2018; Yamane et al., 2013). *T. solium*, on the other hand, is divided into two main clades—the Asian and African/American genotypes. Using the *COI* and *ND1* gene, phylogenetic analysis revealed distinct clustering of *T. solium* from Africa and Latin America and another clade of *T. solium* from Asia (China, Philippines, and India, among others) (Yan et al., 2013). Inferences from molecular data also suggest that *T. solium* probably emerged from ancestors with pig-hyena prey-predator associations, whereas *T. saginata* and *T. asiatica* from bovine-feline prey-predator associations.

In addition to the three species mentioned above, the molecular phylogeny of *Taenia* has been reconstructed. Through the use of three nuclear protein-coding genes (RNA polymerase II subunit, phosphoenolpyruvate carboxykinase, and DNA polymerase delta), Knapp et al. (2011) showed that *Taenia* was divided into two main clades (Knapp et al., 2011). Presented in Fig. 12.2, the first clade consisted of the three human *Taenia*, along with *Taenia serialis*, *Taenia madoquae*, *Taenia multiceps*, and *Taenia ovis*. Within this clade, *T. solium* was sister to the remainder of the species, and a sister relationship resulted with *T. saginata* and *T. asiatica*. The second clade consisted of *Taenia twitchelli*, *Taenia martis*, *Taenia crassiceps*, *Taenia laticollis*, and *Taenia hydatigena*. *Taenia taeniaeformis* and *Taenia parva* were the most basal among the Taenia species sampled. Congruent with the phylogeny obtained from the nuclear protein-coding genes, the nuclear 18S rRNA gene, partial mitochondrial genes, and complete mitochondrial genomes also revealed *T. taeniaeformis* as the earliest diverging species within *Taenia* (Jia et al., 2010, 2012; Liu et al., 2011). Knapp et al. (2011) suggested that the two main clades were also similar to the included species. The close relationship between *T. saginata* and *T. asiatica* was also ascertained through the various genetic markers and, more recently, through whole-genome analysis (Liu et al., 2021). Of the five *Taenia* draft genomes available, Liu et al. (2021) revealed a close relationship between *T. asiatica* and *T. saginata*, with *T. multiceps* sister. *T. solium* was, in turn, sister to the clade of *T. asiatica* + *T. saginata* + *T. multiceps*, and *T. taeniaeformis* was the most basal of the five species (Liu et al., 2021).

Echinococcus

Echinococcus, the other genera of medical importance in family Taeniidae, causes a life-threatening disease known as echinococcosis or hydatidosis. Two types of echinococcosis can occur, either alveolar echinococcosis caused by *Echinococcus multilocularis* or cystic echinococcosis caused by *Echinococcus granulosus* (WHO, 2021a). The definitive hosts are carnivores, while intermediate hosts are herbivores. Humans are accidental hosts where incomplete development of metacestode larva

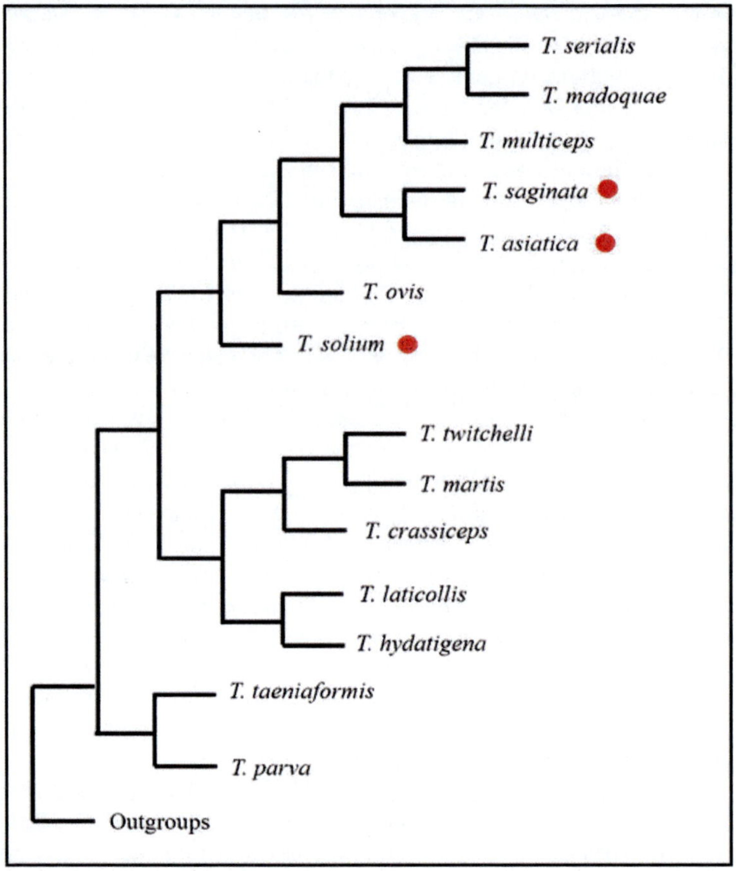

Fig. 12.2 Phylogeny of *Taenia* using the combined nuclear protein-coding genes. The red circle indicates the three *Taenia* species parasitizing humans as definitive hosts. The phylogenetic tree was adapted from Knapp et al. (2011)

occurs (Nakao, Lavikainen, Yanagida, & Ito, 2013). Multiple taxonomic revisions have been performed for this genus, primarily due to the paucity of morphological descriptions and lack of geographical or ecological segregation evidence. Sixteen species have been previously suggested, but molecular evidence has revealed that the genus contains only 11 species with various genotypes (Thompson, 2008). The 11 species are *Echinococcus multilocularis*, *Echinococcus shiquicus*, *Echinococcus vogeli*, *Echinococcus oligarthra*, and the *Echinococcus granulosus sensu lato* (s.l.) species complex, which comprises six species. The *E. granulosus* s.l. species complex comprises of *E. granulosus sensu stricto* (s.s.) (G1, G2, G3), *Echinococcus equinus* (G4), *Echinococcus ortleppi* (G5), *Echinococcus canadensis* (G6/G7), *Echinococcus canadensis* (G8), *Echinococcus canadensis* (G10), and *Echinococcus felidis*. Table 12.1 shows the species and their respective intermediate and definitive hosts.

Phylogenetic relationships within *Echinococcus* have mainly been studied using the mitochondrial protein-coding genes *COI* and *ND1* (Bowles, Blair, & McManus, 1992, 1995; Bowles & McManus, 1993; Thompson et al., 2006). Initial molecular inferences were performed by Bowles et al. (1992) and Bowles and McManus (1993), where *E. granulosus* was divided based on host usage into G1 to G8 genotypes (Bowles et al., 1992; Bowles & McManus, 1993). Subsequent molecular data then confirmed the validity of the species status of the genotypes, rendering the classification of *E. granulosus* s.l. species complex based on Table 12.1 (Nakao, McManus, Schantz, Craig, & Ito, 2007). The non-monophyly of *E. granulosus* s.l. species complex was also observed using mitochondrial genes and complete mitochondrial genomes, with *E. multilocularis* and *E. shiquicus* nested in between the species complex (Nakao et al., 2007, 2013). Figure 12.3 depicts the phylogeny obtained from mitochondrial DNA markers. On the other hand, analyses from nuclear protein-coding genes gave contrasting phylogenies, where Saarma et al. (2009) could recover *E. granulosus* s.l. species complex as monophyletic, while

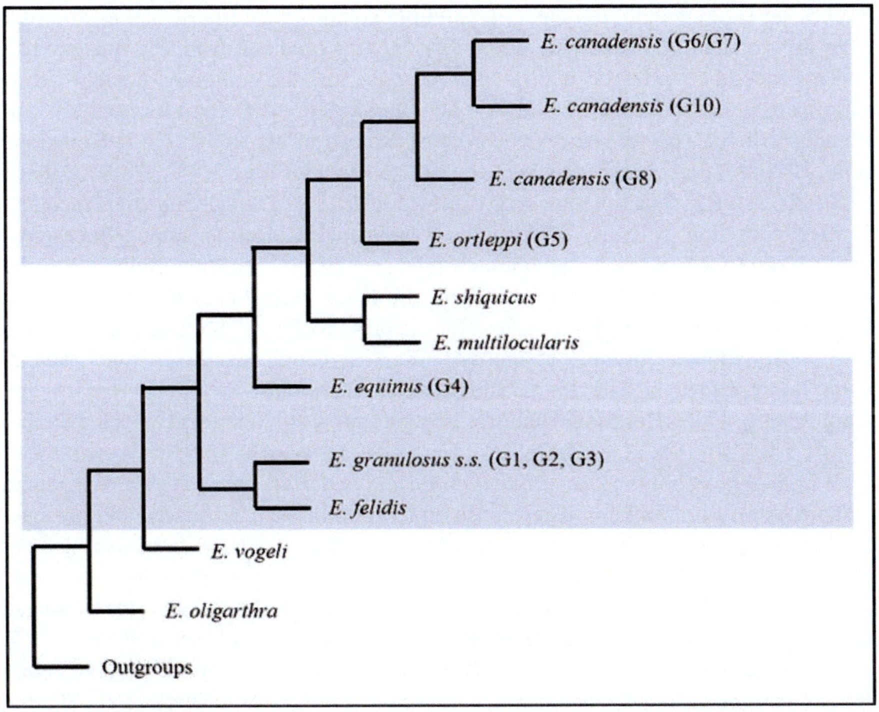

Fig. 12.3 Phylogeny of *Echinococcus* using mitochondrial genes. Species that are part of *E. granulosus* s.l. species complex are shaded in blue, with *E. shiquicus* and *E. multilocularis* nested in the middle

Knapp et al. (2011) showed a phylogeny similar to the mitochondrial phylogenies (Knapp et al., 2011; Saarma et al., 2009).

Aside from molecular phylogenetics, many molecular assays have been developed to detect and identify *Echinococcus* for accurate disease diagnosis. Examples include using PCR restriction fragment length polymorphism (PCR-RFLP) to detect various genotypes among isolates from different hosts, the design of species- and genotype-specific primers for multiplex PCR, and using high-resolution melting PCR for species detection (Boubaker et al., 2016; Chaabane-Banaoues et al., 2016; Espinoza et al., 2014; Santos, Espínola, Ferreira, Margis, & Zaha, 2013). The large repertoire of genetic markers used includes the nuclear rRNA genes and spacer regions, the nuclear protein-coding genes, the mitochondrial rRNA genes, and the mitochondrial protein-coding genes. These studies have made valuable contributions in delineating genetic variation among *Echinococcus*, arising from the sampling of isolates from different hosts and localities.

12.2.2 Diphyllobothriidea

Backed up by molecular evidence, Diphyllobothriidea was established as an order from the division of Pseudophyllidea. Phylogenies obtained from the nuclear 18S rRNA genes, the combined nuclear rRNA genes, and the combined nuclear rRNA and partial mitochondrial DNA all supported the non-monophyly of Pseudophyllidea (Abdel-Gaber et al., 2018; Brabec et al., 2006, 2015; Kodedová et al., 2000; Mariaux, 1998; Olson et al., 2001; Olson & Caira, 1999; Waeschenbach et al., 2007, 2012; Waeschenbach & Littlewood, 2017). Two clear groups resulted: species belonging to Diphyllobothriidea were in basal position among Eucestoda, with Haplobothriidea, Spathebothriidea, and Caryophyllidea, while Bothriocephalidea was the sister group to the tetrafossate lineages. Differences in life cycle features and host usage of species within Diphyllobothriidea and Bothriocephalidea further supported the division of the two orders. Diphyllobothriidea is able to parasitize birds, mammals, and even reptiles and amphibians, while Bothriocephalidea is predominantly restricted to parasitizing teleost fishes (Brabec et al., 2006).

Within Diphyllobothriidea, *Diphyllobothrium* and *Spirometra* are the two genera that are parasites of humans. They cause diphyllobothriasis and sparganosis, respectively, where infection to humans occurs through the consumption of fish or other second intermediate hosts and paratenic hosts containing the infective stage plerocercoid larvae. Human infection can also result from drinking water contaminated with copepods (first intermediate host) containing procercoid larvae (CDC, 2017, 2019; Yamasaki et al., 2021). *Diphyllobothrium latum* is the most geographically widespread, occurring in Asia, Europe, and even the Americas (Park et al., 2007).

Phylogenetic relationships within Diphyllobothriidae have mainly been studied using complete mitochondrial genomes and the nuclear rRNA genes (Hernandez-Orts, Scholz, Brabec, Kuzmina, & Kuchta, 2015; Park et al., 2007; Yamasaki,

Izumiyama, & Nozaki, 2017; Zhang et al., 2017. Using both the primary sequence and secondary structure of the 18S rRNA gene, Zhang et al. (2018) explored the taxonomic position of *Spirometra* with *Schistocephalus*, *Sparganum*, *Adenocephalus*, *Diplogonoporus*, and *Diphyllobothrium* (Zhang, Duan, Wang, et al., 2017). Phylogenetic analysis revealed that *Schistocephalus* was the most basal, with *Sparganum* and *Spirometra* sister. Another clade formed with *Diphyllobothrium*, *Diplogonoporus*, and *Adenocephalus*. *Diphyllobothrium* was not recovered as monophyletic, with *Diphyllobothrium stemmacephalum* forming a sister relationship with *Diplogonoporus grandis*, while the rest of *Diphyllobothrium* species, comprising of *Diphyllobothrium latum*, *Diphyllobothrium nihonkaiense*, and *Diphyllobothrium klebanovskii* formed another clade. *Di. grandis* are frequently found parasitizing whales, but human cases have been reported in Japan, Korea, and Spain (Chung et al., 1995; Clavel, Bargues, Castillo, Rubio, & Mas-Coma, 1997; Yamasaki, Ohmae, & Kuramochi, 2012). Based on morphology, *Diplogonoporus* has a double set of genitalia in a single proglottid, while many *Diphyllobothrium* species have one set of genitalia per proglottid (Yamasaki et al., 2012). The non-monophyly and close relationship between *D. stemmacephalum* and *Diplogonoporus* were also reinforced using the nuclear 28S rRNA genes, the mitochondrial *COI* gene, and complete mitochondrial genomes (Hernandez-Orts et al., 2015; Yamasaki et al., 2012, 2017). Yamasaki et al. (2012) proposed a revision of the taxonomic status and the validity of genera *Diplogonoporus* and *Diphyllobothrium* (Yamasaki et al., 2012). However, additional support to make these findings conclusive should be carried out with more comprehensive sampling and analysis.

12.3 Moving Forward

Progresses have been made using molecular tools in unraveling the puzzle for cestodes molecular systematics, with the circumscription of new orders and new insights using molecular data. However, taxonomic challenges persist. The molecular phylogenetic backbone for Eucestoda has yet to be fully resolved, especially with conflicts in the early branching lineages. Moreover, the discrepancies in sequence coverage between genetic markers and species sampled render the comparison of phylogenies difficult. To date, 11 out of the 15 orders of representative species of Eucestoda have complete mitochondrial genomes available. Although the mitochondrial genome phylogenies are consistent with the phylogeny obtained from the combined nuclear rRNA and partial mitochondrial DNA with the same orders, the absence of representative orders makes it difficult to draw a sound conclusion from the obtained phylogenies.

Like with trematodes and nematodes, DNA barcoding of cestodes has not been smooth sailing. Choosing an appropriate genetic marker, the suitability of primers, and obtaining quality DNA from specimens are some of the challenges associated with DNA barcoding. The road ahead for cestode molecular systematics could lie in the sampling and sequencing of more taxa, along with the ease of using next-generation sequencing (NGS) technologies. These advances in molecular techniques, coupled with the increasing emphasis on integrating morphological and molecular methods for taxonomy, render a bright prospectus for DNA barcoding. Additionally, with 14 whole draft genomes of species in Eucestoda currently available, phylogenomics and transcriptomics can greatly assist in the near future. A more comprehensive picture is needed for Eucestoda, not only in molecular phylogenetic relationships but also in factors and molecules that play a role in parasitism and consequently in developing drug targets for disease control.

References

Abdel-Gaber, R., Alajmi, R., Quraishy, S. A., Morsy, K., & Rasheid, K. A. (2018). Characterization of 28S rRNA sequences of cestoda parasite *Electrotaenia malapteruri* Fritsch, 1886 from the Electric catfish *Malapterurus electricus* (Siluriformes: Malapteruridae). *Acta Tropica, 183*, 153–161. https://doi.org/10.1016/j.actatropica.2018.04.016

Bobes, R. J., Fragoso, G., Fleury, A., Garcia-Varela, M., Sciutto, E., Larralde, C., & Laclette, J. P. (2014). Evolution, molecular epidemiology and perspectives on the research of taeniid parasites with special emphasis on *Taenia solium*. *Infection, Genetics and Evolution, 23*, 150–160. https://doi.org/10.1016/j.meegid.2014.02.005

Boubaker, G., Marinova, I., Gori, F., Hizem, A., Müller, N., Casulli, A., Jerez Puebla, L. E., Babba, H., Gottstein, B., & Spiliotis, M. (2016). A dual PCR-based sequencing approach for the identification and discrimination of *Echinococcus* and *Taenia* taxa. *Molecular and Cellular Probes, 30*(4), 211–217. https://doi.org/10.1016/j.mcp.2016.05.004

Bowles, J., Blair, D., & McManus, D. P. (1992). Genetic variants within the genus *Echinococcus* identified by mitochondrial DNA sequencing. *Molecular and Biochemical Parasitology, 54*(2), 165–173. https://doi.org/10.1016/0166-6851(92)90109-w

Bowles, J., Blair, D., & McManus, D. P. (1995). A molecular phylogeny of the genus *Echinococcus*. *Parasitology, 110*(3), 317–328. https://doi.org/10.1017/s0031182000080902

Bowles, J., & McManus, D. P. (1993). Rapid discrimination of *Echinococcus* species and strains using a polymerase chain reaction-based RFLP method. *Molecular and Biochemical Parasitology, 57*(2), 231–239. https://doi.org/10.1016/0166-6851(93)90199-8

Brabec, J., Kuchta, R., & Scholz, T. (2006). Paraphyly of the Pseudophyllidea (Platyhelminthes: Cestoda): Circumscription of monophyletic clades based on phylogenetic analysis of ribosomal RNA. *International Journal for Parasitology, 36*(14), 1535–1541. https://doi.org/10.1016/j.ijpara.2006.08.003

Brabec, J., Waeschenbach, A., Scholz, T., Littlewood, D. T., & Kuchta, R. (2015). Molecular phylogeny of the Bothriocephalidea (Cestoda): Molecular data challenge morphological classification. *International Journal for Parasitology, 45*(12), 761–771. https://doi.org/10.1016/j.ijpara.2015.05.006

Brooks, D., & McLennan, D. (1993). *Parascript parasites and the language of evolution* (1st ed.). Smithsonian Institution Press.

Caira, J. N., Jensen, K., Waeschenbach, A., Olson, P. D., & Littlewood, D. T. (2014). Orders out of chaos–molecular phylogenetics reveals the complexity of shark and stingray tapeworm relationships. *International Journal for Parasitology, 44*(1), 55–73. https://doi.org/10.1016/j.ijpara.2013.10.004

CDC. (2017). *Sparganosis*. https://www.cdc.gov/dpdx/sparganosis/index.html

CDC. (2019). *Diphyllobothriasis*. https://www.cdc.gov/dpdx/diphyllobothriasis/index.html

Chaabane-Banaoues, R., Oudni-M'rad, M., M'Rad, S., Amani, H., Mezhoud, H., & Babba, H. (2016). A novel PCR-RFLP assay for molecular characterization of *Echinococcus granulosus* sensu lato and closely related species in developing countries. *Parasitology Research, 115*(10), 3817–3824. https://doi.org/10.1007/s00436-016-5143-x

Cheng, T., Liu, G. H., Song, H. Q., Lin, R. Q., & Zhu, X. Q. (2016). The complete mitochondrial genome of the dwarf tapeworm *Hymenolepis nana*–A neglected zoonotic helminth. *Parasitology Research, 115*(3), 1253–1262. https://doi.org/10.1007/s00436-015-4862-8

Chung, D. I., Kong, H. H., Moon, C. H., Choi, D. W., Kim, T. H., Lee, D. W., & Park, J. J. (1995). The first human case of *Diplogonoporus balaenopterae* (Cestoda: Diphyllobothriidae) infection in Korea. *The Korean Journal of Parasitology, 33*(3), 225–230. https://doi.org/10.3347/kjp.1995.33.3.225

Clavel, A., Bargues, M. D., Castillo, F. J., Rubio, M. D., & Mas-Coma, S. (1997). Diplogonoporiasis presumably introduced into Spain: First confirmed case of human infection acquired outside the Far East. *The American Journal of Tropical Medicine and Hygiene, 57*(3), 317–320. https://doi.org/10.4269/ajtmh.1997.57.317

de Chambrier, A., Waeschenbach, A., Fisseha, M., Scholz, T., & Mariaux, J. (2015). A large 28S rDNA-based phylogeny confirms the limitations of established morphological characters for classification of proteocephalidean tapeworms (Platyhelminthes, Cestoda). *Zookeys, 500*, 25–59. https://doi.org/10.3897/zookeys.500.9360

Espinoza, S., Salas, A. M., Vargas, A., Freire, V., Diaz, E., Sánchez, G., & Venegas, J. (2014). Detection of the G3 genotype of *Echinococcus granulosus* from hydatid cysts of Chilean cattle using *COX1* and *ND1* mitochondrial markers. *Parasitology Research, 113*(1), 139–147. https://doi.org/10.1007/s00436-013-3636-4

Feng, Y., Feng, H. L., Fang, Y. H., & Su, Y. B. (2017). Characterization of the complete mitochondrial genome of *Khawia sinensis* belongs among platyhelminths, cestodes. *Experimental Parasitology, 177*, 35–39. https://doi.org/10.1016/j.exppara.2017.04.005

Foronda, P., Casanova, J. C., Valladares, B., Martinez, E., & Feliu, C. (2004). Molecular systematics of several cyclophyllid families (Cestoda) based on the analysis of 18S ribosomal DNA gene sequences. *Parasitology Research, 93*(4), 279–282. https://doi.org/10.1007/s00436-004-1130-8

Guo, A. (2015). The complete mitochondrial genome of *Anoplocephala perfoliata*, the first representative for the family Anoplocephalidae. *Parasites & Vectors, 8*, 549. https://doi.org/10.1186/s13071-015-1172-z

Guo, A. (2016a). Characterization of the complete mitochondrial genome of the cloacal tapeworm *Cloacotaenia megalops* (Cestoda: Hymenolepididae). *Parasites & Vectors, 9*(1), 490. https://doi.org/10.1186/s13071-016-1782-0

Guo, A. (2016b). The complete mitochondrial genome of the tapeworm *Cladotaenia vulturi* (Cestoda: Paruterinidae): Gene arrangement and phylogenetic relationships with other cestodes. *Parasites & Vectors, 9*, 475. https://doi.org/10.1186/s13071-016-1769-x

Guo, A., Wang, L., Zhang, S., Zheng, Y., Georgiev, B. B., Luo, X., Huang, S., & Cai, X. (2019). Mitochondrial genome of *Paruterina candelabraria* (Cestoda: Paruterinidae), with implications for the relationships between the genera *Cladotaenia* and *Paruterina*. *Acta Tropica, 189*, 1–5. https://doi.org/10.1016/j.actatropica.2018.09.016

Healy, C. J., Caira, J. N., Jensen, K., Webster, B. L., & Littlewood, D. T. (2009). Proposal for a new tapeworm order, Rhinebothriidea. *International Journal for Parasitology, 39*(4), 497–511. https://doi.org/10.1016/j.ijpara.2008.09.002

Hernandez-Orts, J. S., Scholz, T., Brabec, J., Kuzmina, T., & Kuchta, R. (2015). High morphological plasticity and global geographical distribution of the Pacific broad tapeworm *Adenocephalus pacificus* (syn. *Diphyllobothrium pacificum*): Molecular and morphological survey. *Acta Tropica, 149*, 168–178. https://doi.org/10.1016/j.actatropica.2015.05.017

Hoberg, E. P., Gardner, S. L., & Campbell, R. A. (1999). Systematics of the Eucestoda: Advances toward a new phylogenetic paradigm, and observations on the early diversification of tapeworms and vertebrates. *Systematic Parasitology, 42*(1), 1–12. https://doi.org/10.1023/a:1006099009495

Hoberg, E. P., Jones, A., & Bray, R. A. (1999). Phylogenetic analysis among the families of the Cyclophyllidea (Eucestoda) based on comparative morphology, with new hypotheses for co-evolution in vertebrates. *Systematic Parasitology, 42*(1), 51–73. https://doi.org/10.1023/a:1006100629059

Hoberg, E. P., Mariaux, J., Justine, J. L., Brooks, D. R., & Weekes, P. J. (1997). Phylogeny of the orders of the Eucestoda (Cercomeromorphae) based on comparative morphology: Historical perspectives and a new working hypothesis. *The Journal of Parasitology, 83*(6), 1128–1147.

Ito, A., Li, T., Wandra, T., Dekumyoy, P., Yanagida, T., Okamoto, M., & Budke, C. M. (2019). Taeniasis and cysticercosis in Asia: A review with emphasis on molecular approaches and local lifestyles. *Acta Tropica, 198*, 105075. https://doi.org/10.1016/j.actatropica.2019.105075

Ito, A., Yanagida, T., & Nakao, M. (2016). Recent advances and perspectives in molecular epidemiology of *Taenia solium* cysticercosis. *Infection, Genetics and Evolution, 40*, 357–367. https://doi.org/10.1016/j.meegid.2015.06.022

Jia, W., Yan, H., Lou, Z., Ni, X., Dyachenko, V., Li, H., & Littlewood, D. T. (2012). Mitochondrial genes and genomes support a cryptic species of tapeworm within *Taenia taeniaeformis*. *Acta Tropica, 123*(3), 154–163. https://doi.org/10.1016/j.actatropica.2012.04.006

Jia, W. Z., Yan, H. B., Guo, A. J., Zhu, X. Q., Wang, Y. C., Shi, W. G., Chen, H. T., Zhan, F., Zhang, S. H., Fu, B. Q., Littlewood, D. T., & Cai, X. P. (2010). Complete mitochondrial genomes of *Taenia multiceps*, *T. hydatigena* and *T. pisiformis*: Additional molecular markers for a tapeworm genus of human and animal health significance. *BMC Genomics, 11*, 447. https://doi.org/10.1186/1471-2164-11-447

Jimenez, J. A., Rodriguez, S., Moyano, L. M., Castillo, Y., & García, H. H. (2010). Differentiating *Taenia* eggs found in human stools: Does Ziehl-Neelsen staining help? *Tropical Medicine & International Health, 15*(9), 1077–1081. https://doi.org/10.1111/j.1365-3156.2010.02579.x

Khalil, L., Jones, A., & Bray, R. (1994). *Keys to the cestode parasite of vertebrates*. CABI Publishing.

Knapp, J., Nakao, M., Yanagida, T., Okamoto, M., Saarma, U., Lavikainen, A., & Ito, A. (2011). Phylogenetic relationships within *Echinococcus* and *Taenia* tapeworms (Cestoda: Taeniidae): An inference from nuclear protein-coding genes. *Molecular Phylogenetics and Evolution, 61*(3), 628–638. https://doi.org/10.1016/j.ympev.2011.07.022

Kodedová, I., Dolezel, D., Broucková, M., Jirků, M., Hypsa, V., Lukes, J., & Scholz, T. (2000). On the phylogenetic positions of the Caryophyllidea, Pseudophyllidea and Proteocephalidea (Eucestoda) inferred from 18S rRNA. *International Journal for Parasitology, 30*(10), 1109–1113. https://doi.org/10.1016/s0020-7519(00)00090-4

Li, W. X., Fu, P. P., Zhang, D., Boyce, K., Xi, B. W., Zou, H., Li, M., Wu, S. G., & Wang, G. T. (2018). Comparative mitogenomics supports synonymy of the genera *Ligula* and *Digramma* (Cestoda: Diphyllobothriidae). *Parasites & Vectors, 11*(1), 324. https://doi.org/10.1186/s13071-018-2910-9

Li, W. X., Zhang, D., Boyce, K., Xi, B. W., Zou, H., Wu, S. G., Li, M., & Wang, G. T. (2017). The complete mitochondrial DNA of three monozoic tapeworms in the Caryophyllidea: A mitogenomic perspective on the phylogeny of eucestodes. *Parasites & Vectors, 10*(1), 314. https://doi.org/10.1186/s13071-017-2245-y

Li, W. X., Zhang, D., Fu, P. P., Song, R., Zou, H., Li, M., Wu, S. G., & Wang, G. T. (2019). Characterization and phylogenomics of the complete mitochondrial genome of the polyzoic cestode *Gangesia oligonchis* (Platyhelminthes: Onchoproteocephalidea). *Journal of Helminthology, 94*, e58. https://doi.org/10.1017/S0022149X19000452

Littlewood, D., Waeschenbach, A., & Nikolov, P. (2008). In search of mitochondrial markers for resolving the phylogeny of cyclophyllidean tapeworms (Platyhelminthes, Cestoda) — a test study with Davaineidae. *Acta Parasitologica, 53*(2), 4. https://doi.org/10.2478/s11686-008-0029-4

Liu, G. H., Korhonen, P. K., Young, N. D., Lu, J., Wang, T., Fu, Y. T., Koehler, A. V., Hofmann, A., Chang, B. C. H., Wang, S., Li, N., Lin, C. Y., Zhang, H., Xiangli, L., Lin, L., Liu, W. M., Li, N., Li, H. W., Gasser, R. B., & Zhu, X. Q. (2021). *Dipylidium caninum* draft genome – A new resource for comparative genomic and genetic explorations of flatworms. *Genomics, 113*(3), 1272–1280. https://doi.org/10.1016/j.ygeno.2021.02.019

Liu, G. H., Lin, R. Q., Li, M. W., Liu, W., Liu, Y., Yuan, Z. G., Song, H. Q., Zhao, G. H., Zhang, K. X., & Zhu, X. Q. (2011). The complete mitochondrial genomes of three cestode species of *Taenia* infecting animals and humans. *Molecular Biology Reports, 38*(4), 2249–2256. https://doi.org/10.1007/s11033-010-0355-0

Mahmud, R., Lim, Y. A. L., & Amir, A. (2017). *Medical parasitology* (1st ed.). Springer.

Mariaux, J. (1998). A molecular phylogeny of the Eucestoda. *The Journal of Parasitology, 84*(1), 114–124.

Morand, S., Krasnov, B., & Poulin, R. (2006). *Micromammals and macroparasites from evolutionary ecology to management* (1st ed.). Springer.

Nakao, M., Lavikainen, A., Iwaki, T., Haukisalmi, V., Konyaev, S., Oku, Y., Okamoto, M., & Ito, A. (2013). Molecular phylogeny of the genus *Taenia* (Cestoda: Taeniidae): Proposals for the resurrection of *Hydatigera* Lamarck, 1816 and the creation of a new genus *Versteria*. *International Journal for Parasitology, 43*(6), 427–437. https://doi.org/10.1016/j.ijpara.2012.11.014

Nakao, M., Lavikainen, A., Yanagida, T., & Ito, A. (2013). Phylogenetic systematics of the genus *Echinococcus* (Cestoda: Taeniidae). *International Journal for Parasitology, 43*(12–13), 1017–1029. https://doi.org/10.1016/j.ijpara.2013.06.002

Nakao, M., McManus, D. P., Schantz, P. M., Craig, P. S., & Ito, A. (2007). A molecular phylogeny of the genus *Echinococcus* inferred from complete mitochondrial genomes. *Parasitology, 134* (Pt 5), 713–722. https://doi.org/10.1017/S0031182006001934

Nakao, M., Yanagida, T., Konyaev, S., Lavikainen, A., Odnokurtsev, V. A., Zaikov, V. A., & Ito, A. (2013). Mitochondrial phylogeny of the genus Echinococcus (Cestoda: Taeniidae) with emphasis on relationships among Echinococcus canadensis genotypes. *Parasitology, 140*(13), 1625–1636. https://doi.org/10.1017/S0031182013000565

Olson, P. D., & Caira, J. N. (1999). Evolution of the major lineages of tapeworms (Platyhelminthes: Cestoidea) inferred from 18S ribosomal DNA and elongation factor-1alpha. *The Journal of Parasitology, 85*(6), 1134–1159.

Olson, P. D., Littlewood, D. T., Bray, R. A., & Mariaux, J. (2001). Interrelationships and evolution of the tapeworms (Platyhelminthes: Cestoda). *Molecular Phylogenetics and Evolution, 19*(3), 443–467. https://doi.org/10.1006/mpev.2001.0930

Olson, P. D., Ruhnke, T. R., Sanney, J., & Hudson, T. (1999). Evidence for host-specific clades of tetraphyllidean tapeworms (Platyhelminthes: Eucestoda) revealed by analysis of 18S ssrDNA. *International Journal for Parasitology, 29*(9), 1465–1476. https://doi.org/10.1016/s0020-7519(99)00106-x

Palm, H. W., Waeschenbach, A., Olson, P. D., & Littlewood, D. T. (2009). Molecular phylogeny and evolution of the Trypanorhyncha Diesing, 1863 (Platyhelminthes: Cestoda). *Molecular Phylogenetics and Evolution, 52*(2), 351–367. https://doi.org/10.1016/j.ympev.2009.01.019

Park, J. K., Kim, K. H., Kang, S., Jeon, H. K., Kim, J. H., Littlewood, D. T., & Eom, K. S. (2007). Characterization of the mitochondrial genome of *Diphyllobothrium latum* (Cestoda: Pseudophyllidea) – Implications for the phylogeny of eucestodes. *Parasitology, 134*(Pt 5), 749–759. https://doi.org/10.1017/S003118200600206X

Saarma, U., Jogisalu, I., Moks, E., Varcasia, A., Lavikainen, A., Oksanen, A., Simsek, S., Andresiuk, V., Denegri, G., Gonzalez, L. M., Ferrer, E., Garate, T., Rinaldi, L., & Maravilla, P. (2009). A novel phylogeny for the genus *Echinococcus*, based on nuclear data, challenges relationships based on mitochondrial evidence. *Parasitology, 136*(3), 317–328. https://doi.org/10.1017/S0031182008005453

Santos, G. B., Espínola, S. M., Ferreira, H. B., Margis, R., & Zaha, A. (2013). Rapid detection of *Echinococcus* species by a highresolution melting (HRM) approach. *Parasites & Vectors, 6*(1), 327. https://doi.org/10.1186/1756-3305-6-327

Sato, M. O., Sato, M., Yanagida, T., Waikagul, J., Pongvongsa, T., Sako, Y., Sanguankiat, S., Yoonuan, T., Kounnavang, S., Kawai, S., Ito, A., Okamoto, M., & Moji, K. (2018). *Taenia solium*, *Taenia saginata*, *Taenia asiatica*, their hybrids and other helminthic infections occurring in a neglected tropical diseases' highly endemic area in Lao PDR. *PLoS Neglected Tropical Diseases, 12*(2), e0006260. https://doi.org/10.1371/journal.pntd.0006260

Sharma, S., Lyngdoh, D., Roy, B., & Tandon, V. (2016). Molecular phylogeny of Cyclophyllidea (Cestoda: Eucestoda): an in-silico analysis based on mtCOI gene. *Parasitology Research, 115*(9), 3329–3335. https://doi.org/10.1007/s00436-016-5092-4

Solano, D., Navarro, J. C., Leon-Reyes, A., Benitez-Ortiz, W., & Rodriguez-Hidalgo, R. (2016). Molecular analyses reveal two geographic and genetic lineages for tapeworms, *Taenia solium* and *Taenia saginata*, from Ecuador using mitochondrial DNA. *Experimental Parasitology, 171*, 49–56. https://doi.org/10.1016/j.exppara.2016.10.015

Thompson, R. C. (2008). The taxonomy, phylogeny and transmission of *Echinococcus*. *Experimental Parasitology, 119*(4), 439–446. https://doi.org/10.1016/j.exppara.2008.04.016

Thompson, R. C., Boxell, A. C., Ralston, B. J., Constantine, C. C., Hobbs, R. P., Shury, T., & Olson, M. E. (2006). Molecular and morphological characterization of *Echinococcus* in cervids from North America. *Parasitology, 132*(3), 439–447. https://doi.org/10.1017/s0031182005009170

von Nickisch-Rosenegk, M., Lucius, R., & Loos-Frank, B. (1999). Contributions to the phylogeny of the Cyclophyllidea (Cestoda) inferred from mitochondrial 12S rDNA. *Journal of Molecular Evolution, 48*(5), 586–596. https://doi.org/10.1007/PL00006501

Waeschenbach, A., & Littlewood, D. T. J. (2017). A molecular framework for the Cestoda. In J. N. Caira & K. Jensen (Eds.), *Planetary biodiversity inventory (2008-2017): Tapeworms from the vertebrate bowels of the earth*. University of Kansas.

Waeschenbach, A., Webster, B. L., Bray, R. A., & Littlewood, D. T. (2007). Added resolution among ordinal level relationships of tapeworms (Platyhelminthes: Cestoda) with complete small and large subunit nuclear ribosomal RNA genes. *Molecular Phylogenetics and Evolution, 45*(1), 311–325. https://doi.org/10.1016/j.ympev.2007.03.019

Waeschenbach, A., Webster, B. L., & Littlewood, D. T. (2012). Adding resolution to ordinal level relationships of tapeworms (Platyhelminthes: Cestoda) with large fragments of mtDNA. *Molecular Phylogenetics and Evolution, 63*(3), 834–847. https://doi.org/10.1016/j.ympev.2012.02.020

WHO. (2012). *Research priorities for helminth infections: technical report of the TDR disease reference group on helminth infections*. World Health Organization.

WHO. (2018). *WHO methods and data sources for global burden of disease estimates 2000–2016*. WHO.

WHO. (2021a). *Echinococcosis*. https://www.who.int/news-room/fact-sheets/detail/echinococcosis

WHO. (2021b). *Taeniasis/cysticercosis*. https://www.who.int/news-room/fact-sheets/detail/taeniasis-cysticercosis

Yamane, K., Yanagida, T., Li, T., Chen, X., Dekumyoy, P., Waikagul, J., Nkouawa, A., Nakao, M., Sako, Y., Ito, A., Sato, H., & Okamoto, M. (2013). Genotypic relationships between *Taenia saginata*, *Taenia asiatica* and their hybrids. *Parasitology, 140*(13), 1595–1601. https://doi.org/10.1017/S0031182013001273

Yamasaki, H., Izumiyama, S., & Nozaki, T. (2017). Complete sequence and characterization of the mitochondrial genome of *Diphyllobothrium stemmacephalum*, the type species of genus *Diphyllobothrium* (Cestoda: Diphyllobothriidae), using next generation sequencing. *Parasitology International, 66*(5), 573–578. https://doi.org/10.1016/j.parint.2017.06.005

Yamasaki, H., Ohmae, H., & Kuramochi, T. (2012). Complete mitochondrial genomes of *Diplogonoporus balaenopterae* and *Diplogonoporus grandis* (Cestoda: Diphyllobothriidae) and clarification of their taxonomic relationships. *Parasitology International, 61*(2), 260–266. https://doi.org/10.1016/j.parint.2011.10.007

Yamasaki, H., Sanpool, O., Rodpai, R., Sadaow, L., Laummaunwai, P., Un, M., Thanchomnang, T., Laymanivong, S., Aung, W. P. P., Intapan, P. M., & Maleewong, W. (2021). *Spirometra* species from Asia: Genetic diversity and taxonomic challenges. *Parasitology International, 80*, 102181. https://doi.org/10.1016/j.parint.2020.102181

Yan, H., Lou, Z., Li, L., Ni, X., Guo, A., Li, H., Zheng, Y., Dyachenko, V., & Jia, W. (2013). The nuclear 18S ribosomal RNA gene as a source of phylogenetic information in the genus Taenia. *Parasitology Research, 112*(3), 1343–1347. https://doi.org/10.1007/s00436-012-3199-9

Zhang, X., Duan, J. Y., Shi, Y. L., Jiang, P., Zeng, J., Wang, Z. Q., & Cui, J. (2017). Comparative mitochondrial genomics among *Spirometra* (Cestoda: Diphyllobothriidae) and the molecular phylogeny of related tapeworms. *Molecular Phylogenetics and Evolution, 117*, 75–82. https://doi.org/10.1016/j.ympev.2017.06.003

Zhang, X., Duan, J. Y., Wang, Z. Q., Jiang, P., Liu, R. D., & Cui, J. (2017). Using the small subunit of nuclear ribosomal DNA to reveal the phylogenetic position of the plerocercoid larvae of *Spirometra* tapeworms. *Experimental Parasitology, 175*, 1–7. https://doi.org/10.1016/j.exppara.2017.01.006

Zhang, Y., Zhao, W., Yang, D., Tian, Y., Zhang, W., & Liu, A. (2018). Genetic characterization of three mitochondrial gene sequences of goat/sheep-derived coenurus cerebralis and cysticercus tenuicollis isolates in Inner Mongolia, China. *Parasite, 25*, 1. https://doi.org/10.1051/parasite/2018002

Chapter 13
DNA Taxonomy of Parasitic Helminths

Abstract Taxonomy, which is fundamental for the delimitation of species, lays the foundation for constructing the tree of life for organisms. Defined as the science of circumscribing, naming, describing, and classifying taxa, the concept of taxonomy was brought about by Carl Linnaeus in 1735, allowing for taxa classification and standardization of taxa names. Molecular advances allowed for the use of DNA for taxonomical studies, bringing about DNA taxonomy and DNA barcoding. Since then, the use of DNA sequences for taxonomical studies has become commonplace for most organisms. Progress in helminth taxonomy is no exception, and in recent decades, DNA sequences have played an increased role in taxonomy. DNA taxonomy, DNA barcoding, and molecular systematics were put into context in this chapter, followed by the progress made in the scientific community. Examples for the molecular identification of the three groups of parasitic helminths, gaps in the progress of helminth DNA barcoding, and practical steps and recommendations for application were also detailed.

13.1 Putting Into Context: DNA Taxonomy, DNA Barcoding, and Molecular Systematics

13.1.1 Definitions

As mentioned in the above paragraphs (Meier, Shiyang, Vaidya, & Ng, 2006; Yamakawa, 2014), taxonomy is the science of circumscribing, naming, describing, and classifying organisms using observable characteristics. These characters include morphological, ecological, and behavioral observations (classical taxonomy) (Convention on Biological Diversity, n.d.). In his attempt to identify every known plant and animal, Carl Linnaeus standardized the way organisms were named by introducing the binomial nomenclature and the hierarchical taxonomic system (Biology Online, n.d.; Convention on Biological Diversity, n.d.). For binomial nomenclature, each scientific name comprises two names—the first describes the genus that the organism belongs to, while the second refers to the species of the organism (Biology Dictionary, n.d.). Binomial nomenclature allowed for the formality of naming an

U. Thaenkham et al., *Molecular Systematics of Parasitic Helminths*,
https://doi.org/10.1007/978-981-19-1786-8_13

organism, where taxonomists globally can speak a common language through the universality of the name of the particular organism. Using the hierarchical taxonomical system (Kingdom, Phyla, Class, Order, etc.), organisms could now be classified based on shared characteristics. These hierarchical groups render the measure of the degree of relatedness among taxa possible and allow for inferences of evolutionary relationships among organisms.

The use of genetic material (DNA sequences) is now no stranger to taxonomists and is widely utilized with species identification and description. The term DNA taxonomy was proposed by Tautz, Arctander, Minelli, Thomas, and Vogler (2003), where they suggested giving DNA a central role in taxonomy (Tautz et al., 2003). The rationale behind the proposal was that DNA sequences could provide new dimensions for the currently available taxonomic knowledge while maintaining taxonomic information obtained through classical taxonomy (Blaxter, 2004; Fontaneto, Flot, & Tang, 2015; Pires & Marinoni, 2010; Tautz et al., 2003; Vogler & Monaghan, 2007). With the speed and convenience of using DNA sequences for taxonomy, the identification of organisms can be performed quickly, in contrast to classical taxonomy methods.

On the other hand, DNA barcoding was proposed by Hebert, Cywinska, Ball, and deWaard (2003) as a method for species identification (Hebert, Cywinska, et al., 2003). The concept of DNA barcoding is based on using a short, standardized region of the DNA (molecular genetic marker), seeking to have a universal molecular identification system for all species (Hebert, Cywinska, et al., 2003; Trivedi, Rehman, Saggu, Panneerselvam, & Ghosh, 2020). Simply put, DNA barcoding can be seen as an application of DNA taxonomy, where a standard genetic marker is used for species identification. The rationale of DNA barcoding stems from the advances in molecular techniques and the increasing ease of obtaining genetic data. It aims to facilitate the identification of species already defined by morphological criteria and describe new species through the interpretation of species diversity through DNA sequences (Dasmahapatra & Mallet, 2006; Frézal & Leblois, 2008; Köhler, 2007; Morand, 2018).

The current DNA barcode for animals is the mitochondrial partial cytochrome *c* oxidase subunit I (*COI*) gene, consisting of approximately 600 base pairs at the 5′ end of the gene (Hebert, Ratnasingham, & deWaard, 2003; Ratnasingham & Hebert, 2007). In the two decades since the concept of DNA barcoding was introduced, it has been widely applied in the molecular identification of species. Applications include species identification of organisms with similar morphological characters, revealing cryptic species, aiding in detecting food products, forensics, and discovering new taxa in biodiversity surveys. Moreover, the confirmation of species through molecular sequences, along with other morphological and ecological information, is integral for the taxonomy field, aiding in the description, circumscription, and classification of species.

Molecular systematics, as previously defined in Chap. 1, encompasses both the classification and the understanding of evolutionary relationships among organisms through molecular methods. With species as the fundamental unit, molecular systematics provides a foundation and framework for constructing the tree of life. Through the use of DNA sequences and molecular phylogenetics, each fundamental unit can be placed in a phylogenetic context, enabling the classification of organisms. Moreover, evolutionary relationships among organisms and the forces driving speciation can be deciphered through the historical evidence left in the genetic material of the organism. Patterns and rates of changes in genetic material are crucial for reconstructing the evolutionary history of organisms. Taxonomy, through the naming and description of species, allows for the proper recognition of each fundamental unit, and species identification can be performed via DNA barcoding. Thus, identifying species (alpha taxonomy) through DNA taxonomy and applying DNA barcoding falls under the big umbrella of molecular systematics. The nexuses between molecular systematics, DNA taxonomy, and DNA barcoding are intertwined and are essential ingredients in building up the tree of life for all organisms.

13.1.2 DNA Barcoding Advances

DNA barcoding was first proposed by Hebert, Cywinska, et al. (2003), where they suggested that the mitochondrial *COI* gene can be used as a universal region of the genome for species identification of animals (Hebert, Cywinska, et al., 2003). The proposal for using the mitochondrial *COI* gene as a global species identification system for animals was based on the genetic marker's resolution for reciprocal monophyly and the presence of barcode gap (interspecies divergence higher than intraspecies divergence) (Hebert, Ratnasingham, & deWaard, 2003). Additionally, voucher specimens, which are preserved specimens archived in a collection facility from which the DNA sequences are obtained, are an essential component of DNA barcoding. The Barcode of Life project was set up as a public collection of reference sequences tasked to speed up species identification using DNA sequences and as a tool to support the Convention of Biological Diversity (http://cbd.int).

In 2004, the Consortium for the Barcode of Life (CBOL) was established as an international initiative devoted to DNA barcoding. Thereafter, the International Barcode of Life project was launched in 2010 as a research alliance of various research groups across countries to construct a DNA barcode library for animals (https://ibol.org). To have a centralized DNA database, the Barcode of Life Systems Database (BOLD) (http://boldsystems.org) was established as a platform for DNA sequences and voucher specimen information to be deposited (Centre for Biodiversity Genomics, 2021; Ratnasingham & Hebert, 2007). With a centralized database linking DNA sequences, voucher specimens, and DNA barcoding projects across various research groups and countries, researchers can now easily tap into the BOLD platform as a resource for their respective projects and organisms of interest.

In animals, the mitochondrial *COI* gene is the DNA barcode of choice and has been successfully utilized for various groups such as mammals, arthropods, birds, and fishes (Trivedi et al., 2020). Although the mitochondrial *COI* gene is the most popular DNA barcode, other genetic markers have been used for DNA barcoding for other groups of organisms. In plants, the *COI* gene could not provide sufficient species-level resolution due to lower levels of sequence variation. The plastid maturase K (*matK*) and the ribulose bisphosphate carboxylase (*rbcL*) genes are used instead for DNA barcoding and have been shown to provide sufficient species-level resolution for plant identification (Group, 2009; Kress, 2017).

Similarly, the mitochondrial *COI* gene in fungi is unsuitable as a DNA barcode, as it is only suitable for a few genera. Thus, the use of multiple loci such as the nuclear internal transcribed spacer (ITS) region and the beta-tubulin gene has thus been suggested for fungi DNA barcoding (Hoang, Irinyi, Chen, Sorrell, & Meyer, 2019; Schoch et al., 2012; Trivedi et al., 2020). For species identification of bacteria, the 16S ribosomal RNA (rRNA) gene is widely utilized and has also been used to quantify the bacteria community from environmental samples (Trivedi et al., 2020). Currently, the BOLD server can support *COI* barcodes for animals, *rbcL* and *matK* for plants, and ITS for fungi. To date, there are close to 1,000,000 specimens with the respective barcode sequences on the server, representing approximately 300,000 species. A quick search on PubMed (https://pubmed.ncbi.nlm.nih.gov/) using "DNA barcoding" as a keyword revealed close to 9000 full texts. The potential of DNA barcodes as a universal identification tool is clearly advantageous for taxonomy.

The advantages of DNA barcoding as an application for DNA taxonomy are abundant. Firstly, DNA barcodes provide a link between voucher specimens (previously identified by morphology) and their respective DNA sequences (Hebert, Cywinska, et al., 2003). The retainment of voucher specimens and DNA provides an avenue for re-checking species identity through morphological characters and DNA sequences. Secondly, the use of a standard gene and region allows for comparing DNA sequences across species and studies (Frézal & Leblois, 2008; Hebert, Cywinska, et al., 2003). Thirdly, DNA barcodes allow for species identification in the event where morphological characters are not able to fully resolve a specimen's identity (Pires & Marinoni, 2010). Morphologically similar species can now be accurately identified through DNA sequences.

Moreover, the issue of morphological plasticity can now be overcome. Fourthly, immatures from various hosts are able to be identified through DNA barcoding (Morand, 2018). Lastly, the speed and convenience for analyzing the large number of samples through molecular-based techniques give DNA barcoding an edge over identification using classical taxonomy (Teletchea, 2010; Vogler & Monaghan, 2007).

However, despite the advantages of DNA barcoding, the full extent of the usefulness of DNA barcoding has been hotly debated. We will not go into the details of the debates but will explain why the DNA barcoding concept is an overpromise for taxonomy and caution its potential drawbacks. One of the main reasons for not fully agreeing with DNA barcoding is that DNA sequences on their own without reference sequences or prior morphological identification to compare to is essentially just a string of "As, Ts, Cs, and Gs" (Köhler, 2007; Vogler & Monaghan, 2007).

DNA barcoding has been promised to be useful for species discovery. However, matching the unknown sequence to species identity is impossible without reference sequences or initial morphological identification. Moreover, species delimitation using DNA barcodes can be problematic, as species boundaries (defined by interspecies and intraspecies variation) are not a fixed value among different groups of organisms (Darling & Blum, 2007; Krishnamurthy, Francis, & Robert, 2012; Taylor & Harris, 2012). A universal threshold for genetic variation encompassing all organisms to define what constitutes sufficient variation or one species from another is not feasible. Another reason is due to the use of the sole use of the mitochondrial *COI* gene for animal barcodes (Teletchea, 2010). Although the *COI* gene can provide robust species identification of animals due to their high rate of evolution (part of the mitochondrial genome), non-supporters argue that it is limited in its utility. The presence of possible pseudogenes and introgression can create confusion about species boundaries and their identity (Rubinoff, Cameron, & Will, 2006). Relying on a single gene from only the mitochondrial loci might not be sufficient to prevent confounders.

To circumvent these limitations that have been addressed, the concept of integrative taxonomy was introduced. Integrative taxonomy was proposed by Dayrat (2005), where it is defined as the science that aims to delimit the units of life diversity from multiple and complementary perspectives, which include phylogeography, comparative morphology, population genetics, ecology, development, behavior, etc. (Dayrat, 2005). The rationale behind an integrative approach to taxonomy is that the complexities of species biology require that species boundaries be studied from multiple perspectives. With integrative taxonomy, both classical and DNA taxonomy can be used together, serving as additional information to aid species identification (Goldstein & DeSalle, 2011; Padial, Miralles, De la Riva, & Vences, 2010). Integrative taxonomy thus allows for the comparison of information from various dimensions, strengthening hypotheses on species boundaries and delimitation. Additionally, the use of multiple loci for species identification has been promoted in recent years, and many publications on species identification now include both mitochondrial and nuclear genetic markers for increased support.

13.2 Molecular Identification and DNA Barcoding of Helminths

13.2.1 Progress

As we have seen in Chaps. 10–12, various genetic markers have been used for the molecular systematics of helminths. Similarly, studies utilizing genetic markers for the molecular identification of helminths have rapidly increased in the past two decades. Progress has been made, as evidenced by the numerous studies using genetic markers for species identification of helminths (the utilities of genetic markers were discussed in Chap. 6). Molecular-based techniques and markers such as randomly amplified polymorphic DNA (RAPD), restriction fragment length polymorphism (RFLP), amplified fragment length polymorphism (AFLP), and simple sequence repeats (SSR) have also supported the identification and assessment of genetic variation among helminths. The reliance on molecular data for helminth taxonomy to aid classical taxonomy to describe species has gradually gained momentum in the past decades. Additionally, the use of both morphological and molecular information is now commonplace for helminth molecular identification.

The nuclear 18S and 28S rRNA genes were the first genetic markers to be used for the molecular identification of helminths. Floyd, Abebe, Papert, and Blaxter (2002) used 52 segments of the 18S rRNA gene for species identification of soil nematodes using a molecular operational taxonomic unit (MOTU) scheme (Floyd et al., 2002). Morphology and molecular approaches were also explored in a survey of marine nematodes, where the D2–D3 segment of the 28S rRNA gene was utilized to match specimens with their respective molecular sequences (Pereira, Fonseca, Mundo-Ocampo, Guilherme, & Rocha-Olivares, 2010). The ITS region, possessing higher sequence variability than the nuclear rRNA genes, is another popular marker for helminth species identification. Among nematodes, ITS has shown to be promising, with high interspecific variability and low intraspecific variability. These differences in genetic variation between species and within species render the ITS region useful for species identification. Evidence from plant-parasitic nematodes and strongylid nematodes support the usefulness of ITS as a genetic marker for molecular identification (Kuchboev et al., 2020; Palomares-Rius, Cantalapiedra-Navarrete, Archidona-Yuste, Subbotin, & Castillo, 2017). The ITS region has also been utilized for species identification among trematodes. Examples of their utility include the molecular identification of morphologically similar immatures and adults and linking different life stages found in various hosts (Chibwana, Nkwengulila, Locke, McLaughlin, & Marcogliese, 2015; Georgieva et al., 2013). Additionally, both the ITS region and the *COI* gene have been used in unison as complementary markers for additional support for molecular identification (Locke et al., 2015).

With the idea of DNA barcoding using the mitochondrial *COI* gene as a universal DNA barcode for animals, as proposed by Hebert, Cywinska, et al. (2003), the utility of *COI* was also explored among helminths (Hebert, Cywinska, et al., 2003). Among *Schistosoma*, the usefulness of the *COI* gene was supported through substantial genetic diversity (Webster et al., 2013). Additionally, using the popular *COI* primers developed by Bowles, Blair, and McManus (1992), species identification using the mitochondrial *COI* gene has also been successful among cestodes (Bowles et al., 1992). Among free-living marine nematodes, the utility of *COI* was explored for 41 species. Derycke, Vanaverbeke, Rigaux, Backeljau, and Moens (2010) demonstrated that the *COI* gene could differentiate almost all species tested and there was high congruence between morphology and molecular identification (Derycke et al., 2010). Similar to other groups of organisms, the ability of the *COI* gene for species-level resolution of helminths cannot be questioned.

Albeit the large number of studies that have been performed for helminth molecular identification and the use of both morphological and molecular data for helminth identification, DNA barcoding of helminths has not been without complications. From the examples, we can see that different genetic markers have been used for the molecular identification of helminths. To be utilized as a successful genetic marker for DNA barcoding, some factors will need to be considered. These factors include a standardized genetic marker with sufficient variation for species-level resolution, primers that can amplify a broad range of species, a comprehensive reference database of sequences, and voucher specimens with relevant information (Köhler, 2007). Currently, although the *COI* gene is the standard DNA barcode of choice for animals (encompassing helminths), a suitable genetic marker has not been found for helminths. We will further elaborate on these challenges for helminth DNA barcoding in the following subsection.

13.2.2 Challenges

The challenges of accurate species identification and delimitations among helminths were discussed thoroughly in Chap. 5. Briefly, the huge diversity of helminths and their ability to infect various hosts are just factors that could affect the morphological identification of helminths. Additionally, the presence of morphologically plastic characters due to convergent evolution further complicates identification. Due to the small size and lack of diagnostic morphological characters, molecular identification is a welcoming tool to aid in the taxonomy of helminths. However, molecular-based identification also presents its own set of challenges.

Firstly, obtaining high-quality DNA with good yield could present a problem for helminth specimens (Ondrejicka, Locke, Morey, Borisenko, & Hanner, 2014). Helminths are generally small, and adult specimens might not always be present in the collected samples. Immatures or egg stages can be found, depending on the sample collection and the host surveyed. The reagents for the staining and preservation of helminths also affect the DNA quality, and specimens that have been stained for morphological identification cannot be used for downstream molecular analysis (Sepulveda & Kinsella, 2013). In fecal or soil samples, the presence of inhibitors can affect molecular analysis, and protocols will have to be thoroughly optimized to obtain maximum DNA from the specimens. Due to the DNA yield and quality, primers for DNA barcoding should be robust to amplify small amounts of DNA and DNA that have been degraded due to preservation techniques.

Secondly, high levels of sequence variation in the *COI* gene are present for helminths. Moszczynska, Locke, McLaughlin, Marcogliese, and Crease (2009) revealed that the *COI* alignment of diverse species within Platyhelminthes had high variation, making it difficult to design primers to successfully amplify species for the entire phylum (Moszczynska et al., 2009). PCR success for parasitic nematodes has also been evaluated using the universal *COI* primers for invertebrates by Folmer, Black, Hoeh, Lutz, and Vrijenhoek (1994), where only 65% of sequences were successfully recovered (Folmer et al., 1994; Prosser, Velarde-Aguilar, Leon-Regagnon, & Hebert, 2013). Using 41 marine nematode species, Derycke et al. (2010) compared PCR success with the HCO2198/LCO1490 primers (henceforth named “Folmer primers”) and JB3/JB4.5 primers (Derycke et al., 2010). PCR success was 65.8% and 87.8%, respectively, and the results revealed that the *COI* gene is able to successfully identify almost all nematode species tested, with congruence between morphology. Despite the higher PCR success using the JB3/JB4.5 primers, some studies have noted poor amplification success, and genus- or species-specific primers usually had to be designed for successful PCR amplification (Lv et al., 2016; Morise, Miyazaki, Yoshimitsu, & Eki, 2012). Using both the 18S rRNA gene and the *COI* gene for Trichuridae, the JB3/JB4.5 *COI* primers had to be modified in order to be able to amplify species belonging to *Capillaria*, *Trichuris*, and *Eucoleus* (Guardone, Deplazes, Macchioni, Magi, & Mathis, 2013). Similarly, lower PCR amplification success was observed for the *COI* gene compared to the 12S rRNA gene for filarial nematodes (Ferri et al., 2009). Thus, the high sequence variability of the helminth *COI* gene limits the use of the universal *COI* primers for DNA barcoding.

Thirdly, although the highly conserved 18S sequences allow for the design of primers that can amplify many species across nematodes and Platyhelminthes, this genetic marker is limited by its ability to discriminate against some closely related species successfully. The conserved sequence of the 18S rRNA gene can result in higher PCR success than the *COI* gene. Examples of primers include the 18S primers used for Platyhelminthes by Olson and Caira (1999) and nematodes by Blaxter et al. (1998). Despite the robust phylogenetic resolution at higher taxonomic levels, species-level resolution is low, especially for distinguishing closely related species. In an assessment of ten genetic markers for helminths by Chan, Chaisiri, Saralamba, Morand, and Thaenkham (2021), the low sequence variation among the selected closely related species was observed for the 18S rRNA sequences across nematodes, trematodes, and cestodes (Chan et al., 2021).

Fourthly, as stated by proponents of classical taxonomy, DNA sequences on their own are just DNA sequences without any species identity (Krishnamurthy et al., 2012). Prior identification based on morphology should ideally be performed to species level to gauge the unknown specimen. DNA-based identification is further complicated if the unknown DNA sequence cannot be matched to any species present in the reference database. A comprehensive reference database with the target sequences from the genetic marker will be advantageous to increase the probability of matching unknown sequences (Hajibabaei et al., 2005). For helminths, the nuclear 18S rRNA gene has the highest proportion of sequences and species in the NCBI database relative to the other genetic markers due to its popularity as a genetic marker for helminth molecular-based studies. However, with the low species-level resolution, the full utility of the 18S rRNA gene for DNA barcoding of helminths remains limited despite having a comprehensive database of sequences in NCBI. There are approximately 1600 nematode species with *COI* barcodes for the mitochondrial *COI* gene, while approximately 600 platyhelminth species have *COI* barcodes based on the BOLD database. The proportion of helminth species with *COI* barcode is in stark contrast to the species diversity present.

Lastly, species delimitation of helminths is complex due to cryptic diversity. Large amounts of cryptic species have been revealed with molecular data, making species delimitation of helminths difficult (Morand, 2018). Questions include the species boundaries and how much intra- and interspecies variation is needed to define one species from another.

13.2.3 Possible Solutions

Although a suitable genetic marker for helminth has not yet been found, various studies have attempted to overcome the challenges in search of an optimal solution. Prosser et al. (2013) reported the development of a primer cocktail (a mixture of three primer sets) to amplify *COI* barcodes for representative parasitic nematodes belonging to Spirurida, Rhabditida, and Panagrolamida (Prosser et al., 2013). Their study revealed similar PCR amplification success of the cocktail primers with the Folmer primers but had a lower percentage of successful sequences recovered. Using the mitochondrial 12S and 16S rRNA genes, primers encompassing the four clades of animal parasitic nematodes were developed by Chan, Chaisiri, Morand, Saralamba, and Thaenkham (2020). The potential utility of the mitochondrial rRNA genes for molecular systematics and DNA barcoding was revealed, with robust resolution at high and low taxonomic levels. Additionally, the ability of the mitochondrial rRNA genes to discriminate closely related species among nematodes provided support that they might be a suitable alternative to the *COI* gene as a DNA barcode marker for nematodes (Chan et al., 2020).

The high *COI* sequence divergence in digenetic trematodes among platyhelminths made it challenging to design primers for trematodes. Moszczynska et al. (2009) initially developed degenerate primers attempting to amplify the *COI* gene from representatives of three digenean families, two cestode families, and one turbellarian family (Moszczynska et al., 2009). However, the degenerate primers could not successfully amplify the majority of the representative species, with a 40% and 5% success rate for Digenea and Eucestoda, respectively. Another set of primers specific to family Diplostomatidae was designed instead. The authors suggested that amplification of the 18S rRNA gene could be first performed to screen representative trematodes to the family level, and the family-specific *COI* primers were then used for species identification. In another study, degenerate primers targeting the *COI* gene for digeneans and cestodes showed an overall success of 91%, showing an improvement compared to the primers by Moszczynska et al. (2009) and Van Steenkiste, Locke, Castelin, Marcogliese, and Abbott (2015). However, the degeneracy of primers targeting both Digenea and Eucestoda was high, with a degeneracy of 4608 and 62,208 for the forward and reversed primers, respectively. The high degeneracy can limit the utility of the primers for successful PCR amplification. As an alternative to the *COI* gene, primers targeting three orders in Digenea (Plagiorchiida, Echinostomatida, and Striegda) were developed using the mitochondrial rRNA genes (Chan, Saralamba, Saralamba, Ruangsittichai, & Thaenkham, 2022). Successful amplification occurred with all representative species at various life stages, and closely related species were successfully discriminated.

In addition to searching for a suitable and standardized genetic marker, a more coordinated approach will be needed for successful helminth DNA barcoding studies. The importance of a consolidated and accurate database of sequences, along with accurate species identity, is pivotal to DNA barcoding. With more advanced morphological and molecular technology, species identification through integrative taxonomy allows for comparing multiple and complementary data sources. Integrative taxonomy paves the road ahead for DNA barcoding and will undoubtedly be integral in laying the foundations for the DNA taxonomy of helminths.

13.2.4 Examples

Various genetic markers have been utilized for molecular identification of parasitic helminths, and this section will provide examples for each group of helminths in relation to the genetic marker's utilities and pitfalls for molecular identification.

Nematodes

Molecular identification of nematodes was first performed using the 18S rRNA gene as a genetic marker. The popularity of the 18S rRNA gene was further enhanced with the development of broadly applicable primers for the phylum (Blaxter et al., 1998). Early studies attempted to use a MOTU scheme, where the 18S rRNA gene was used as a molecular barcode for soil nematode identification (Blaxter et al., 2005; Floyd et al., 2002). Through the MOTU scheme, DNA sequences could be grouped into taxonomic units based on their similarity, allowing for a rapid assessment of nematode taxon diversity without a correlation to a database of sequences from known species. Species diversity among nematodes of various habitats has been assessed using the 18S rRNA gene, and the habitats include agricultural soils, rainforests, marine, and invertebrates (Griffiths, Donn, Neilson, & Daniell, 2016; Morise et al., 2012; Pereira et al., 2010; Powers et al., 2009).

The easy PCR amplification of the 18S rRNA gene, along with the availability of universal primers and the existence of an 18S rRNA gene phylogenetic tree, is a factor contributing to the popularity of the 18S rRNA gene for nematode species identification. The increased number of studies utilizing the 18S rRNA gene as a genetic marker for nematodes led to increased availability of sequences in databases, thus propagating its use. Despite the popularity of the 18S rRNA gene, species-level resolution might not be possible in some nematode groups. The low sequence variation of the 18S rRNA gene has been reported in several studies, where closely related species cannot be differentiated. For example, low to no sequence variation has been reported among *Trichuris* spp., *Haemonchus* spp. for animal parasitic nematodes, and *Meloidogyne* spp. for plant-parasitic nematodes (Callejon et al., 2013; Powers, Mullin, Harris, Sutton, & Higgins, 2005; Zarlenga, Stringfellow, Nobary, & Lichtenfels, 1994). In another study comparing the 18S rRNA gene and the mitochondrial *COI* gene for the species composition of nematodes present in agricultural field soils, the 18S rRNA gene provided only half the number of operational taxonomic units (OTUs) as compared to the *COI* gene (Morise et al., 2012).

The nuclear 28S rRNA gene, especially the D2–D3 region, has been utilized, especially among plant nematodes. The usefulness of the 28S rRNA gene for species identification was supported in studies looking at the diversity of plant-parasitic nematodes and free-living nematodes, with higher sequence divergences between species compared to the 18S rRNA gene (Palomares-Rius et al., 2017; Pereira et al., 2010). Similarly, with their high sequence variation between species, the ITS regions have been used in the molecular identification of closely related species. Examples include the successful differentiation of the human hookworms *Necator americanus* and *Ancylostoma duodenale* and between the human whipworm *Trichuris trichiura* and pig whipworm *Trichuris suis* (Cutillas et al., 2009; Nissen et al., 2012; Yong et al., 2007). However, the high sequence variation hinders the development of universal primers, and no universal nematode primers for ITS have been developed. Species-specific ITS primers are used to differentiate the species of interest, depending on the research.

As mentioned earlier, the mitochondrial *COI* gene is a popular genetic marker with its robust species-level resolution and fame as a DNA barcode. Derycke et al. (2010) explored the use of the *COI* gene for DNA barcoding of free-living marine nematodes and showed that most of the species tested were able to be differentiated (Derycke et al., 2010). Closely related and cryptic species have been successfully differentiated to assess nematode species diversity and population structure (Callejon et al., 2013; Monteiro et al., 2019; Setsuda et al., 2016). However, despite being a reliable genetic marker that is robust for species identification, the unavailability of a primer pair that is able to work across the entire phylum successfully limits its utility as a genetic marker. Although degenerate primers and primer cocktails have been suggested, the primers often perform poorly, and there have been instances where family or genus level primers had to be specially designed to yield successful amplicons (Ferri et al., 2009; Guardone et al., 2013; Lv et al., 2016).

Recently, the use of the mitochondrial rRNA genes as a genetic marker was supported, with the development of broad phylum primers that can amplify parasitic nematodes of medical importance for four clades belonging to orders Trichocephalida, Oxyurida, Ascaridia, Spirurida, Rhabditida, and Strongylida (Chan, Chaisiri, Morand, et al., 2020). The same primers were applied to the closely related *Angiostrongylus cantonensis* and *Angiostrongylus malaysiensis*, revealing their robustness for species identification (Chan, Chaisiri, Dusitsittipon, et al., 2020). The availability of phylum-wide primers, along with their robustness for species-level identification, could render the mitochondrial rRNA genes a suitable genetic marker for molecular identification and DNA barcoding of nematodes.

Trematodes

The nuclear 18S and 28S rRNA genes were first widely utilized as genetic markers for trematodes, owing to the study by Olson, Cribb, Tkach, Bray, and Littlewood (2003) in providing a comprehensive phylogeny for Digenea (Olson et al., 2003).

However, the low sequence variation made cryptic species identification challenging, considering that trematodes have higher species diversity than the other group of helminths. Circumventing that, PCR-RFLP has been used to differentiate between closely related species. Using the 18S rRNA gene, common foodborne trematode species belonging to *Clonorchis sinensis*, *Metagonimus yokogawai*, *Pygidiopsis summa*, *Heterophyopsis continua*, and *Gymnophalloides seoi* metacercariae in fish and oysters were successfully differentiated through the RFLP band patterns (Pyo et al., 2013). Thaenkham et al. (2011) also showed successful differentiation using PCR-RFLP with the 28S rRNA gene for six species of heterophyid intestinal flukes endemic to Asia (Thaenkham et al., 2011). The species are *Haplorchis taichui*, *Haplorchis pumilio*, *Haplorchis yokogawai*, *Procerovum varium*, *Stellantchasmus falcatus*, and *Centrocestus formosanus*. Although molecular identification of closely related species is possible through PCR-RFLP with the nuclear rRNA genes, the low sequence variation still limits their use as genetic markers for DNA barcoding.

With their high sequence variability, the nuclear ITS regions render them helpful to differentiate between closely related species. Species-specific primers targeting the species of interest are commonly used. Examples include examining the prevalence of fish-borne trematodes using the ITS2 as a genetic marker, molecular identification for human-associated liver fluke species through ITS1, and the successful differentiation of *Fasciola hepatica* and *Fasciola gigantica* with both ITS1 and ITS2 (Mahami-Oskouei, Dalimi, Forouzandeh-Moghadam, & Rokni, 2011). Additionally, Le et al. (2008) revealed the ability of the ITS2 region to detect hybrids of *F. hepatica* and *F. gigantica* in Vietnam through comparison with the mitochondrial *COI* and *ND1* genes (Le et al., 2008). In recent years, trematode molecular studies have heeded the call to use more than one genetic marker, in addition to classical taxonomy. The mitochondrial *COI* gene has been frequently utilized alongside the nuclear ITS regions as complementary information to support species identity and hypotheses.

The mitochondrial *COI* gene was initially widely utilized to assess genetic diversity among *Schistosoma* species. Studies have been performed for *Schistosoma haematobium* and *Schistosoma mansoni* across the African continent and revealed substantial diversity and geographical separation of genotypes (Webster et al., 2013; Webster, Webster, et al., 2013). The development of *COI* primers specific for Diplostomidae by Moszczynska et al. (2009) aimed to build an extensive DNA barcode library for diplostomids enhanced the use of the *COI* gene for molecular identification (Moszczynska et al., 2009). The discovery of cryptic species, the identification of developmental stages in different hosts to elucidate the parasite's life cycle, and the identification of problematic taxa are some examples of studies that were performed for Diplostomidae using the *COI* gene (Chibwana et al., 2015; Georgieva et al., 2013; Locke et al., 2015). However, there is still no truly universal primer targeting the *COI* gene for a broad range of digeneans to date, restricting its potential as a DNA barcode marker.

Targeting representative species belonging to Plagiorchiida, Echinostomatida, and Striegda, the potential of the mitochondrial 12S and 16S rRNA genes as a DNA barcode marker was explored (Chan et al., 2022). The study revealed 100%

PCR amplification success with the representative trematodes and the primers' ability to successfully amplify the different life stages of the parasite (egg, cercaria, metacercaria, adult). Additionally, closely related species were able to be successfully discriminated, supporting the potential of the mitochondrial rRNA genes for DNA barcoding.

Cestodes

Molecular identification of cestodes frequently utilizes mitochondrial genetic markers. The mitochondrial *COI* gene and nicotinamide adenine dinucleotide 1 (*ND1*) genes are popular genetic markers used for cestode molecular identification. The first few studies involved the use of the JB3/JB4.5 (for *COI*) and the JB11/JB12 (for *ND1*) primers developed by Bowles et al. (1992) and Bowles and McManus (1993) through the characterization of species and strains within the genus *Echinococcus* (Bowles et al., 1992; Bowles & McManus, 1993). Inter- and intraspecific variations were observed, supporting the use of the mitochondrial genes for species and strain identification. Gasser, Zhu, and McManus (1999) successfully employed these primers for nine species in *Taenia* (Gasser et al., 1999). Compared to nematodes and trematodes, the *COI* JB3/JB4.5 primers are more widely utilized in cestodes, especially among cestodes of medical importance to humans. A possible reason could be the lesser sequence variation among cestodes, allowing for higher PCR success comparatively. Among other examples, the mitochondrial *COI* gene, with its robustness for species-level resolution, has provided support for cryptic species among *Hymenolepis*, species identification among *Spirometra*, and the circumscription of *Taenia asiatica* to a distinct species (Almeida, Coscarelli, Melo, Melo, & Pinto, 2016; Gasser et al., 1999; Macnish, Morgan-Ryan, Monis, Behnke, & Thompson, 2002). However, the utility of the *COI* gene for cestodes is not without challenges. Nyman et al. (2021) investigated cestode species in marine and freshwater ringed seals using the *COI* gene and found that the degenerate *COI* primers for Platyhelminthes by Setsuda et al. (2016) produced variable results (Nyman et al., 2021).

In addition to the *ND1* and *COI* genes, other genetic markers have also been used for cestode molecular identification. A multi-locus approach targeting the nuclear 18S rRNA gene, ITS1 and ITS2 regions, mitochondrial *COI*, and cytochrome b (*cytB*) genes was undertaken to assess inter- and intraspecies variation among *Diphyllobothrium* by Wicht et al. (2010). They showed that the nuclear genes are unsuitable for inter- and intraspecific characterization between *Diphyllobothrium dendriticum* and *Diphyllobothrium ditremum*, while the mitochondrial genes could discriminate between them. Molecular identification on cestode communities in carnivores was also performed using the nuclear 18S rRNA gene, the mitochondrial 12S rRNA, and the *COI* gene (Bajer et al., 2020). Higher levels of sequence divergence among *Mesocestoides* spp. were noted for the mitochondrial genetic markers compared to the 18S rRNA gene, supporting the use of mitochondrial genes for cestode molecular identification.

13.3 Applying DNA Barcoding for Helminth Taxonomy: Strategy and Steps

Even though there is no suitable genetic marker for helminth DNA barcoding yet, general management and practice can still be applied for helminth molecular identification and taxonomy. The general overview of the steps includes specimen collection and management, molecular analysis, molecular sequence analysis, and database management (Fig. 13.1). Application of these steps will be helpful for taxonomic studies to link voucher specimens to DNA sequences.

13.3.1 Management of Voucher Specimens

In Chap. 9, guidelines for sample preparation of helminths for morphological and molecular analysis were detailed. It is undeniably the most important in influencing further downstream steps. Applying appropriate procedures during specimen collection should be performed. Voucher specimens, which are preserved specimens achieved in a collection facility, can allow for either morphological or molecular re-examination and the replication and validation of past analysis. DNA-friendly preservation methods such as keeping in 96–100% ethanol at −20 °C or direct freezing (−20 °C) of the specimens are highly recommended. DNA degrading chemicals such as formalin should be avoided. Proper labeling of the specimens with the collection details should be recorded. Generally, details recorded can include host type, organ isolated from, stage of helminth, locality, date of collection, specimen identity, and specimen code. If the specimens isolated from a particular host and organ are of the same stage and species, they can be kept together in the same vial (voucher specimen stock).

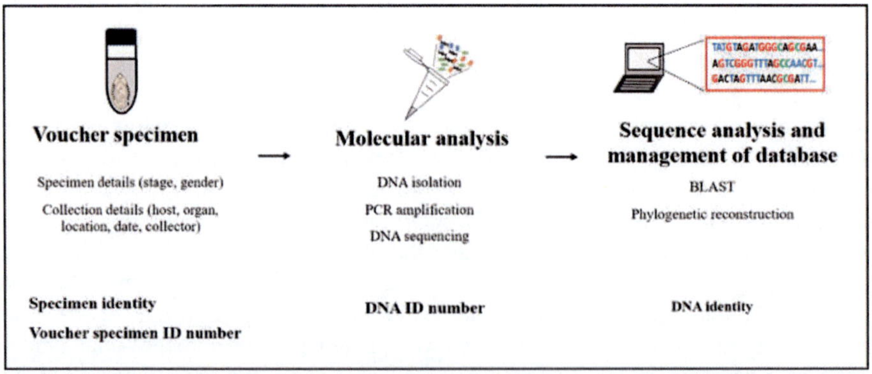

Fig. 13.1 Major steps for DNA barcoding linking voucher specimens to DNA sequences

Depending on the helminth size, a portion of the helminth can be cut and used for downstream molecular analysis. The remainder of the helminth not used for molecular analysis will be kept in another vial (individual voucher specimen) and preserved for future re-examination. The vial containing the individual helminth should be labeled appropriately, with details tracing back to the original collection (voucher specimen stock). If the whole helminth has to be used for molecular analysis, the isotype specimens in the voucher specimen stock vial can be used if there is a need for re-examination.

13.3.2 Molecular Analysis

The molecular analysis includes the isolation of DNA, PCR amplification of the target gene, and DNA sequencing. Generally, the age of the specimen and the quality of DNA isolated are inversely proportional. Fresher and more recent specimens usually result in higher DNA yield and quality than older specimens. Before isolating DNA, if the voucher specimen was preserved in 96–100% ethanol, the specimen should be washed thoroughly in distilled water to remove any excess ethanol. If the helminth is too small to be washed in distilled water, ensure that the majority of the ethanol is removed before DNA isolation. For the isolation of DNA, it can be performed with either a commercial kit through silica-column-based extraction or the traditional phenol-chloroform extraction. A series of steps include the mechanical and chemical lysis of the cells to release the DNA, remove contaminants such as proteins through enzymes, and finally wash DNA out in the elution step. For mechanical lysis of the specimen, a tissue lyser homogenizer through bead beating can be used to break down the cuticle for nematodes. Quantification and quality checking of the isolated DNA should be ideally performed before using it for downstream analysis. For long-term storage, the DNA should be kept at −20 °C and freeze-thawing avoided. For short-term storage, the DNA can be kept at 4 °C. The isolated DNA should be given a unique identifier that can be traced back to the details of the identity of the voucher specimen.

Depending on the PCR protocol and optimized conditions, the DNA concentration should be modified accordingly before PCR amplification. Evaluation of the specific genetic marker for its robustness for DNA barcoding and species-level resolution should be undertaken for the specific helminth of interest. Chan et al. (2021) provided some helpful guidelines and genetic distance cut-off values specific for parasitic nematodes, trematodes, and cestodes and can be valuable in deciding which genetic marker to choose for helminths (Chan et al., 2021). In Chap. 7, the principles, steps, and recommendations for PCR and DNA sequencing were already detailed. It is important to ensure that the DNA sequence obtained can be traced back to the details of the voucher specimen.

13.3.3 Sequence Analysis and Management of Database

Using NCBI BLAST (https://blast.ncbi.nlm.nih.gov/Blast.cgi), the DNA sequence identity can be revealed based on the closest match on the NCBI database through comparison with all of the sequences that are available in NCBI. If an internal reference DNA sequence database for the helminth of interest and a genetic marker is available, the query DNA sequence can also be compared against the internal database. However, in instances where there is no reference sequence available or matching to the unknown DNA sequence and where the exact species identity was unable to be determined through morphology, the exact species identity of the specimen is still unknown. To circumvent this, phylogenetic analysis can be performed to investigate where the query DNA sequence cluster to. If a phylogenetic tree with comprehensive reference sequences of species is available, the phylogenetic position of the unknown sequence can be known. A neighbor-joining tree is usually constructed for DNA barcoding to determine how phylogenetically related the query DNA sequence is to other species through clustering based on similarity. In the absence of an exact reference species sequence, the identity of the unknown can be inferred through its phylogenetic placement in relation to other species.

Additionally, genetic distance analysis can also be performed to determine inter- and intraspecies genetic variation between the selected DNA sequences. Details about how DNA sequence analysis can be performed are in Chap. 8, along with the various types of phylogenetic trees that can be reconstructed. Finally, the DNA sequence obtained can be uploaded onto a reference database (e.g., NCBI or BOLD for the *COI* gene), along with their corresponding voucher specimen details.

Throughout the whole process, from helminth specimen collection to DNA sequence analysis, it is important to ensure that the details linking the voucher specimen to the DNA sequence are recorded properly at each stage. Good data management is essential for a successful DNA barcoding project, ensuring that the DNA sequence obtained can be traced back to the voucher specimen.

13.4 Recommendations and Outlook for DNA Taxonomy of Parasitic Helminths

For successful DNA barcoding, three main ingredients are vital. They are the availability of broadly applicable primers to amplify a wide range of helminth species, the robustness of the genetic marker for species-level resolution, and a comprehensive database of DNA sequences. With the high sequence variability for helminths, the mitochondrial *COI* gene hinders the availability of a broadly appropriate primer, albeit its robustness for species-level resolution. The nuclear ITS regions present similar challenges to the mitochondrial *COI* gene. The nuclear 18S rRNA gene, on the other hand, does not have enough sequence variability for species-level discrimination of some species. However, broadly applicable primers are available, and it has the highest proportion of helminth species sequences available (in comparison to the other genetic markers for helminths). The potential of the mitochondrial rRNA genes as a DNA barcode marker should not be disregarded, with primers that are now broadly applicable for a wide range of parasitic helminths and robust species-level resolution. The number of sequences available in the public database is a limiting factor, restricting its use as a DNA barcode marker.

However, the outlook for DNA taxonomy and barcoding of helminths is promising through the combination of multi-locus markers, integrative taxonomy, and next-generation sequencing (NGS). Recently, studies involving the detection of ancient DNA and DNA metabarcoding of helminths in various habitats have been made possible through NGS. Additionally, species identification and discovery of helminths in the field of biodiversity is even more critical now in the face of a changing climate and habitat loss. Climate changes can drastically reduce habitats and hosts for helminths, potentially leading to a race in discovering species before they are gone. Furthermore, the process of deforestation and urbanization reduces the interface between humans and animals, increasing the chance of zoonoses and cross-species transmission of helminthic diseases. Clearly, DNA taxonomy is pivotal for helminth biodiversity and in guiding disease control and strategies.

References

Almeida, G. G., Coscarelli, D., Melo, M. N., Melo, A. L., & Pinto, H. A. (2016). Molecular identification of *Spirometra* spp. (Cestoda: Diphyllobothriidae) in some wild animals from Brazil. *Parasitology International, 65*(5), 428–431. https://doi.org/10.1016/j.parint.2016.05.014

Bajer, A., Alsarraf, M., Dwuznik, D., Mierzejewska, E. J., Kolodziej-Sobocinska, M., Behnke-Borowczyk, J., Banasiak, L., Grzybek, M., Tolkacz, K., Kartawik, N., Stanczak, L., Opalinska, P., Krokowska-Paluszak, M., Gorecki, G., Alsarraf, M., & Behnke, J. M. (2020). Rodents as intermediate hosts of cestode parasites of mammalian carnivores and birds of prey in Poland, with the first data on the life-cycle of *Mesocestoides melesi*. *Parasites & Vectors, 13*(1), 95. https://doi.org/10.1186/s13071-020-3961-2

Biology Dictionary. (n.d.). *Binomial nomenclature*. https://biologydictionary.net/binomial-nomenclature/

Biology Online. (n.d.). *Binomial nomenclature*. https://www.biologyonline.com/dictionary/binomial-nomenclature

Blaxter, M., Mann, J., Chapman, T., Thomas, F., Whitton, C., Floyd, R., & Abebe, E. (2005). Defining operational taxonomic units using DNA barcode data. *Philosophical Transactions of the Royal Society of London. Series B, Biological Sciences, 360*(1462), 1935–1943. https://doi.org/10.1098/rstb.2005.1725

Blaxter, M. L. (2004). The promise of a DNA taxonomy. *Philosophical Transactions of the Royal Society of London. Series B, Biological Sciences, 359*(1444), 669–679. https://doi.org/10.1098/rstb.2003.1447

Blaxter, M. L., De Ley, P., Garey, J. R., Liu, L. X., Scheldeman, P., Vierstraete, A., Vanfleteren, J. R., Mackey, L. Y., Dorris, M., Frisse, L. M., Vida, J. T., & Thomas, W. K. (1998). A molecular evolutionary framework for the phylum Nematoda. *Nature, 392*(6671), 71–75. https://doi.org/10.1038/32160

Bowles, J., Blair, D., & McManus, D. P. (1992). Genetic variants within the genus *Echinococcus* identified by mitochondrial DNA sequencing. *Molecular and Biochemical Parasitology, 54*(2), 165–173. https://doi.org/10.1016/0166-6851(92)90109-w

Bowles, J., & McManus, D. P. (1993). NADH dehydrogenase 1 gene sequences compared for species and strains of the genus *Echinococcus*. *International Journal for Parasitology, 23*(7), 969–972. https://doi.org/10.1016/0020-7519(93)90065-7

Callejon, R., Nadler, S., De Rojas, M., Zurita, A., Petrasova, J., & Cutillas, C. (2013). Molecular characterization and phylogeny of whipworm nematodes inferred from DNA sequences of cox1 mtDNA and 18S rDNA. *Parasitology Research, 112*(11), 3933–3949. https://doi.org/10.1007/s00436-013-3584-z

Centre for Biodiversity Genomics. (2021). *The global taxonomy initiative 2020: A step-by-step guide for DNA barcoding*. Technical Series No. 94, 66.

Chan, A. H. E., Chaisiri, K., Dusitsittipon, S., Jakkul, W., Charoennitiwat, V., Komalamisra, C., & Thaenkham, U. (2020). Mitochondrial ribosomal genes as novel genetic markers for discrimination of closely related species in the *Angiostrongylus cantonensis* lineage. *Acta Tropica, 211*, 105645. https://doi.org/10.1016/j.actatropica.2020.105645

Chan, A. H. E., Chaisiri, K., Morand, S., Saralamba, N., & Thaenkham, U. (2020). Evaluation and utility of mitochondrial ribosomal genes for molecular systematics of parasitic nematodes. *Parasites & Vectors, 13*(1), 364. https://doi.org/10.1186/s13071-020-04242-8

Chan, A. H. E., Chaisiri, K., Saralamba, S., Morand, S., & Thaenkham, U. (2021). Assessing the suitability of mitochondrial and nuclear DNA genetic markers for molecular systematics and species identification of helminths. *Parasites & Vectors, 14*(1), 233. https://doi.org/10.1186/s13071-021-04737-y

Chan, A. H. E., Saralamba, N., Saralamba, S., Ruangsittichai, J., & Thaenkham, U. (2022). The potential use of mitochondrial ribosomal genes (12S and 16S) in DNA barcoding and phylogenetic analysis of trematodes. *BMC Genomics, 23*, 104.

Chibwana, F. D., Nkwengulila, G., Locke, S. A., McLaughlin, J. D., & Marcogliese, D. J. (2015). Completion of the life cycle of *Tylodelphys mashonense* (Sudarikov, 1971) (Digenea: Diplostomidae) with DNA barcodes and rDNA sequences. *Parasitology Research, 114*(10), 3675–3682. https://doi.org/10.1007/s00436-015-4595-8

Convention on Biological Diversity. (n.d.) *What is taxonomy?* https://www.cbd.int/gti/taxonomy.shtml

Cutillas, C., Callejon, R., de Rojas, M., Tewes, B., Ubeda, J. M., Ariza, C., & Guevara, D. C. (2009). *Trichuris suis* and *Trichuris trichiura* are different nematode species. *Acta Tropica, 111* (3), 299–307. https://doi.org/10.1016/j.actatropica.2009.05.011

Darling, J. A., & Blum, M. J. (2007). DNA-based methods for monitoring invasive species: A review and prospectus. *Biological Invasions, 9*(7), 751–765. https://doi.org/10.1007/s10530-006-9079-4

Dasmahapatra, K. K., & Mallet, J. (2006). Taxonomy: DNA barcodes: Recent successes and future prospects. *Heredity, 97*(4), 254–255. https://doi.org/10.1038/sj.hdy.6800858

Dayrat, B. (2005). Towards integrative taxonomy. *Biological Journal of the Linnean Society, 85*(3), 407–415. https://doi.org/10.1111/j.1095-8312.2005.00503.x

Derycke, S., Vanaverbeke, J., Rigaux, A., Backeljau, T., & Moens, T. (2010). Exploring the use of cytochrome oxidase c subunit 1 (*COI*) for DNA barcoding of free-living marine nematodes. *PLoS One, 5*(10), e13716. https://doi.org/10.1371/journal.pone.0013716

Ferri, E., Barbuto, M., Bain, O., Galimberti, A., Uni, S., Guerrero, R., Ferte, H., Bandi, C., Martin, C., & Casiraghi, M. (2009). Integrated taxonomy: Traditional approach and DNA barcoding for the identification of filarioid worms and related parasites (Nematoda). *Frontiers in Zoology, 6*, 1. https://doi.org/10.1186/1742-9994-6-1

Floyd, R., Abebe, E., Papert, A., & Blaxter, M. (2002). Molecular barcodes for soil nematode identification. *Molecular Ecology, 11*(4), 839–850. https://doi.org/10.1046/j.1365-294x.2002.01485.x

Folmer, O., Black, M., Hoeh, W., Lutz, R., & Vrijenhoek, R. (1994). DNA primers for amplification of mitochondrial cytochrome c oxidase subunit I from diverse metazoan invertebrates. *Molecular Marine Biology and Biotechnology, 3*(5), 294–299.

Fontaneto, D., Flot, J.-F., & Tang, C. Q. (2015). Guidelines for DNA taxonomy, with a focus on the meiofauna. *Marine Biodiversity, 45*(3), 433–451. https://doi.org/10.1007/s12526-015-0319-7

Frézal, L., & Leblois, R. (2008). Four years of DNA barcoding: Current advances and prospects. *Infection, Genetics and Evolution, 8*(5), 727–736. https://doi.org/10.1016/j.meegid.2008.05.005

Gasser, R. B., Zhu, X., & McManus, D. P. (1999). NADH dehydrogenase subunit 1 and cytochrome c oxidase subunit I sequences compared for members of the genus *Taenia* (Cestoda). *International Journal for Parasitology, 29*(12), 1965–1970. https://doi.org/10.1016/s0020-7519(99)00153-8

Georgieva, S., Soldánová, M., Pérez-Del-Olmo, A., Dangel, D. R., Sitko, J., Sures, B., & Kostadinova, A. (2013). Molecular prospecting for European *Diplostomum* (Digenea: Diplostomidae) reveals cryptic diversity. *International Journal for Parasitology, 43*(1), 57–72. https://doi.org/10.1016/j.ijpara.2012.10.019

Goldstein, P. Z., & DeSalle, R. (2011). Integrating DNA barcode data and taxonomic practice: Determination, discovery, and description. *Bioessays, 33*(2), 135–147. https://doi.org/10.1002/bies.201000036

Griffiths, B., Donn, S., Neilson, R., & Daniell, T. (2016). Molecular sequencing and morphological analysis of a nematode community. *Applied Soil Ecology, 32*(3), 325–337.

Group, C. P. W. (2009). A DNA barcode for land plants. *Proceedings of the National Academy of Sciences of the United States of America, 106*(31), 12794–12797. https://doi.org/10.1073/pnas.0905845106

Guardone, L., Deplazes, P., Macchioni, F., Magi, M., & Mathis, A. (2013). Ribosomal and mitochondrial DNA analysis of Trichuridae nematodes of carnivores and small mammals. *Veterinary Parasitology, 197*(1-2), 364–369. https://doi.org/10.1016/j.vetpar.2013.06.022

Hajibabaei, M., deWaard, J. R., Ivanova, N. V., Ratnasingham, S., Dooh, R. T., Kirk, S. L., Mackie, P. M., & Hebert, P. D. (2005). Critical factors for assembling a high volume of DNA barcodes. *Philosophical Transactions of the Royal Society of London. Series B, Biological Sciences, 360*(1462), 1959–1967. https://doi.org/10.1098/rstb.2005.1727

Hebert, P. D., Cywinska, A., Ball, S. L., & deWaard, J. R. (2003). Biological identifications through DNA barcodes. *Proceedings of the Biological Sciences, 270*(1512), 313–321. https://doi.org/10.1098/rspb.2002.2218

Hebert, P. D., Ratnasingham, S., & deWaard, J. R. (2003). Barcoding animal life: cytochrome c oxidase subunit 1 divergences among closely related species. *Proceedings of the Biological Sciences, 270*(Suppl 1), 96–99. https://doi.org/10.1098/rsbl.2003.0025

Hoang, M. T. V., Irinyi, L., Chen, S. C. A., Sorrell, T. C., & Meyer, W. (2019). Dual DNA barcoding for the molecular identification of the agents of invasive fungal infections. *Frontiers in Microbiology, 10*, 1647. https://doi.org/10.3389/fmicb.2019.01647

Köhler, F. (2007). From DNA taxonomy to barcoding - How a vague idea evolved into a biosystematic tool. *Zoosystematics and Evolution, 83*(S1), 44–51. https://doi.org/10.1002/mmnz.200600025

Kress, W. J. (2017). Plant DNA barcodes: Applications today and in the future. *Journal of Systematics and Evolution, 55*(4), 291–307. https://doi.org/10.1111/jse.12254

Krishnamurthy, K., Francis, P., & Robert, A. (2012). A critical review on the utility of DNA barcoding in biodiversity conservation. *Biodiversity and Conservation, 21*(8), 1901–1919. https://doi.org/10.1007/s10531-012-0306-2

Kuchboev, A., Sobirova, K., Karimova, R., Amirov, O., von Samson-Himmelstjerna, G., & Krücken, J. (2020). Molecular analysis of polymorphic species of the genus *Marshallagia* (Nematoda: Ostertagiinae). *Parasites & Vectors, 13*(1), 411. https://doi.org/10.1186/s13071-020-04265-1

Le, T. H., De, N. V., Agatsuma, T., Thi Nguyen, T. G., Nguyen, Q. D., McManus, D. P., & Blair, D. (2008). Human fascioliasis and the presence of hybrid/introgressed forms of *Fasciola hepatica* and *Fasciola gigantica* in Vietnam. *International Journal for Parasitology, 38*(6), 725–730. https://doi.org/10.1016/j.ijpara.2007.10.003

Locke, S. A., Al-Nasiri, F. S., Caffara, M., Drago, F., Kalbe, M., Lapierre, A. R., McLaughlin, J. D., Nie, P., Overstreet, R. M., Souza, G. T., Takemoto, R. M., & Marcogliese, D. J. (2015). Diversity, specificity and speciation in larval Diplostomidae (Platyhelminthes: Digenea) in the eyes of freshwater fish, as revealed by DNA barcodes. *International Journal for Parasitology, 45*(13), 841–855. https://doi.org/10.1016/j.ijpara.2015.07.001

Lv, J., Zhang, Y., Feng, C., Yuan, X., Sun, D., Deng, J., Wang, C., Wu, S., & Lin, X. (2016). Species discrimination in the subfamily Ostertagiinae of Northern China: Assessment of DNA barcode in a taxonomically challenging group. *Parasitology Research, 115*(3), 987–996. https://doi.org/10.1007/s00436-015-4826-z

Macnish, M. G., Morgan-Ryan, U. M., Monis, P. T., Behnke, J. M., & Thompson, R. C. (2002). A molecular phylogeny of nuclear and mitochondrial sequences in *Hymenolepis nana* (Cestoda) supports the existence of a cryptic species. *Parasitology, 125*(6), 567–575. https://doi.org/10.1017/s0031182002002366

Mahami-Oskouei, M., Dalimi, A., Forouzandeh-Moghadam, M., & Rokni, M. (2011). Molecular identification and differentiation of *Fasciola* isolates using PCR-RFLP method based on internal transcribed spacer (ITS1, 5.8S rDNA, ITS2). *Iranian Journal of Parasitology, 6*(3), 35–42.

Meier, R., Shiyang, K., Vaidya, G., & Ng, P. K. (2006). DNA barcoding and taxonomy in Diptera: A tale of high intraspecific variability and low identification success. *Systematic Biology, 55*(5), 715–728. https://doi.org/10.1080/10635150600969864

Monteiro, K. J. L., Jaeger, L. H., Nunes, B. C., Calegar, D. A., Reis, E., Bacelar, P. A. A., Santos, J. P. D., Boia, M. N., & Carvalho-Costa, F. A. (2019). Mitochondrial DNA reveals species composition and phylogenetic relationships of hookworms in northeastern Brazil. *Infection, Genetics and Evolution, 68*, 105–112. https://doi.org/10.1016/j.meegid.2018.11.018

Morand, S. (2018). Advances and challenges in barcoding of microbes, parasites, and their vectors and reservoirs. *Parasitology, 145*(5), 537–542. https://doi.org/10.1017/S0031182018000884

Morise, H., Miyazaki, E., Yoshimitsu, S., & Eki, T. (2012). Profiling nematode communities in unmanaged flowerbed and agricultural field soils in Japan by DNA barcode sequencing. *PLoS One, 7*(12), e51785. https://doi.org/10.1371/journal.pone.0051785

Moszczynska, A., Locke, S. A., McLaughlin, J. D., Marcogliese, D. J., & Crease, T. J. (2009). Development of primers for the mitochondrial cytochrome c oxidase I gene in digenetic trematodes (Platyhelminthes) illustrates the challenge of barcoding parasitic helminths. *Molecular Ecology Resources, 9*(1), 75–82. https://doi.org/10.1111/j.1755-0998.2009.02634.x

Nissen, S., Al-Jubury, A., Hansen, T. V., Olsen, A., Christensen, H., Thamsborg, S. M., & Nejsum, P. (2012). Genetic analysis of *Trichuris suis* and *Trichuris trichiura* recovered from humans and pigs in a sympatric setting in Uganda. *Veterinary Parasitology, 188*(2), 68–77. https://doi.org/10.1016/j.vetpar.2012.03.004

Nyman, T., Papadopoulou, E., Ylinen, E., Wutke, S., Michell, C. T., Sromek, L., Sinisalo, T., Andrievskaya, E., Alexeev, V., & Kunnasranta, M. (2021). DNA barcoding reveals different cestode helminth species in northern European marine and freshwater ringed seals. *International Journal for Parasitology: Parasites and Wildlife, 15*, 255–261. https://doi.org/10.1016/j.ijppaw.2021.06.004

Olson, P. D., & Caira, J. N. (1999). Evolution of the major lineages of tapeworms (Platyhelminthes: Cestoidea) inferred from 18S ribosomal DNA and elongation factor-1alpha. *The Journal of Parasitology, 85*(6), 1134–1159.

Olson, P. D., Cribb, T. H., Tkach, V. V., Bray, R. A., & Littlewood, D. T. J. (2003). Phylogeny and classification of the Digenea (Platyhelminthes: Trematoda). *International Journal for Parasitology, 33*(7), 733–755. https://doi.org/10.1016/s0020-7519(03)00049-3

Ondrejicka, D. A., Locke, S. A., Morey, K., Borisenko, A. V., & Hanner, R. H. (2014). Status and prospects of DNA barcoding in medically important parasites and vectors. *Trends in Parasitology, 30*(12), 582–591. https://doi.org/10.1016/j.pt.2014.09.003

Padial, J. M., Miralles, A., De la Riva, I., & Vences, M. (2010). The integrative future of taxonomy. *Frontiers in Zoology, 7*, 16. https://doi.org/10.1186/1742-9994-7-16

Palomares-Rius, J. E., Cantalapiedra-Navarrete, C., Archidona-Yuste, A., Subbotin, S. A., & Castillo, P. (2017). The utility of mtDNA and rDNA for barcoding and phylogeny of plant-parasitic nematodes from Longidoridae (Nematoda, Enoplea). *Scientific Reports, 7*(1), 10905. https://doi.org/10.1038/s41598-017-11085-4

Pereira, T. J., Fonseca, G., Mundo-Ocampo, M., Guilherme, B. C., & Rocha-Olivares, A. (2010). Diversity of free-living marine nematodes (Enoplida) from Baja California assessed by integrative taxonomy. *Marine Biology, 157*(8), 1665–1678. https://doi.org/10.1007/s00227-010-1439-z

Pires, A., & Marinoni, L. (2010). DNA barcoding and traditional taxonomy unified through integrative taxonomy: A view that challenges the debate questioning both methodologies. *Biota Neotropica, 10*(2), 35.

Powers, T. O., Mullin, P. G., Harris, T. S., Sutton, L. A., & Higgins, R. S. (2005). Incorporating molecular identification of *Meloidogyne* spp. into a large-scale regional nematode survey. *Journal of Nematology, 37*(2), 226–235.

Powers, T. O., Neher, D. A., Mullin, P., Esquivel, A., Giblin-Davis, R. M., Kanzaki, N., Stock, S. P., Mora, M. M., & Uribe-Lorio, L. (2009). Tropical nematode diversity: Vertical stratification of nematode communities in a Costa Rican humid lowland rainforest. *Molecular Ecology, 18*(5), 985–996. https://doi.org/10.1111/j.1365-294X.2008.04075.x

Prosser, S. W., Velarde-Aguilar, M. G., Leon-Regagnon, V., & Hebert, P. D. (2013). Advancing nematode barcoding: a primer cocktail for the cytochrome c oxidase subunit I gene from vertebrate parasitic nematodes. *Molecular Ecology Resources, 13*(6), 1108–1115. https://doi.org/10.1111/1755-0998.12082

Pyo, K. H., Kang, E. Y., Hwang, Y. S., Jun, H. C., Sohn, W. M., Cho, S. H., Lee, W. J., Chai, J. Y., & Shin, E. H. (2013). Species identification of medically important trematodes in aquatic food samples using PCR-RFLP targeting 18S rRNA. *Foodborne Pathogens and Disease, 10*(3), 290–292. https://doi.org/10.1089/fpd.2012.1299

Ratnasingham, S., & Hebert, P. D. (2007). The barcode of life data system. *Molecular Ecology Notes, 7*(3), 355–364. https://doi.org/10.1111/j.1471-8286.2007.01678.x

Rubinoff, D., Cameron, S., & Will, K. (2006). A genomic perspective on the shortcomings of mitochondrial DNA for "barcoding" identification. *The Journal of Heredity, 97*(6), 581–594. https://doi.org/10.1093/jhered/esl036

Schoch, C. L., Seifert, K. A., Huhndorf, S., Robert, V., Spouge, J. L., Levesque, C. A., & Chen, W. (2012). Nuclear ribosomal internal transcribed spacer (ITS) region as a universal DNA barcode marker for Fungi. *Proceedings of the National Academy of Sciences of the United States of America, 109*(16), 6241–6246. https://doi.org/10.1073/pnas.1117018109

Sepulveda, M. S., & Kinsella, J. M. (2013). Helminth collection and identification from wildlife. *Journal of Visualized Experiments, 82*, e51000. https://doi.org/10.3791/51000

Setsuda, A., Da, N., Hasegawa, H., Behnke, J. M., Rana, H. B., Dhakal, I. P., & Sato, H. (2016). Intraspecific and interspecific genetic variation of *Gongylonema pulchrum* and two rodent *Gongylonema* spp. (*G. aegypti* and *G. neoplasticum*), with the proposal of *G. nepalensis* n. sp. for the isolate in water buffaloes from Nepal. *Parasitology Research, 115*(2), 787–795. https://doi.org/10.1007/s00436-015-4806-3

Tautz, D., Arctander, P., Minelli, A., Thomas, R. H., & Vogler, A. P. (2003). A plea for DNA taxonomy. *Trends in Ecology & Evolution, 18*(2), 70–74. https://doi.org/10.1016/S0169-5347(02)00041-1

Taylor, H. R., & Harris, W. E. (2012). An emergent science on the brink of irrelevance: a review of the past 8 years of DNA barcoding. *Molecular Ecology Resources, 12*(3), 377–388. https://doi.org/10.1111/j.1755-0998.2012.03119.x

Teletchea, F. (2010). After 7 years and 1000 citations: comparative assessment of the DNA barcoding and the DNA taxonomy proposals for taxonomists and non-taxonomists. *Mitochondrial DNA, 21*(6), 206–226. https://doi.org/10.3109/19401736.2010.532212

Thaenkham, U., Phuphisut, O., Pakdee, W., Homsuwan, N., Sa-nguankiat, S., Waikagul, J., Nawa, Y., & Dung do, T. (2011). Rapid and simple identification of human pathogenic heterophyid intestinal fluke metacercariae by PCR-RFLP. *Parasitology International, 60*(4), 503–506. https://doi.org/10.1016/j.parint.2011.09.004

Trivedi, S., Rehman, H., Saggu, S., Panneerselvam, C., & Ghosh, S. K. (2020). *DNA barcoding and molecular phylogeny* (2nd ed.). Springer.

Van Steenkiste, N., Locke, S. A., Castelin, M., Marcogliese, D. J., & Abbott, C. L. (2015). New primers for DNA barcoding of digeneans and cestodes (Platyhelminthes). *Molecular Ecology Resources, 15*(4), 945–952. https://doi.org/10.1111/1755-0998.12358

Vogler, A. P., & Monaghan, M. T. (2007). Recent advances in DNA taxonomy. *Journal of Zoological Systematics and Evolutionary Research, 45*(1), 1–10. https://doi.org/10.1111/j.1439-0469.2006.00384.x

Webster, B. L., Culverwell, C. L., Khamis, I. S., Mohammed, K. A., Rollinson, D., & Stothard, J. R. (2013). DNA barcoding of *Schistosoma haematobium* on Zanzibar reveals substantial genetic diversity and two major phylogenetic groups. *Acta Tropica, 128*(2), 206–217. https://doi.org/10.1016/j.actatropica.2012.06.002

Webster, B. L., Webster, J. P., Gouvras, A. N., Garba, A., Lamine, M. S., Diaw, O. T., Seye, M. M., Tchuem Tchuente, L. A., Simoonga, C., Mubila, L., Mwanga, J. R., Lwambo, N. J., Kabatereine, N. B., Lange, C. N., Kariuki, C., Mkoji, G. M., Rollinson, D., & Stothard, J. R. (2013). DNA 'barcoding' of *Schistosoma mansoni* across sub-Saharan Africa supports substantial within locality diversity and geographical separation of genotypes. *Acta Tropica, 128*(2), 250–260. https://doi.org/10.1016/j.actatropica.2012.08.009

Wicht, B., Ruggeri-Bernardi, N., Yanagida, T., Nakao, M., Peduzzi, R., & Ito, A. (2010). Inter- and intra-specific characterization of tapeworms of the genus *Diphyllobothrium* (Cestoda: Diphyllobothriidea) from Switzerland, using nuclear and mitochondrial DNA targets. *Parasitology International, 59*(1), 35–39. https://doi.org/10.1016/j.parint.2009.09.002

Yamakawa, S. (2014). *Taxonomy is revamped in DNA-based system*. https://www.jhunewsletter.com/article/2014/03/taxonomy-is-revamped-in-dna-based-system-87011

Yong, T. S., Lee, J. H., Sim, S., Lee, J., Min, D. Y., Chai, J. Y., Eom, K. S., Sohn, W. M., Lee, S. H., & Rim, H. J. (2007). Differential diagnosis of *Trichostrongylus* and hookworm eggs via PCR using ITS-1 sequence. *The Korean Journal of Parasitology, 45*(1), 69–74. https://doi.org/10.3347/kjp.2007.45.1.69

Zarlenga, D. S., Stringfellow, F., Nobary, M., & Lichtenfels, J. R. (1994). Cloning and characterization of ribosomal RNA genes from three species of *Haemonchus* (Nematoda: Trichostrongyloidea) and identification of PCR primers for rapid differentiation. *Experimental Parasitology, 78*(1), 28–36. https://doi.org/10.1006/expr.1994.1003

Chapter 14
Implementation of Genetic Markers from Molecular Systematics to DNA Taxonomy and Field Application

Abstract The role of genetic markers for molecular systematics, DNA taxonomy, and DNA barcoding has been thoroughly discussed in the previous chapters. In recent years, a shift towards using next-generation sequencing (NGS) technologies has been prominent, and these molecular advances are playing a pivotal role in species identification, monitoring, and disease diagnosis for a vast variety of organisms. The shift towards NGS in helminth research is no exception. Progress has been made in utilizing NGS for helminth whole genome sequencing and transcriptomics. In the WormBase ParaSite database (https://parasite.wormbase.org/index.html), 157 nematodes and 45 platyhelminths whole genomes have been annotated serving as a valuable resource for researchers. Additionally, the use of genetic markers in the context of DNA metabarcoding of helminths through targeted amplicon sequencing is also gaining traction. In view of these NGS advances, this chapter presented some of the applications of genetic markers and the types of research that have been applied through NGS. A focus on the scaling up of DNA barcoding to DNA metabarcoding and towards field-based monitoring, especially in the context of helminths, was also discussed.

14.1 Applications of Genetic Markers Through Next-Generation Sequencing

Defined as massively parallel sequencing through the use of high-throughput sequencing, NGS has revolutionized the field of molecular biology (Koboldt, Steinberg, Larson, Wilson, & Mardis, 2013). Building on the basics of Sanger sequencing, second- and third-generation sequencing technologies (also known as NGS) are now at the forefront of molecular research. These technologies have increased the ability to obtain sequence information from minute amounts of DNA and even from single molecules (Metzker, 2010). Commonly used platforms for NGS include those manufactured by Illumina and Thermo Fisher Scientific. Each of these platforms is different in its chemistries, advantages, and limitations. For example, platforms manufactured by Illumina (Illumina HiSeq and MiSeq) are based on sequencing by synthesis principle, while the Ion Torrent from Thermo

U. Thaenkham et al., *Molecular Systematics of Parasitic Helminths*,
https://doi.org/10.1007/978-981-19-1786-8_14

Fisher Scientific is based on emulsion polymerase chain reaction (PCR) (Quail et al., 2012). More recently, the PacBio and MinION platforms have been developed (Plesivkova, Richards, & Harbison, 2018). PacBio utilizes the detection of released fluorophores as each nucleotide is incorporated in the sequencing process. The MinION, on the other hand, is a nanopore sequencer utilizing nanopore technology. In Chap. 7, the mechanisms behind the NGS platforms for second- and third-generation sequencing were detailed. Deciding which platform to use for the respective application highly depends on the user's specifications. These specifications include cost per run, read length, sequences yield per run, accuracy, and run time.

Initially, NGS was used to study whole genomes and portions of the different regions of the genomes. Generally, applications of NGS can be divided into the sequencing of whole genomes for subsequent analysis and applied through targeted amplicon sequencing of portions of the gene of interest. For sequencing of whole genomes, whole genomes and transcriptomes can be characterized, in addition to gene expression profiling of organisms (Ekblom & Galindo, 2011;Koboldt et al., 2013 ; Metzker, 2010). Examples include whole sequencing genomes of *Homo sapiens* and Neanderthals to unravel our evolutionary history (Callaway, 2014; Green et al., 2006). Whole genome sequencing projects were able to gain momentum through the aid of NGS. Our understanding of factors causing human diseases such as cancer has also increased, facilitated through NGS. Identifying candidate genes for drug targets and cellular processes relating to gene regulation is now possible, and these findings can provide valuable strategies for the treatment of diseases (Borsting & Morling, 2015; Koboldt et al., 2013).

Targeted amplicon sequencing through NGS is also highly popular, especially in the field of biodiversity and conservation. Targeted amplicon sequencing utilizes molecular genetic markers, amplifying a targeted portion of the genome (Koboldt et al., 2013; Metzker, 2010). Molecular identification and surveys of species in various types of habitats can be performed. With the effect of climate change and biodiversity loss, conservation genetics have received increasing attention. Conservation genetics is the application of genetics in the preservation of species where knowledge of relatedness among organisms and the discovery of new species is crucial in uncovering species biodiversity in a changing climate (Sabir et al., 2014). The application of DNA metabarcoding is one such example widely used for conservation genetics, allowing for the move from classical taxonomy to relying less on taxonomic expertise and saving cost and time. Studies applying DNA metabarcoding on environmental samples in agricultural landscapes allow monitoring and examining ecosystems (Hajibabaei, Shokralla, Zhou, Singer, & Baird, 2011; Tedersoo, Drenkhan, Anslan, Morales-Rodriguez, & Cleary, 2019; Tillmar, Dell'Amico, Welander, & Holmlund, 2013). These include identifying agricultural pests and investigating trophic interactions (prey-predator relationships) through diet analysis within the ecosystem. Detection of traces of environmental DNA (eDNA) of organisms present in the environment due to the high sensitivity of the NGS offers an opportunity to detect minute amounts of DNA in the environment without a sampling of the actual organism present (Shokralla, Spall, Gibson, & Hajibabaei, 2012; Valentini et al., 2016).

Additionally, microbial assessment of water quality using NGS can also allow for monitoring microbial species that serve as indicators of water contamination (Tan et al., 2015). Biosecurity monitoring programs for detecting invasive species in soil and marine environments also provide an avenue for predicting health threats in the environment. The monitoring of food products for food safety and quality is another example of studies that have utilized NGS through genetic markers (Littlefair & Clare, 2016; Staats et al., 2016). Thus, the implementation of genetic markers using NGS is inevitably valuable for monitoring ecosystems, surveying habitats, and species identification of organisms across multiple disciplines and organizations. The application of DNA metabarcoding through NGS will be discussed in more detail in Sect. 14.2.

The use of genetic markers unifies molecular systematics, taxonomy, and molecular identification through DNA barcoding. As mentioned in Chap. 13, evolutionary relationships among organisms and the historical evidence can be deciphered through molecular systematics, and species identification using DNA sequences enhances alpha taxonomy through DNA barcoding. Additionally, studying species diversity within and among populations enhances our understanding of the evolutionary forces driving speciation. With NGS platforms, it offers not only an avenue for sequencing of whole genomes and transcriptomes, but it is also vastly applicable at the species and population level, from biodiversity to the monitoring of pathogens and biosecurity or food safety. Applications through NGS contribute to and play a key role in various industries.

14.2 DNA Metabarcoding

The previous section showed that DNA metabarcoding utilizes genetic markers coupled with NGS technologies. This section will provide an overview of the general concept of DNA metabarcoding, followed by its relevance in the context of genetic markers and helminths.

14.2.1 General Overview

DNA metabarcoding, which is an extension of DNA barcoding, can be defined as the use of NGS to massively sequence and identify large amounts of species in a bulk sample (Cristescu, 2014; Taberlet, Coissac, Pompanon, Brochmann, & Willerslev, 2012). Figure 14.1 illustrates the general workflow and main steps of a DNA metabarcoding approach. Taberlet, Coissac, Pompanon, et al. (2012) first introduced the term DNA metabarcoding to designate high-throughput multispecies identification, either from bulk samples of entire organisms or from an environment sample (Taberlet, Coissac, Pompanon, et al., 2012). The terms DNA metabarcoding, metagenomics, and eDNA metabarcoding have been used somewhat

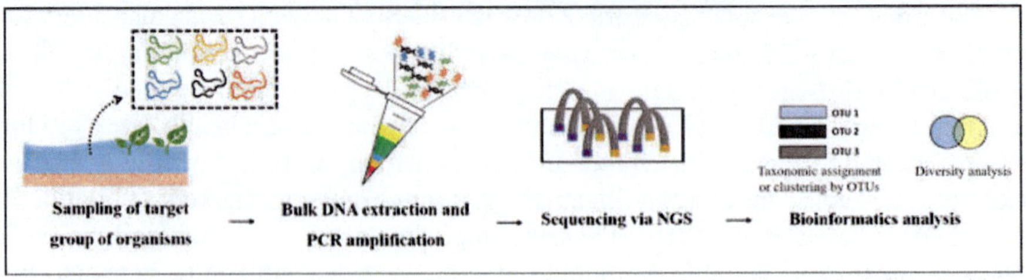

Fig. 14.1 Main steps of the DNA metabarcoding approach

interchangeably in different publications. However, we will clarify the use of these terminologies for easy reference in the rest of the chapter. DNA metabarcoding refers to high-throughput sequencing technologies such as NGS for the multispecies identification of DNA extraction from an environment sample or bulk samples of organisms (Taberlet, Coissac, Pompanon, et al., 2012). Organisms can either be isolated prior to analysis or have their DNA directly extracted from the environment. eDNA metabarcoding, on the other hand, refers to DNA that is extracted from environmental samples without first isolating any target organism (Taberlet, Coissac, Hajibabaei, & Rieseberg, 2012). These environmental samples can include soil, water, air, fecal, or gut contents. The term DNA metabarcoding is thus an umbrella term encompassing eDNA metabarcoding. Metagenomics differs from DNA metabarcoding as it refers to the study of all genomes within a particular ecosystem instead of a targeted set of genes (Pavan-Kumar, Gireesh-Babu, & Lakra, 2015). Initially, DNA metabarcoding was applied in the field of microbiology, where a first study was published, and they assessed the microbial diversity in a marine community using the prokaryotic 16S ribosomal RNA (rRNA) gene through Roche 454 pyrosequencing (Hugerth & Andersson, 2017; Sogin et al., 2006).

The success of the DNA metabarcoding approach to assessing microbial communities was a game-changer in microbial ecology, where various communities can now be assessed through NGS. Subsequently, instead of DNA metabarcoding studies, metagenomic studies gained popularity for microbial ecology by characterizing complete bacterial genomes in the ecosystem instead of targeting a specific gene (Hugerth & Andersson, 2017). The application of DNA metabarcoding was thereafter expanded to target eukaryotes, following the success of utilizing NGS for community-level analysis of bulk samples. NGS has revolutionized molecular-based identification and significantly transformed research in areas like ecology, conservation, medicine, and biosecurity.

14.2.2 Advantages of DNA Metabarcoding

The advantages of DNA metabarcoding for species identification and discovery are many folds compared to traditional taxonomy (using morphological characters) and molecular identification using Sanger sequencing. Firstly, compared to the traditional methods of species identification using morphological characters, the use of NGS can greatly accelerate the detection and identification of species in a shorter span of time (Deiner et al., 2017). Moreover, DNA metabarcoding decreases the reliance on taxonomic expertise, circumventing the limitations of insufficiently trained taxonomists for morphological identification. Secondly, the use of NGS allows for the amplification and detection of more than one species in a single reaction, compared to Sanger sequencing, accounting for more than 1000 samples in a single Illumina sequencing run (Tedersoo et al., 2019). The pooling of samples through parallel sequencing runs is more cost-effective than Sanger sequencing. Mena et al. (2021), through the comparison of established inventory techniques for biodiversity monitoring (live-trapping, camera trapping, mist-netting) and eDNA metabarcoding sampling, revealed that eDNA metabarcoding is a helpful tool that saves effort and sampling costs (Mena et al., 2021). Thirdly, the high sensitivity of NGS platforms can allow for the detection of minute amounts of DNA that is present in a mixture of species. The high sensitivity of NGS compared to Sanger sequencing is advantageous, especially to detect trace amounts of the target organism's DNA in the environment. Studies have revealed that DNA metabarcoding was able to detect a greater number of species present in a sample than morphological identification (Porter & Hajibabaei, 2018). Different taxa were detected through DNA metabarcoding and morphological identification, supporting its use to complement traditional taxonomy in species discovery and diversity. Lastly, DNA metabarcoding can be applied noninvasive for surveillance and monitoring endangered species (De Barba et al., 2014; Deiner et al., 2017). Analysis of traces of DNA in environmental samples can allow for assessment without harming the individual.

14.2.3 Applications and Examples of DNA Metabarcoding

Among eukaryotes, the mitochondrial cytochrome c oxidase subunit I (*COI*) gene, with its status as a DNA barcode region, is a popular genetic marker for DNA metabarcoding. The availability of primers for various groups of organisms, along with the plethora of sequences available on the database for a taxonomic assignment, is a factor contributing to the *COI*'s gene popularity for DNA metabarcoding (Andujar, Arribas, Yu, Vogler, & Emerson, 2018). Arthropods are a common group of eukaryotes that frequently utilize the *COI* gene for DNA metabarcoding (Liu, Clarke, Baker, Jordan, & Burridge, 2019). Bulk detection of terrestrial arthropods through DNA metabarcoding showed the effectiveness of the application (Braukmann et al., 2019). Elbrecht et al. (2019) compared the performances of commonly used and newly designed *COI* primer sets to recover species in a bulk DNA sample from more than 300 insect species collected from Malaise traps (Elbrecht et al., 2019). Mosquito surveillance through DNA metabarcoding was performed by Batovska et al. (2018), where mosquitoes were pooled together for species identification and virus detection (Batovska et al., 2018). Biodiversity assessment targeting sediment communities using the *COI* gene showed an abundance of phyla Arthropoda, Nematoda, and Cnidaria. The authors also concluded that the *COI* gene could be used to accurately characterize communities with high resolution (Atienza et al., 2020).

Additionally, DNA mini-barcodes have been developed for the *COI* gene as the universal 650 bp fragment may be challenging to amplify from DNA obtained from archival or environmental samples (Meusnier et al., 2008). Characterizing fish diets using modified Folmer primers (targeting shorter amplicon size) was performed for three species of planktivorous and benthivorous coral reef fishes, revealing DNA metabarcoding as a valuable asset for large-scale biodiversity assessments and food web ecological studies (Leray et al., 2013). Clearly, the *COI* gene is an ideal choice for DNA metabarcoding of eukaryotes with its species-level resolution, a large number of sequences available on the reference database, and primers targeting most groups available.

The nuclear 18S rRNA gene is also another popular genetic marker for DNA metabarcoding of organisms. Faria et al. (2018) aimed to disentangle meiofauna diversity patterns using the 18S rRNA gene through DNA metabarcoding from a small tidal flat bay (Faria et al., 2018). DNA metabarcoding studies have also been performed to target various soil and marine organisms for ecological surveys and monitoring of ecosystems. Through the design of primers targeting the 18S rRNA gene to explore metazoan biodiversity in soil, the study performed by Capra et al. (2016) supports the use of 18S DNA metabarcoding as a powerful and repeatable approach for uncovering the composition and structure of soil metazoan communities (Capra et al., 2016). Biodiversity levels in the Antarctic peninsula's sediment communities targeting meiofauna were also conducted using the 18S rRNA gene (Fonseca et al., 2017). The results revealed higher than expected diversity with an expected similarity to temperate regions. Also, the phyla Nematoda, Arthropoda, Platyhelminthes, and Annelida were the four dominant eukaryotic groups.

In marine environments, cephalopods were sampled for DNA metabarcoding using the hypervariable V2 region of the 18S rRNA gene (de Jonge et al., 2021). The highly conserved 18S rRNA gene enables the accessible design of universal primers to target many groups of organisms, thus rendering it suitable and successful for broad-scale DNA metabarcoding studies in various environments and habitats, even in difficult to reach places. However, the true potential of the 18S rRNA gene for DNA metabarcoding has been hotly debated. The low species-level resolution of the 18S rRNA gene limits its utility for DNA metabarcoding. Studies comparing the 18S rRNA gene and the mitochondrial *COI* gene showed that the *COI* gene could generate more operational taxonomic units (OTUs) compared to the 18S rRNA gene. Also, although similar taxonomic coverage was observed using the *COI* gene and the 18S rRNA gene among zooplanktons, the mitochondrial *COI* gene resolved three folds more taxa to the species level compared to the 18S rRNA gene (Clarke, Beard, Swadling, & Deagle, 2017).

Aside from the mitochondrial *COI* gene and the nuclear 18S rRNA gene, the mitochondrial 12S and 16S rRNA genes are also popular genetic markers for DNA metabarcoding, especially among fishes and mammals. A set of universal primers for fishes using the mitochondrial 12S rRNA gene have been developed and utilized for DNA metabarcoding of fishes, not only in enclosures but also in open seas and river systems (Brys et al., 2021; Milan et al., 2020; Miya et al., 2015; Morey, Bartley, & Hanner, 2020). Kelly, Port, Yamahara, and Crowder (2014) estimated vertebrate fauna in a 4.5-million-liter aquarium tank through eDNA metabarcoding. They showed that it was possible to detect eDNA of fishes from the water using a 106 bp fragment of the 12S rRNA gene, indicating its potential for marine environmental monitoring through DNA metabarcoding (Kelly et al., 2014).

Sampling kelp forest ecosystems to target marine fishes and mammals was conducted by Port et al. (2016), where concordance between the species present using DNA metabarcoding and spatial dive surveys was observed (Port et al., 2016). Insights into understanding predator diets and their variation between habitats were discovered using the droppings of the African leopard. Targeting the 16S rRNA gene, species of prey of the leopard at various habitats were confirmed through DNA metabarcoding (Havmøller et al., 2021). Diet analysis of the Australian sea lion, which is an opportunistic apex predator, through their droppings also supports DNA metabarcoding as a noninvasive method for insights into predator-prey relationships (Berry et al., 2017). The ability of species-level resolution and availability of primers targeting various groups of eukaryotes (useful mainly for diet analysis) contribute to the popularity of the mitochondrial rRNA genes for DNA metabarcoding studies, not only to understand predator-prey relationships in the ecosystem but also biodiversity monitoring of habitats.

14.3 Helminth DNA Metabarcoding

DNA metabarcoding has also been applied for helminths, significantly contributing to various domains. Firstly, the characterization of helminth communities contributes to the field of helminth biodiversity and conservation. With a projected estimate of 100,000 helminth species, only approximately 15% are currently described (Carlson, Dallas, Alexander, Phelan, & Phillips, 2020). It is thus of paramount importance to rapidly uncover the helminth species that are currently present and even more so in the face of changing habitat and habitat loss. Secondly, DNA metabarcoding of helminths has been applied as a biomonitoring tool in agricultural settings, contributing to commercial and non-commercial stakeholders. Helminths play many roles in the proper functioning of an ecosystem, and their community structure in the environment can serve as indicators of pollution (Porazinska et al., 2009). Thirdly, the detection and monitoring of parasitic helminth communities in hosts contribute to the field of medical parasitology for human and animal health. DNA metabarcoding allows for a noninvasive method of helminth sampling in hosts, useful for detection (Aivelo & Medlar, 2018). The following paragraphs will describe these applications of helminth DNA metabarcoding with examples through the implementation of genetic markers in detail.

14.3.1 Biodiversity and Conservation of Helminths

Initial studies for helminth DNA metabarcoding targeted the soil and marine habitats, aiming to characterize plant and free-living nematodes. Through the sampling of nematodes in the soil, litter, and canopy habitats of the lowland tropical rainforest of Costa Rica, Porazinska et al. (2010) revealed high species richness and diversity in these habitats (Porazinska et al., 2010). Using the nuclear 18S rRNA gene, they revealed that only six species were shared across the three different habitats, showing the vast species diversity of nematodes. However, only 40% of the sequences belonged to Nematoda, while 26% belonged to other eukaryotic taxa such as Acari, fungi, and protozoa. The non-specificity of the 18S rRNA gene for nematodes could be a limitation of utilizing this genetic marker for helminth DNA metabarcoding. Targeting the V4 and V9 region of the 18S rRNA gene for eukaryotes, the meiofaunal diversity of the soil in the Atlantic Forest was investigated (Müller et al., 2019). The results revealed a considerable number of newly or still uncharacterized nematodes in the community. Sapkota and Nicolaisen (2015) developed a primer set for the 18S rRNA gene that was predicted to amplify only DNA from nematodes (Sapkota & Nicolaisen, 2015). By sampling soils from agricultural fields, a taxonomic assignment from DNA metabarcoding revealed that the taxa represented most of the nematode orders defined in Blaxter et al. (1998) except Strongylida, Spirurida, and Oxyurida. Additionally, the primer set classified 64.4% of the sequences belonging to Nematoda, improving the 40% obtained from Porazinska et al. (2010) and Blaxter et al. (1998).

In marine habitats, free-living nematodes from coastal and deep-sea sediments were sampled to assess the biodiversity of deep-sea nematode assemblages (Dell'Anno, Carugati, Corinaldesi, Riccioni, & Danovaro, 2015). The suitability of the DNA metabarcoding approach for marine nematodes was tested using the 18S rRNA gene. Congruence between morphological identification and DNA metabarcoding analysis was obtained, but only at the taxonomic order level to family level. Additionally, a large portion of the OTUs remained unassigned. The low resolution of the 18S rRNA gene for taxonomic assignment can be partly attributed to the low sequence variation of the genetic marker. Comparing the 18S and 28S rRNA genes for freshwater nematode communities, Schenk, Kleinbölting, and Traunspurger (2020) also noted that many OTUs have a low taxonomic resolution, thus hindering accurate taxonomic assignments. A comparison between the 18S rRNA gene and the mitochondrial *COI* gene for DNA metabarcoding was performed on marine nematode communities in three types of land use areas—aquaculture, industry, and nature reserve (Tytgat et al., 2019). The results revealed that the 18S rRNA gene outperforms the *COI* gene in terms of PCR amplification success and the comprehensiveness of reference database sequences. As mentioned in the previous chapter, the high sequence variability of the helminth *COI* gene limits its utility as a genetic marker as only selected species are successfully amplified due to PCR amplification bias.

Another interesting area utilizing DNA metabarcoding for helminths involves characterizing helminth communities in preserved vertebrate specimens. The high sensitivity of NGS to detect minute amounts of DNA and shorter or degraded DNA sequences is extremely valuable in this aspect. The feasibility of utilizing DNA metabarcoding to identify parasite biodiversity and community structure within museum-archived gastrointestinal tracts was tested by Greiman et al. (2018). Targeting the cestode mitochondrial 12S and 16S rRNA genes and the nuclear 28S rRNA gene, assignment of many helminth species was possible through DNA metabarcoding.

14.3.2 Biomonitoring Through Helminths

Helminths serve as important indicators of the health of the ecosystem and have been used as biological indicators in agricultural soils and water bodies. Nematodes, especially, play an important role in the ecosystem for the process of decomposition and nutrient mineralization. They have diverse roles in soils such as fungal feeders, bacterial feeders, omnivore-predators, or plant parasites (Bongers & Ferris, 1999; Neher, 2001; Porazinska et al., 2009; Treonis et al., 2018). Changes in the community structure of soil nematodes can provide valuable insights into the soil ecosystem and relationships among various organisms. Taxonomic profiling of nematodes from copse soil was successfully performed to examine the types of nematodes present (Kenmotsu, Uchida, Hirose, & Eki, 2020). Through DNA metabarcoding using the 18S rRNA gene, fungivores, plant parasites, bacterivores, and omnivore-predator nematodes were present. Fungivores and plant-parasitic were the most abundant group of nematodes. In addition to the diverse roles of nematodes in soil, the presence of plant-parasitic nematodes in the soil is also an indicator of soil and crop health and can have economic impacts on agriculture.

Bell, Namaganda, Urwin, and Atkinson (2021) applied DNA metabarcoding to evaluate the soil quality of banana plantations in Uganda, with the final aim of establishing a method for future assessments of transgenic crops and improving cropping systems (Bell et al., 2021). Using the 18S rRNA gene, nematode community indices were consistent with other soils in Uganda, and the soils were successfully assessed as having a mature soil web condition with high nutrient enrichment and balanced decomposition channels. The DNA metabarcoding approach was also successfully applied to characterize soil nematode communities in three cropping systems (Treonis et al., 2018). The diversity and abundance of each group of nematodes (omnivore-predator, bacterivores, fungivores) were compared between the three cropping systems using the 18S rRNA gene. The characterization of soil nematode communities can provide insights into improving cropping systems and optimizing crop management strategies.

In water bodies, water quality is commonly assessed to detect pathogens and species that can serve as bioindicators to assess the extent of pollution. Schenk et al. (2020) evaluated an index based on DNA metabarcoding of nematode communities in aquatic sediments, revealing the potential of DNA metabarcoding as a biomonitoring tool in aquatic environments.

14.3.3 Noninvasive Detection and Monitoring of Parasitic Helminths

DNA metabarcoding has greatly aided in detecting medically important parasitic helminths for humans and animals. As mentioned in the previous chapters, detection of helminths is usually performed through morphological identification (from fecal or host tissue samples) or molecular analysis through Sanger sequencing or species-specific PCR. Where possible, hosts will have to be dissected, and helminths can then be isolated. DNA metabarcoding is thus an attractive alternative as a sensitive noninvasive method for species detection and monitoring through host fecal samples. Additionally, the noninvasive approach offers an opportunity to analyze helminth communities in difficult to capture hosts (Aivelo & Medlar, 2018). Using the DNA extracted from fecal samples of individual rats, DNA metabarcoding using the 18S rRNA gene was performed, and helminths belonging to Nematoda and Cestoda were successfully detected (Tanaka et al., 2014).

Greater varieties of helminths were detected through DNA metabarcoding compared to morphological examination. Helminth communities of animals such as the red kangaroo and frogs have been investigated through the noninvasive approach of DNA metabarcoding with the internal transcribed spacer 2 (ITS2) region and the 18S rRNA gene, respectively (Aivelo, Harris, Cadle, & Wright, 2018; Lott, Hose, & Power, 2015). Aivelo, Medlar, Loytynoja, Laakkonen, and Jernvall (2015) used DNA metabarcoding of fecal samples from 11 host species targeting the 18S rRNA gene for nematodes and found similar helminth assemblages in distantly related host species with different ecological niches (Aivelo, Medlar, Löytynoja, Laakkonen, & Jernvall, 2018). The success of the noninvasive approach using fecal samples for DNA metabarcoding shows its potential as a tool for species identification of helminths. DNA metabarcoding also allows longitudinal studies and tracking helminth communities in host species over time. The intestinal nematode community of mouse lemurs was investigated for at least 2 years and revealed that the nematode community structure was consistent among hosts over time (Aivelo et al., 2015).

The DNA metabarcoding approach has also been applied to gastrointestinal helminths of medical importance to livestock. Avramenko et al. (2015) introduced a concept of "nemabiome," where they developed and applied the DNA metabarcoding approach using the ITS2 region on parasitic nematode communities belonging to order Strongylida in cattle fecal samples (Avramenko et al., 2015). They demonstrated that relative quantification of gastrointestinal nematode communities from fecal samples could be achieved using ITS2-based DNA metabarcoding. The impact of routine antiparasitic treatment (macrocyclic lactone) was examined on the gastrointestinal nematode communities of cattle through the ITS2 DNA metabarcoding approach, illustrating the power of DNA metabarcoding in monitoring the emergence of antiparasitic treatment resistance in livestock (Avramenko et al., 2017). More recently, the ITS2-based DNA metabarcoding approach was validated and applied in a large-scale survey of sheep farms across the United Kingdom (Redman et al., 2019). The distribution and abundance of gastrointestinal nematode communities across 99 sheep flocks revealed eight different species of nematodes belonging to *Trichostrongylus*, *Teladorsagia*, *Haemonchus*, *Cooperia*, *Oesophagostomum*, and *Chabertia*. The power of utilizing DNA metabarcoding was demonstrated through a bulk sampling of fecal samples and can be appropriate for large-scale surveillance and epidemiological studies of gastrointestinal nematodes of livestock.

14.3.4 Limitations and Considerations for Helminth DNA Metabarcoding

Various factors will have to be in place for a successful DNA metabarcoding approach. Firstly, the target organism group should select an appropriate genetic marker with sufficient species-level resolution. Secondly, primers that can successfully amplify a broad range of species within the target group with limited PCR amplification bias are ideal. Thirdly, primers should be specific to the target group to reduce non-specific amplification of non-target organisms. This factor is especially crucial for eDNA metabarcoding. Fourthly, the reference database of sequences for the selected genetic marker should be comprehensive for taxonomic assignment. Lastly, bioinformatics pipelines and cut-off values for threshold similarity index should be evaluated, as the reliability of taxonomic assignment is affected by the clustering-based OTUs in instances where no reference sequences are available.

To date, DNA metabarcoding studies have mainly been performed on nematodes, especially on soil and marine environments, utilizing the nuclear 18S rRNA gene. DNA metabarcoding primers specifically for soil nematodes (NF1-18Sr2b) and marine nematodes (SSU04F-SSUR22) have been developed and widely used (Creer et al., 2010; Fonseca et al., 2014; Holovachov, Haenel, Bourlat, & Jondelius, 2017; Porazinska et al., 2010). Additionally, the detection of parasitic nematodes in fecal samples is usually performed either with the primers targeting the 18S rRNA gene of eukaryotes or the primers targeting soil and marine nematodes. However, the low species-level resolution of the 18S rRNA gene remains a limitation. Taxonomic assignments for nematodes using the 18S rRNA gene are usually only at the family to genera level.

Moreover, the conserved sequence of the 18S rRNA gene results in low specificity for nematodes, where DNA of other eukaryotes gets amplified as well. Porazinska et al. (2010) revealed that only 40% of reads belonged to nematodes using the NF1-18Sr2b primer set (Porazinska et al., 2010). However, despite their limitations, the 18S rRNA gene remains a popular genetic marker for DNA

Table 14.1 Considerations for helminth DNA metabarcoding

Main DNA metabarcoding workflow steps	Considerations
Sampling	• How should the helminths of interest be collected for maximum representation in the environment? • How should the collected samples be kept and transported?
DNA extraction	• Should the helminths be isolated prior to DNA extraction? • Which is the optimal isolation method for maximum species recovery? • Which DNA extraction method should be used? • Will there be inhibitors from the environment?
PCR amplification	• Which genetic marker is most suitable for species-level resolution? • Which primers can target a broad species range? • How can we ensure low PCR amplification bias? • How sensitive and specific are the primers?
NGS platforms	• Which NGS platform is most appropriate (number of reads, length of reads, read quality and accuracy, cost per sample)?
Bioinformatics	• Is the reference database for the selected genetic marker comprehensive? • Will Sanger sequencing be needed to build up the reference library? • What kind of analysis should be performed? • What are the species boundaries/cut-off similarity threshold for the selected helminths?

metabarcoding of nematodes. The availability of the 18S rRNA gene reference sequences for nematodes on the NCBI database and SILVA database, along with primer sets that are frequently utilized, is a factor influencing the popularity of this genetic marker for DNA metabarcoding of nematodes.

On the other hand, the mitochondrial *COI* gene has a higher species-level resolution than the nuclear 18S rRNA gene, making it valuable for DNA metabarcoding. However, for helminths, the high sequence variability of the *COI* gene limits the design of primers that can amplify a broad number of helminth species (Moszczynska, Locke, McLaughlin, Marcogliese, & Crease, 2009). Moreover, PCR amplification bias is prominent, as only selected taxa get amplified. Ahmed et al. (2019) evaluated the 18S primers (NF1-18Sr2b, SSU04F-SSUR22) against the *COI* primers (JB3-JB5GED) for nematodes and revealed that the *COI* primer had the lowest number of taxa recovered (Ahmed et al., 2019). Although the mitochondrial *COI* gene is the DNA barcode marker of choice for other groups of organisms, the limitations of using the *COI* gene for DNA barcoding and metabarcoding of helminths outweigh its utility.

Albeit the progress made, DNA metabarcoding of helminths is still in the infancy stage. Table 14.1 depicts some considerations that should be taken. Appropriate genetic markers, in terms of their resolution for species identification, and the development of primers targeting broad species range remains a challenge for helminth DNA metabarcoding. Careful evaluation of primer sets and suitable

DNA extraction methods, especially for eDNA metabarcoding, should be performed to fully utilize the potential of DNA metabarcoding. Finally, once a suitable genetic marker is selected, reference sequences should be obtained via Sanger sequencing before DNA metabarcoding studies to build a comprehensive database for accurate taxonomic assignments. Thus far, the most comprehensive genetic marker for helminths is the 18S rRNA gene, available on the NCBI and SILVA databases.

Additionally, cut-off threshold values for OTUs and suitable bioinformatic pipelines should also be evaluated, as genetic variations for species boundaries might not be the same across helminths. Chan, Chaisiri, Saralamba, Morand, and Thaenkham (2021) previously evaluated ten genetic markers for their suitability of helminth molecular-based studies and revealed the high sequence variability among helminths despite being at the same taxonomic level (Chan et al., 2021). Thus, care will have to be taken when using fixed cut-off threshold values for helminths to interpret OTUs.

14.3.5 Future of Helminth DNA Metabarcoding

Among helminths, DNA metabarcoding has mainly been conducted on nematodes in various environments, including nematodes of medical importance. Examples have been discussed in the above paragraphs. However, for cestodes, DNA metabarcoding studies are limited. Cestodes belonging to family Hymenolepdidae were detected in rats using primers targeting the 18S rRNA gene for eukaryotes and nematodes belonging to Ascarididae, Heterakoidea, Oxyuroidea, Panagrolaimoidea, Rhabditoidea, Strongylida, and Acuarioidea (Tanaka et al., 2014). Greiman et al. (2018) used the mitochondrial 16S rRNA gene targeting cestodes for DNA metabarcoding of helminths present in the GI tract of museum-preserved specimens (Greiman et al., 2018). To date, DNA metabarcoding targeting trematodes has not been performed.

The applications for helminth DNA metabarcoding are vast. Aside from environmental sampling and detection of helminths, parasitic helminths in human and animal hosts such as meat and vegetables through DNA metabarcoding can also be applied. The monitoring and surveillance of helminths in food products are important for food quality and safety checking, especially for imported products. Additionally, routine surveillance for the detection of helminths in potable water in addition to microbial and chemical contents can be added to the parameters for increased water quality. DNA metabarcoding surveillance of helminths can also be included in disease response measures (e.g., flooding event), where increased risk of helminthic disease transmission is probable.

With the potential applications for DNA metabarcoding for helminths, especially for the surveillance of medically important helminths, the selection and evaluation of a suitable genetic marker for DNA metabarcoding of helminths remain high on the priority list. A suitable genetic marker and the development of universal primers targeting parasitic helminths across nematodes, trematodes, and cestodes will facilitate progress for helminth DNA metabarcoding. Evaluation with mock communities of known species abundance should be conducted to test the sensitivity of the primers, along with the selection of appropriate methods for helminth isolation and eDNA extraction. These evaluations will aid in determining the most suitable methodology for helminth DNA metabarcoding to reduce biasness from DNA extraction and PCR amplification. As presented in Table 14.1, the buildup of reference sequences and evaluation of bioinformatics pipelines for analysis can be performed once appropriate genetic markers have been thoroughly evaluated and primers developed. Without a doubt, research in helminth DNA metabarcoding will accelerate at an even more incredible pace in the future, contributing to biodiversity, disease detection and surveillance, and even security on a global scale.

14.4 Other Field Applications for Helminths

Aside from applying DNA metabarcoding of helminths for biodiversity, species detection, and biomonitoring, advancements in NGS or other sequencing technologies can also allow faster response in disease outbreaks and diagnosis, especially in the field setting. For example, portable laboratories and devices can qualify for laboratory techniques to be performed directly in the field instead of transporting the specimens back to a central laboratory (Liu et al., 2019). This will bring about a more rapid response system in instances where timing is crucial for disease control. A possibility in the future is the use of Bento Lab, which is an easy-to-use DNA analysis kit created by Bethan Wolfenden and Philipp Boeing at the University College London (Bento Lab, 2016). Consisting of a PCR thermocycler, a centrifuge, and a gel electrophoresis box with an LED transilluminator, the Bento Lab is a portable and convenient laboratory that can be suitable for a field setting for quick diagnosis. Using the Bento Lab system, Knot, Zouganelis, Weedall, Wich, and Rae (2020) tested field samples for DNA barcoding of nematodes (Knot et al., 2020). Additionally, sequencing was performed using the MinION sequencing platform and compared to Sanger sequencing data to assess the quality of results obtained via the MiniION platform. They showed that nematodes were successfully identified using the 18S rRNA gene through Bento Lab and the MinION platforms, revealing exciting possibilities for monitoring helminths directly in the field.

The outlook for helminths using NGS is promising, with DNA metabarcoding as one of its main applications. Helminth DNA metabarcoding can complement the traditional gold standard for morphological identification and has enormous potential in a field setting. Moreover, rapid quantification and diagnosis of helminths in livestock can be performed via noninvasive sampling of fecal samples (Avramenko et al., 2015, 2017; Redman et al., 2019). With eDNA, metabarcoding provides a stepping stone for sampling habitats, hosts, and helminths that are difficult to obtain, potentially bringing new depths towards our understanding of host-parasite interactions. For genomics, the sequencing of helminth whole genomes has made genomic data publicly available that allow for wide genomic approaches to target candidate genes for anthelminthic therapy and to study the genetics of anthelmintic resistance

(McVeigh, 2020). Additionally, transcriptomics can provide insights at the cellular level for helminthic disease control. Undoubtedly, the use of NGS will contribute tremendously to our understanding of helminths, and helminth research in the genomics era is an exciting area to delve into.

References

Ahmed, M., Back, M. A., Prior, T., Karssen, G., Lawson, R., Adams, I., & Sapp, M. (2019). Metabarcoding of soil nematodes: the importance of taxonomic coverage and availability of reference sequences in choosing suitable marker(s). *Metabarcoding and Metagenomics, 3*, 36408. https://doi.org/10.3897/mbmg.3.36408

Aivelo, T., Harris, K., Cadle, J., & Wright, P. (2018). Exploring non-invasive sampling of parasites by metabarcoding gastrointestinal nematodes in Madagascar frog species. *Basic and Applied Herpetology, 32*, 29–40.

Aivelo, T., & Medlar, A. (2018). Opportunities and challenges in metabarcoding approaches for helminth community identification in wild mammals. *Parasitology, 145*(5), 608–621. https://doi.org/10.1017/S0031182017000610

Aivelo, T., Medlar, A., Loytynoja, A., Laakkonen, J., & Jernvall, J. (2015). Tracking year-to-year changes in intestinal nematode communities of rufous mouse lemurs (*Microcebus rufus*). *Parasitology, 142*(8), 1095–1107. https://doi.org/10.1017/S0031182015000438

Aivelo, T., Medlar, A., Löytynoja, A., Laakkonen, J., & Jernvall, J. (2018). Metabarcoding gastrointestinal nematodes in sympatric endemic and nonendemic species in Ranomafana National Park, Madagascar. *International Journal of Primatology, 39*(1), 49–64. https://doi.org/10.1007/s10764-017-0010-x

Andujar, C., Arribas, P., Yu, D. W., Vogler, A. P., & Emerson, B. C. (2018). Why the *COI* barcode should be the community DNA metabarcode for the metazoa. *Molecular Ecology, 27*(20), 3968–3975. https://doi.org/10.1111/mec.14844

Atienza, S., Guardiola, M., Præbel, K., Antich, A., Turon, X., & Wangensteen, O. S. (2020). DNA metabarcoding of deep-sea sediment communities using *COI*: Community assessment, spatio-temporal patterns and comparison with 18S rDNA. *Diversity, 12*(4), 123.

Avramenko, R. W., Redman, E. M., Lewis, R., Bichuette, M. A., Palmeira, B. M., Yazwinski, T. A., & Gilleard, J. S. (2017). The use of nemabiome metabarcoding to explore gastro-intestinal nematode species diversity and anthelmintic treatment effectiveness in beef calves. *International Journal for Parasitology, 47*(13), 893–902. https://doi.org/10.1016/j.ijpara.2017.06.006

Avramenko, R. W., Redman, E. M., Lewis, R., Yazwinski, T. A., Wasmuth, J. D., & Gilleard, J. S. (2015). Exploring the gastrointestinal "nemabiome": deep amplicon sequencing to quantify the species composition of parasitic nematode communities. *PLoS One, 10*(12), e0143559. https://doi.org/10.1371/journal.pone.0143559

Batovska, J., Lynch, S. E., Cogan, N. O. I., Brown, K., Darbro, J. M., Kho, E. A., & Blacket, M. J. (2018). Effective mosquito and arbovirus surveillance using metabarcoding. *Molecular Ecology Resources, 18*(1), 32–40. https://doi.org/10.1111/1755-0998.12682

Bell, C. A., Namaganda, J., Urwin, P. E., & Atkinson, H. J. (2021). Next-generation sequencing of the soil nematode community enables the sustainability of banana plantations to be monitored. *Applied Soil Ecology, 166*, 103999. https://doi.org/10.1016/j.apsoil.2021.103999

Bento Lab. (2016). *Nature Biotechnology, 34*(5), 455–455. https://doi.org/10.1038/nbt0516-455

Berry, T. E., Osterrieder, S. K., Murray, D. C., Coghlan, M. L., Richardson, A. J., Grealy, A. K., Stat, M., Bejder, L., & Bunce, M. (2017). DNA metabarcoding for diet analysis and biodiversity: A case study using the endangered Australian sea lion (*Neophoca cinerea*). *Ecology and Evolution, 7*(14), 5435–5453. https://doi.org/10.1002/ece3.3123

Blaxter, M. L., De Ley, P., Garey, J. R., Liu, L. X., Scheldeman, P., Vierstraete, A., Vanfleteren, J. R., Mackey, L. Y., Dorris, M., Frisse, L. M., Vida, J. T., & Thomas, W. K. (1998). A molecular evolutionary framework for the phylum Nematoda. *Nature, 392*(6671), 71–75. https://doi.org/10.1038/32160

Bongers, T., & Ferris, H. (1999). Nematode community structure as a bioindicator in environmental monitoring. *Trends in Ecology & Evolution, 14*(6), 224–228. https://doi.org/10.1016/s0169-5347(98)01583-3

Borsting, C., & Morling, N. (2015). Next generation sequencing and its applications in forensic genetics. *Forensic Science International. Genetics, 18*, 78–89. https://doi.org/10.1016/j.fsigen.2015.02.002

Braukmann, T. W. A., Ivanova, N. V., Prosser, S. W. J., Elbrecht, V., Steinke, D., Ratnasingham, S., de Waard, J. R., Sones, J. E., Zakharov, E. V., & Hebert, P. D. N. (2019). Metabarcoding a diverse arthropod mock community. *Molecular Ecology Resources, 19*(3), 711–727. https://doi.org/10.1111/1755-0998.13008

Brys, R., Haegeman, A., Halfmaerten, D., Neyrinck, S., Staelens, A., Auwerx, J., & Ruttink, T. (2021). Monitoring of spatiotemporal occupancy patterns of fish and amphibian species in a lentic aquatic system using environmental DNA. *Molecular Ecology, 30*(13), 3097–3110. https://doi.org/10.1111/mec.15742

Callaway, E. (2014). Oldest-known human genome sequenced. *Nature, 514*(7523), 413. https://doi.org/10.1038/514413a

Capra, E., Giannico, R., Montagna, M., Turri, F., Cremonesi, P., Strozzi, F., Leone, P., Gandini, G., & Pizzi, F. (2016). A new primer set for DNA metabarcoding of soil Metazoa. *European Journal of Soil Biology, 77*, 53–59. https://doi.org/10.1016/j.ejsobi.2016.10.005

Carlson, C. J., Dallas, T. A., Alexander, L. W., Phelan, A. L., & Phillips, A. J. (2020). What would it take to describe the global diversity of parasites? *Proceedings of the Biological Sciences, 287*(1939), 20201841. https://doi.org/10.1098/rspb.2020.1841

Chan, A. H. E., Chaisiri, K., Saralamba, S., Morand, S., & Thaenkham, U. (2021). Assessing the suitability of mitochondrial and nuclear DNA genetic markers for molecular systematics and species identification of helminths. *Parasites & Vectors, 14*(1), 233. https://doi.org/10.1186/s13071-021-04737-y

Clarke, L. J., Beard, J. M., Swadling, K. M., & Deagle, B. E. (2017). Effect of marker choice and thermal cycling protocol on zooplankton DNA metabarcoding studies. *Ecology and Evolution, 7*(3), 873–883. https://doi.org/10.1002/ece3.2667

Creer, S., Fonseca, V. G., Porazinska, D. L., Giblin-Davis, R. M., Sung, W., Power, D. M., Packer, M., Carvalho, G. R., Blaxter, M. L., Lambshead, P. J., & Thomas, W. K. (2010). Ultrasequencing of the meiofaunal biosphere: practice, pitfalls and promises. *Molecular Ecology, 19*(Suppl 1), 4–20. https://doi.org/10.1111/j.1365-294X.2009.04473.x

Cristescu, M. E. (2014). From barcoding single individuals to metabarcoding biological communities: towards an integrative approach to the study of global biodiversity. *Trends in Ecology & Evolution, 29*(10), 566–571. https://doi.org/10.1016/j.tree.2014.08.001

De Barba, M., Miquel, C., Boyer, F., Mercier, C., Rioux, D., Coissac, E., & Taberlet, P. (2014). DNA metabarcoding multiplexing and validation of data accuracy for diet assessment: Application to omnivorous diet. *Molecular Ecology Resources, 14*(2), 306–323. https://doi.org/10.1111/1755-0998.12188

de Jonge, D. S. W., Merten, V., Bayer, T., Puebla, O., Reusch, T. B. H., & Hoving, H.-J. T. (2021). A novel metabarcoding primer pair for environmental DNA analysis of Cephalopoda (Mollusca) targeting the nuclear 18S rRNA region. *Royal Society Open Science, 8*(2), 201388. https://doi.org/10.1098/rsos.201388

Deiner, K., Bik, H. M., Machler, E., Seymour, M., Lacoursiere-Roussel, A., Altermatt, F., Creer, S., Bista, I., Lodge, D. M., de Vere, N., Pfrender, M. E., & Bernatchez, L. (2017). Environmental DNA metabarcoding: Transforming how we survey animal and plant communities. *Molecular Ecology, 26*(21), 5872–5895. https://doi.org/10.1111/mec.14350

Dell'Anno, A., Carugati, L., Corinaldesi, C., Riccioni, G., & Danovaro, R. (2015). Unveiling the biodiversity of deep-sea nematodes through metabarcoding: are we ready to bypass the classical taxonomy? *PLoS One, 10*(12), e0144928. https://doi.org/10.1371/journal.pone.0144928

Ekblom, R., & Galindo, J. (2011). Applications of next generation sequencing in molecular ecology of non-model organisms. *Heredity, 107*(1), 1–15. https://doi.org/10.1038/hdy.2010.152

Elbrecht, V., Braukmann, T. W. A., Ivanova, N. V., Prosser, S. W. J., Hajibabaei, M., Wright, M., Zakharov, E. V., Hebert, P. D. N., & Steinke, D. (2019). Validation of *COI* metabarcoding primers for terrestrial arthropods. *PeerJ, 7*, e7745. https://doi.org/10.7717/peerj.7745

Faria, L. C., Di Domenico, M., Andrade, S. C. S., Santos, M. C. D., Fonseca, G., Zanol, J., & Amaral, A. C. Z. (2018). The use of metabarcoding for meiofauna ecological patterns assessment. *Marine Environmental Research, 140*, 160–168. https://doi.org/10.1016/j.marenvres.2018.06.013

Fonseca, V. G., Carvalho, G. R., Nichols, B., Quince, C., Johnson, H. F., Neill, S. P., Lambshead, J. D., Thomas, W. K., Power, D. M., & Creer, S. (2014). Metagenetic analysis of patterns of distribution and diversity of marine meiobenthic eukaryotes. *Global Ecology and Biogeography, 23*(11), 1293–1302. https://doi.org/10.1111/geb.12223

Fonseca, V. G., Sinniger, F., Gaspar, J. M., Quince, C., Creer, S., Power, D. M., Peck, L. S., & Clark, M. S. (2017). Revealing higher than expected meiofaunal diversity in Antarctic sediments: A metabarcoding approach. *Scientific Reports, 7*(1), 6094. https://doi.org/10.1038/s41598-017-06687-x

Green, R. E., Krause, J., Ptak, S. E., Briggs, A. W., Ronan, M. T., Simons, J. F., Du, L., Egholm, M., Rothberg, J. M., Paunovic, M., & Pääbo, S. (2006). Analysis of one million base pairs of Neanderthal DNA. *Nature, 444*(7117), 330–336. https://doi.org/10.1038/nature05336

Greiman, S. E., Cook, J. A., Tkach, V. V., Hoberg, E. P., Menning, D. M., Hope, A. G., Sonsthagen, S. A., & Talbot, S. L. (2018). Museum metabarcoding: A novel method revealing gut helminth communities of small mammals across space and time. *International Journal for Parasitology, 48*(13), 1061–1070. https://doi.org/10.1016/j.ijpara.2018.08.001

Hajibabaei, M., Shokralla, S., Zhou, X., Singer, G. A., & Baird, D. J. (2011). Environmental barcoding: A next-generation sequencing approach for biomonitoring applications using river benthos. *PLoS One, 6*(4), e17497. https://doi.org/10.1371/journal.pone.0017497

Havmøller, R. W., Jacobsen, N. S., Havmøller, L. W., Rovero, F., Scharff, N., & Bohmann, K. (2021). DNA metabarcoding reveals that African leopard diet varies between habitats. *African Journal of Ecology, 59*(1), 37–50. https://doi.org/10.1111/aje.12817

Holovachov, O., Haenel, Q., Bourlat, S. J., & Jondelius, U. (2017). Taxonomy assignment approach determines the efficiency of identification of OTUs in marine nematodes. *Royal Society Open Science, 4*(8), 170315. https://doi.org/10.1098/rsos.170315

Hugerth, L. W., & Andersson, A. F. (2017). Analysing microbial community composition through amplicon sequencing: from sampling to hypothesis testing. *Frontiers in Microbiology, 8*, 1561. https://doi.org/10.3389/fmicb.2017.01561

Kelly, R. P., Port, J. A., Yamahara, K. M., & Crowder, L. B. (2014). Using environmental DNA to census marine fishes in a large mesocosm. *PLoS One, 9*(1), e86175. https://doi.org/10.1371/journal.pone.0086175

Kenmotsu, H., Uchida, K., Hirose, Y., & Eki, T. (2020). Taxonomic profiling of individual nematodes isolated from copse soils using deep amplicon sequencing of four distinct regions of the 18S ribosomal RNA gene. *PLoS One, 15*(10), e0240336. https://doi.org/10.1371/journal.pone.0240336

Knot, I. E., Zouganelis, G. D., Weedall, G. D., Wich, S. A., & Rae, R. (2020). DNA barcoding of nematodes using the MinION. *Frontiers in Ecology and Evolution, 8*, 100. https://doi.org/10.3389/fevo.2020.00100

Koboldt, D. C., Steinberg, K. M., Larson, D. E., Wilson, R. K., & Mardis, E. R. (2013). The next-generation sequencing revolution and its impact on genomics. *Cell, 155*(1), 27–38. https://doi.org/10.1016/j.cell.2013.09.006

Leray, M., Yang, J. Y., Meyer, C. P., Mills, S. C., Agudelo, N., Ranwez, V., Boehm, J. T., & Machida, R. J. (2013). A new versatile primer set targeting a short fragment of the mitochondrial *COI* region for metabarcoding metazoan diversity: Application for characterizing coral reef fish gut contents. *Frontiers in Zoology, 10*(1), 34. https://doi.org/10.1186/1742-9994-10-34

Littlefair, J. E., & Clare, E. L. (2016). Barcoding the food chain: From Sanger to high-throughput sequencing. *Genome, 59*(11), 946–958. https://doi.org/10.1139/gen-2016-0028

Liu, M., Clarke, L. J., Baker, S. C., Jordan, G. J., & Burridge, C. P. (2019). A practical guide to DNA metabarcoding for entomological ecologists. *Ecological Entomology, 45*(3), 373–385. https://doi.org/10.1111/een.12831

Lott, M. J., Hose, G. C., & Power, M. L. (2015). Parasitic nematode communities of the red kangaroo, *Macropus rufus*: Richness and structuring in captive systems. *Parasitology Research, 114*(8), 2925–2932. https://doi.org/10.1007/s00436-015-4494-z

McVeigh, P. (2020). Post-genomic progress in helminth parasitology. *Parasitology, 147*(8), 835–840. https://doi.org/10.1017/s0031182020000591

Mena, J. L., Yagui, H., Tejeda, V., Bonifaz, E., Bellemain, E., Valentini, A., Tobler, M. W., Sánchez-Vendizú, P., & Lyet, A. (2021). Environmental DNA metabarcoding as a useful tool for evaluating terrestrial mammal diversity in tropical forests. *Ecological Applications, 31*(5), e02335. https://doi.org/10.1002/eap.2335

Metzker, M. L. (2010). Sequencing technologies - The next generation. *Nature Reviews. Genetics, 11*(1), 31–46. https://doi.org/10.1038/nrg2626

Meusnier, I., Singer, G. A., Landry, J. F., Hickey, D. A., Hebert, P. D., & Hajibabaei, M. (2008). A universal DNA mini-barcode for biodiversity analysis. *BMC Genomics, 9*, 214. https://doi.org/10.1186/1471-2164-9-214

Milan, D. T., Mendes, I. S., Damasceno, J. S., Teixeira, D. F., Sales, N. G., & Carvalho, D. C. (2020). New 12S metabarcoding primers for enhanced Neotropical freshwater fish biodiversity assessment. *Scientific Reports, 10*(1), 17966. https://doi.org/10.1038/s41598-020-74902-3

Miya, M., Sato, Y., Fukunaga, T., Sado, T., Poulsen, J. Y., Sato, K., Minamoto, T., Yamamoto, S., Yamanaka, H., Araki, H., Kondoh, M., & Iwasaki, W. (2015). MiFish, a set of universal PCR primers for metabarcoding environmental DNA from fishes: Detection of more than 230 subtropical marine species. *Royal Society Open Science, 2*(7), 150088. https://doi.org/10.1098/rsos.150088

Morey, K. C., Bartley, T. J., & Hanner, R. H. (2020). Validating environmental DNA metabarcoding for marine fishes in diverse ecosystems using a public aquarium. *Environmental DNA, 2*(3), 330–342. https://doi.org/10.1002/edn3.76

Moszczynska, A., Locke, S. A., McLaughlin, J. D., Marcogliese, D. J., & Crease, T. J. (2009). Development of primers for the mitochondrial cytochrome c oxidase I gene in digenetic trematodes (Platyhelminthes) illustrates the challenge of barcoding parasitic helminths. *Molecular Ecology Resources, 9*, 75–82. https://doi.org/10.1111/j.1755-0998.2009.02634.x

Müller, C. A., de Mattos Pereira, L., Lopes, C., Cares, J., dos Anjos Borges, L. G., Giongo, A., Graeff-Teixeira, C., & Loureiro Morassutti, A. (2019). Meiofaunal diversity in the Atlantic Forest soil: A quest for nematodes in a native reserve using eukaryotic metabarcoding analysis. *Forest Ecology and Management, 453*, 117591. https://doi.org/10.1016/j.foreco.2019.117591

Neher, D. A. (2001). Role of nematodes in soil health and their use as indicators. *Journal of Nematology, 33*(4), 161–168.

Pavan-Kumar, A., Gireesh-Babu, P., & Lakra, W. (2015). DNA metabarcoding: A new approach for rapid biodiversity assessment. *Cell Science & Molecular Biology, 2*(1), 111.

Plesivkova, D., Richards, R., & Harbison, S. (2018). A review of the potential of the MinION™ single-molecule sequencing system for forensic applications. *WIREs Forensic Science, 1*(1), 1323. https://doi.org/10.1002/wfs2.1323

Porazinska, D. L., Giblin-Davis, R. M., Esquivel, A., Powers, T. O., Sung, W., & Thomas, W. K. (2010). Ecometagenetics confirm high tropical rainforest nematode diversity. *Molecular Ecology, 19*(24), 5521–5530. https://doi.org/10.1111/j.1365-294X.2010.04891.x

Porazinska, D. L., Giblin-Davis, R. M., Faller, L., Farmerie, W., Kanzaki, N., Morris, K., Powers, T. O., Tucker, A. E., Sung, W., & Thomas, W. K. (2009). Evaluating high-throughput sequencing as a method for metagenomic analysis of nematode diversity. *Molecular Ecology Resources, 9*(6), 1439–1450. https://doi.org/10.1111/j.1755-0998.2009.02611.x

Port, J. A., O'Donnell, J. L., Romero-Maraccini, O. C., Leary, P. R., Litvin, S. Y., Nickols, K. J., Yamahara, K. M., & Kelly, R. P. (2016). Assessing vertebrate biodiversity in a kelp forest ecosystem using environmental DNA. *Molecular Ecology, 25*(2), 527–541. https://doi.org/10.1111/mec.13481

Porter, T. M., & Hajibabaei, M. (2018). Scaling up: A guide to high-throughput genomic approaches for biodiversity analysis. *Molecular Ecology, 27*(2), 313–338. https://doi.org/10.1111/mec.14478

Quail, M. A., Smith, M., Coupland, P., Otto, T. D., Harris, S. R., Connor, T. R., Bertoni, A., Swerdlow, H. P., & Gu, Y. (2012). A tale of three next generation sequencing platforms: Comparison of ion torrent, pacific biosciences and illumina MiSeq sequencers. *BMC Genomics, 13*, 341. https://doi.org/10.1186/1471-2164-13-341

Redman, E., Queiroz, C., Bartley, D. J., Levy, M., Avramenko, R. W., & Gilleard, J. S. (2019). Validation of ITS-2 rDNA nemabiome sequencing for ovine gastrointestinal nematodes and its application to a large scale survey of UK sheep farms. *Veterinary Parasitology, 275*, 108933. https://doi.org/10.1016/j.vetpar.2019.108933

Sabir, J., Mutwakil, M., El-Hanafy, A., Al-Hejin, A., Sadek, M., Abou-Alsoud, M., Wuresh, M., Saini, K., & Ahmed, M. (2014). Applying molecular tools for improving livestock performance from dna markers to next generation sequencing technologies. *Journal of Food, Agriculture and Environment, 12*(2), 351–363.

Sapkota, R., & Nicolaisen, M. (2015). High-throughput sequencing of nematode communities from total soil DNA extractions. *BMC Ecology, 15*(3), 34. https://doi.org/10.1186/s12898-014-0034-4

Schenk, J., Hoss, S., Brinke, M., Kleinbolting, N., Bruchner-Huttemann, H., & Traunspurger, W. (2020). Nematodes as bioindicators of polluted sediments using metabarcoding and microscopic taxonomy. *Environment International, 143*, 105922. https://doi.org/10.1016/j.envint.2020.105922

Schenk, J., Kleinbölting, N., & Traunspurger, W. (2020). Comparison of morphological, DNA barcoding, and metabarcoding characterizations of freshwater nematode communities. *Ecology and Evolution, 10*(6), 2885–2899. https://doi.org/10.1002/ece3.6104

Shokralla, S., Spall, J. L., Gibson, J. F., & Hajibabaei, M. (2012). Next-generation sequencing technologies for environmental DNA research. *Molecular Ecology, 21*(8), 1794–1805. https://doi.org/10.1111/j.1365-294X.2012.05538.x

Sogin, M. L., Morrison, H. G., Huber, J. A., Welch, D. M., Huse, S. M., Neal, P. R., Arrieta, J. M., & Herndl, G. J. (2006). Microbial diversity in the deep sea and the underexplored "rare biosphere". *Proceedings of the National Academy of Sciences, 103*(32), 12115–12120. https://doi.org/10.1073/pnas.0605127103

Staats, M., Arulandhu, A. J., Gravendeel, B., Holst-Jensen, A., Scholtens, I., Peelen, T., Prins, T. W., & Kok, E. (2016). Advances in DNA metabarcoding for food and wildlife forensic species identification. *Analytical and Bioanalytical Chemistry, 408*(17), 4615–4630. https://doi.org/10.1007/s00216-016-9595-8

Taberlet, P., Coissac, E., Hajibabaei, M., & Rieseberg, L. H. (2012). Environmental DNA. *Molecular Ecology, 21*(8), 1789–1793. https://doi.org/10.1111/j.1365-294X.2012.05542.x

Taberlet, P., Coissac, E., Pompanon, F., Brochmann, C., & Willerslev, E. (2012). Towards next-generation biodiversity assessment using DNA metabarcoding. *Molecular Ecology, 21*(8), 2045–2050. https://doi.org/10.1111/j.1365-294X.2012.05470.x

Tan, B., Ng, C., Nshimyimana, J. P., Loh, L. L., Gin, K. Y., & Thompson, J. R. (2015). Next-generation sequencing (NGS) for assessment of microbial water quality: Current progress, challenges, and future opportunities. *Frontiers in Microbiology, 6*, 1027. https://doi.org/10.3389/fmicb.2015.01027

Tanaka, R., Hino, A., Tsai, I. J., Palomares-Rius, J. E., Yoshida, A., Ogura, Y., Hayashi, T., Maruyama, H., & Kikuchi, T. (2014). Assessment of helminth biodiversity in wild rats using 18S rDNA based metagenomics. *PLoS One, 9*(10), e110769. https://doi.org/10.1371/journal.pone.0110769

Tedersoo, L., Drenkhan, R., Anslan, S., Morales-Rodriguez, C., & Cleary, M. (2019). High-throughput identification and diagnostics of pathogens and pests: Overview and practical recommendations. *Molecular Ecology Resources, 19*(1), 47–76. https://doi.org/10.1111/1755-0998.12959

Tillmar, A. O., Dell'Amico, B., Welander, J., & Holmlund, G. (2013). A universal method for species identification of mammals utilizing next generation sequencing for the analysis of DNA mixtures. *PLoS One, 8*(12), e83761. https://doi.org/10.1371/journal.pone.0083761

Treonis, A. M., Unangst, S. K., Kepler, R. M., Buyer, J. S., Cavigelli, M. A., Mirsky, S. B., & Maul, J. E. (2018). Characterization of soil nematode communities in three cropping systems through morphological and DNA metabarcoding approaches. *Scientific Reports, 8*(1), 2004. https://doi.org/10.1038/s41598-018-20366-5

Tytgat, B., Nguyen, D. T., Nguyen, T. X. P., Pham, T. M., Long, P. K., Vanreusel, A., & Derycke, S. (2019). Monitoring of marine nematode communities through 18S rRNA metabarcoding as a sensitive alternative to morphology. *Ecological Indicators, 107*, 105554. https://doi.org/10.1016/j.ecolind.2019.105554

Valentini, A., Taberlet, P., Miaud, C., Civade, R., Herder, J., Thomsen, P. F., Bellemain, E., Besnard, A., Coissac, E., Boyer, F., Gaboriaud, C., Jean, P., Poulet, N., Roset, N., Copp, G. H., Geniez, P., Pont, D., Argillier, C., Baudoin, J. M., . . . Dejean, T. (2016). Next-generation monitoring of aquatic biodiversity using environmental DNA metabarcoding. *Molecular Ecology, 25*(4), 929–942. https://doi.org/10.1111/mec.13428

Chapter 15
Concluding Remarks and Further Prospective

Abstract Aimed at providing fundamental knowledge and information for research on molecular systematics, the molecular systematics of parasitic helminths was thoroughly presented in the previous chapters. The link between specimens and the phenotypic characters of helminths of medical importance, to evolutionary processes leading up to phenotypic and genotypic variations to species diversity of helminths, and subsequently to the use of molecular information to build phylogenies for enhanced understanding of relationships among helminth groups for molecular systematics was discussed. To close this book, we focused on pressing issues like climate change and its impact on helminths and technological advances that are gaining momentum in the scientific community in relation to helminths.

15.1 Climate Change: Helminths in a Changing World

Accelerating changes to the climate and environment due to anthropogenic activity and global warming is a concerning area and has implications for both helminth biodiversity and the impacts of helminthic infection on human health. Climate change refers to changes in climate properties such as temperature, precipitation, and extreme weather events that persist for a prolonged period (Ellwanger et al., 2020; Liang & Gong, 2017). In the recent assessment by the Intergovernmental Panel on Climate Change (IPCC) performed in 2021, findings revealed that climate change is already affecting every region globally, with humans contributing largely to the many observed changes in weather extremes (IPCC, 2021). Additionally, with every increment of warming temperatures, climate changes are projected to be larger in frequency and intensity. A 1.5 °C to 2 °C increase in global surface temperature is predicted to occur within this century unless major reductions in carbon dioxide, greenhouse gas emissions, and human anthropogenic activity occur. The association between alterations in the climate and their effect on animal and plant ecosystems is closely interlinked. Associations exist on multiple levels, where ecosystems and habitats are responding to climate changes, which are linked to vertebrates living in these habitats having to adapt to changes in their environment and, in turn, allied with parasites that reside in them (either vertebrates or environment) having to

U. Thaenkham et al., *Molecular Systematics of Parasitic Helminths*,
https://doi.org/10.1007/978-981-19-1786-8_15

evolve to new conditions in their host (Yeh et al., 2020). As helminths are reliant on both their external environment (in some parts of the life cycle) and the internal environment of their hosts, changes with either environment will inevitably influence survival, population dynamics, evolution, and impact on human health (Petney et al., 2021). As mentioned previously in Chaps. 3 and 4, changes at the genotypic and phenotypic levels can occur as helminths coevolve with their host and adapt to new environmental conditions over time.

Firstly, with respect to helminth biodiversity in a changing world, potential effects of climate change include reduced diversity leading to possible extinction. With hosts being affected by climate change, host extinctions or invasions in certain areas change the possible habitats of helminths that reside in their hosts. As previously mentioned in Chap. 3, specialist parasites and those with complex life cycles (requiring more than one host) are more likely to go extinct as compared to generalist parasites (Colles et al., 2009; Petney et al., 2021; Rohr et al., 2011). The impacts of direct and indirect effects of climate seasonality on helminths infecting amphibians were modeled by Martins et al. (2021). They found that helminth species diversity was directly and indirectly affected by temperature, precipitation, and host richness, reinforcing the close interplay of relationships among climate, hosts, and helminths (Martins et al., 2021). In another study, Rubenina et al. (2021) demonstrated variation in the species composition of helminths in frogs by investigating environmental and anthropogenic factors (Rubenina et al., 2021). The effect of wet and dry season rice paddy plantation was demonstrated for *Bithynia siamensis*, the first intermediate snail host of *Opisthorchis viverrini*. A population collapse was observed for *B. siamensis* due to dry rice paddy conditions, which led to a decrease in *O. viverrini* found, showing the close association between hosts and helminths (Chaiyasaeng et al., 2019). The possible decrease of hosts due to changes in the environment can thus lead to a decline in the proportion of helminths. The loss of helminths due to environmental changes, especially in the context of helminths that have not yet been discovered in many host species (which can become extinct soon), is a critical issue for biodiversity. Five to 10% of the 457 parasite species included in a model by Carlson et al. (2017) are predicted to become extinct by 2070 due to habitat loss from climate change. Additionally, 30% of parasitic worms are predicted to be extinct, considering other factors like host-driven coextinctions (Carlson et al., 2017).

Secondly, the impacts on human health with helminths in a changing world can lead to a greater ability of parasites to host shift due to a decline in the preferred host. Host switching can also lead to the possibility of increasing the pathogenicity of helminthic diseases (Carlson et al., 2017; Petney et al., 2021; Rohr et al., 2011). Additionally, higher global temperatures can potentially speed up the reproduction time for helminths, leading to more generations per year. Higher global temperatures can also be associated with faster reproduction time for certain host species, increasing the proportion of available habitats for helminths (Petney et al., 2021). The effect of climate change can also shift the dominance of one helminth species to another. For example, in Africa, hookworm larvae have been reported to remain viable at a temperature of 40 °C or higher, unlike *Ascaris lumbricoides* and *Trichuris trichiura* eggs that stop developing at 38 °C (Blum & Hotez, 2018; Brooker et al., 2004, 2006; Okulewicz, 2017). An increase in global temperatures can lead to a shift in favor of dominance and survival for hookworms among the soil-transmitted helminths. Additionally, with greater environmental extremes, helminths that can arrest their development in human tissues and remain shielded from the external climate may possess an advantage over those with developmental stages in the external climate (Blum & Hotez, 2018; Pullan & Brooker, 2012). Animal health also suffers the impacts of climate change. The impact of climate on anthelmintic resistance in helminths infecting equines was investigated through modeling mechanisms by Sauermann et al. (2020) (Sauermann et al., 2020). The model replicates the dynamics of cyathostomin, the administration of anthelmintics, and the buildup of resistance genotypes. The results revealed that climate change will influence the level of parasitism and the development of anthelmintic resistance, and the severity differs based on the magnitude of climate change (Sauermann et al., 2020).

Thirdly, in the context of tools for identifying and detecting helminths, morphological and molecular detection methods will have to be updated, with changes in the genotypic and phenotypic levels occurring at an increased pace (Petney et al., 2021; Yeh et al., 2020). At the phenotypic level, morphological variations due to phenotypic plasticity can occur over short periods of time in the face of a changing environment (Hoberg & Brooks, 2015). While at the genotypic level, genetic markers and molecular methods for accurate species identification will have to complement the pace at which the genetic material of the host and parasites coevolve with the environment (Yeh et al., 2020). The possibility of adaptive radiation and cryptic species adapting to new environmental conditions is increased.

15.2 Keeping Pace: Technological Advances and Helminths

Advances in technological advances both in the field of molecular biology and computer science have greatly accelerated research in helminths. Although a large part of helminthology still comprises traditional methods for species identification and disease diagnosis, advances in science are playing an integral role and will be a significant part of research in helminths in the future. In Chap. 14, NGS technologies were exemplified, and DNA metabarcoding through NGS and bioinformatics was emphasized. In addition to DNA metabarcoding, other new and promising research frontiers include studies on the microbiome, genomics including genome editing, drug and vaccine developments, and machine learning with artificial intelligence.

Studies involving the interaction between helminths and the microbiota in the host have gained attention in recent years. Interactions can exist between host-microbiota, helminth-host, and helminth-microbiota, illustrating the complex interrelationships that can occur (Walshe et al., 2020). Questions include the role that the host-microbiome plays in the development and fitness of helminths, mechanisms behind the interactions, and control strategies against drug resistance that can potentially be harnessed from the interaction between helminths and the microbiome

(Jenkins et al., 2019; Rapin & Harris, 2018; Sharpton et al., 2020). The role of the host-microbiome in helminth development was demonstrated by Hayes et al. (2010), where the egg hatching of *Trichuris muris* was triggered by the presence of the fimbriae (protein structures located on the surface of cells) of the bacteria (Hayes et al., 2010). In field-based studies, a positive association was found between the presence of *T. trichiura* and *A. lumbricoides* with bacterial diversity when children naturally infected with *T. trichiura* and *A. lumbricoides* were compared with those that were not infected (Cooper et al., 2013). In Sri Lanka, differences in bacterial communities were detected in individuals infected with soil-transmitted helminths and those not infected but subjected to anthelminthic treatment (Jenkins et al., 2017). Additionally, the potential of microbes being a useful source for the discovery of novel anthelmintic drugs can be enhanced by understanding the interaction between host-microbiome and helminths (Glendinning et al., 2014; Sharpton et al., 2020).

The sequencing of whole genomes has provided researchers with a wealth of genetic information online for target organisms. To date, 157 nematodes and 45 platyhelminths whole genomes have been sequenced and annotated in the WormBase ParaSite database (https://parasite.wormbase.org/index.html). Access to helminth genomic data has accelerated the pace of comparative approaches in evolutionary studies (mechanisms and parasitism strategies) and functional genomics and transcriptomics (gene functions and interactions) (McVeigh, 2020). By analyzing 81 genomes of parasitic and nonparasitic helminths, gene families that modulate immune responses related to parasitism were revealed (International Helminth Genomes Consortium, 2019). RNA interference (RNAi) technology, a gene-silencing method to investigate transcriptional changes in organisms, has been adopted among helminths. Examples include the implementation of RNAi in *Schistosoma mansoni*, where gene functions were manipulated in the various developmental stages of the trematode's life cycle (Da'dara & Skelly, 2015; Rinaldi et al., 2009). Clustered regularly interspaced short palindromic repeats, more commonly recognized as CRISPR, are rapidly becoming the forefront of genetic research. The CRISPR/Cas9 technology enables genes to be edited using the Cas9 enzyme to cut a targeted location in the genome. The use of CRISPR/Cas9 was performed on *Strongyloides stercoralis*, where mutations were introduced, and these mutations were shown to be heritable from F1 to F3 following host passage (Bryant et al., 2018; Gang et al., 2017). New genetic information can be introduced in the organism for the selection process or even the introduction of specific genes that can potentially be useful for helminth control (McVeigh, 2020). In addition to genomes, progress has been made in developing drugs against helminths and the availability of tools for helminth molecular diagnosis and detection. Improvements to our ability to identify regions in the genome that correlate with drug resistance using quantitative trait loci (QTL) can benefit efforts aiming to overcome anthelmintic drug resistance. Doyle et al. (2019) detected QTL associated with ivermectin resistance in *Haemonchus contortus*, suggesting the potential to understand the genetics associated with drug resistance (Doyle & Cotton, 2019; Doyle et al., 2019). The use of more sensitive PCR detection methods like quantitative PCR, digital droplet PCR, and NGS technology contributed to progress for helminth species detection.

Helminthic infections can now be quantified, and greater resolution for detecting polymorphisms in helminths can now be achieved (Khurana et al., 2021; Kumar et al., 2020).

Another exciting area is using artificial intelligence (AI) to support helminth species identification through machine learning. Artificial intelligence is defined as the ability of a machine to perform intelligent functions such as learning, decision-making, and perception. With a digital image system (Helminth Egg Automatic Detector), helminth eggs obtained from wastewater systems were successfully discriminated from other types of artifacts and identified based on images taken under a microscope (Jiménez et al., 2016). A web-based application was subsequently developed based on the Helminth Egg Automatic Detector system, presenting a helpful tool reaching a wider audience through online systems (Jiménez et al., 2020). Using image processing and AI prediction models, another integrative and interactive cloud-based tool, was developed by Lee et al. (2021) for species identification and quantification of helminth eggs (Lee et al., 2021).

Additionally, the tool serves as a database of helminth egg image collection that is freely accessible. The digitalization of microscopy, assisted by AI, also allows easing the burden on experienced taxonomists required for helminth identification. Moreover, web-based tools and mobile applications enable species identification to be performed remotely. A system was developed for remote and automatic detection and quantification of *T. trichiura* eggs in fecal samples, with just a use of a 3D printed adaptor and a smartphone (Dacal et al., 2021).

15.3 Closing Remarks

The development and use of additional tools will undoubtedly improve current species detection and identification methods for helminths. As we have seen throughout this book, various mechanisms and driving forces are at work to shape the biodiversity of helminths and are intricately connected to human and animal health. Studying the molecular systematics of parasitic nematodes, trematodes, and cestodes allowed for the elucidation of relationships between and within them and increased our understanding of evolutionary forces driving parasitism. Moreover, molecular tools have significantly contributed to helminth systematics in recent decades, and as technological advances firmly take root in helminthology, the road ahead will be transformed in the next era of genomics and AI.

References

Blum, A. J., & Hotez, P. J. (2018). Global "worming": Climate change and its projected general impact on human helminth infections. *PLoS Neglected Tropical Diseases, 12*(7), e0006370. https://doi.org/10.1371/journal.pntd.0006370

Brooker, S., Clements, A. C., & Bundy, D. A. (2006). Global epidemiology, ecology and control of soil-transmitted helminth infections. *Advances in Parasitology, 62*, 221–261. https://doi.org/10.1016/s0065-308x(05)62007-6

Brooker, S., Kabatereine, N. B., Tukahebwa, E. M., & Kazibwe, F. (2004). Spatial analysis of the distribution of intestinal nematode infections in Uganda. *Epidemiology and Infection, 132*(6), 1065–1071. https://doi.org/10.1017/s0950268804003024

Bryant, A. S., Ruiz, F., Gang, S. S., Castelletto, M. L., Lopez, J. B., & Hallem, E. A. (2018). A critical role for thermosensation in host seeking by skin-penetrating nematodes. *Current Biology, 28*(14), 2338–2347–e.6. https://doi.org/10.1016/j.cub.2018.05.063

Carlson, C. J., Burgio, K. R., Dougherty, E. R., Phillips, A. J., Bueno, V. M., Clements, C. F., Castaldo, G., Dallas, T. A., Cizauskas, C. A., Cumming, G. S., Doña, J., Harris, N. C., Jovani, R., Mironov, S., Muellerklein, O. C., Proctor, H. C., & Getz, W. M. (2017). Parasite biodiversity faces extinction and redistribution in a changing climate. *Science Advances, 3*(9), e1602422. https://doi.org/10.1126/sciadv.1602422

Chaiyasaeng, M., Pechdee, P., Sereewong, C., Suwannatrai, A., Laha, T., & Tesana, S. (2019). Effects of aestivation on survival of *Bithynia siamensis goniomphalos* snails and the infection of *Opisthorchis viverrini* in the irrigation area of wet- and dry-season rice paddy. *Acta Tropica, 192*, 55–60. https://doi.org/10.1016/j.actatropica.2019.01.014

Colles, A., Liow, L. H., & Prinzing, A. (2009). Are specialists at risk under environmental change? Neoecological, paleoecological and phylogenetic approaches. *Ecology Letters, 12*(8), 849–863. https://doi.org/10.1111/j.1461-0248.2009.01336.x

Cooper, P., Walker, A. W., Reyes, J., Chico, M., Salter, S. J., Vaca, M., & Parkhill, J. (2013). Patent human infections with the whipworm, *Trichuris trichiura*, are not associated with alterations in the faecal microbiota. *PLoS One, 8*(10), e76573. https://doi.org/10.1371/journal.pone.0076573

Da'dara, A. A., & Skelly, P. J. (2015). Gene suppression in schistosomes using RNAi. In C. Peacock (Ed.), *Parasite genomics protocols. Methods in molecular biology*. Humana Press.

Dacal, E., Bermejo-Peláez, D., Lin, L., Álamo, E., Cuadrado, D., Martínez, Á., Mousa, A., Postigo, M., Soto, A., Sukosd, E., Vladimirov, A., Mwandawiro, C., Gichuki, P., Williams, N. A., Muñoz, J., Kepha, S., & Luengo-Oroz, M. (2021). Mobile microscopy and telemedicine platform assisted by deep learning for the quantification of *Trichuris trichiura* infection. *PLoS Neglected Tropical Diseases, 15*(9), e0009677. https://doi.org/10.1371/journal.pntd.0009677

Doyle, S. R., & Cotton, J. A. (2019). Genome-wide approaches to investigate anthelmintic resistance. *Trends in Parasitology, 35*(4), 289–301. https://doi.org/10.1016/j.pt.2019.01.004

Doyle, S. R., Illingworth, C. J. R., Laing, R., Bartley, D. J., Redman, E., Martinelli, A., Holroyd, N., Morrison, A. A., Rezansoff, A., Tracey, A., Devaney, E., Berriman, M., Sargison, N., Cotton, J. A., & Gilleard, J. S. (2019). Population genomic and evolutionary modelling analyses reveal a single major QTL for ivermectin drug resistance in the pathogenic nematode, *Haemonchus contortus*. *BMC Genomics, 20*(1), 218. https://doi.org/10.1186/s12864-019-5592-6

Ellwanger, J. H., Kulmann-Leal, B., Kaminski, V. L., Valverde-Villegas, J. M., Veiga, A., Spilki, F. R., Fearnside, P. M., Caesar, L., Giatti, L. L., Wallau, G. L., Almeida, S. E. M., Borba, M. R., Hora, V. P. D., & Chies, J. A. B. (2020). Beyond diversity loss and climate change: Impacts of Amazon deforestation on infectious diseases and public health. *Anais da Academia Brasileira de Ciências, 92*(1), e20191375. https://doi.org/10.1590/0001-3765202020191375

Gang, S. S., Castelletto, M. L., Bryant, A. S., Yang, E., Mancuso, N., Lopez, J. B., Pellegrini, M., & Hallem, E. A. (2017). Targeted mutagenesis in a human-parasitic nematode. *PLoS Pathogens, 13*(10), e1006675. https://doi.org/10.1371/journal.ppat.1006675

Glendinning, L., Nausch, N., Free, A., Taylor, D. W., & Mutapi, F. (2014). The microbiota and helminths: Sharing the same niche in the human host. *Parasitology, 141*(10), 1255–1271. https://doi.org/10.1017/S0031182014000699

Hayes, K. S., Bancroft, A. J., Goldrick, M., Portsmouth, C., Roberts, I. S., & Grencis, R. K. (2010). Exploitation of the intestinal microflora by the parasitic nematode *Trichuris muris*. *Science, 328*(5984), 1391–1394. https://doi.org/10.1126/science.1187703

Hoberg, E. P., & Brooks, D. R. (2015). Evolution in action: Climate change, biodiversity dynamics and emerging infectious disease. *Philosophical Transactions of the Royal Society of London. Series B, Biological Sciences, 370*(1665). https://doi.org/10.1098/rstb.2013.0553

International Helminth Genomes Consortium. (2019). Comparative genomics of the major parasitic worms. *Nature Genetics, 51*(1), 163–174. https://doi.org/10.1038/s41588-018-0262-1

IPCC. (2021). Summary for policymakers: Climate change 2021: The physical science basis. Contribution of working group I to the sixth assessment report of the intergovernmental panel on climate change.

Jenkins, T. P., Brindley, P. J., Gasser, R. B., & Cantacessi, C. (2019). Helminth microbiomes - a hidden treasure trove? *Trends in Parasitology, 35*(1), 13–22. https://doi.org/10.1016/j.pt.2018.10.007

Jenkins, T. P., Rathnayaka, Y., Perera, P. K., Peachey, L. E., Nolan, M. J., Krause, L., Rajakaruna, R. S., & Cantacessi, C. (2017). Infections by human gastrointestinal helminths are associated with changes in faecal microbiota diversity and composition. *PLoS One, 12*(9), e0184719. https://doi.org/10.1371/journal.pone.0184719

Jiménez, B., Maya, C., Velásquez, G., Barrios, J. A., Perez, M., & Román, A. (2020). Helminth Egg Automatic Detector (HEAD): Improvements in development for digital identification and quantification of helminth eggs and their application online. *Experimental Parasitology, 217*, 107959. https://doi.org/10.1016/j.exppara.2020.107959

Jiménez, B., Maya, C., Velásquez, G., Torner, F., Arambula, F., Barrios, J. A., & Velasco, M. (2016). Identification and quantification of pathogenic helminth eggs using a digital image system. *Experimental Parasitology, 166*, 164–172. https://doi.org/10.1016/j.exppara.2016.04.016

Khurana, S., Singh, S., & Mewara, A. (2021). Diagnostic techniques for soil-transmitted helminths - recent advances. *Research and Reports in Tropical Medicine, 12*, 181–196. https://doi.org/10.2147/RRTM.S278140

Kumar, S., Gupta, S., Mohmad, A., Fular, A., Parthasarathi, B. C., & Chaubey, A. K. (2020). Molecular tools-advances, opportunities and prospects for the control of parasites of veterinary importance. *International Journal of Tropical Insect Science, 1-10*. https://doi.org/10.1007/s42690-020-00213-9

Lee, C.-C., Huang, P.-J., Yeh, Y.-M., Li, P.-H., Chiu, C.-H., Cheng, W.-H., & Tang, P. (2021). Helminth egg analysis platform (HEAP): An opened platform for microscopic helminth egg identification and quantification based on the integration of deep learning architectures. *Journal of Microbiology, Immunology and Infection*. https://doi.org/10.1016/j.jmii.2021.07.014

Liang, L., & Gong, P. (2017). Climate change and human infectious diseases: A synthesis of research findings from global and spatio-temporal perspectives. *Environment International, 103*, 99–108. https://doi.org/10.1016/j.envint.2017.03.011

Martins, P. M., Poulin, R., & Gonçalves-Souza, T. (2021). Integrating climate and host richness as drivers of global parasite diversity. *Global Ecology and Biogeography, 30*(1), 196–204. https://doi.org/10.1111/geb.13213

McVeigh, P. (2020). Post-genomic progress in helminth parasitology. *Parasitology, 147*(8), 835–840. https://doi.org/10.1017/s0031182020000591

Okulewicz, A. (2017). The impact of global climate change on the spread of parasitic nematodes. *Annals of Parasitology, 63*(1), 15–20. https://doi.org/10.17420/ap6301.79

Petney, T. N., Sithithaworn, P., & Andrews, R. H. (2021). Parasite diversity, dynamics, and climate change. In T. N. Petney, P. Sithithaworn, & H. Mehlhorn (Eds.), *Biodiversity of southeast Asian parasites and vectors causing human disease* (pp. 183–204). Springer.

Pullan, R. L., & Brooker, S. J. (2012). The global limits and population at risk of soil-transmitted helminth infections in 2010. *Parasites & Vectors, 5*(1), 81. https://doi.org/10.1186/1756-3305-5-81

Rapin, A., & Harris, N. L. (2018). Helminth-bacterial interactions: Cause and consequence. *Trends in Immunology, 39*(9), 724–733. https://doi.org/10.1016/j.it.2018.06.002

Rinaldi, G., Morales, M. E., Alrefaei, Y. N., Cancela, M., Castillo, E., Dalton, J. P., Tort, J. F., & Brindley, P. J. (2009). RNA interference targeting leucine aminopeptidase blocks hatching of *Schistosoma mansoni* eggs. *Molecular and Biochemical Parasitology, 167*(2), 118–126. https://doi.org/10.1016/j.molbiopara.2009.05.002

Rohr, J. R., Dobson, A. P., Johnson, P. T., Kilpatrick, A. M., Paull, S. H., Raffel, T. R., Ruiz-Moreno, D., & Thomas, M. B. (2011). Frontiers in climate change-disease research. *Trends in Ecology & Evolution, 26*(6), 270–277. https://doi.org/10.1016/j.tree.2011.03.002

Rubenina, I., Kirjusina, M., Ceirans, A., Gravele, E., Gavarane, I., Pupins, M., & Krasnov, B. R. (2021). Environmental, anthropogenic, and spatial factors affecting species composition and species associations in helminth communities of water frogs (*Pelophylax esculentus* complex) in Latvia. *Parasitology Research, 120*(10), 3461–3474. https://doi.org/10.1007/s00436-021-07303-8

Sauermann, C. W., Leathwick, D. M., Lieffering, M., & Nielsen, M. K. (2020). Climate change is likely to increase the development rate of anthelmintic resistance in equine cyathostomins in New Zealand. *International Journal for Parasitology: Drugs and Drug Resistance, 14*, 73–79. https://doi.org/10.1016/j.ijpddr.2020.09.001

Sharpton, T. J., Combrink, L., Arnold, H. K., Gaulke, C. A., & Kent, M. (2020). Harnessing the gut microbiome in the fight against anthelminthic drug resistance. *Current Opinion in Microbiology, 53*, 26–34. https://doi.org/10.1016/j.mib.2020.01.017

Walshe, N., Mulcahy, G., Hodgkinson, J., & Peachey, L. (2020). No worm is an island; the influence of commensal gut microbiota on cyathostomin infections. *Animals (Basel), 10*(12). https://doi.org/10.3390/ani10122309

Yeh, K. B., Fair, J. M., Smith, W., Martinez Torres, T., Lucas, J., Monagin, C., Winegar, R., & Fletcher, J. (2020). Assessing climate change impact on ecosystems and infectious disease: Important roles for genomic sequencing and a one health perspective. *Tropical Medicine and Infectious Disease, 5*(2). https://doi.org/10.3390/tropicalmed5020090

Glossary

Acetabula The plural form of "Acetabulum" refers to the muscular suckers being a saucer-shaped organ in an invertebrate, for example, in some annelid worms and flatworms. This organ is the ventral sucker in trematodes located at the anterior end of the body for penetrating and gripping host tissue. The acetabulum is one of the diagnostic characters for morphological identification among trematodes.

Adaptive radiation Process of rapid diversification from an ancestral species resulting in a multitude of new forms to fit in different environments. Adaptive radiation leads to ecological and phenotypic diversity.

Allele frequency A measure of the relative frequency of an allele on a genetic locus and shows a frequency of variant form within a population.

Alleles Alleles are variants of a gene and are inherited from each parent. If both alleles are the same for the individual, they are homozygous for that gene, whereas if the alleles are different, they are heterozygous.

Allopatric speciation The speciation process, also known as geographic speciation, occurs due to geographical barriers. Gene flow is reduced, promoting genetic divergence.

Amplified fragment length polymorphism (AFLP) A PCR-based technique amplifying a subset of digested DNA fragments with restriction enzymes to generate and compare unique DNA fingerprints for a genome of interest.

Analogous Similar biological structures have corresponding functions but do not share common evolutionary origins.

Antagonistic interactions An interaction between organisms beneficial for one and detrimental for the other.

Anthelmintic resistance Genetically transmitted loss of sensitivity of an anthelmintic drug in helminth populations that was previously susceptible to the same anthelmintic.

Antigenic proteins A molecular structure binding to a specific antibody resulting in an immune response.

U. Thaenkham et al., *Molecular Systematics of Parasitic Helminths*,
https://doi.org/10.1007/978-981-19-1786-8

Autopolyploidy It occurs when an individual has more than 2 sets of chromosomes (multiple sets) derived from an original taxon.

Balancing selection Balancing selection occurs when multiple alleles maintain genetic diversity in populations over long evolutionary periods.

Barcode gap The difference between inter- and intraspecific genetic distances within a group of organisms important in determining whether the genetic marker is robust for DNA barcoding.

Basal relationship/clade The direction of the base of a rooted phylogenetic tree referring to nodes adjacent to the root diverging early in the evolutionary history of the clade.

Bet-hedging The theory addresses how individuals should optimize their fitness in a variable and unpredictable environment, which occurs when organisms suffer decreased fitness in their typical conditions in exchange for increased fitness in stressful conditions.

Bilateral symmetry A body plan where the left and right sides of the organism can be divided into mirror images on a median axis.

Binomial nomenclature It is the formal way of naming a species. Each scientific name comprises two names—the first describes the genus that the organism belongs to, while the second refers to the species of the organism.

Bothria The plural form of "Bothrium" characteristic of Pseudophyllidea. They are elongate, dorsal, or ventral longitudinal grooves on the scolex of Cestoda with weak muscles.

Buccal capsule A nematode structure, functioning for the initial uptake of food into the alimentary tract and connects the opening at the anterior end of the nematode to the beginning of the esophagus. This morphological character can be modified depending on the food and feeding mechanism.

Caudal bursa The posterior portion of the male adult nematode in Strongylida.

Caudal end The caudal end refers to the posterior section of nematodes. Examples of morphological characters at the caudal end include caudal bursa, caudal papillae, and spicules.

Cephalic end The cephalic end refers to the anterior section of nematodes. Examples of morphological characters at the cephalic end include cephalic alae, buccal capsule, and various types of lips.

Cercaria The immature (larval) stage of Digenea usually possesses a tail. In most digeneans, the cercaria is shed from the first intermediate host and excysted in the second intermediate host as metacercaria.

Cholangiocarcinoma (CCA) Cholangiocarcinoma is a type of cancer in the bile duct. The liver fluke *Opisthorchis viverrini* can cause cholangiocarcinoma, and the disease pathway consists of mechanical damage and immunopathologic reactions. Due to their damage caused to the host, *Opisthorchis viverrini* is thus classified as a group 1 carcinogen.

Chromosomal mutation Changes such as deletions and insertions, inversions, translocations, or duplications that occur on the chromosomes. These mutations can result in the rearrangement, duplication, or loss of parts of the chromosome.

Chromosomal mutations can result in drastic effects, with Down syndrome as an example of a disease caused by it.

Coalescence process The process is the model that indicates all descent lines eventually converge onto a single gene copy, assuming neither recombination nor natural selection occurs and no gene flow or population structure.

Coevolution A reciprocal evolutionary change and a dynamic microevolutionary process concerned with changes in allele frequencies within a population/among populations due to major evolutionary forces.

Concerted evolution The process converts copies of a gene in a multigene family into the same copy, resulting in paralogous genes that are more closely related than homologous genes.

Conservation genetics The interdisciplinary subfield of population genetics combining ecology, systematics, biology, molecular methods, and evolutionary analysis to solve problems related to conservation and provides insights using genetic tools and genomic information, particularly the dynamics of genes in populations to avoid extinction.

Conserved region/site Refers to nucleotide sites between two or more DNA sequences that are identical to each other.

Contig The contiguous sequence is generated from both the forward and reverse strands of the DNA sequence after sequencing, creating an overlap region.

Convergent evolution Process where distantly related organisms evolve similar traits in similar environments. Convergence in phylogenies can result from the grouping of organisms that are not evolutionarily related as they do not share a recent common ancestor.

Coracidium The stage of development of cestodes in order Pseudophyllidea/ Diphyllobothridea that hatch out from the egg.

Cryptic lineage Lineages that are morphologically similar and closely related but are genotypically different. Species that are part of cryptic lineages are cryptic species. Molecular analysis is usually required to resolve the species status among cryptic species.

Cysticercoid The larva form of cestodes with the scolex filling the enclosing cyst. Cysticercoid develops in the intestinal mucosa or other tissues in intermediate hosts when the oncosphere larva hatch.

Dead-end host A host that harbors an organism that is unable to further develop in their life-cycle stage. It is also called "accidental host." For example, humans are dead-end hosts of *Angiostrongylus cantonensis*, where the infective third-stage larvae are unable to develop to an adult stage in humans.

Definitive host The final host of the parasite where it develops into an adult stage and can reproduce. For example, humans are the definitive host of *Ascaris lumbricoides*, where adults develop and reproduce in the intestinal tract of humans, and eggs are then passed out through feces.

Degenerate primers Primers contain a mixture of possible nucleotide bases in some positions, giving a population of primers with similar sequences (not identical) covering all possible nucleotide combinations.

Deleterious mutation A harmful mutation that results in a decrease in the individual's fitness compared to the normal.

Direct life cycle Helminths requiring one host for transmission in their life cycle.

Disruptive selection It is also called "diversifying selection," an evolutionary force driving a population apart. Disruptive selection occurs due to the alleles for the extreme traits being favored over intermediate ones, resulting in an increase in allele frequency of the extreme traits. The resulting population is then divided into two distinct groups.

Divergence time The time taken for lineages to change and separate from each other and can be measured by the proportion of changes in the DNA sequence.

DNA barcoding An application of molecular identification where a short segment of the gene is used as a universal genetic marker to identify organisms. The short, standardized gene segment (approximately 400 to 800 base pairs) is a DNA barcode, serving as a molecular identifier for species identification. The concept of DNA barcoding is based on the unique DNA sequence that each species possesses, allowing for species identification.

DNA marker A gene or DNA sequence with a known location on the chromosome that can identify individuals or species. DNA marker is also called "Molecular Genetic Marker" or "Genetic marker."

DNA metabarcoding An extension of DNA barcoding to the parallel detection and sequencing of bulk samples using next-generation sequencing technology. A subset of DNA metabarcoding is environmental DNA (eDNA) metabarcoding, performed on eDNA without first isolating any target organism.

DNA mini-barcode Shorter sequence of a DNA barcode that is usually between 100 and 200 base pairs. DNA mini-barcodes have been proposed as a strategy for DNA metabarcoding due to the difficulties of amplifying minute amounts of DNA from environmental DNA.

DNA taxonomy The classification of organisms using DNA sequences.

Dorsoventrally flattened A term describing the shape of the cestodes and trematodes indicating that the body is flat from the dorsal and ventral surfaces.

Earliest diverged Refers to the taxa that first split off from the most recent common ancestor.

Ecdysozoa Superphylum Ecdysozoa includes animals with a cuticle and develops by molting its cuticle. Phylum Nematoda is classified within Ecdysozoa.

Ecological niche The position of a species in an ecosystem describing both the range of conditions necessary for species persistence and ecological role in the ecosystem.

Ecological process Ecological processes influence the state of an ecosystem that allow for the measurement of the health of an ecosystem. The four ecological processes are the water cycle, mineral cycle, energy cycle, and community dynamics. These processes directly impact the environment, thus having an indirect impact on hosts and parasites and their interactions.

Effective population size (N_e) The number of individuals contributes to the next generation of offspring. It is a key parameter in population genetics and determines the rate of change in the composition of a population.

Elevate/erect A taxonomy term used to raise the taxonomic status of a particular species/group.

Encapsulated/non-encapsulated The presence or absence of a host-derived collagen sheath around the infected muscle cell. This characteristic is used for differentiating species within *Trichinella*.

Environmental gradient A change in abiotic factors through space or time is related to altitude, depth, temperature, soil humidity, and precipitation.

Evolutionary arms race A series of selective sweeps occur when coevolving species develop a positive selection for antagonized traits.

Evolutionary factors Factors leading to evolution in a population over time which include mutations, genetic drift, natural selection, and gene flow (migration).

Evolutionary forces The forces of evolution include founder effects, genetic drift, mutation, migration, and selection.

Evolutionary history Refers to how extinct and extant species have evolved since the emergence of life on earth.

Evolutionary process Indicates processes or mechanisms that give rise to genetic variation in a population over time. These forces contributing to genetic variation include natural selection, genetic drift, mutations, and gene flow.

Evolutionary rate Speed of mutation that occurs on the DNA sequence over time. The evolutionary rate of the genetic marker refers to the speed whereby mutation occurs, and a higher evolutionary rate indicates a higher degree of sequence variation of the genetic marker.

Evolutionary trade-off theory The situation in evolution occurs when a change in one trait increases fitness, but a simultaneous change in another trait reduces fitness, thus preventing the organism from optimizing both changes.

Excretory-secretory product A product released by helminths may regulate the immune response in their host. These products can be used for immunodetection of helminthic infection.

Fixation probability The probability that all the alleles will eventually be the ancestor within a population at that locus.

Fixation time The time it takes for a new allele to become fixed in the population.

Free-living nematodes The nematodes present in the environment are not parasitic to any group of organisms, and they feed on bacteria, algae, dead organisms, and living tissues.

Furcocercous cercariae A fork-shaped tail cercaria characteristic of members in Schistosomatidae.

Gene expression profiling Measurement of the activity of thousands of genes at once to create a global picture of cellular function to determine the patterns of genes expressed by using mRNA levels during transcription.

Gene family A set of genes formed by duplication events that originate from a common ancestor and have similar biochemical functions.

Gene flow/migration rate The process of gene flow refers to the migration of genetic material from one population to another. It is an evolutionary force allowing novel alleles to be introduced into populations and increases genetic variation. Populations with gene flow occurring can help to maintain diversity.

Gene pool The set of all alleles at all loci existing in a population.

Gene tree Phylogenetic tree represents the true evolutionary relationship on the gene set that best reflects the evolution of the sequences showing a genealogy of a gene. It is constructed from DNA sequences for a genetic locus, which is not necessarily congruent with the species tree representing the actual evolutionary pathway of the species involved.

Gene-for-gene model The type of interaction represents the situation where the host needs to recognize specific effectors of the parasites to launch its defense.

Generalist species The species that can grow and develop in a wide variety of environmental conditions and can make use of a variety of different resources

Genetic distance A measurement of sequence variation and divergence between individuals and is measured by the proportion of nucleotide sites different between individuals. It is a parameter determining how genetically similar or different are the individuals in question.

Genetic diversity The range of genetic traits or alleles within a species or population allowing certain traits to be selected for and increases the population's survival in instances where there are external environmental stressors. Factors affecting genetic diversity include population size and evolutionary forces acting on the population.

Genetic incompatibility A major cause of genetic incompatibility is postzygotic reproductive isolation. Genetic incompatibility results when the offspring of two different species are not viable or unable to reproduce.

Genetic polymorphism An alteration in DNA sequence found in the general population at a frequency greater than 1% can be used to determine the diversity of individuals.

Genetic variation The variety of differences in the DNA sequence between different organisms providing the basis for speciation, divergence, and molecular evolutionary relationships. Genetic variation can arise from mutations, recombination, crossing over, gene flow, and natural selection.

Geometric-mean fitness The elimination of well-adapted genotypes invoked only within the limited context of microevolution in response to environmental variance.

Gradient PCR A type of PCR where the primer annealing temperature is adjusted to make a temperature gradient. Gradient PCR is performed to determine the optimal primer annealing temperature for PCR amplification.

Group 1 carcinogen Standard International Agency for Research on Cancer (IARC) classification of substances that have cancer-causing effects on humans. A substance classified as Group 1 indicates sufficient evidence of carcinogenicity in humans when exposed. *Opisthorchis viverrini* is classified as a group 1 carcinogen.

Hairpin formation It occurs when the primer can fold back on itself (resulting in a hairpin-like loop) due to complementary nucleotides pairing and forming hydrogen bonds with each other.

Heterogeneous environment Host types using different resources/niches within a local environment.

Heterozygous allele The two different alleles at a particular gene locus.

Heuristic guide tree A quick tree that is generated based on the distance matrix of the pairwise sequence alignment scores.

High-resolution melting PCR A quantitative PCR assay utilizes high-resolution melting curve analysis to quantify amplicons generated. Data for the melting curve is collected at 0.2 °C increments instead of the usual 0.5 °C increments.

Homogenous environment Parasites have similar probabilities to transmit to each host type existing in the same geographical area.

Homologous gene/sequence Genes or sequences are inherited by sharing of a common ancestor.

Homoplastic morphological characters Morphological characters similar between two genetically different species, limiting the use of these characters for morphological identification. Homoplastic characters between two genetically different species can result from convergent evolution and being in similar environments.

Homozygous allele The two sequences at a given locus in an individual are copies of the same allele.

Horizontal gene transfer (HGT) Horizontal gene transfer is the transmission of genetic material between different genomes. It is also called "Lateral gene transfer or LGT."

Host specificity Refers to helminth species that are specific only to certain types of hosts.

Host–parasite association The interaction between a host and a parasite, whereby the association between both organisms can either be mutually benefiting or one-sided.

Host–parasite relationship The relationship between the parasite and its host, where the host is the habitat that the parasite resides in.

Host–parasites coevolution The reciprocal evolution between host and parasite where both organisms continually adapt to each other.

Host–parasite interaction Refers to the interaction between host and parasite through being together in the same environment.

Hyperinfection Hyperinfection results when *Strongyloides* larvae proliferate and disseminate to multiple organs in humans, resulting in organ failure and systemic sepsis.

Hypervariable region Region of the segment of DNA that has higher levels of sequence variation.

Immunoblotting Immunoblotting techniques refer to assays that utilize antibodies to detect target proteins of interest. It is commonly used as a diagnostic tool to diagnose human helminthic infections through antigen-antibody recognition.

Indirect life cycle At least one or two intermediate hosts are required to complete the parasite life cycle. *Gnathostoma* is an example that utilizes the indirect life cycle, where copepods are the first intermediate host and fishes or aquatic animals are the second intermediate hosts.

Infective stage Refers to the parasitic stage, where the parasite can initiate a reaction in the host. For nematodes like *Ascaris lumbricoides*, *Trichuris trichiura*, and *Enterobius vermicularis*, the egg stage is the infective stage, whereas for hookworms and *Strongyloides stercoralis*, the filariform larva (L3) is the infective stage. For trematodes, the infective stage is usually the metacercaria (except for *Schistosoma,* where the infective stage is the cercaria). For cestodes, the infective stage is usually the cysticercus, embryonated egg, or gravid proglottid.

Integrated taxonomy Refers to using morphological, ecological, and molecular information for taxonomy.

Intermediate host Refers to host where the immature stage of the parasite can either reproduce asexually or develop to another immature stage. For example, in most trematodes, snails are the first intermediate hosts where asexual reproduction occurs (miracidium to cercaria), and fishes are the second intermediate hosts where the cercaria develops to metacercaria stage.

Interspecific variation Refers to the amount of sequence variation between organisms (species level and higher). Genetic distances can measure variation through the proportion of difference between nucleotides.

Intraspecies variation Refers to the amount of sequence variation within the same species and can be used to measure genetic variation among populations.

Introgression The process relates to incorporating genetic material from one organism to another through repeated backcrossing events. Introgression can play an essential role in the diversification of genetic material and the evolution of the organism. Among helminths, introgression has been suggested to occur between *Schistosoma* species.

Invasive species Refers to any species that are not native to a particular area and can cause harm to the environment and other groups of organisms living in the area.

Isotype Refers to the duplicate specimen collected from the same habitat and host.

Life-cycle complexity Life cycle refers to an organism's growth, development, and course of events. Life-cycle complexity indicates the requirement of two or more hosts being involved in the organism's life cycle.

Local adaptation The ability of parasites to perform better (survive and thrive) in or hosts from the same environment/niches.

Local maladaptation It occurs when a more harmful trait to the organism is selected for.

Locus A genetic locus refers to the location of the particular gene. For example, the 18S rRNA gene (in the nuclear DNA) is different from the *COI* gene (in the mitochondrial DNA).

Long branch attraction It occurs when two phylogenetically distantly related organisms are closely related on a phylogenetic tree, skewing true evolutionary interpretations.

Longitudinal studies Studies that take place over an extended period with the same individual.

Long-PCR amplification PCR amplification of DNA targets over 5Kb in length and typically cannot be amplified using routine PCR.

Macrocyclic lactone A class of drug used in the control of intestinal and filarial nematodes.

Macroparasite Organisms that survive on or within another organism have higher genomic complexity and life cycles that require either one or several hosts to be completed. Parasitic helminths such as nematodes, trematodes, and cestodes, and parasitic arthropods including parasitoids and ectoparasites are examples of such.

Marine nematodes Free-living nematodes that reside in the marine sediments. Examples include nematodes belonging to Clade II.

Matching-alleles model Interactions between the host immune system and antigenic parasite explicitly matching the host for infection.

Maximum parsimony (MP) Character-based phylogenetic inference based on the tree that requires the least number of changes. The most parsimonious tree (least number of changes) is most likely to be the best.

Meiofauna A diverse assemblage of eukaryotic organisms found in both freshwater ecosystems and the marine realm, including small microscopic animals and protists.

Meiosis A process occurring in sexually reproducing organisms, with two rounds of cellular division, produces four cells containing half the amount of genetic information.

Melting temperature (T_m) The temperature where half of the double-stranded DNA will dissociate to become a single DNA strand.

Metacercaria The immature stage of trematodes that is encysted. The metacercaria develops from the cercaria and develops into an adult trematode. The metacercaria stage is usually the infective stage of the trematode.

Metacestode The larval form of cestode that can be in a cyst form. It is usually the infective stage where definitive hosts get infected through ingestion of metacestode. An example is *Echinococcus granulosus* metacestode that can be found in humans.

Microhabitat Small part of a habitat possessing specific conditions conducive for an organism.

Microsatellite Microsatellites are also known as single sequence repeats and are repetitive DNA motifs distributed throughout the genome, usually occurring in the noncoding regions of nuclear DNA. They are made up of tandem repeats of 1 to 6 base pairs and are often used as genetic markers for genotyping, parental testing, and studies on genetic variation.

Minimum evolution (ME) Distance-based phylogenetic inference based on the assumption that the tree with the smallest sum of branch length is most likely to be the best.

Miracidium The larva stage of Digenea after hatching out from egg. After hatching, the miracidium can penetrate a suitable first intermediate snail host and develop to sporocyst, redia, and cercaria.

Mock communities An artificial mix of species created *in vitro* to simulate the composition of an actual sample for DNA metabarcoding.

Model organism Nonhuman species that are extensively studied and can serve as models to understand molecular and cellular processes in science.

Molecular identification Species identification using molecular methods by detecting variations in the organism's genotype.

Molecular operational taxonomic unit (MOTU) The grouping of similar or closely related individuals based on molecular sequences.

Monophyletic clade A clade consists of all the descendants of the most common ancestor.

Morphological characters Observable and phenotypic characters of an organism which is used for morphological identification. The use of morphological characters is the gold standard for the taxonomy for helminths.

Morphological disparity The measure of morphological variation among species and higher taxa. This has been used to characterize and investigate the evolution of variation in the organism's anatomy, function, and ecology.

Morphological identification Species identification using observable and phenotypic characters of the organism. Morphological identification also includes other ecological and behavioral traits. It is usually performed through morphological identification keys.

Natural selection The process where individuals with certain genotypes are more likely than individuals with other genotypes to survive and reproduce to pass on their alleles to the next generation.

Negative frequency-dependent selection Refers to the decreased fitness of a phenotype or genotype as it becomes more common. It is the most powerful selective force maintaining balanced polymorphisms.

Negative selection Selective removal of deleterious mutations from the population, keeping the frequency of deleterious mutations to low levels over many generations.

Neutral theory Theory indicating that variations in the genetic material are due to chance implying that evolution can occur without natural selection.

Non-monophyletic clade The descendants of the most common ancestor not in the same clade.

Nuclear sequence of mitochondrial origin (NUMTs) Insertions of mitochondrial DNA sequences into the nuclear genome. NUMTs can lead to misidentification and incorrect phylogenetic inferences.

Nucleotide substitution Nucleotide substitution refers to a change in the nucleotide base in the DNA. Transition and transversion are the two types of nucleotide substitution, where a transition occurs when a purine base is substituted to another purine base, while transversion occurs when a purine base is substituted to a pyrimidine base.

Obligate parasite A parasite living in a host and cannot complete its life cycle without it.

Oncosphere The larva stage of the tapeworm with a hexacanth embryo.

Ontology Formal representation of concepts and knowledge. Refers to the formal representation that species concepts can be applied and tested.

Operational taxonomic units (OTUs) A classification term to group similar organisms.

Orthologous sequence Refers to sequences that are related to a common ancestor.

Ovijector The muscular portion of the oviduct of nematodes where eggs exit from the female.

Oviparous A reproductive term refers to females that lay eggs. Examples of nematodes that are oviparous include *Ascaris lumbricoides* and *Trichuris trichiura.*

Ovoviviparous A reproductive term referring to females that have eggs with the larvae that hatch in the reproductive tract before passing out. An example of a nematode that is ovoviviparous is *Strongyloides stercoralis*, where the rhabditiform larva is hatched out from eggs deposited in the intestinal mucosa of hosts.

Oxidative phosphorylation A metabolic process occurs in the mitochondria to form Adenosine Triphosphate (ATP) through the transfer of electrons.

Paralogous sequence Genes are similar due to gene copies resulting from gene duplication events and not from sharing a common ancestor.

Parasite virulence Parasite virulence refers to the effects or damage that the parasite causes to its host. It can be assessed as reducing host fitness following infection by a parasite. For parasitic helminths, a balance between exploiting hosts for resources for survival and reproduction and avoiding killing the hosts is usually the strategy influencing their virulence.

Parasitic burden Parasitic burden refers to the intensity of parasite infection in the host and can be categorized either as mild, moderate, or heavy. The intensity of helminthic infection can be quantified based on the number of eggs or adults detected in fecal samples.

Paratenic host A host that the parasite does not develop but serves as a vehicle for parasite transmission. For example, in *Anisakis*, animals (such as large fishes or other aquatic animals) that prey on smaller fishes that are the second intermediate host of *Anisakis* are paratenic hosts. The L3 stage does not develop in paratenic hosts, and development to adults will only proceed when definitive hosts predate on the paratenic hosts.

Parthenogenetic Process refers to the generation of offspring without requiring sperm from the male. *Strongyloides stercoralis* is an example of a nematode that can reproduce via parthenogenesis.

PCR amplification bias Refers to the process when selective species are amplified through PCR. Amplification bias can occur due to the primers binding only to some species' DNA targets or the abundance of a particular DNA target.

Phenotype/phenotypic The observable properties or characters of an organism including its appearance and behavior.

Phenotypic plasticity Variations in morphological characters in genetically similar species can arise in similar environments. These characters are known as "plastic."

Phylogenetic framework A term refers to the phylogeny that is the most comprehensive and accurate representation to date.

Phylogenetic inference/relationship It refers to the depiction of how related groups of organisms or species are to each other through phylogenetic trees.

Phylogenetic resolution A term refers to the degree of robustness of the phylogenetic tree. A resolved phylogeny is usually a phylogenetic tree congruent with other phylogenies and has strong bootstrap support.

Phylogenetically informative sites A term refers to nucleotide sites that are useful for distinguishing one sequence from another.

Plant-parasitic Nematodes that feed on parts of the plant (stem, root, leaves, flowers, seed) and affect their health. They can parasitize on the outside of the plant (ectoparasites), parasitize within the plant (migratory and sedentary endoparasites), or both (semi-endoparasites). Examples include nematodes belonging to Clades II and IV.

Plerocercoid The next stage of development of cestodes in order Pseudophyllidea after procercoid. The plerocercoid is the infective stage.

Polyembryony mechanism A mechanism where more than one embryo develops from a fertilized egg.

Polymerase chain reaction-restriction fragment length polymorphism (PCR-RFLP) A molecular technique uses restriction enzymes for genotyping variations between individuals or species. Differentiation relies on the different amplicon sizes (band patterns) produced after digestion with specific restriction enzymes. In PCR-RFLP, a PCR is first performed targeting the gene of interest. Target amplicons are then digested with restriction enzymes to give the specific band patterns corresponding to the individual/species.

Polymorphism Polymorphisms are produced by changes in the DNA sequences, resulting in variations in DNA sequence among individuals. Examples of changes include mutations at the nucleotide or chromosomal level. Genetic polymorphism within and between populations can be used to investigate evolutionary forces in the population.

Polyphyletic clade/assemblage A clade is considered polyphyletic if it consists of unrelated organisms which do not share a recent common ancestor.

Polyzoic Cestode with more than one set of reproductive organs.

Population aggregation analysis Species delimitation based on morphological apparent-fixed differences, which systematically compares character state distributions among populations.

Population genetics A field of study dealing with the genetic variation within and among populations and the evolutionary factors and forces explaining the genetic variation present.

Positive frequency-dependent selection The favoring of common alleles, accelerating the loss of rare alleles and decreasing genetic variation causing the fitness of a phenotype or genotype to increase as it becomes more common.

Positive selection The selection of new advantageous mutations in the population and may increase the fitness of individuals in the population.

Procercoid The first larva stage of development of cestodes in order Pseudophyllidea/Diphyllobothridea that comes after coracidium.

Protoscolex larva The stage of the cestode larva formed from the hydatid metacestode. Protoscolices are produced in intermediate hosts of cestodes (e.g., for *Echinococcus granulosus*), and the protoscolices further develop to adults in definitive hosts.

Pseudogenes Nonfunctional DNA sequences similar to functional genes due to duplication events or a loss of function.

Purifying selection The negative selection removing deleterious variations, where deleterious mutations that are produced in each generation are constantly swept away.

Purine Base consisting of a six-membered and a five-membered nitrogen-containing ring. The nucleotide bases of Adenine and Guanine are examples.

Pyrimidine Base consisting of a six-membered nitrogen-containing ring. The nucleotide bases of Thymine, Cytosine, and Uracil are examples.

Radio-labeled nucleotide The nucleotide modified with radioisotopes. Radio-labeled nucleotides were traditionally used in Maxam-Gilbert sequencing to determine DNA sequences.

Random amplified polymorphic DNA (RAPD) A PCR-based technique using primers of arbitrary nucleotide sequence to amplify random segments of the DNA.

Random genetic drift Due to chance, the process of changes in allele frequencies from generation to generation may cause gene variants to reduce genetic variation and cause an initially increasing number of rare alleles and fix them in the population.

Red Queen hypothesis It refers to the hypothesis whereby species must adapt and evolve for reproductive advantage and survival because competing organisms are also evolving.

Redia The larva stage of Digenea after developing from sporocyst in the first intermediate host. Redia then develops to cercaria by asexual reproduction, which is shed out from the first intermediate host.

Reference sequence/database A database containing genetic sequences from many species useful for comparisons with sequences obtained from unknown species.

Relaxed selection Refers to environmental change that often eliminates or weakens a source of selection formerly important for the maintenance of a particular trait.

Replication slippage Misalignment of DNA strands during DNA replication due to mispairing between neighboring repeats.

Reproductive isolation It refers to processes that led to the prevention of the population from successfully producing offspring. Two populations are considered reproductively isolated when they are unable to produce offspring.

Reservoir It refers to the natural habitat of the parasite. The host is not affected by the presence of the parasite and serves as a source for parasite transmission to other organisms.

Resinous mounting agent A solution that is commonly used to embed a specimen for permanent mounts in taxonomy. Examples of resinous mounting agents used for helminth taxonomy include Canada Balsam and Permount.

Restriction fragment length polymorphism (RFLP) A molecular-based technique using restriction enzymes to generate different band patterns in the DNA, allowing studies involving polymorphism and variations in the DNA sequence.

Reticulate genetic processes Processes resulting in the transfer of genetic material and leading to the lack of independence between two evolutionary lineages. Examples include hybridization and horizontal gene transfer.

Rhabditiform larva A term used to refer to the larva stages of some soil-transmitted helminths. Rhabditiform larva is stages one and two, and is the noninfective stage.

Rostellum The anterior end of a cestode that is part of the scolex. The rostellum can be unarmed or armed with rows of hooks.

Scolex It refers to the anterior end (head) of cestodes. The scolex serves as attachment organs to attach to the host intestinal wall. Various forms of scolex exist, including those with four muscular suckers, sucking grooves, and rostellum.

Secondary structure A secondary structure of DNA forms when the primary DNA sequence strand forms complementary base-pairing through hydrogen bonding.

Selective pressure External agents or environmental factors acting on the population that influences the organism's ability to survive.

Selective sweep A process where a new advantageous mutation increases in frequency in the population and becomes fixed over time.

Semi-nested PCR Refers to the nested PCR assay where the second round of PCR utilizes one primer from the first PCR round instead of two different primers.

Sexual selection Any selection that arises from fitness differences is associated with nonrandom success in the competition for access to gamete for fertilization.

Sodium-dodecyl-sulfate-polyacrylamide gel electrophoresis (SDS-PAGE) SDS-PAGE is used for the separation of proteins on a polyacrylamide gel. Proteins are separated based on their molecular weight using an electric current, and the gel is subsequently stained and visualized.

Soil-transmitted helminths Helminths transmitted to humans through contact with soil contaminated with helminths. Examples include *Ascaris lumbricoides* and *Trichuris trichiura*.

Specialist species A species growing and developing only in a narrow range of environmental conditions or has a limited diet.

Speciation mechanism Processes leading to the formation of new/distinct species. These include allopatric, sympatric, parapatric, and peripatric mechanisms.

Species boundaries A term used to imply how different one species is from another and to be distinct from each other.

Species circumscription The taxonomic term refers to species delimitation (determining one species from another).

Species complex A group of closely related species whereby their species boundaries are unclear. Examples of a species complex in trematodes include *Paragonimus westermani*, *Paragonimus heterotremus*, and *Echinostoma revolutum*. Examples of a species complex in cestodes include *Echinococcus granulosus*.

Species concept Concept and theories proposed which define species as individuals. They are the criteria that are closely tied to species delimitation and speciation. Examples of species concepts include the Biological Species Concept, Phylogenetic Species Concept, and Evolutionary Species Concept.

Species delimitation A process where new species boundaries are determined and new species discovered.

Species diversity It refers to the amount (richness) and abundance of species found in the ecosystem. Higher levels of species diversity relate to greater ecosystem stability as the species present are more likely to withstand external environmental stresses such as climate change.

Species tree A phylogenetic tree represents the evolutionary history of a set of species (or populations). It defines the pattern of branching of species lineage via the speciation process.

Speciesness It is about the reality of species, types of species, and their diversity.

Sporocyst The larva stage of Digenea after developing from miracidium through asexual reproduction in the first intermediate host.

Standard curve A graph obtained through calibration allows for quantifying a particular concentration in a sample used in quantitative PCR to measure the amount of DNA present. The curve is obtained through serial dilutions of known concentrations of DNA.

Stop codon A codon (trinucleotides) signals the end of the DNA translation process. UAA, UAG, and UGA serve as stop codons in the standard genetic code for most eukaryotes and prokaryotes. For mitochondrial genomes, AGG serves as a stop codon.

Strobila The strobila comprises the neck and the body portion of cestodes, where a long chain of segments makes up the body.

Substitution saturation Substitution saturation results when multiple nucleotide substitutions occur at the same site in a sequence. The process leads to decreased phylogenetic information as the resulting substitutions might not be an accurate indicator of evolutionary relationships. A genetic marker that is saturated might not be a suitable genetic marker for molecular analysis.

Sympatric speciation Refers to the speciation process without geographical barriers where one population becomes two (reproductive isolation).

Tapeworm The common name for cestodes is their long, flat, and segmented bodies. The adult tapeworm consists of a head and strobila section, where the strobila consists of a neck portion and a long chain of segments. They are parasites of human and animal hosts and mostly reside in the gastrointestinal tract of hosts.

Taxonomic assignment A process of grouping organisms with their respective species or closely related species.

Taxonomic categorical/hierarchy level It refers to categorizing organisms into relative levels (Kingdom, Phylum, Class, Order, Family, Genus, Species), with Kingdom as the highest rank and Species as the lowest rank.

Taxonomic profiling It refers to identifying taxa and measuring the relative abundance of organisms present in the sample.

Taxonomic status The taxonomic status of an organism refers to its current official consensus.

Taxonomy The science of classifying organisms based on shared characters allows for the determination of relatedness among organisms. These characters can either be phenotypic or genotypic. Traditional and classical taxonomy utilizes morphological and observable characters for classification, while DNA taxonomy utilizes DNA sequences to classify organisms.

Thermo-resistant enzyme An enzyme that is resistant to inactivation at high temperatures. An example of a thermo-resistant enzyme is Taq polymerase used in PCR.

Topology It refers to the general structure and shape of a phylogenetic tree. It summarizes and depicts the pattern of relatedness among organisms present on the leaves of the phylogenetic tree. Phylogenetic trees with similar topology are indicative of similar relationships among organisms.

Transcriptomics Transcriptomics is the study of transcriptomes, the set of messenger RNA molecules that are expressed. Transcriptome analysis allows for identifying and comparing genes and their expression levels in a cell in response to varying treatments. Transcriptomics utilizes high-throughput sequencing methods such as microarray analysis, and an example is RNA-Seq.

Transgressive segregation Formation of extreme phenotypes or transgressive phenotypes observed in segregated hybrid populations compared to phenotypes observed in the parental lines.

Trophic interactions Associations between species in an ecosystem. An example of trophic interaction is the association between predator and prey.

Universal primer A primer can be used to amplify DNA from broad groups of organisms. Examples of universal primers include the HCO2198-LCO1490 Folmer primers targeting the mitochondrial *COI* gene of invertebrates and the primers targeting the 16S rRNA gene in bacteria.

Ventrogenital sac The modified ventral sucker present in Heterophyidae.

Viviparous A reproduction strategy that allows females to forth their live young directly. An example is *Trichinella* species, where females give birth to live young in the digestive tract of human hosts.

Voucher specimen Voucher specimens refer to preserved specimens that are archived in a collection facility. They are important, especially for DNA barcoding studies where the corresponding DNA of the voucher specimen is available for species identification.

Whole-genome sequencing Whole-genome sequencing is a method to analyze the complete DNA sequence of organisms and utilizes next-generation sequencing technology. It allows for a high-resolution and high-throughput method to generate large volumes of genetic information to support the assembly of the complete genome of an organism. The first helminth to have its genome sequenced is *Caenorhabditis elegans*.

Zoonosis Infectious disease caused by a pathogen that has jumped from an animal to human. These pathogens can be parasitic, viral, or bacterial and spread to humans through direct contact or food, water, or the environment. Zoonoses represent a large proportion of new and re-emerging diseases in humans. An example of zoonosis includes Trichinellosis, where humans get infected by consuming undercooked meat from domestic or wildlife.

Index

U. Thaenkham et al., *Molecular Systematics of Parasitic Helminths*,
https://doi.org/10.1007/978-981-19-1786-8

Q

R

S